Beyond Two: Theory and Applications of Multiple-Valued Logic

Studies in Fuzziness and Soft Computing

Further volumes of this series can be found at our homepage.

Vol. 93. V.V. Cross and T.A. Sudkamp
Similarity and Compatibility in Fuzzy Set Theory, 2002
ISBN 3-7908-1458-X

Vol. 94. M. MacCrimmon and P. Tillers (Eds.)
The Dynamics of Judicial Proof, 2002
ISBN 3-7908-1459-8

Vol. 95. T.Y. Lin, Y.Y. Yao and L.A. Zadeh (Eds.)
Data Mining, Rough Sets and Granular Computing, 2002
ISBN 3-7908-1461-X

Vol. 96. M. Schmitt, H.-N. Teodorescu, A. Jain, A. Jain, S. Jain and L.C. Jain (Eds.)
Computational Intelligence Processing in Medical Diagnosis, 2002
ISBN 3-7908-1463-6

Vol. 97. T. Calvo, G. Mayor and R. Mesiar (Eds.)
Aggregation Operators, 2002
ISBN 3-7908-1468-7

Vol. 98. L.C. Jain, Z. Chen and N. Ichalkaranje (Eds.)
Intelligent Agents and Their Applications, 2002
ISBN 3-7908-1469-5

Vol. 99. C. Huang and Y. Shi
Towards Efficient Fuzzy Information Processing, 2002
ISBN 3-7908-1475-X

Vol. 100. S.-H. Chen (Ed.)
Evolutionary Computation in Economics and Finance, 2002
ISBN 3-7908-1476-8

Vol. 101. S.J. Ovaska and L.M. Sztandera (Eds.)
Soft Computing in Industrial Electronics, 2002
ISBN 3-7908-1477-6

Vol. 102. B. Liu
Theory and Practice of Uncertain Programming, 2002
ISBN 3-7908-1490-3

Vol. 103. N. Barnes and Z.-Q. Liu
Knowledge-Based Vision-Guided Robots, 2002
ISBN 3-7908-1494-6

Vol. 104. F. Rothlauf
Representations for Genetic and Evolutionary Algorithms, 2002
ISBN 3-7908-1496-2

Vol. 105. J. Segovia, P.S. Szczepaniak and M. Niedzwiedzinski (Eds.)
E-Commerce and Intelligent Methods, 2002
ISBN 3-7908-1499-7

Vol. 106. P. Matsakis and L.M. Sztandera (Eds.)
Applying Soft Computing in Defining Spatial Relations, 2002
ISBN 3-7908-1504-7

Vol. 107. V. Dimitrov and B. Hodge
Social Fuzziology, 2002
ISBN 3-7908-1506-3

Vol. 108. L.M. Sztandera and C. Pastore (Eds.)
Soft Computing in Textile Sciences, 2003
ISBN 3-7908-1512-8

Vol. 109. R.J. Duro, J. Santos and M. Graña (Eds.)
Biologically Inspired Robot Behavior Engineering, 2003
ISBN 3-7908-1513-6

Vol. 110. E. Fink
Changes of Problem Representation, 2003
ISBN 3-7908-1523-3

Vol. 111. P.S. Szczepaniak, J. Segovia, J. Kacprzyk and L.A. Zadeh (Eds.)
Intelligent Exploration of the Web, 2003
ISBN 3-7908-1529-2

Vol. 112. Y. Jin
Advanced Fuzzy Systems Design and Applications, 2003
ISBN 3-7908-1537-3

Vol. 113. A. Abraham, L.C. Jain and J. Kacprzyk (Eds.)
Recent Advances in Intelligent Paradigms and Applications, 2003
ISBN 3-7908-1538-1

Melvin Fitting
Ewa Orłowska
Editors

Beyond Two: Theory and Applications of Multiple-Valued Logic

With 25 Figures
and 19 Tables

Springer-Verlag Berlin Heidelberg GmbH

Professor Dr. Melvin Fitting
Lehmann College
Department of Mathematics
and Computer Science
250 Bedford Park Boulevard West
Bronx, NY 10468-1589
USA
fitting@lehmann.cuny.edu

Professor Dr. Ewa Orłowska
National Institute
of Telecommunications
ul. Szachowa 1
04-894 Warsaw
Poland
orlowska@itl.waw.pl

ISSN 1434-9922

DOI 10.1007/978-3-7908-1769-0

Bibliographic information published by Die Deutsche Bibliothek
Die Deutsche Bibliothek lists this publication in the Deutsche Nationalbibliografie; detailed bibliographic data is available in the Internet at <http://dnb.ddb.de>.

Originally published by Physica-Verlag Heidelberg in 2003
MyCopy version of the original edition 2003

www.springer.com/mycopy

Preface

If Bertrand Russell had been more inclined to the short answer question instead of the essay, he might have asked, "True or false, logic is two-valued." And if he had asked this of Alfred North Whitehead, we all know what the answer would have been. But if Russell had asked a wider variety of people, again we all know what would have happened. One person might have answered, "I don't know," another "Well, it is and it isn't," and yet a third might not have replied at all. If Russell had pondered these responses, a variety of many-valued logics might have had earlier births than they did.

Actually, the make-believe anecdote about Russell is a little misleading. While the two classical truth values are quite explicit in the writings of Boole, they are not particularly so in those of Russell. In *Principia Mathematica*, for instance, truth values are implicit in the axiomatization. The guiding principle behind Russell's choice of axioms seems to have been to assume enough to get the job done. Later, the work of Post and Łukasiewicz (and Wittgenstein) extracted truth values and generalized them. The generalizations were both for their own sake – mathematicians exploring the possibilities – and with philosophical applications in mind. Today a make-believe dialog of the Russell/Whitehead sort might take quite a different form: "How many truth values are there?" Answer: "How many do you need?" We live in a more consumer-oriented society.

Of course, nothing that was said above is quite exactly the truth (or quite exactly falsehood either). Logics with more than two truth values have a lineage that predates the twentieth century, and they are still not as accepted as they should be. Let us give a few examples to illustrate this.

In Buddhist writings there is a famous passage wherein a disciple, Malunkyaputta, asks the Buddha a series of questions that he strongly feels are of importance, concerning the existence of a soul, the afterlife, and so on. The questions have a certain ritualized form. Somewhat condensed, here is a typical question: "Is it the case that the saint exists after death; is it the case that the saint does not exist after death; is it the case that the saint both exists and does not exist after death; is it the case that the saint neither exists nor does not exist after death?" In effect, Malunkyaputta and the Buddha are using a four-valued logic long, long before it was made explicit by Nuel Belnap.

The Jain religion was founded more-or-less contemporarily with Buddhism, and there is a system of Jain logic known as Syadvada. In this, an observation might be true, or false, or indeterminate in various ways. This leads to a seven-valued logic whose values are: true, false, true or false, indeterminate, true or indeterminate, false or indeterminate, true or false or indeterminate. It is not clear quite what system this relates to today, but it is clear that the subject of many-valued logic has a longer history than one might have supposed.

But now let us turn to the current state of things. Even though a well-known store catalog (Williams-Sonoma) offers for sale a "fuzzy logic rice cooker" (at $199) acceptance and application of many-valued logics is not as widespread as it should be. Here is a concrete example.

The programming language Java has two conjunction connectives, written & and &&, and two disjunction connectives, written | and ||. How do the two sets of connectives differ? In writing programs, connectives typically occur in commands like: "if (*A*&&*B*)|*C*; else *D*;" where *C* and *D* are actions to be executed. But *A* and *B* themselves may involve computations, such as $x/y > 0$, or method invocations (procedure calls, function calls) that return boolean values. Something like $x/y > 0$ is meaningless if y is 0, and a method call might enter an infinite loop and never terminate. In Java textbooks the difference between & and && is always explained (if it is explained at all) procedurally – under what circumstances will an instruction result in a crash or a hang. But if we think of the underlying logic as three-valued (true, false, undefined) simpler explanations can be offered. The logic of Java's & and | is Kleene's weak three-valued logic. The logic of Java's && and || is an asymmetric logic sometimes known as Lisp logic, or McCarthy's logic.

Instead of giving straightforward three-valued truth tables, programming textbooks show students two-valued truth tables and then teach them how they apply in a setting that is not two-valued. This makes thinking difficult, and much that could be a problem is simply ignored. It is, for instance, taken for granted that (*A*&&(*B*||*C*)) as a condition in an if/else statement will behave the same as ((*A*&&*B*)||(*A*&&*C*)). In fact this is the case, but it is so because the two expressions are equivalent in McCarthy's logic, and not because their equivalence is a classical tautology. These complaints should not be associated with Java alone – the textbook treatment of every programming language follows a similar pattern. Full acceptance of many-valued logics, even in situations where they are quite natural, is still some years away.

But, even though the past is fuzzier than we thought, and the present is not quite what we would want, still the subject of many-valued logics is entering a kind of golden age. It is no longer the provence of the few – its value has been seen by a substantial number, if not yet by the many. Interesting mathematical investigations are under way. Significant applications in computer science and in philosophy have been made. All this is part of a general trend in formal logic, the shift from one-logic-to-rule-them-all to a rich and teeming universe of possibilities. If one looks at current developments in modal logics, in temporal logics, in substructural logics, in quantum logics, one sees the many-valued logic experience repeated over and over. It has become a rich and varied world indeed.

The present volume collects together expanded versions of papers originally presented at ISMVL 2001 and the accompanying workshop, and represents the state of the art for much current research in many-valued logics. Here is a brief summary of the contents.

The papers in Part I present algebraic approaches to multiple valued logics and their applications. *Algebras for Hazard Detection* by J. Brzozowski, Z. Ésik, and Y. Iland studies multi-valued algebras which are capable of modeling digital circuits with possible hazard occurrences. A general algebraic theory is presented which provides a framework for simulation of gate circuits and for hazard detection, identification, and counting. In the paper *An Abstract Algebraic Logic View of Some Multiple-valued Logics* by J. Font, an algebraic framework is developed for presenting and studying algebraic behaviour of a broad family of Łukasiewicz-style logics. The paper *Representation Theorems and the Semantics of Non-classical Logics, and the Applications to Automated Theorem Proving* by V. Sofronie-Stokkermans develops a representability theory based on relationships between an algebraic and a Kripke-style semantics. General representation theorems are presented for a broad class of non-classical logics, including several multiple-valued logics. The results of the paper lead to a method of constructing resolution theorem proving systems for the respective logics. The paper *An Algebraic Approach to Entropy and its Generalizations - A Survey* by D. Simovici examines various concepts of entropy that may be associated to algebraic objects such as functions, partitions, and set collections. For each of these notions of entropy an axiomatization is presented.

The papers in Part II deal with proof theory of multiple-valued logics. *Classical Gentzen-type Methods in Propositional Many-valued Logics* discusses methods of developing Gentzen-style deduction systems for multiple-valued logics and presents several such systems. Cut elimination theorems are proved for those systems. The paper *Sequents of Relations Caluli: A Framework for Analytic Deduction in Many-valued Logics* by M. Baaz, A. Ciabattoni, and Ch. Fermüller presents Gentzen-style deduction systems for Gödel logics. The authors point out some general features of those systems that provide guidelines for development of such systems for a broader class of logics. The paper *Polarity-based Stochastic Local Search Algorithms for Non-classical Satisfiability* by Z. Stachniak presents algorithms for checking satisfiability of formulas of a class of multiple-valued logics. The algorithms are based on a method of polarity. Originally, polarity was introduced as a strategy for resolution-style systems. In this paper polarity is extended to the many-valued setting. In *Model Checking for a Multi-valued Computation Tree*

Logic by B. Konikowska and W. Penczek a multiple-valued extension of the computation tree logic CTL* is introduced and a translation of model checking for classical CTL* to model checking for this extension is developed. The paper *Complexity of Many-valued Logics* by R. Hähnle presents a survey of computational complexity results in several classes of multiple-valued logics. Two kinds of problems are considered: first, the complexity of the satisfiability and/or validity problem and, second, the size of mixed integer programming representations of the propositional connectives and quantifiers.

The papers in Part III are devoted to fuzzy logics, their methodology, and applications. *Ternary Kleenean Non-additive Measures* by T. Araki, M. Mukaidono, and F. Yamamoto studies some extensions of fuzzy measure and their algebraic properties. The extensions enable us to relate the concept of vagueness treated in fuzzy logic with the concept of ambiguity expressible with the fuzzy measure. *The Hierarchy of t-norm Based Residuated Fuzzy Logics* by F. Esteva, L. Godo, and A. García-Cerdaña presents a survey of fuzzy logics based on residuated lattices with an additional axiom of pre-linearity. In the paper *A Development of Set Theory in Fuzzy Logic* by P. Hájek and Z. Haniková an axiomatic theory of fuzzy sets is developed. It is shown that the classical set theory ZF is interpretable in fuzzy set theory. The paper *A Fuzzy Generalisation of Information Relations* by A. Radzikowska and E. Kerre presents a class of fuzzy relations derived from an information system and studies their properties. Correspondence theorems are presented that relate properties of the relations with the properties of fuzzy information operators.

Finally Part IV presents applications of multiple-valued logics to control theory and rational belief. In *Weierstrass Approximation Theorem and Łukasiewicz Formulas with one Quantifiable Variable* by S. Aguzzoli and D. Mundici it is shown that an infinite-valued Łukasiewicz logic is expressive enough to approximately represent any continuous control function. Complexity of the decision problem for the logic is given. The paper *A Łukasiewicz-style Many-valued Similarity Reasoning. Review* by E. Turunen presents applications of fuzzy similarity relations to the representation of and reasoning about control of signalized pedestrian crossing and control of the water level in a reservoir. In the paper *Two Values, Three values, Many Values, No Values* by Charles Morgan a standard formal semantics for logical systems is compared with a probabilistic semantics defined in terms of the notion of a degree of rational belief.

April 2002 Melvin Fitting and Ewa Orłowska

Table of Contents

Part I

Algebras of Multiple-valued Logics and Their Applications

Chapter 1
Algebras for Hazard Detection

Janusz Brzozowski[1], Zoltán Ésik[2], and Yaacov Iland[1]

[1] School of Computer Science, University of Waterloo, Waterloo, ON, Canada N2L 3G1
{brzozo,yiland}@uwaterloo.ca
http://maveric.uwaterloo.ca
[2] Department of Computer Science, University of Szeged, Árpád tér 2, 6720 Szeged, Hungary
esik@inf.u-szeged.hu
http://www.inf.u-szeged.hu/~esik

Abstract. Hazards pulses are undesirable short pulses caused by stray delays in digital circuits. Such pulses not only may cause errors in the circuit operation, but also consume energy, and add to the computation time. It is therefore very important to detect hazards in circuit designs. Two-valued Boolean algebra, which is commonly used for the analysis and synthesis of digital circuits, cannot detect hazard conditions directly. To overcome this limitation several multi-valued algebras have been proposed for hazard detection. This paper surveys these algebras, and studies their mathematical properties. Also, some recent results unifying most of the multi-valued algebras presented in the literature are described. Our attention in this paper is restricted to the study of static and dynamic hazards in gate circuits.

1 Introduction

The two-element Boolean algebra, $\mathbf{A_2}$, has been the standard algebra for circuit analysis and design, since Shannon's pioneering work in 1938 [40]. The problem of hazards has been recognized very early; hazards were already discussed in 1951 in a book by Keister, Ritchie and Washburn [28]. In 1957 Huffman [27] proposed informal definitions of static and dynamic hazards, and provided some characterizations of these hazards. Roughly speaking, a static hazard is one or more pulses occurring in a signal which should be constant, and a dynamic hazard is one or more pulses in a signal that is changing from one binary value to the other. A more formal treatment of hazards was given by McCluskey [33] in 1962, and further results were obtained by Unger [42] in 1969. These early works used the two-valued Boolean algebra for gate circuits.

Although the above-mentioned binary methods were successfully applied to a number of problems related to hazards, they are indirect, and require the use of such tools as Karnaugh maps, or detailed knowledge of the circuit structure including wire delays. Consequently, researchers turned their attention to algebras using more than two values, with the hope of finding a more direct method for hazard detection.

The first nonbinary hazard algebra was the three-valued ternary algebra, which we call $\mathbf{A_3}$, introduced by Goto [22] in 1948. In 1953 a four-valued algebra, $\mathbf{A_4}$, was described by Metze [34]. A five-valued algebra, $\mathbf{A_5}$, was presented by Lewis [32] in 1972. A six-valued algebra, $\mathbf{A_6}$, was derived by Hayes [25] in 1986 for the detection of static hazards. An eight-valued algebra, $\mathbf{A_8}$, is implicit in the 1974 work of Breuer and Harrison [5]. The same algebra is also implicit in the 1974 work of Fantauzzi [20], who presented a nine-valued algebra, $\mathbf{A_9}$. In 1986 Hayes [25] studied the then-known multivalued algebras. He presented general methods for constructing new algebras. Using these methods he obtained a thirteen-valued algebra, $\mathbf{A_{13}}$. Breuer and Harrison [5] proposed a 27-valued algebra, $\mathbf{A_{27}}$, for eliminating hazards in test generation.

In several papers mentioned above, little attention has been paid to the mathematical structure of hazard algebras. This paper summarizes the mathematical properties of hazard algebras, and examines their completeness and usefulness for hazard detection.

Some very recent results by Brzozowski and Ésik [8, 9] generalize and unify most of the previous work on hazard algebras. An infinite algebra, $\mathbf{C}$, and an infinite number of finite algebras, $\mathbf{C_k}$, were introduced for counting signal transitions and hazard pulses. A characterization of the results of simulation in algebra $\mathbf{C}$ for a restricted class of feedback-free circuits has been given by Gheorghiu [21].

The remainder of the paper is structured as follows. Section 2 introduces the algebraic laws which occur in many of the hazard algebras. Sections 3–10 examine the mathematical and hazard-detecting properties of the hazard algebras with three, four, five, six, eight, nine, thirteen and 27 elements. Section 11 summarizes the recent work by Brzozowski and Ésik on the infinite hazard algebra, and its application to the classification of hazard algebras.

2 Laws of hazard algebras

Logic circuits are often viewed as being constructed with OR gates, AND gates and inverters. Thus, algebras for such circuits naturally have three operations, $+$, $*$, and $^-$, corresponding to these gates. The normal logic values are denoted by 0 and 1, and an unknown value (if present in the algebra) is represented by Φ. The values 0, 1, and Φ are constants in the algebras. For these reasons, we are dealing with general algebraic systems having the form $P = \langle \mathcal{A}, +, *, ^-, 0, 1 \rangle$, where $\mathcal{A}$ is a set of elements, + and $*$ are binary operations on $\mathcal{A}$, $^-$ is a unary operation on $\mathcal{A}$, and 0 and 1 are constants in $\mathcal{A}$. We take the liberty of considering Φ as a constant of the algebra whenever appropriate.

Various subsets of the following set of laws have been used to describe algebraic structures defined for the study of hazard algebras. We first recall the

Table 1. Laws pertaining to hazard algebras

	For all $a, b, c \in \mathcal{A}$:	
Idempotence	L1 $a + a = a$	L1′ $a * a = a$
Commutativity	L2 $a + b = b + a$	L2′ $a * b = b * a$
Associativity	L3 $a + (b + c) = (a + b) + c$	L3′ $a * (b * c) = (a * b) * c$
Absorption	L4 $a + (a * b) = a$	L4′ $a * (a + b) = a$
Identity	L5 $a + 0 = a$	L5′ $a * 1 = a$
Bounding	L6 $a + 1 = 1$	L6′ $a * 0 = 0$
Distributivity	L7 $a + (b * c) = (a + b) * (a + c)$	L7′ $a * (b + c) = (a * b) + (a * c)$
Involution	L8 $\overline{(\overline{a})} = a$	
De Morgan's laws	L9 $\overline{(a + b)} = \overline{a} * \overline{b}$	L9′ $\overline{(a * b)} = \overline{a} + \overline{b}$
Complement laws	L10 $a + \overline{a} = 1$	L10′ $a * \overline{a} = 0$
Ternary laws	L11 $(a + \overline{a}) + \Phi = a + \overline{a}$	L11′ $(a * \overline{a}) * \Phi = a * \overline{a}$
Self-complement	L12 $\overline{\Phi} = \Phi$	

definitions of several algebraic systems. A *semigroup* is a system $S = \langle \mathcal{A}, + \rangle$ satisfying L3. A *bisemigroup* [8, 9] is a pair of semigroups, $S_+ = \langle \mathcal{A}, + \rangle$ and $S_* = \langle \mathcal{A}, * \rangle$, with the same underlying set. A bisemigroup is *commutative* if both of its operations are commutative, *i.e.*, if it satisfies L2 and L2′. A bisemigroup is *bounded* if it has two constants 0 and 1 satisfying L5, L5′, L6, and L6′. A commutative bisemigroup is *de Morgan* [8, 9, 19] if it is bounded and has a unary operation $^-$ satisfying L8, L9 and L9′. A *semilattice* is a semigroup satisfying L1 and L2. A *bisemilattice* [6] is a pair of semilattices $S_+ = \langle \mathcal{A}, + \rangle$ and $S_* = \langle \mathcal{A}, * \rangle$ which have the same underlying set. A bisemilattice is *de Morgan* [6] if it is bounded and satisfies L8, L9 and L9′. A *lattice* is a bisemilattice satisfying L4 and L4′. A lattice is said to be *bounded* if L5, L5′, L6 and L6′ hold. A lattice is said to be *distributive* if L7 and L7′ hold. A *de Morgan algebra* is a bounded, distributive lattice equipped with a $^-$ operation satisfying L8, L9 and L9′. A de Morgan algebra is a *Boolean algebra* if L10 and L10′ hold. A de Morgan algebra is a *ternary algebra* [13] if it contains Φ, and L11, L11′ and L12 hold.

A binary relation on $\mathcal{A}$ which is reflexive, antisymetric and transitive is called a partial order on $\mathcal{A}$. For $a, b, c \in \mathcal{A}$, c is the least upper bound (*lub*) of a and b if $a \leq c$, $b \leq c$, and for any d such that $a \leq d$ and $b \leq d$, we have $c \leq d$. If the *lub* exists for every pair of elements in a partially ordered set (poset), let $a + b = lub\{a, b\}$. Then $\langle \mathcal{A}, + \rangle$ is a semilattice. Conversely, given a semilattice operation $+$, define $a \leq b$ iff $a + b = b$. Then $\leq$ is a partial order such that $a + b$ is in fact the least upper bound of a and b, for any $a, b \in \mathcal{A}$. A Hasse diagram representing such a partial order is then a convenient way of defining the operation $+$.

In a bisemilattice [6] $\langle \mathcal{A}, +, * \rangle$ there are two partial orders:

$$a \leq b \text{ iff } a + b = b,$$

and

$$a \sqsubseteq b \text{ iff } a * b = a.$$

In case the bisemilattice is a lattice, these partial orders coincide.

3 The three-valued algebra

Three-valued algebra has been introduced by Goto [22, 23] (1948). Other early work using three-valued algebra for hazard detection includes the papers by Moisil [35] (1956), Roginskii [38] (1959), Muller [36] (1959), Yoeli and Rinon [43] (1964), and Eichelberger [17] (1965).

In three-valued algebra, 0 and 1 are logic 0 and logic 1 respectively, while Φ represents an unknown value. The operations of the three-valued algebra are shown in Table 2. It is easily verified that the three-valued algebra is a ternary

Table 2. Operations of three-valued algebra

$+$	0	Φ	1
0	0	Φ	1
Φ	Φ	Φ	1
1	1	1	1

$*$	0	Φ	1
0	0	0	0
Φ	0	Φ	Φ
1	0	Φ	1

a	$\overline{a}$
1	0
Φ	Φ
0	1

algebra (defined above). It is not a Boolean algebra, since $\Phi + \overline{\Phi} = \Phi + \Phi = \Phi \neq 1$, which violates the complement law. The partial order corresponding to the $+$ and $*$ operations in the three-valued algebra is given in Fig. 1(a). We can read off the result of either binary operation from this figure. For example, $\Phi + 1 = 1$ since the least upper bound of $\{\Phi, 1\}$ is 1, and $\Phi * 0 = 0$ since the greatest lower bound of $\{0, \Phi\}$ is 0.

At this point it is convenient to describe some of the results in the 1986 paper of Hayes [25], in which the author surveys the algebras that have been used for hazard detection and other simulation tasks. He also clarifies

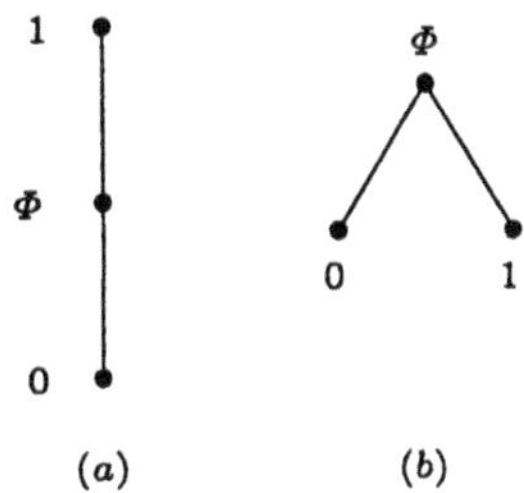

Fig. 1. Partial orders for $\mathbf{A_3}$: (a) $\leq$ and $\sqsubseteq$; (b) uncertainty order

a number of issues, introduces a unified notation, and describes two methods for generating new hazard algebras from smaller ones. Here we digress to describe the first method of Hayes, which sheds further light on algebra $\mathbf{A_3}$.

The first method constructs a new algebra $\mathbf{A}'$ from a given algebra $\mathbf{A}$. The underlying set $\mathcal{A}'$ of $\mathbf{A}'$ is a subset of the power set of the underlying set $\mathcal{A}$ of $\mathbf{A}$. The operations are extended from $\mathbf{A}$ to $\mathbf{A}'$ as follows. If $X_1, \ldots, X_n$ are in $2^{\mathcal{A}}$ and $f(x_1, \ldots, x_n)$ is an n-ary operation in $\mathbf{A}$, then f is extended to $\mathbf{A}'$ by defining $f(X_1, \ldots, X_n) = \{f(x_1, \ldots, x_n) \mid x_i \in X_i, i = 1, \ldots, n\}$.

For example, let the given algebra be $\mathbf{A_2}$. We can choose the following subset of $2^{\{0,1\}}$: $\{\{0\}, \{1\}, \{0,1\}\}$ as the underlying set of a new algebra, $\mathbf{A}_2'$. Hayes observes that $\mathbf{A}_2'$ is isomorphic to $\mathbf{A_3}$, as the reader can easily verify. The correspondence is $\{0\} \mapsto 0$, $\{1\} \mapsto 1$, and $\{0,1\} \mapsto \Phi$.

Eichelberger [17] introduces an efficient method for hazard and race detection. His ternary simulation consists of two algorithms, A and B, which use the uncertainty partial order shown in Fig. 1(b). This partial order puts values with more uncertainty higher in the order. We want to analyze what happens when some inputs change from their initial binary values to their final binary values. We first apply Algorithm A, in which the changing inputs are set to the uncertain value Φ. We then simulate the circuit using the three-valued algebra to determine whether the uncertainty introduced by the changing inputs can spread to other gates. All unstable variables are changed at the same time. It can be shown that in Algorithm A a variable can only change from a binary value to Φ, or it can remain unchanged. Thus, Algorithm A stops in at most n steps, if there are n gates in the circuit.

Algorithm B uses the output of Algorithm A as the initial state, and now simulates the circuit using the final values of the changing inputs. This time a variable can only change from Φ to a binary value, or it can remain constant. Algorithm B also terminates in at most n steps.

A static hazard is detected if an output has the same value in the initial state of Algorithm A and in the final state of Algorithm B, but has value Φ in the final state of Algorithm A. Ternary simulation cannot detect dynamic hazards, but it can detect oscillations in sequential circuits. The method is described in detail in [13].

Ternary simulation was shown to be correct with respect to binary analysis by Brzozowski and Seger [11–13]. A more general version, which does not require that the initial state of the circuit be stable, was described in [39]. Brzozowski, Lou and Negulescu [10] gave a set-theoretic characterization of finite ternary algebras, and this result was generalized to the infinite case by Ésik [18]. Free ternary algebras were recently characterized by Balbes [2].

The three-valued algebra has been used in circuit simulators, for example, in the TEGAS simulator of Thompson and Szygenda [41]. Three-valued simulation is also described in a 1976 book by Breuer and Friedman [4].

Summary: Simulation in the three-valued algebra, $\mathbf{A_3}$, is an efficient method for detecting static hazards, but cannot handle dynamic hazards. Its correctness with respect to binary analysis has been proven. Algebra $\mathbf{A_3}$ is the smallest example of ternary algebra; the mathematical properties of ternary algebra are well understood. We return to $\mathbf{A_3}$ in Section 11.

4 The four-valued algebra

The four-valued algebra of Metze has the form $P = \langle \mathcal{A}, +, *, ^{-}, 0, 1\rangle$, where $\mathcal{A}$ = {0, 1/0, 0/1, 1}, + and $*$ are defined in Table 3, and $\{0,1\}$ and $\{0/1, 1/0\}$ are complementary pairs. The values 0 and 1 are the usual logic values, and 0/1 and 1/0 represent transitions from 0 to 1 and 1 to 0, respectively. The same partial order corresponds to both operations + and $*$; it is in fact a total order: $0 \leq 1/0 \leq 0/1 \leq 1$.

Table 3. Operations + and $*$ in $\mathbf{A_4}$

+	0	1/0	0/1	1
0	0	1/0	0/1	1
1/0	1/0	1/0	0/1	1
0/1	0/1	0/1	0/1	1
1	1	1	1	1

*	0	1/0	0/1	1
0	0	0	0	0
1/0	0	1/0	1/0	1/0
0/1	0	1/0	0/1	0/1
1	0	1/0	0/1	1

Metze observes that addition and multiplication in $\mathbf{A_4}$ are idempotent, commutative and associative, and that the distributive, involution, and de Morgan's laws hold.

From Table 3 it is clear that Metze's algebra satisfies also the absorption, identity, and bounding laws. Hence $\mathbf{A_4}$ is a de Morgan algebra. It is not a Boolean algebra, since

$$0/1 + \overline{0/1} = 0/1 + 1/0 = 0/1 \neq 1.$$

Since $\mathbf{A_4}$ does not have a self-complementary element, it is not a ternary algebra.

Metze's algebra is at least as powerful as $\mathbf{A_3}$, since the mapping $h : \mathbf{A_4} \to \mathbf{A_3}$, defined below is a surjective homomorphism:

$$h(1) = 1, \quad h(0) = 0, \quad h(0/1) = h(1/0) = \Phi.$$

Thus, if we perform a simulation in $\mathbf{A_4}$ and apply this homomorphism to the results, we obtain the results of ternary simulation.

Metze presents some examples in which one can detect hazards by examining sequences of values that can occur in a circuit output. For example, if an output sequence is $0, 1/0, 1$, then there is dynamic hazard. However, these methods are rather involved, and no general algorithm is described.

Summary: From the hazard point of view, $\mathbf{A_4}$ is a step in the right direction, but has a flaw, as we discuss in the next section. Algebra $\mathbf{A_4}$ is a de Morgan algebra. A set-theoretic characterization of de Morgan algebras was recently discovered by Brzozowski [7].

5 The five-valued algebra

In 1972 Lewis presents a successful generalization of $\mathbf{A_3}$ [32], and corrects the flaw in $\mathbf{A_4}$. He notes that $\mathbf{A_4}$ has no element to represent a signal for which neither the actual value (0 or 1) nor the direction of a transition (if there is one) is known. Also, since $\mathbf{A_3}$ has no values to represent transitions without hazards, it cannot differentiate between such transitions and the corresponding dynamic hazards. Thus, Lewis adds Φ to Metze's algebra, and modifies the operations so that $0/1$ and $1/0$ now represent transitions without hazards. He defines a five-valued algebra, $\mathbf{A_5}$, with values (in our notation) 0, 1, 01 (representing a 0-to-1 transition without hazards), 10 (a 1-to-0 transition without hazards), and Φ (an unknown value). Table 4 shows the operations $+$ and $*$. The pairs $\{0, 1\}$ and $\{01, 10\}$ are complementary, and $\overline{\Phi} = \Phi$. Five-valued simulation is described by Breuer and Friedman [4].

Table 4. $+$ and $*$ operations for $\mathbf{A_5}$

$+$	0	01	Φ	10	1
0	0	01	Φ	10	1
01	01	01	Φ	Φ	1
Φ	Φ	Φ	Φ	Φ	1
10	10	Φ	Φ	10	1
1	1	1	1	1	1

$*$	0	01	Φ	10	1
0	0	0	0	0	0
01	0	01	Φ	Φ	01
Φ	0	Φ	Φ	Φ	Φ
10	0	Φ	Φ	10	10
1	0	01	Φ	10	1

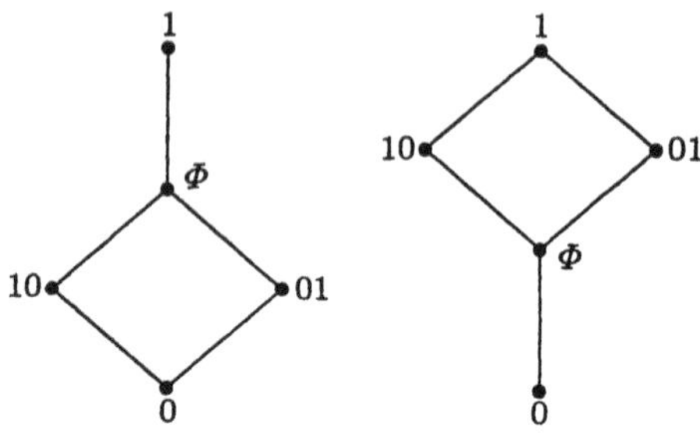

Fig. 2. Partial orders for $\mathbf{A_5}$: (a) $\leq$; (b) $\sqsubseteq$

Algebra $\mathbf{A_5}$ is not a lattice, since the absorption law fails:

$$01 * (01 + 10) = 01 * \Phi = \Phi \neq 01.$$

Distributivity does not hold because

$$01 * (10 + 1) = 01 \neq \Phi = \Phi + 01 = (01 * 10) + (01 * 1).$$

The five-valued algebra satisfies the laws of a de Morgan bisemilattice. Details of the verification are given in [3]. De Morgan bisemilattices were studied by Brzozowski [6], who showed that $\mathbf{A_5}$ is even more special, since it is a locally distributive de Morgan bilattice. He also provided a set-theoretic characterization for locally distributive de Morgan bilattices.

The partial orders for the five-valued algebra are given in Fig. 2 (a) and (b).

Hayes [25] uses $\mathbf{A_5}$ to give a second example of the application of his power-set method. Algebra $\mathbf{A_5}$ can be constructed using five elements of $2^{\mathcal{A}_2 \times \mathcal{A}_2}$, with $\mathcal{A}_2 = \{0, 1\}$, and with the following correspondence:

$$\begin{aligned} 0 &\mapsto \{(0,0)\} \\ 1 &\mapsto \{(1,1)\} \\ 01 &\mapsto \{(0,0),(0,1),(1,1)\} \\ 10 &\mapsto \{(1,1),(1,0),(0,0)\} \\ \Phi &\mapsto \{(0,0),(0,1),(1,0),(1,1)\} \end{aligned}$$

Summary: Algebra $\mathbf{A_5}$ is capable of detecting both static and dynamic hazards. It is a de Morgan bisemilattice and a locally distributive de Morgan bilattice [6], but not a de Morgan algebra. We return to $\mathbf{A_5}$ in Section 11.

6 The six-valued algebra

In this section it is convenient to discuss the second method of Hayes since the six-valued algebra is generated by this technique. This is the subdirect product [24] method. We obtain the underlying set $\mathcal{A}$ of a new algebra by taking a subset $\mathcal{S}$ of the direct product $\mathcal{P}$ of the underlying sets $\mathcal{A}_1, \ldots, \mathcal{A}_n$ of some n algebras, such that $\mathcal{S}$ determines a subalgebra of $\mathcal{P}$, and each element of each $\mathcal{A}_i$ appears as the ith component of some element of $\mathcal{S}$. We define the operations on $\mathcal{A}$ componentwise, using the operations on the n algebras. For example, algebra $\mathbf{A_6}$ is defined using $\mathbf{A_3}$ and two copies of $\mathbf{A_2}$. More specifically, $\mathbf{A_6} = \langle \mathcal{A}_6, +, *, ^-, 0, 1\rangle$, where

$$\mathcal{A}_6 = \{(0,0,0),(0,\Phi,0),(0,\Phi,1),(1,\Phi,0),(1,\Phi,1),(1,1,1)\}$$

is a subset of $\mathcal{A}_2 \times \mathcal{A}_3 \times \mathcal{A}_2$, $\mathcal{A}_2 = \{0,1\}$, and $\mathcal{A}_3 = \{0,\Phi,1\}$. For $+$, the first and third components are added in $\mathbf{A_2}$, and the second component, in $\mathbf{A_3}$; the other two operations are handled similarly. For example,

$$(0,0,0) + (1,\Phi,0) = (0+1, 0+\Phi, 0+0) = (1,\Phi,0),$$

and

$$\overline{(0,\Phi,1)} = (\overline{0},\overline{\Phi},\overline{1}) = (1,\Phi,0).$$

In an algebra constructed by the subdirect product method, a law holds iff it holds in all the component algebras. Thus, since both $\mathbf{A_2}$ and $\mathbf{A_3}$ are de Morgan algebras, we know that $\mathbf{A_6}$ is also a de Morgan algebra. It is not a Boolean algebra since the complement laws do not hold in $\mathbf{A_3}$, and it is not a ternary algebra, since there is no self-complemented element in $\mathbf{A_2}$.

Hayes interprets the ordered triples (a,b,c) in $\mathcal{A}_6$ as follows: a is the initial binary value of a signal, b, the transient value, and c, the final binary value. Thus, $(0,0,0)$ represents a constant signal with value 0, and $(0,\Phi,0)$ represents a static hazard, since the transient component has the unknown value. The value $(0,\Phi,1)$ represents a 0-to-1 transition, with or without hazards. The remaining values are similarly interpreted. It is clear that $\mathbf{A_6}$ is capable of representing static hazards, but not dynamic hazards.

In our notation we represent the elements of $\mathbf{A_6}$ as 0, $0\Phi0$, 1, $1\Phi1$, $0\Phi1$, and $1\Phi0$. The partial order corresponding to both binary operations in $\mathbf{A_6}$ is given in Fig. 3.

Summary: Algebra $\mathbf{A_6}$ is capable of representing static, but not dynamic hazards. It is a de Morgan algebra, but not a Boolean or ternary algebra. We return to $\mathbf{A_6}$ in Section 11.

7 The eight-valued algebra

The recent work of Brzozowski and Ésik discussed in Section 11 shows that there is a natural seven-valued algebra, but this algebra has not been considered before. Thus, the next value used in the past is eight.

An eight-valued algebra, $\mathbf{A_8}$, is discussed by Hayes [25] in 1986 for representing static and dynamic hazards. Hayes points out that this algebra

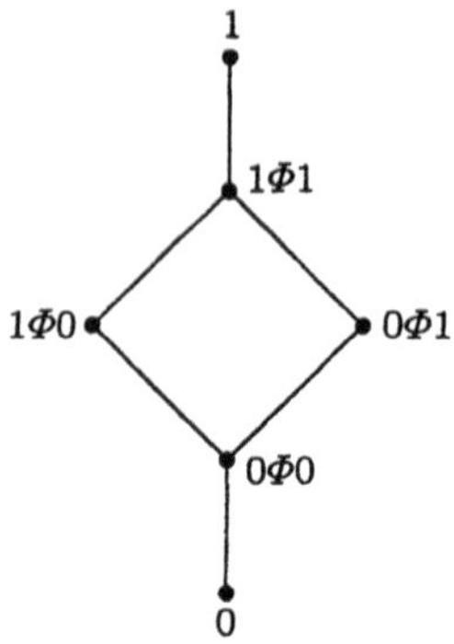

Fig. 3. Partial order for $\mathbf{A_6}$

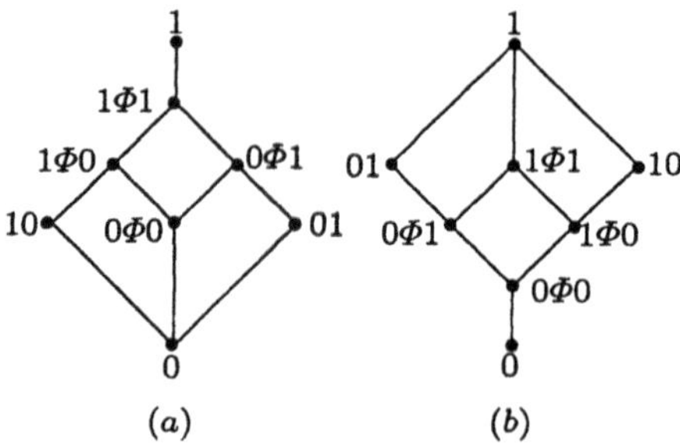

Fig. 4. Partial orders for $\mathbf{A_8}$: (a) $\leq$; (b) $\sqsubseteq$

appears already in the 1974 paper by Breuer and Harrison [5]. In that work, however, this algebra is not explicitly used, but is part of a 27-element algebra, to which we return later. In [5] the values consist of the pairs $(0,0),(0,1),(1,0),(1,1)$, which represent the levels of a signal before and after a change, and each pair is accompanied by a hazard status bit indicating whether or not a hazard occurs during the transition, for a total of eight values. Hayes also points out that the eight-valued algebra appears implicitly in the 1974 work of Fantauzzi [20], but there it is a part of a nine-valued algebra. Fantauzzi uses a different set of symbols, but the basic idea is the same. We return to his algebra in Section 8. Eight-valued simulation is described by Breuer and Friedman [4]. Algebra $\mathbf{A_8}$ is also used in [1, 26].

Using the subdirect product construction, Hayes generates the underlying set $\mathcal{A}_8$ of $\mathbf{A_8}$ as a subset of $\mathcal{A}_2 \times \mathcal{A}_5 \times \mathcal{A}_2$, or equivalently of $\mathcal{A}_2 \times (2^{\mathcal{A}_2 \times \mathcal{A}_2}) \times \mathcal{A}_2$. Algebra $\mathbf{A_2}$ is a Boolean algebra, and thus a de Morgan bisemilattice. Algebra $\mathbf{A_5}$ is a de Morgan bisemilattice, but is not a lattice. It is therefore immediate that $\mathbf{A_8}$ is a de Morgan bisemilattice. The partial orders for $\mathbf{A_8}$ are given, in our notation, in Fig. 4.

Summary: Algebra $\mathbf{A_8}$ is capable of representing both static and dynamic hazards. It is a de Morgan bisemilattice, but not a lattice. We return to $\mathbf{A_8}$ in Section 11.

8 The nine-valued algebra

As we have mentioned above, a nine-valued algebra, $\mathbf{A_9}$, was introduced by Fantauzzi [20] in 1974. The values are those of $\mathbf{A_8}$ together with a value called "ambiguous" by the author; we represent this value by Φ.

Like all hazard algebras, $\mathbf{A_9}$ has two binary operations, $+$ and $*$, and a unary operation $^-$. Hayes [25] argues that the set $\mathcal{A}_9 = \mathcal{A}_8 \cup \{\Phi\}$ is not closed under the operations $*$ and $+$. When $\mathcal{A}_9$ is viewed as consisting of ordered triples, it is easy to see that this is indeed the case. While $(1, \Phi, 0) * (\Phi, \Phi, \Phi)$ should give $(\Phi, \Phi, 0)$, this last value is not in $\mathcal{A}_9$. Fantauzzi defines $(1, \Phi, 0) * (\Phi, \Phi, \Phi)$ to be (Φ, Φ, Φ), and this results in operations $*$ and $+$ which are not associative.

Algebra $\mathbf{A_9}$ is also used in [31].

Summary: Algebra $\mathbf{A_9}$ is flawed, since the binary operations are ill defined. The subalgebra $\mathbf{A_8}$ of $\mathbf{A_9}$ is well defined, as we have stated above. There does exist a useful hazard algebra with nine elements, as we show in Section 11.[3]

9 The 13-valued algebra

As shown in Section 11, there is a hazard algebra with eleven elements, but the next value considered in the literature is thirteen.

Algebra $\mathbf{A_{13}}$ is due to Hayes [25]. It is generated by the subdirect product construction, and its underlying set is a subset of $\mathcal{A}_3 \times \mathcal{A}_5 \times \mathcal{A}_3$. The laws of a ternary algebra are a superset of those of a de Morgan bisemilattice. Algebra $\mathcal{A}_3$ is a ternary algebra, and $\mathcal{A}_5$ is a de Morgan bisemilattice. Hence, $\mathbf{A_{13}}$ is a de Morgan bisemilattice. This was first observed by Beare and Brzozowski [3]. The partial order $\leq$ corresponding to $+$ for $\mathbf{A_{13}}$ is shown in Fig. 5 in our notation.

The thirteen-valued algebra can represent all of the states used in the eight-valued algebra. In addition, there are values to represent a completely unknown signal, (Φ, Φ, Φ) (which we denote simply by Φ), signals $(0, \Phi, \Phi)$ and $(1, \Phi, \Phi)$ starting at 0 or 1 and becoming unknown (we denote them by

[3] A nine valued algebra is used by Muth [37] to generalize the D-algorithm for test generation, but has not been applied to hazard detection. It is the direct product $\mathbf{A_3} \times \mathbf{A_3}$, and so is a ternary algebra. A nine-valued simulator is also used by Knudsen for NMOS circuits [29, 30]. The nine values are 0, 1, three rising values (below threshold, critically near threshold, and above threshold), three falling values, and the unknown value. The simulator has some hazard-detecting capabilities, but NMOS circuits are outside the scope of this paper.

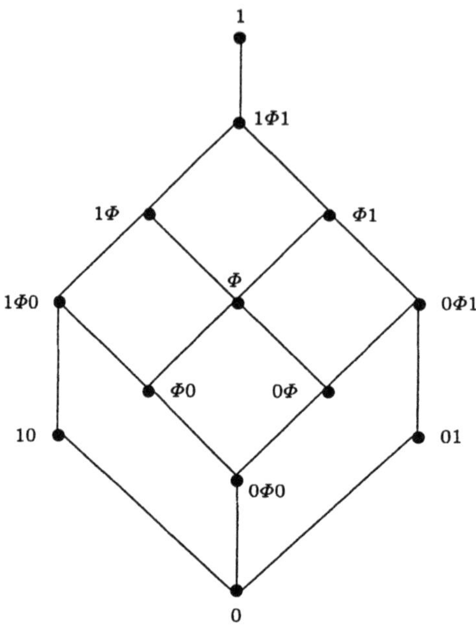

Fig. 5. Partial order $\leq$ for $\mathbf{A_{13}}$

0Φ and 1Φ), and signals $(\Phi, \Phi, 0)$ and $(\Phi, \Phi, 1)$ (denoted by $\Phi 0$ and $\Phi 1$), which begin in an unknown state and change to 0 or 1. In our notation each element is a word over the alphabet $\{0, 1, \Phi\}$, as is further explained in Section 11.

Chakraborty, Agrawal and Bushnell [14] rediscovered the thirteen-valued algebra independently in 1992. They give an alternate method of constructing $\mathbf{A_{13}}$. They examine all triples in $\mathbf{A_3} \times \mathbf{A_2} \times \mathbf{A_3}$. The first coordinate is the initial value, the third coordinate is the final value and the second coordinate indicates whether or not a hazard occurs in the transition. The eighteen elements so obtained are reduced to thirteen by noting that for signals with unknown initial or final states, it is not useful to specify the presence or absence of hazards. Hence, any two signals $(a, b, 0)$ and $(a, b, 1)$ are identified, if either a or b or both are Φ. Algebra $\mathbf{A_{13}}$ is also used in [15, 16]. Precise simulation algorithms (like Algorithms A and B for ternary simulation) for general sequential circuits have not been studied for $\mathbf{A_{13}}$.

Summary: Algebra $\mathbf{A_{13}}$ is the most complete of the algebras discussed above for analyzing circuits with respect to hazards and unknown signals, provided that counting hazard pulses (see Section 11) is not important. Algebra $\mathbf{A_{13}}$ is capable of representing static and dynamic hazards and unknown values. It is de Morgan bisemilattice, but not a lattice.

10 The 27-valued algebra

Next, we briefly consider the 27-valued algebra introduced by Breuer and Harrison [5] for eliminating static and dynamic hazards in test generation.[4] This algebra is the direct product $\mathbf{A_3} \times \mathbf{A_3}$, with an additional three-valued component, and the following interpretation. Each element of the algebra is an ordered triple (a, b, c), where a and b are the initial and final values of a transition, each from the set $\{0, \Phi, 1\}$ (in our notation), and c is the hazard status component taking its values from $\{hf, hsu, hp\}$ representing "hazard-free," "hazard status unknown," and "hazard present." The algebra is flawed, as $(0,1,hp) * (1,1,hsu)$ should give $(0,1,hp)$, but instead is defined as $(0,1,hsu)$. This results in a non-associative algebra. For example,

$$((0,1,hp) * (1,1,hsu)) * (0,1,hsu) \neq (0,1,hp) * ((1,1,hsu) * (0,1,hsu)).$$

Summary: The 27-valued algebra is flawed, since it labels an output as "hazard status unknown" when there should be a hazard. Consequently, it is not an associative algebra. As we have mentioned above, the eight-valued subalgebra $\mathbf{A_8}$ of $\mathbf{A_{27}}$ is well defined and useful.

11 Change-counting algebras

This section is a summary of the recent work of Brzozowski and Ésik [8, 9] on an infinite hazard algebra.

11.1 The infinite algebra C

Consider waveforms with a constant initial value, a transient period involving a finite number of changes, and a constant final value. Waveforms of this type are called *transients*. Figure 6 gives four examples of transients. With each such transient is associated a binary word, *i.e.*, a sequence of 0s and 1s, in a natural way. In this binary word a 0 (1) represents a maximal interval during which the signal has the value 0 (1). Such an interval is called a *0-interval* (*1-interval*).

In this section we use $\vee$, $\wedge$, and $^-$ for the Boolean OR, AND, and NOT operations, respectively. For the present, assume that circuits are constructed with 2-input OR gates, 2-input AND gates and inverters. We examine how transients are processed by such gates. The case of the inverter is the easiest

[4] Hławiczka and Badura [26] describe a 16-valued algebra for analyzing flow tables of asynchronous circuits in order to detect various types of conditions, such as critical races, essential hazards, and D-trios [42]. A discussion of their work is outside the scope of this paper. The authors observe that their algebra satisfies laws L1-L3, L5, L8, L9, and their primed versions. Since this algebra also satisfies L6, it is, in fact, a de Morgan bisemilattice.

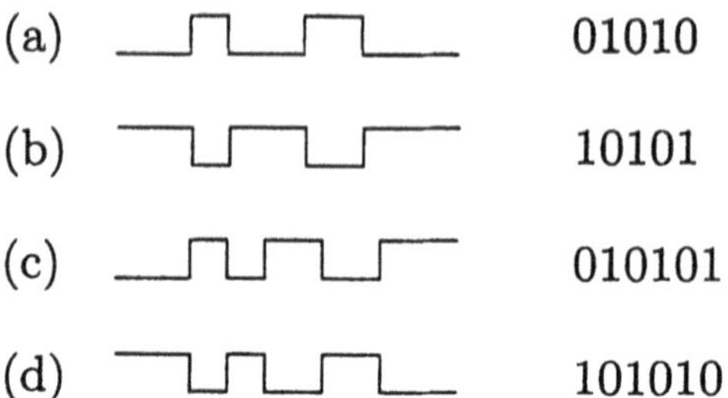

Fig. 6. Transients: (a) constant 0 with static hazards; (b) constant 1 with static hazards; (c) change from 0 to 1 with dynamic hazards; (d) change from 1 to 0 with dynamic hazards

one. If $t = a_1 \dots a_j$ is the binary word of a transient at the input of an inverter, then its output has the transient $\overline{t} = \overline{a_1} \dots \overline{a_j}$. For example, in Fig. 6, the first two transients are complementary, as are the last two. The following propositions show the largest number of changes possible at the output of an OR gate and an AND gate.

Proposition 1. *If the inputs of an* OR *gate have m and n 0-intervals respectively, then the maximum number of 0-intervals in the output signal is 0 if $m = 0$ or $n = 0$, and is $m + n - 1$, otherwise.*

Proposition 2. *If the inputs of an* AND *gate have m and n 1-intervals respectively, then the maximum number of 1-intervals in the output signal is 0 if $m = 0$ or $n = 0$, and is $m + n - 1$, otherwise.*

Let

$$T = \{0, 1, 01, 10, 010, 101, 0101, 1010, 01010, \dots\}$$

be the set of all binary words of alternating 0s and 1s. In regular expression notation,

$$T = 0(10)^* \cup 1(01)^* \cup 0(10)^*1 \cup 1(01)^*0.$$

This is the set of all transients. The *change-counting algebra* is defined as $\mathbf{C} = (T, +, *, ^-, 0, 1)$, where $+$, $*$ and $^-$ are defined below. For any $t \in T$ define $z(t)$ (z for zeros) and $u(t)$ (u for units) to be the number of 0s in t and the number of 1s in t, respectively. Let $\alpha(t)$ and $\omega(t)$ be the first and last letters of t, and let $l(t)$ denote the length of t. For example, if $t = 10101$, then $z(t) = 2$, $u(t) = 3$, $\alpha(t) = \omega(t) = 1$, and $l(t) = 5$.

Operations $+$ and $*$ are binary operations on T intended to represent the worst-case OR-ing and AND-ing of two transients at the inputs of a gate. We assume that input changes specified by the transients occurring at the inputs of a gate can take place at any time. For example, consider an OR gate with inputs X_1 and X_2 and transients 01 and 010, respectively. The input changes can occur in any one of the following orders: $X_1X_2X_2$, $X_2X_1X_2$, or $X_2X_2X_1$. One verifies that, in the first two cases, the output transient is 01, whereas in the last case, that transient is 0101. Therefore, we define $01 + 010 = 0101$. In general, the $+$ and $*$ operations are defined as follows:

$$t + 0 = 0 + t = t, \quad t + 1 = 1 + t = 1,$$

for any $t \in T$. If w and w' are words in T of length > 1, and their *sum* is denoted by $t = w + w'$, then t is that word in T that begins with $\alpha(w) \vee \alpha(w')$, ends with $\omega(w) \vee \omega(w')$, and has $z(t) = z(w) + z(w') - 1$, by Proposition 1. For example, $010 + 1010 = 101010$.

Next, define

$$t * 1 = 1 * t = t, \quad t * 0 = 0 * t = 0,$$

for any $t \in T$. Consider now the *product* of two words $w, w' \in T$ of length > 1, and denote it by $t = w * w'$. Then t is that word in T that begins with $\alpha(w) \wedge \alpha(w')$, ends with $\omega(w) \wedge \omega(w')$, and has $u(t) = u(w) + u(w') - 1$, by Proposition 2. For example, $0101 * 10101 = 01010101$.

The *(quasi-)complement* $\bar{t}$ of a word $t \in T$ is obtained by complementing each letter in t. For example, $\overline{1010} = 0101$. The constants 0 and 1 are the words 0 and 1 of length 1.

Proposition 3. *Algebra* $\mathbf{C} = (T, +, *, ^-, 0, 1)$, *is a commutative de Morgan bisemigroup.*

11.2 Counting changes to a threshold

Since the underlying set T of algebra $\mathbf{C}$ is infinite, an arbitrary number of changes can be counted. An alternative is to count only up to some threshold $k \geq 1$, and consider all transients with length k or more as equivalent.

Relation $\sim_k$ in algebra $\mathbf{C} = (T, +, *, ^-, 0, 1)$ is defined as follows: For $t, s \in T$, $t \sim_k s$ if either $t = s$ or t and s are both of length $\geq k$.

Proposition 4. *Relation* $\sim_k$ *is a congruence relation on* $\mathbf{C}$, *by which we mean that it is an equivalence relation on* T *such that for all* $t, s, w \in T$, $t \sim_k s$ *implies* $(w + t) \sim_k (w + s)$, *and* $\bar{t} \sim_k \bar{s}$; *this then implies that* $(w * t) \sim_k (w * s)$ *whenever* $t \sim_k s$.

The equivalence classes of the quotient algebra $\mathbf{C_k} = \mathbf{C}/\sim_k$ are of two types. Each transient t with $l(t) < k$ is in a class by itself, and all the words of length $\geq k$ constitute a class, which is denoted by Φ. The operations on equivalence classes are as follows. The complement of the class containing t is the class containing $\bar{t}$. The sum (product) of the class containing t and the class containing t' is the class containing $t + t'$ $(t * t')$. Thus, the quotient algebra $\mathbf{C_k}$ is a commutative de Morgan bisemigroup with $2k - 1$ elements.

Example 1. The following are examples of quotient algebras:

- For $k = 2$, the operations $+$, $*$, and $^{-}$ are those of the 3-element ternary algebra $\mathbf{A_3}$. Hence $\mathbf{A_3}$ is isomorphic to $\mathbf{C_2}$.
- For $k = 3$, the operations are those of the 5-element ternary algebra $\mathbf{A_5}$. Hence $\mathbf{A_5}$ is isomorphic to $\mathbf{C_3}$.
- In general, for each $k \geq 2$ there is an algebra $\mathbf{C_k}$ with $2k-1$ elements. □

11.3 Circuit simulation in algebras C and $\mathbf{C_k}$

The following example illustrates how static hazards are detected by ternary simulation.

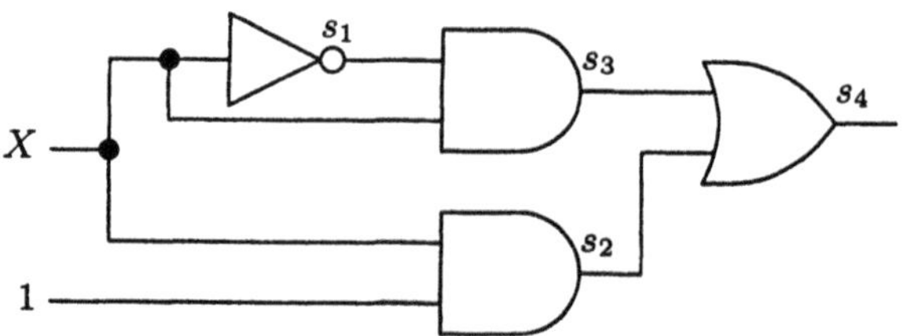

Fig. 7. Circuit with hazards

Table 5. Ternary simulation

	X	s_1	s_2	s_3	s_4
initial state	0	1	0	0	0
	Φ	1	0	0	0
	Φ	Φ	Φ	Φ	0
result A	Φ	Φ	Φ	Φ	Φ
	1	Φ	Φ	Φ	Φ
	1	0	1	Φ	Φ
result B	1	0	1	0	1

Example 2. Consider the circuit shown in Fig. 5. Refer now to Table 5. The values of the input variable X and the state variables $s_1, \ldots, s_4$ are shown in rows as the simulation progresses. We begin in the initial state 01000, which is stable. We wish to study the behavior of the circuit when the input changes from 0 to 1 and is kept constant at 1.

Ternary simulation consists of two algorithms, A and B. In Algorithm A, the input changes to the uncertain or unknown value Φ. Instead of Boolean functions, we now use the ternary functions as defined in Section 3. Thus we use the excitation equations

$$S_1 = \overline{X},\ S_2 = 1 * X,\ S_3 = X * s_1,\ S_4 = s_2 + s_3.$$

After X changes to Φ, Gates 1, 2, and 3, become unstable in the ternary model. All unstable variables are changed at the same time. This results in the second row of Algorithm A. Now Gate 4 becomes unstable and changes, to yield the third row of Algorithm A.

In Algorithm A we introduce uncertainty in the circuit inputs and we see how this uncertainty spreads throughout the circuit. In Algorithm B we start in the state produced by Algorithm A, but now we set the changing inputs from Φ to their final values, thus reducing uncertainty. In our example, X becomes 1. We again use the ternary excitation functions to see whether the uncertainty will be removed from any gates. The final result is the vector 10101, showing that each gate reaches a binary value after the transient is over.

It is clear from the circuit diagram that s_1 changes from 1 to 0 and s_2 changes from 0 to 1 without any hazards. By Boolean analysis, it can be verified [13] that a dynamic hazard is present in s_4.

This example illustrates that ternary simulation is capable of detecting static hazards. Gate s_3 is 0 at the beginning and 0 at the end, but it is Φ at the end of Algorithm A, and this indicates a static hazard [13]. Our example also clearly shows that ternary simulation is not capable of detecting dynamic hazards. Gates s_2 and s_4 both change from 0 to Φ to 1, yet s_2 has no dynamic hazard, while s_4 has one. □

In the examples that follow, we show how the accuracy of the simulation improves when we use algebra $\mathbf{C_k}$ as k increases.

Example 3. In Table 6 we repeat the simulation, this time using five-valued gate functions as defined in $\mathbf{C_3}$. Instead to changing X to Φ, we change it to 01 in Algorithm A, since this is the change we wish to study. From the result of Algorithm A we see that s_1 changes from 1 to 0 and s_2 changes from 0 to 1, both without hazards. The static hazard in s_3 is detected as before, as is the dynamic hazard in s_4. This example shows that quinary simulation is capable of detecting both static and dynamic hazards. □

Example 4. We repeat the simulation now using algebra $\mathbf{C_4}$ with seven values. Refer to Table 7. This time Algorithm A not only reveals that there is a static hazard in s_3, but also identifies it as 010. Note, however, that the dynamic hazard is still not identified. □

Example 5. We repeat the simulation using algebra $\mathbf{C_5}$ with nine values. Refer to Table 8. This time Algorithm A identifies the dynamic hazard as 0101.

Table 6. Quinary simulation

	X	s_1	s_2	s_3	s_4
initial state	0	1	0	0	0
	01	1	0	0	0
	01	10	01	01	0
	01	10	01	Φ	01
result A	01	10	01	Φ	Φ
	1	10	01	Φ	Φ
	1	0	1	10	Φ
result B	1	0	1	0	1

Table 7. Septenary simulation

	X	s_1	s_2	s_3	s_4
initial state	0	1	0	0	0
	01	1	0	0	0
	01	10	01	01	0
	01	10	01	010	01
result A	01	10	01	010	Φ
	1	10	01	010	Φ
	1	0	1	10	Φ
result B	1	0	1	0	1

Observe that the same table results if we simulate our circuit in algebra $\mathbf{C_k}$ with any $k \geq 5$, or in algebra $\mathbf{C}$. Note also, that for $k \geq 4$ Algorithm B is no longer necessary, since the entire history of worst-case signal changes is recorded on each wire. □

A characterization of the results of simulation in algebra $\mathbf{C}$ has been recently obtained by Gheorghiu [21] for feedback-free circuits consisting of 2-input AND gates, 2-input OR gates and inverters.

11.4 Simulation with initial, transient, and final values

In a number of simulators [14, 25], the signal values are ordered triples containing the initial, transient, and final values of a signal. We now show how such algebras can be described in our framework.

Relation $\approx_k$ in the algebra $\mathbf{C} = (T, +, *, ^-, 0, 1)$ is defined as follows. For $t, s \in T$, $t \approx_k s$ if either $t = s$ or $\alpha(t) = \alpha(s)$, $\omega(t) = \omega(s)$, and t and s are both of length $\geq k$. Denote by λ (for left) and ρ (for right) the congruences defined by $t \,\lambda\, s$ iff $\alpha(t) = \alpha(s)$, $t \,\rho\, s$ iff $\omega(t) = \omega(s)$. Then $\approx_k = \lambda \cap \sim_k \cap \rho$.

Proposition 5. *Relation $\approx_k$ is a congruence relation on* $\mathbf{C}$.

Table 8. Nonary simulation

	X	s_1	s_2	s_3	s_4
initial state	0	1	0	0	0
	01	1	0	0	0
	01	10	01	01	0
	01	10	01	010	01
result A	01	10	01	010	0101
	1	10	01	010	0101
	1	0	1	10	0101
result B	1	0	1	0	1

The quotient algebra $\mathbf{C}'_\mathbf{k} = \mathbf{C}/\approx_k$ is a commutative de Morgan bisemigroup with $2(k-1)+4 = 2k+2$ elements. Each word $t \in T$ with $l(t) < k$ determines a singleton congruence class. In addition, for any $b_1, b_2 \in \{0,1\}$, the words $t \in T$ with $l(t) \geq k$, $\alpha(t) = b_1$, and $\omega(t) = b_2$ determine a congruence class that we denote by $b_1 \Phi b_2$. Since $\approx_k \subseteq \sim_k$, $\mathbf{C_k}$ is a quotient of $\mathbf{C}'_\mathbf{k}$. It can be constructed from $\mathbf{C}'_\mathbf{k}$ by identifying the four elements $0\Phi0$, $0\Phi1$, $1\Phi0$, and $1\Phi1$. Also, if $k \leq m$, then $\mathbf{C}'_\mathbf{k}$ is a quotient of $\mathbf{C}'_\mathbf{m}$.

Proposition 6. *$\mathbf{C}'_\mathbf{k}$ is isomorphic to a subdirect product of two copies of the 2-element Boolean algebra $\mathbf{A_2}$ and algebra $\mathbf{C_k}$.*

Example 6. The following are examples of algebras $\mathbf{C}'_\mathbf{k}$:

- For $k = 2$, there is a six-element algebra $\mathbf{C}'_\mathbf{2}$. It is isomorphic to $\mathbf{A_6}$.
- For $k = 3$, there is a eight-element algebra $\mathbf{C}'_\mathbf{3}$. It is isomorphic to $\mathbf{A_8}$.
- In general, for any $k \geq 2$ there is a $(2k+2)$-element algebra $\mathbf{C}'_\mathbf{k}$. □

It is shown in [9] that simulation in $\mathbf{C}'_\mathbf{k}$ may not terminate for circuits with feedback; hence this approach is not suitable for such circuits.

11.5 Simulation with unknown values

As Hayes [25] points out, most digital simulators include an unknown value. Algebra $\mathbf{A_{13}}$ is an example of an algebra that allows us to represent a value that is completely unknown, as well as a value that is unknown initially, but becomes known at the end, and one that is known initially, but unknown at the end. This example can be generalized as follows. Consider $\mathbf{C}'_\mathbf{k} = (T'_k, +, *, ^-, 0, 1)$, where

$$T'_k = T_k \setminus \{\Phi\} \cup \{0\Phi0, 0\Phi1, 1\Phi0, 1\Phi1\},$$

and the operations of $\mathbf{C}'_\mathbf{k}$ are defined as above. We define a new algebra, $\mathbf{C}''_\mathbf{k}$, by adjoining five elements to $\mathbf{C}'_\mathbf{k}$. Let

$$T''_k = T'_k \cup \{0\Phi, 1\Phi, \Phi0, \Phi1, \Phi\}.$$

We consider all the elements of T''_k to be words over the alphabet $\{0, 1, \Phi\}$. Complementation in $\mathbf{C}''_k$ is letter by letter complementation in the three-valued algebra $\mathbf{A_3}$. For example, $\overline{0\Phi1} = 1\Phi0$. Addition is defined as follows. First,

$$t + 0 = 0 + t = t, \quad t + 1 = 1 + t = 1,$$

for any $t \in T''_k$. Next, if w and w' are words in T''_k of length > 1, then their *sum* is denoted by $t = w + w'$, and is a word in T''_k that begins with $\alpha(t) = \alpha(w) \vee \alpha(w')$, and ends with $\omega(t) = \omega(w) \vee \omega(w')$. It remains to compute the middle portion of t, if any. We have $t = \Phi$, if $\alpha(t) = \omega(t) = \Phi$, $t = \Phi a$, if $\alpha(t) = \Phi, \omega(t) = a \in \{0, 1\}$, and $t = a\Phi$, if $\alpha(t) = a \in \{0, 1\}, \omega(t) = \Phi$.

There remain only the cases where both $\alpha(t)$ and $\omega(t)$ are in $\{0, 1\}$, *i.e.*, where both w and w' are in T'_k. Here the rules of $\mathbf{C}'_k$ apply.

Example 7. The following are examples of algebras $\mathbf{C}''_k$:

- For $k = 3$, there is a thirteen-element algebra $\mathbf{C}''_3$. It is isomorphic to $\mathbf{A_{13}}$.
- In general, for any $k \geq 2$ there is a $(2k + 7)$-element algebra $\mathbf{C}''_k$. □

Summary: Algebra $\mathbf{C}$ leads to a general theory of simulation of gate circuits for the purpose of hazard detection, identification, and counting. The same simulation algorithms can be used to count the number of signal changes during a given input change of a circuit. This provides an estimate of the worst-case energy consumption of that input change. If a circuit has m inputs and n gates, the simulation algorithms run in $O(m + n^2)$ time. By choosing the value of the threshold k one can count signal changes and hazards to any degree of accuracy. For further properties of these algebras see [8, 9].

12 Conclusions

We have presented a survey of the algebras that have been designed for the detection of static and dynamic hazards. We have generalized the simulation algorithms, previously used only in the three-valued algebra, to the change-counting algebra $\mathbf{C}$ and its related algebras $\mathbf{C_k}$. We have provided a single framework which includes all the successful hazard algebras as special cases.

Acknowledgments: This research was supported by the Natural Sciences and Engineering Research Council of Canada under grant No. OGP0000871, and Grant No. T30511 from the National Foundation of Hungary for Scientific Research.

References

1. Andrew, R.: An Algorithm for Eight-Valued Simulation and Hazard Detection in Gate Networks. Proc. 16th Int. Symp. on Multiple-Valued Logic, IEEE (1986) 273–280

2. Balbes, R.: Free Ternary Algebras. Int. J. Algebra and Computation, vol. 10, no. 6 (2000) 739–749
3. Beare, B.D. and Brzozowski, J.A.: An Exploration of the Properties of a Thirteen-Valued Algebra. Unpublished (1999)
4. Breuer, M.A. and Friedman, A.D.: Diagnosis & Reliable Design of Digital Systems. Computer Science Press (1976)
5. Breuer, M.A. and Harrison, R.L.: Procedures for Eliminating Static and Dynamic Hazards in Test Generation. IEEE Trans. Comput., vol. C-23, no. 10, Oct. (1974) 1069–1078
6. Brzozowski, J.A.: De Morgan Bisemilattices. Int. Symp. on Multiple-Valued Logic. IEEE (2000) 173–178
7. Brzozowski, J.A.: A Characterization of de Morgan Algebras. Int. Journal of Algebra and Computation. Vol. 11, No. 5, Oct. (2001) 525–527
8. Brzozowski, J.A. and Ésik, Z.: Hazard Algebras (Extended Abstract). A Half-Century of Automata Theory, Salomaa, A., Wood, D. and Yu, S. eds., World Scientific, Singapore (2001) 1–19
9. Brzozowski, J.A. and Ésik, Z.: Hazard Algebras, Maveric Report 00-2, University of Waterloo, Waterloo, ON, Canada, 27 pp., July (2000); revised December (2001). http://maveric.uwaterloo.ca/publication.html
10. Brzozowski, J.A., Lou, J.J. and Negulescu, R.: A Characterization of Finite Ternary Algebras. Int. J. Algebra and Computation, vol. 7, no. 6 (1997) 713–721
11. Brzozowski, J.A. and Seger, C-J.: Correspondence between Ternary Simulation and Binary Race Analysis in Gate Networks. Proc. Coll. Automata, Languages and Programming, Kott, L. ed., Springer-Verlag, Berlin, July (1986) 69–78
12. Brzozowski, J.A. and Seger, C-J.: A Characterization of Ternary Simulation of Gate Networks. IEEE Trans. Computers, vol. C-36, no. 11, November (1987), 1318–1327
13. Brzozowski, J.A. and Seger, C-J.: Asynchronous Circuits, Springer-Verlag, Berlin (1995)
14. Chakraborty, T., Agrawal, V. and Bushnell, M.: Delay Fault Models and Test Generation for Random Logic Sequential Circuits. Proc. Design Automation Conf., IEEE, June (1992) 165–172
15. Chakraborty, S. and Dill, D.L.: More Accurate Polynomial-Time Min-Max Timing Simulation. Proc. Int. Symp. Advanced Research in Asynchronous Circuits and Systems, April (1997) 112–123
16. Chakraborty, S., Dill, D.L. and Yun, K.Y.: Min-Max Timing Analysis and An Application to Asynchronous Circuits. Proc. IEEE (1999) 1–13
17. Eichelberger, E.B.: Hazard Detection in Combinational and Sequential Switching Circuits. IBM J. Res. and Dev., vol. 9, (1965) 90–99
18. Ésik, Z.: A Cayley Theorem for Ternary Algebras. Int. J. Algebra and Computation, vol. 8, no. 3, (1998) 311–316
19. Ésik, Z.: Free De Morgan bisemigroups and bisemilattices. Algebra Colloquium, to appear
20. Fantauzzi, G.: An Algebraic Model for the Analysis of Logical Circuits. IEEE Trans. Computers, vol. C–23, no. 6, June (1974) 576–581
21. Gheorghiu, M.: Circuit Simulation Using a Hazard Algebra. MMath Thesis, Department of Computer Science, University of Waterloo, Waterloo, Ontario, Canada N2L 3G1, December (2001) http://maveric.uwaterloo.ca/publication.html

22. Goto, M.: Application of Three-Valued Logic to Construct the Theory of Relay Networks (in Japanese). Proc. Joint Meeting IEE, IECE, and I. of Illum. E. of Japan, (1948)
23. Goto, M.: Application of Logical Mathematics to the Theory of Relay Networks (in Japanese). J. Inst. Elec. Eng. of Japan, vol. 69, no. 729, (1949)
24. Grätzer, G.: Universal Algebra. Second Edition, Springer-Verlag (1979)
25. Hayes, J.P.: Digital Simulation with Multiple Logic Values. IEEE Trans. CAD, vol. CAD–5, no. 2 (1986) 274–283
26. Hławiczka, A. and Badura, D.: The Method of Recognition of Critical Hazards, Critical Races, Essential Hazards and D-Trio. Proc. Int. Symp. Multiple-Valued Logic, IEEE, (1982) 298–311
27. Huffman, D.A.: The Design and Use of Hazard-Free Switching Circuits. J. ACM, vol. 4 (1957) 47–62
28. Keister, W., Ritchie, A.E. and Washburn, S.H.: The Design of Switching Circuits. D. Van Nostrand, New York (1951)
29. Knudsen, M.S.: A Compact Nine-Valued Logic Simulation Algorithm. Proc. Int. Symp. Circuits and Systems, vol. 3, IEEE, (1982) 1190–1193
30. Knudsen, M.S.: A Nine-Valued Logic Simulator for Digital N-MOS Circuits. Proc. Int. Symp. Multiple-Valued Logic. IEEE, (1982) 293–297
31. Kung, D.S.: Hazard-Non-Increasing Gate-Level Optimization Algorithms. Proc. Int. Conf. Computer-Aided Design, IEEE, (1992) 631–634
32. Lewis, D.W.: Hazard Detection by a Quinary Simulation of Logic Devices with Bounded Propagation Delays, MSc Thesis, University of Syracuse (1972)
33. McCluskey, E.J.: Transient Behavior of Combinational Logic Circuits. In Redundancy Techniques for Computing Systems. Wilcox, R.H. and Mann, W.C. eds., Spartan Books, Washington, DC (1962) 9–46
34. Metze, G.A.: Many-Valued Logic and the Design of Switching Circuits, MSc Thesis, University of Illinois, Urbana (1953)
35. Moisil, Gr.C.: Sur l'application des logiques à trois valeurs à l'étude des schémas à contacts et relais. Actes proc. congr. intern. automatique, (1956) 48
36. Muller, D.E.: Treatment of Transition Signals in Electronic Switching Circuits by Algebraic Methods. IRE Trans. Electronic Computers, vol. EC–8, no. 3, September (1959) 401
37. Muth, P.: A Nine-Valued Circuit Model for Test Generation. IEEE Trans. Computers, vol. C–25, no. 6, June (1976)
38. Roginskii, V.N.: The Operation of Relay Networks in Transitional Periods. Avtomatika i Telemekhanika, vol. 20, no. 10, October (1959) 1408–1416
39. Seger, C-J. and Brzozowski, J.A.: Generalized Ternary Simulation of Sequential Circuits. Theoretical Informatics and Applications, vol. 28, No. 3–4, (1994) 159–186
40. Shannon, C.E.: A Symbolic Analysis of Relay and Switching Circuits. Trans. AIEE, vol. 57 (1938) 713–723
41. Thompson, E.W. and Szygenda, S.A.: Three Levels of Accuracy for the Simulation of Different Fault Types in Digital Systems. Proc. Design Automation Conference, IEEE (1975) 105-113
42. Unger, S.H.: Asynchronous Sequential Switching Circuits. Wiley-Interscience, New York (1969)
43. Yoeli, M. and Rinon, S.: Application of Ternary Algebra to the Study of Static Hazards. J. ACM, vol. 11, no. 1, January (1964) 84–97

Chapter 2
An Abstract Algebraic Logic View of Some Mutiple-valued Logics

Josep Maria Font

Faculty of Mathematics, University of Barcelona
Gran Via 585, E-08007 Barcelona, Spain
font@mat.ub.es

Abstract. Abstract Algebraic Logic is a general theory of the algebraization of deductive systems arising as an abstraction of the well-known Lindenbaum-Tarski process. The notions of logical matrix and of Leibniz congruence are among its main building blocks. Its most successful part has been developed mainly by BLOK, PIGOZZI and CZELAKOWSKI, and obtains a deep theory and very nice and powerful results for the so-called protoalgebraic logics. I will show how the idea (already explored by WÓJCKICI and NOWAK) of defining logics using a scheme of "preservation of degrees of truth" (as opposed to the more usual one of "preservation of truth") characterizes a wide class of logics which are not necessarily protoalgebraic and provide another fairly general framework where recent methods in Abstract Algebraic Logic (developed mainly by JANSANA and myself) can give some interesting results. After the general theory is explained, I apply it to an infinite family of logics defined in this way from subalgebras of the real unit interval taken as an MV-algebra. The general theory determines the algebraic counterpart of each of these logics without having to perform any computations for each particular case, and proves some interesting properties common to all of them. Moreover, in the finite case the logics so obtained are protoalgebraic, which implies they have a "strong version" defined from their Leibniz filters; again, the general theory helps in showing that it is the logic defined from the same subalgebra by the truth-preserving scheme, that is, the corresponding finite-valued logic in the most usual sense. However, for infinite subalgebras the obtained logic turns out to be the same for all such subalgebras and is not protoalgebraic, thus the ordinary methods do not apply. After introducing some (new) more general abstract notions for non-protoalgebraic logics I can finally show that this logic too has a strong version, and that it coincides with the ordinary infinite-valued logic of Łukasiewicz.

1 On abstract algebraic logic

In papers on the algebraic study of a specific logic it is common to read sentences like "[such and such class of algebras] plays in relation to [such and such logic] a role similar to that played by Boolean algebras in relation to classical logic". Many works in the Algebraic Logic literature are devoted to the study of particular logics and the particular associated class of algebras, and often the said "role" amounts to very little more than the completeness theorem. Thus most of the benefits of having an algebraic counterpart of a logic were usually obtained by suitably devised ad-hoc procedures. ***Abstract Algebraic Logic*** is the branch or part of Algebraic Logic where the emphasis is put on the process of algebraization itself rather than on its results for this or that logic, and where the process is analysed and described at a truly *abstract* level. The *general* theories it develops account for the algebraization of particular logics, and can be used to obtain properties of the logics from those of the algebras or vice-versa, once their connection has been established. It also identifies, either by metalogical or by algebraic conditions, some classes of logics where certain methods can be used with particular success, the connection logic-algebras acquiring varying degrees of intensity.

One of the distinctive features of Abstract Algebraic Logic is the very definition of what a *logic* is; not taking this into account may lead to some misunderstandings. TARSKI's conception of logic as *consequence* is adopted: A ***logic*** or ***deductive system*** $\mathcal{S}$ is here identified with a finitary and substitution-invariant consequence relation on the set Fm of formulas; that is, a binary relation $\vdash_{\mathcal{S}} \subseteq \mathcal{P}(Fm) \times Fm$ such that:

1. $\Gamma \vdash_{\mathcal{S}} \varphi$ whenever $\varphi \in \Gamma$.
2. $\Gamma \vdash_{\mathcal{S}} \varphi$ whenever $\Delta \vdash_{\mathcal{S}} \varphi$ and $\Delta \subseteq \Gamma$.
3. $\Gamma \vdash_{\mathcal{S}} \varphi$ whenever $\Delta \vdash_{\mathcal{S}} \varphi$ and $\Gamma \vdash_{\mathcal{S}} \beta$ for every $\beta \in \Delta$.
4. $\Gamma \vdash_{\mathcal{S}} \varphi$ implies $\sigma\Gamma \vdash_{\mathcal{S}} \sigma\varphi$ for every substitution σ.
5. $\Gamma \vdash_{\mathcal{S}} \varphi$ implies $\Gamma_0 \vdash_{\mathcal{S}} \varphi$ for some finite $\Gamma_0 \subseteq \Gamma$.

For very general theoretical studies it may be useful to drop the finitarity condition 5, as is done in [11, 15, 45]; in such cases it is better to speak of ***consequence relations***. Note that Abstract Algebraic Logic departs from the approaches where a logic is identified with a *set of formulas* closed under some conditions or rules. It also departs from those where a logic is understood as a *proof system* of a certain kind, and from those requiring that each logic should always have both a semantics and a proof theory. While any of these devices can *define* a logic, they are not regarded as part of the notion of logic itself, but as *properties* a logic may or may not have.

The second distinctive feature of Abstract Algebraic Logic (or of Algebraic Logic in general) is that its models are taken on *algebras* of the same similarity type as the language of the formulas, and that the interpretations or evaluations are the *homomorphisms* from the formula algebra $\boldsymbol{Fm}$ to the algebra where the model resides. The model itself can be some kind of structure over the algebra, such as a subset or a family of subsets.

The most classical and best developed part of Abstract Algebraic Logic uses *subsets* as models; its central notions are those of logical matrix and of Leibniz operator. It is well-known that the origin of the notion of logical matrix can be traced back to the twenties, or even before. The general theory of matrix semantics was established through the work of WÓJCICKI [43], CZELAKOWSKI [9], RASIOWA [34] and many others, and completed its maturity after the introduction of the notion of Leibniz operator by BLOK and PIGOZZI [2] and its systematic study by themselves and other people. A (logical) ***matrix*** is a pair $\langle \boldsymbol{A}, F \rangle$ where $\boldsymbol{A}$ is an algebra of suitable similarity type, and $F \subseteq A$ is the set of so-called *designated elements*, which represent *truth* in the model. One says that $\langle \boldsymbol{A}, F \rangle$ is a ***matrix for*** $\mathcal{S}$ (briefly, an ***$\mathcal{S}$-matrix***) when for every $\Gamma \subseteq Fm$ and every $\varphi \in Fm$ such that $\Gamma \vdash_{\mathcal{S}} \varphi$ the following holds:

$$\text{For every } v \in \operatorname{Hom}(\boldsymbol{Fm}, \boldsymbol{A}), \text{ if } v[\Gamma] \subseteq F \text{ then } v(\varphi) \in F. \qquad (1)$$

The set F is then called an ***$\mathcal{S}$-filter***. For each algebra $\boldsymbol{A}$ the family of all the $\mathcal{S}$-filters on $\boldsymbol{A}$ is denoted by $\mathcal{F}i_{\mathcal{S}}\boldsymbol{A}$.

The ***Leibniz congruence*** of a matrix $\langle A, F \rangle$ is defined as

$$\boldsymbol{\Omega}_{\boldsymbol{A}}(F) = \max\{\theta \in \operatorname{Co}\boldsymbol{A} : \text{ if } \langle a, b \rangle \in \theta \text{ and } a \in F \text{ then } b \in F\},$$

where $\operatorname{Co}\boldsymbol{A}$ denotes the set of (algebraic) congruences of the algebra $\boldsymbol{A}$. A matrix is ***reduced*** when its Leibniz congruence is the identity. The first class of algebras naturally associated with a logic, denoted by $\mathsf{Alg}^*\mathcal{S}$, is the class of the algebraic reducts of the reduced matrices of $\mathcal{S}$, that is, the class of algebras $\boldsymbol{A}$ such that there is $F \in \mathcal{F}i_{\mathcal{S}}\boldsymbol{A}$ with $\langle \boldsymbol{A}, F \rangle$ reduced; notice that this F need not be unique. It sometimes happens that the Leibniz congruence and the reduced matrices of a given logic can be nicely represented; for instance in the implicative logics studied in [34] $\langle a, b \rangle \in \boldsymbol{\Omega}_{\boldsymbol{A}}(F)$ if and only if $a \to b, b \to a \in F$, and the algebras in $\mathsf{Alg}^*\mathcal{S}$, there called "$\mathcal{S}$-algebras", are determined by the axioms and rules of the logic plus the familiar condition "if $a \to b = b \to a = 1$ then $a = b$". But in less well-behaved cases things can be much different.

The historical development of Abstract Algebraic Logic can be identified with the process of extending some paradigms of the algebraization of logic, which had proven successful in the study of the best-behaved logics, to wider and wider classes of logics; the extension has been performed in a way that obtains deep and meaningful results and as powerful and nice a theory as possible, while keeping the old results in the already studied cases. The generalization of the Lindenbaum-Tarski process to ***implicative*** logics [34] has been extended and specialised for other (increasingly larger) classes of logics: the ***algebraizable*** ones [3, 12], the ***equivalential*** ones [10], and the ***protoalgebraic*** ones [2, 4]. Each of these three classes of logics can be characterized by a certain aspect of the behaviour of ***the Leibniz operator***: the mapping $\boldsymbol{\Omega}_{\boldsymbol{A}} : F \mapsto \boldsymbol{\Omega}_{\boldsymbol{A}}(F)$ when F ranges over $\mathcal{F}i_{\mathcal{S}}\boldsymbol{A}$. The result has been a powerful, complex and multifaceted theory, which by now forms the established core of Abstract Algebraic Logic, as developed in [4, 5, 11]; [17] is a compact, yet comprehensive survey of recent work in the area. I am just going to give the definitions and properties I will use in the paper.

The classes of logics mentioned in the previous paragraph, together with a few others, form the so-called ***hierarchy*** of logics, also called the *protoalgebraic hierarchy*, the *algebraic hierarchy* or the *Leibniz hierarchy*. A logic $\mathcal{S}$ is ***protoalgebraic*** when for every $\boldsymbol{A}$, the Leibniz operator $\boldsymbol{\Omega}_{\boldsymbol{A}}$ is monotonic on $\mathcal{F}i_{\mathcal{S}}\boldsymbol{A}$, that is, $F \subseteq G$ implies $\boldsymbol{\Omega}_{\boldsymbol{A}}(F) \subseteq \boldsymbol{\Omega}_{\boldsymbol{A}}(G)$ for all $F, G \in \mathcal{F}i_{\mathcal{S}}\boldsymbol{A}$. $\mathcal{S}$ is ***equivalential*** when there is a set $\Delta(p, q)$ of formulas in two variables that defines the Leibniz congruence on $\mathcal{S}$-filters in the following sense: For any $\boldsymbol{A}$, if $F \in \mathcal{F}i_{\mathcal{S}}\boldsymbol{A}$ and $a, b \in A$, then

$$\langle a, b\rangle \in \boldsymbol{\Omega}_{\boldsymbol{A}}(F) \iff \Delta^{\boldsymbol{A}}(a, b) \subseteq F. \tag{2}$$

The set Δ is the set of ***equivalence formulas*** of $\mathcal{S}$. When this set can be taken finite then $\mathcal{S}$ is called ***finitely equivalential***. $\mathcal{S}$ is ***weakly algebraizable*** when the Leibniz operator in injective on $\mathcal{S}$-filters; it is ***algebraizable*** when it is both equivalential and weakly algebraizable. The original definition of this notion by BLOK and PIGOZZI in [3] is now called ***finitely algebraizable***, and corresponds to the logics that are both weakly algebraizable (or algebraizable) and finitely equivalential; for these logics the class $\mathsf{Alg}^*\mathcal{S}$ is a quasivariety, which is called ***the equivalent algebraic semantics*** for $\mathcal{S}$. The links between the logic and this class of algebras are very strong, for instance the Leibniz operator $\boldsymbol{\Omega}_{\boldsymbol{A}}$ becomes an *isomorphism* between the lattices $\mathcal{F}i_{\mathcal{S}}\boldsymbol{A}$ and $\mathrm{Co}_{\mathsf{Alg}^*\mathcal{S}}\boldsymbol{A}$ (the set of congruences of $\boldsymbol{A}$ yielding a quotient in $\mathsf{Alg}^*\mathcal{S}$). Further kinds of algebraizability are denoted by an additional adjective to each of the just mentioned classes: ***regularly*** means that each algebra $\boldsymbol{A} \in \mathsf{Alg}^*\mathcal{S}$ has a special element 1 such that $\mathcal{S}$ is complete with respect to the class of matrices $\{\langle \boldsymbol{A}, \{1\}\rangle : \boldsymbol{A} \in \mathsf{Alg}^*\mathcal{S}\}$; ***strongly*** means that the class $\mathsf{Alg}^*\mathcal{S}$ is a *variety*. Figure 1 shows the organisation of the main classes of the hierarchy appearing in this paper.

Protoalgebraic logics, the largest class in the hierarchy, include the vast majority of logics usually considered in the literature (classical and intuitionistic logics, modal logics, many-valued logics, etc.). They are considered to be the largest class of logics to which the standard model-theoretic methods of the theory of logical matrices can be successfully applied, beyond the most general completeness theorems. The following characterization is of a special interest: A logic is protoalgebraic if and only if there is a set $E(p, q)$ of formulas in two variables satisfying the following two (minimal) requirements, for all formulas φ, ψ:

$$\begin{array}{ll} \text{(Law of Identity)} & \vdash_{\mathcal{S}} E(\varphi, \varphi) \\ \text{(Modus Ponens)} & E(\varphi, \psi) \cup \{\varphi\} \vdash_{\mathcal{S}} \psi \end{array}$$

From this it results that non-protoalgebraic logics must have no implication at all, or at most a rather strange one; formerly it was believed that only

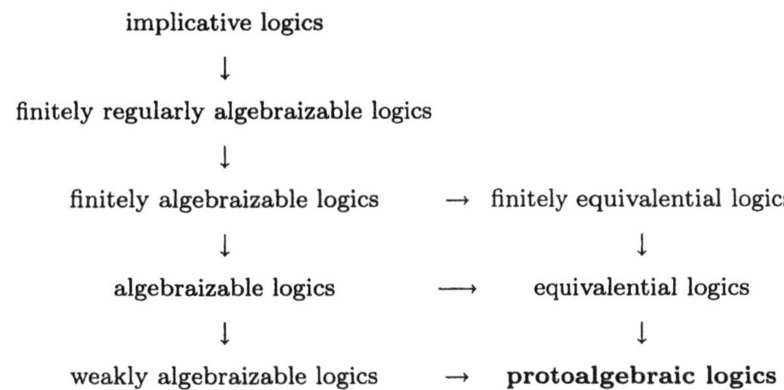

Fig. 1. Some of the main classes of logics in the hierarchy. $\rightarrow$ means $\subseteq$.

very pathological logics would be non-protoalgebraic, but recently their interest has been recognized, in parallel to the identification of several families of (natural) examples: the conjunction-disjunction and the implication-less fragments of intuitionistic logic [20, 35], some subintuitionistic logics [1, 7, 36, 40, 42], BELNAP's four-valued logic [13], and the weak version of system $\mathcal{R}$ of relevance logic, defined by following WÓJCICKI's suggestions in [45, p. 165] and algebraically studied in [19]. And, as I will show in the final section, there is also an infinite multiple-valued logic in this group. The algebraic treatment of these logics clearly calls for another framework with a wider scope.

This recent branch of Abstract Algebraic Logic has grown around the notions of generalized matrix [43], of Tarski congruence, and of full model. A ***generalized matrix*** (called ***abstract logic*** in [6] and in [14]) is a pair $\langle \boldsymbol{A}, \mathcal{C} \rangle$ where $\mathcal{C}$ is a closed-set system (i.e., a family of subsets closed under arbitrary intersections and containing the whole universe) on an algebra $\boldsymbol{A}$. It is a ***model of a logic*** $\mathcal{S}$ when for every $v \in \mathrm{Hom}(\boldsymbol{Fm}, \boldsymbol{A})$, if $\Gamma \vdash_{\mathcal{S}} \varphi$ then $v(\varphi) \in C(v[\Gamma])$, where C is the closure operator associated with the closed-set system $\mathcal{C}$. Obviously, $\langle \boldsymbol{A}, \mathcal{C} \rangle$ is a model of $\mathcal{S}$ if and only if $\mathcal{C} \subseteq \mathcal{F}i_{\mathcal{S}}\boldsymbol{A}$; thus on any algebra, an arbitrary collection of $\mathcal{S}$-filters constitutes a model. This means that not much can be said about models in general, but some can be selected as behaving in more interesting ways. Observe that on any algebra there is a "largest" model $\langle \boldsymbol{A}, \mathcal{F}i_{\mathcal{S}}\boldsymbol{A} \rangle$; and it turns out that models that are "like" these are seen to have more interest. Models of this kind are called ***basic full models***, and a generalized matrix is a ***full model of*** $\mathcal{S}$ when it is the inverse image of a basic full model of $\mathcal{S}$ under a strict surjective homomorphism between generalized matrices; a surjective $h \in \mathrm{Hom}(\boldsymbol{B}, \boldsymbol{A})$ is ***strict*** between $\langle \boldsymbol{B}, \mathcal{D} \rangle$ and $\langle \boldsymbol{A}, \mathcal{C} \rangle$ when $\mathcal{D} = h^{-1}[\mathcal{C}]$. The ***Tarski congruence*** of a generalized matrix $\langle \boldsymbol{A}, \mathcal{C} \rangle$ is defined as

$$\widetilde{\Omega}_{\boldsymbol{A}}(\mathcal{C}) = \bigcap \{ \Omega_{\boldsymbol{A}}(F) : F \in \mathcal{C} \}$$

and $\langle \boldsymbol{A}, \mathcal{C} \rangle$ is ***reduced*** when its Tarski congruence is the identity. Then the class of ***$\mathcal{S}$-algebras***, the second class of algebras canonically associated with a logic, is defined as the class of algebraic reducts of reduced models of $\mathcal{S}$; it is denoted by $\mathbf{Alg}\mathcal{S}$, and it happens to coincide with the class of algebraic reducts of reduced full models of $\mathcal{S}$. The study of these notions has been developed mainly in [14], where the thesis is maintained (by developing a consistent general theory and by analysing many examples) that they account for the algebraization of arbitrary logics in a faithful and meaningful way. In particular, it seems that $\mathbf{Alg}\mathcal{S}$ deserves the title of ***the algebraic counterpart of a logic*** much better than $\mathbf{Alg}^*\mathcal{S}$, specially in the non-protoalgebraic cases; moreover, if $\mathcal{S}$ is protoalgebraic then $\mathbf{Alg}\mathcal{S} = \mathbf{Alg}^*\mathcal{S}$, hence in the best-behaved and well-known cases the classical theory is recovered.

In the quest for new and more encompassing general theories the *study of examples*, either taken individually or in groups that share certain features, is essential. They are needed to test the theory against practice, to confirm (or, in some cases, surprisingly reject) the intuitive ideas or the results obtained by ad-hoc constructions not conforming to any precise methodology, to compare with existing paradigms, to help identify and sort the key notions from the peripheral ones, etc.

In this paper I try to show how the application of a certain general framework, developed in [16] in the context just introduced, helps in understanding and describing the algebraic behaviour of a certain family of multiple-valued logics defined from truth-value algebras that are subalgebras of the real unit interval endowed with Łukasiewicz's familiar operations. The way these logics are defined from the algebras is a particular instance of the so-called ***semilattice-based logics***[1], a general procedure devised in order to formalise a specific view of logics as inference systems preserving degrees of truth. I devote Section 2 to an informal exposition of this idea, and Section 3 to summarize the main elements of the general theory of [16] that will be used in the sequel. Then Sections 4 and 5 contain the detailed treatment of Łukasiewicz-based logics.

As general references on Abstract Algebraic Logic I recommend [3, 5, 11, 14, 17, 34, 45]; for multiple-valued logics the recent survey monographs [8, 24, 25] contain a lot of information.

2 Logics preserving degrees of truth

Non-standard truth values are the basis of any rationale behind the setting up of a multiple-valued logic on semantical grounds. When a logic is defined through the use of a set of more than two values, these 'values' are presented

[1] The term "(semi)lattice-based" has been used in [38], in a non-technical way, to describe a large class of logics having both algebraic and relational semantics linked by representation theorems of different kinds.

as encoding different *forms, ways* or *kinds* of *being true*, and are called 'truth-values'; the classical 0 and 1 representing the extreme cases of *absolute truth* and *absolute falsity.* However, the way these truth-values are actually used in the definition of the logic may point to a different interpretation, and supports the study of an alternative way of using them.

The most common framework for definition of multiple-valued logics is to start from some *set of truth-values* A and among them to select or *designate* a certain element $1 \in A$ as representing *truth.* Then, given some set Val of *evaluations*, that is, functions $Fm \to A$, one can define a consequence relation $\vdash^1$ in the following way: For any $\Gamma \subseteq Fm$ and any $\varphi \in Fm$,

$$\Gamma \vdash^1 \varphi \iff v(\varphi) = 1 \text{ whenever for all } \beta \in \Gamma,\ v(\beta) = 1, \text{ for all } v \in Val. \tag{3}$$

Logics defined in this way are usually said to follow a ***truth-preserving*** scheme. However, if one wants to really believe that all elements of A represent some kind of truth, of which 1 represents absolute truth, then (3) should be rather regarded as a "preservation of absolute truth" scheme, as it does not guarantee the preservation of any other truth-value than absolute truth: the elements of A are used as possible values for the computation of the value $v(\varphi)$ of non-atomic formulas φ from the values of their atomic parts (variables), but then only those evaluations giving final truth-value 1 to the formulas are taken into account in order to define consequence. This way of *using* the truth-values induces one to think that the other values do not carry any kind of truth in themselves, and that they rather represent "kinds of falsity", or simply "values" which are not "truth-values".

The same comment can be made in case one selects or designates a subset $D \subseteq A$ instead of a single element: If the definition is

$$\Gamma \vdash^D \varphi \iff v(\varphi) \in D \text{ whenever for all } \beta \in \Gamma,\ v(\beta) \in D, \text{ for all } v \in Val. \tag{4}$$

then only final values inside D count, and the "truth-content" of values outside D seems not to be relevant for the logic so defined.

An alternative way of using a set A of 'truth-values' as truly representing different kinds of truth, is to think of them as *degrees of truth*, and to *understand consequence in the following sense*: That whenever all premisses attain at least a certain degree of truth, the conclusion should have at least that degree of truth too. This means assuming that there is some (partial) *ordering relation* $\leq$ among the elements of A, and defining a consequence relation $\vdash^{\leq}$ in the following way:

$$\Gamma \vdash^{\leq} \varphi \iff v(\varphi) \geq t \text{ whenever for all } \beta \in \Gamma,\ v(\beta) \geq t, \text{ for all } v \in Val \text{ and all } t \in A. \tag{5}$$

This scheme is referred to as ***preservation of degrees of truth.***

In the case where one wants the logic to be truth-functional, one assumes that the set of truth-values A has an algebraic structure $\boldsymbol{A} = \langle A, \{\lambda^{\boldsymbol{A}} : \lambda \in \mathcal{L}\}\rangle$ of the similarity type $\mathcal{L}$ of the formulas, and takes $Val = \mathrm{Hom}(\boldsymbol{Fm}, \boldsymbol{A})$ as the set of evaluations. Then, both definition schemes can be smoothly represented in a more algebraic-logic style by means of *matrices*, in the technical sense described in Section 1: Schemes (3) and (4) correspond to the logic defined by the logical matrices

$$\langle \boldsymbol{A}, \{1\} \rangle \qquad \text{or} \qquad \langle \boldsymbol{A}, D \rangle \tag{6}$$

respectively, while (5) corresponds to the logic defined by the family of matrices

$$\{ \langle \boldsymbol{A}, [t) \rangle : t \in A \} \tag{7}$$

where $[t) = \{a \in A : t \leq a\}$. Equivalently, (5) corresponds to the logic defined by the generalized matrix $\langle \boldsymbol{A}, \mathcal{C} \rangle$ where $\mathcal{C}$ is the closed-set system generated by the family of sets $\{[t) : t \in A\}$.

These definitions might seem too algebraic, and be considered too restrictive as general schemes for defining logics. However, WÓJCICKI has shown in [45, Chapter 5] that any logic defined either locally or globally by a relational semantics of the most general kind, and hence apparently non-truth-functional, can also be defined by a class of matrices or of generalized matrices; therefore, it happens to be truth-functional with respect to convenient algebraic structures, which are obtained through suitable representation constructions from the relational structures. This issue has also been reviewed, for a large class of particular cases, in [39].

The idea of preservation of the degree of truth is not new. It has been surely discussed in a variety of works related to multiple-valued logics or even more in general, in connection with matrix semantics, as in [45, p. 191][2]. Its specific algebraic side has been less studied, though. It is applied to finite subalgebras of the real unit interval in [21, 45] (see Section 4 for details) and it is studied in general by NOWAK in [33]; in NOWAK's paper the set of truth-values is supposed to have (or to be embedded in) a complete lattice structure with maximum 1, hence (5) can be rephrased as

$$\Gamma \vdash^{\leq} \varphi \iff v(\varphi) \geq \inf\big(\{v(\beta) : \beta \in \Gamma\} \cup \{1\}\big)\,, \quad \text{for all } v \in \mathrm{Hom}(\boldsymbol{Fm}, \boldsymbol{A}). \tag{8}$$

Notice that this is just one of the several possible notions of "preserving degrees of truth" that are introduced and characterised in [33]. However, it turns out that the consequence relations defined in any of the preceding ways cannot in general be guaranteed to be *finitary*, unless the set A is finite, see

[2] Warning: in this book the term "truth-preserving" is used in the sense explained above, except on page 345 where it means "preserving degrees of truth" (and at that place "preserving validity" is used as a replacement for "truth-preserving").

Theorem 20. I am going to show that, by restricting (3) and (5) to finite Γ, one obtains a reasonably smooth general framework to be exploited with the tools of Abstract Algebraic Logic, and which includes many usual logics, while it does not require them to be even protoalgebraic. Moreover, it is not necessary to assume a complete lattice structure, as in [33]; just assuming an *inf-semilattice with maximum* is enough. After these two changes, if as usual the semilattice operation is denoted by $\wedge$ and its maximum is denoted by 1, then condition (8) can be split into the following two:

$$\varphi_0, \dots, \varphi_{n-1} \vdash^{\leq} \psi \iff v(\varphi_0) \wedge \cdots \wedge v(\varphi_{n-1}) \leq v(\psi) \quad \text{for all } v \in \mathrm{Hom}(\boldsymbol{Fm}, \boldsymbol{A}) \tag{8'}$$

$$\vdash^{\leq} \psi \iff v(\psi) = 1 \quad \text{for all } v \in \mathrm{Hom}(\boldsymbol{Fm}, \boldsymbol{A}) \tag{8''}$$

One typical feature of these schemes, which becomes even clearer now, is that the theorems of the logic $\vdash^{\leq}$ will be the same as those of the logic $\vdash^{1}$. It is clear that any possible interest of the present proposal must lie in the *inferential* aspect of logics, rather than in their *assertional* aspect. The problem of how to associate a consequence, or *entailment relation* with a given set of "theorems" or "tautologies" is surely a non-trivial one, and has been discussed many times and from many points of view. I want to highlight here the discussion in [45, Section 2.10] because of its connection with the more specific part of the present paper: there, an operation $\rightarrow$ of "implication" is assumed to exist, which establishes a strong connection between the two key elements at work, namely the ordering relation and the maximum truth-value; applied to the present case this would become:

$$\text{For all } a, b \in A, \quad a \leq b \iff a \rightarrow b = 1. \tag{9}$$

Any logic defined through the schemes (8') and (8") from a truth-value algebra where (9) holds, satisfies

$$\varphi_0, \dots, \varphi_{n-1} \vdash^{\leq} \psi \iff \vdash^{\leq} \varphi_0 \wedge \cdots \wedge \varphi_{n-1} \rightarrow \psi. \tag{10}$$

Viewed the other way round: Imagine that a set of "tautologies" has been defined from $\boldsymbol{A}$ and 1 through (8"). Then if one wants to have connectives $\wedge$ and $\rightarrow$ representing respectively conjunction and inference *inside the language* so that (10) holds, then the natural way is to go for (8') and find the logic $\vdash^{\leq}$ defined by preservation of degrees of truth. However, having (9) is a rather particular property that may not be present in many cases, for instance in all those cases where simply there is no such implication in the language, or when one precisely wants to deal with an implication-less fragment of a richer logic; in those cases (8') and (8") still offer a sound definition of a logic preserving degrees of truth.

Condition (8') or its more general form (5) are often paraphrased as stating that "the conclusion must have at least the degree of truth of the premisses". Since in logical inference premises act collectively, it is generally acknowledged that a reasonable evaluation of the "collective" degree of truth of the set of premisses is the infimum of their degrees of truth. It is interesting to notice that a similar intuition is present in the characterizations of the notions of a fuzzy subset being a fuzzy subalgebra of a (crisp) algebra [31] and of canonical fuzzy numbers [32, 30].

3 Semilattice-based logics

The material in this section is excerpted from [15] and [16].

Let $\mathbf{K}$ be a class of algebras of some (arbitrary but fixed from now on) similarity type having an upper-bounded inf-semilattice reduct; this means there is a partial ordering relation $\leq$ on each $\boldsymbol{A} \in \mathbf{K}$ having a maximum 1, and a binary connective $\wedge$ (which can be either a primitive one or defined by a term in two variables) such that $a \wedge b = \inf\{a, b\}$ for all $a, b \in A$ and all $\boldsymbol{A} \in \mathbf{K}$. There is no harm in assuming that the maximum is represented as a constant $\top$ of the language. It is well-known that this situation can be expressed equationally as the satisfaction in $\mathbf{K}$ of the four following equations

$$x \wedge x \approx x$$
$$x \wedge y \approx y \wedge x$$
$$x \wedge (y \wedge z) \approx (x \wedge y) \wedge z$$
$$x \wedge \top \approx x$$

together with the requirement that for all $a, b \in A$ (for any $\boldsymbol{A} \in \mathbf{K}$),

$$a \leq b \iff a = a \wedge b.$$

Definition 1 *Let $\mathcal{S} = \langle \boldsymbol{Fm}, \vdash_{\mathcal{S}} \rangle$ be a finitary sentential logic. It is said to be* ***semilattice-based with respect to*** $\mathbf{K}$ ***through*** $\wedge$ *when for any $\varphi_0, \ldots, \varphi_{n-1}, \psi \in Fm$ the following hold:*

$$\varphi_0, \ldots, \varphi_{n-1} \vdash_{\mathcal{S}} \psi \iff v(\varphi_0) \wedge \cdots \wedge v(\varphi_{n-1}) \leq v(\psi) \quad \textit{for all } v \in \mathrm{Hom}(\boldsymbol{Fm}, \boldsymbol{A}) \textit{ and all } \boldsymbol{A} \in \mathbf{K} \tag{11}$$

$$\vdash_{\mathcal{S}} \psi \iff v(\psi) = 1 \quad \textit{for all } v \in \mathrm{Hom}(\boldsymbol{Fm}, \boldsymbol{A}) \textit{ and all } \boldsymbol{A} \in \mathbf{K}. \tag{12}$$

Since I am assuming that all the algebras have a maximum and it is denoted by a constant in the language, it follows from (12) that all these logics will have theorems. A slightly more general setting can be obtaining by deleting these assumptions, as is done in [16]; however for the purpose of introducing the multiple-valued cases I want to deal with, one can safely assume such properties and the exposition is somehow simplified.

Elementary properties

1. Independence from $\wedge$: If $\mathcal{S}$ is also semilattice-based with respect to another class $\mathbf{K}'$ through another binary term $\wedge'$ then the varieties generated by $\mathbf{K}$ and by $\mathbf{K}'$ are equal, and modulo this variety the terms $\wedge$ and $\wedge'$ are equivalent.
2. If $\mathcal{S}$ is semilattice based with respect to $\mathbf{K}$ then it is also so with respect to $\mathbf{V}(\mathbf{K})$, and this variety is the only variety with respect to which $\mathcal{S}$ can be semilattice-based.
3. Two formulas φ and ψ are ***interderivable*** modulo $\mathcal{S}$ (a relation I denote by $\varphi \dashv\vdash_{\mathcal{S}} \psi$ and define as "$\varphi \vdash_{\mathcal{S}} \psi$ and $\psi \vdash_{\mathcal{S}} \varphi$") if and only if the equation $\varphi \approx \psi$ is true in $\mathbf{K}$.
4. The interderivability relation $\dashv\vdash_{\mathcal{S}}$ is a congruence of the formula algebra $\boldsymbol{Fm}$, and the quotient algebra $\boldsymbol{Fm}/\dashv\vdash_{\mathcal{S}}$ generates the variety $\mathbf{V}(\mathbf{K})$.
5. The term $\wedge$ is a ***conjunction*** for $\mathcal{S}$, that is, it satisfies the three Hilbert-style rules

$$\varphi \wedge \psi \vdash \varphi \quad , \quad \varphi \wedge \psi \vdash \psi \quad \text{and} \quad \varphi, \psi \vdash \varphi \wedge \psi .$$

6. The logic $\mathcal{S}$ is entirely determined from its interderivability relation $\dashv\vdash_{\mathcal{S}}$, plus condition (12) for theorems. That is (given the result in item 3 above), the equational theory of $\mathbf{K}$ completely determines the inferential part of $\mathcal{S}$.

A logic $\mathcal{S}$ is called ***selfextensional*** when its interderivability relation $\dashv\vdash_{\mathcal{S}}$ is a congruence of the formula algebra $\boldsymbol{Fm}$. The logics with this property enjoy a strong *substitutivity property*: If $\alpha \dashv\vdash_{\mathcal{S}} \beta$ then for every $\varphi(p) \in Fm$, $\varphi(\alpha) \dashv\vdash_{\mathcal{S}} \varphi(\beta)$. This notion was highlighted and studied by WÓJCICKI, see [45]. The first important result about semilattice-based logics is:

Theorem 2 *A logic $\mathcal{S}$ is semilattice-based if and only if it is selfextensional, has theorems, and has a conjunction.* □

This result might be considered to be implicit in [33], although under weakening of some parts and strengthening of others, as has already been discussed in Section 2.

Examples

The preceding result characterizes by three metalogical properties the logics admitting a definition in terms of preserving degrees of truth in the semilattice case. Despite its perhaps non-standard or lesser-known phrasing, I want to emphasise that *the class of logics covered by these properties is very large*: Conjunction is a very weak and common requirement, and WÓJCICKI showed that selfextensional logics are exactly the *local consequences* defined by any possible-world or frame semantics in a very general sense of the word, see

[11, Section 6.7] or [45, Chapter 5]. From this it follows, for instance, that a very large group of modal logics can be studied under this framework.

Moreover, one can prove that all fragments of classical or intuitionistic logic containing conjunction $\wedge$ belong to this group. As is shown in the next section, also a big family of multiple-valued logics belongs to it. And many logics being a strengthening or an expansion of these will also belong to the same group, for instance, all modal logics referred to above (however, notice that not every expansion of a logic in this group belongs to it, as being selfextensional is a property that is not automatically preserved under strengthenings or under expansions).

This group contains many protoalgebraic logics, such as most of the just-mentioned fragments (more precisely, all those fragments containing the implication or equivalence connectives besides conjunction), but also many that are not (and normally these are much less known). Among the non-protoalgebraic examples are all those cited in Section 1, including an infinite multiple-valued logic, according to Theorem 23.

Algebraic models

It is clear from their definition that the logics I am considering bear a special relationship to the class K of algebras and also to the generated variety $\mathsf{V}(\mathsf{K})$. The application of the general notions and tools of Abstract Algebraic Logic show that these relations are not just the expression of the logic's definition, but something more: They conform to the general framework of algebraization of logic put forward in [14], whose particular interest arises in the treatment of logics that are not necessarily protoalgebraic.

For any algebra $\boldsymbol{A} \in \mathsf{V}(\mathsf{K})$ I denote by $\mathcal{F}ilt(\boldsymbol{A})$ the set of all ***(semilattice) filters of*** $\boldsymbol{A}$, that is, those $F \subseteq A$ satisfying:

1. $1 \in F$.
2. If $a \in F$ and $b \in F$ then $a \wedge b \in F$.
3. If $a \in F$ and $a \leq b$ then $b \in F$.

In case $\boldsymbol{A}$ is a lattice then these are just the ordinary lattice filters of $\boldsymbol{A}$. Notice that condition 3 amounts to the converse of 2. One can show:

Theorem 3 *If $\mathcal{S}$ is semilattice-based with respect to K then for each $\boldsymbol{A} \in \mathsf{V}(\mathsf{K})$, $\mathcal{F}i_{\mathcal{S}}\boldsymbol{A} = \mathcal{F}ilt(\boldsymbol{A})$, the generalized matrix $\langle \boldsymbol{A}, \mathcal{F}ilt(\boldsymbol{A}) \rangle$ is reduced, and $\mathsf{Alg}\mathcal{S} = \mathsf{V}(\mathsf{K})$.* □

Hence $\mathsf{V}(\mathsf{K})$ is the class of algebras canonically associated with the logic by the abstract framework, that is, it is *the algebraic counterpart of $\mathcal{S}$*. This result, besides its practical applications for particular logics, has some theoretical significance, for it answers in the affirmative a recurrent question in Abstract Algebraic Logic; namely, it identifies a large class of logics whose algebraic counterpart is a *variety*, something in general not guaranteed by the theory: even for finitely and regularly algebraizable logics $\mathcal{S}$, the general theory establishes that the class $\mathbf{Alg}\mathcal{S}$ is a quasivariety, and not necessarily a variety.

Also the class of *full models* of $\mathcal{S}$ can be characterized with respect to $\mathbf{K}$:

Theorem 4 *Let $\mathcal{S}$ be semilattice-based with respect to $\mathbf{K}$ and let $\langle \boldsymbol{A}, \mathcal{C} \rangle$ be a generalized matrix. Then $\langle \boldsymbol{A}, \mathcal{C} \rangle$ is a full model of $\mathcal{S}$ if and only if there is a strict surjective homomorphism from $\langle \boldsymbol{A}, \mathcal{C} \rangle$ onto some generalized matrix $\langle \boldsymbol{B}, \mathcal{D} \rangle$ such that $\boldsymbol{B} \in \mathbf{V}(\mathbf{K})$ and $\mathcal{D} = \mathcal{F}ilt(\boldsymbol{B})$.* □

The protoalgebraic case: the strong version

Although the main virtue of the semilattice-based framework is its independence of protoalgebraicity, nevertheless one can take also advantage from the power of the classical theory of protoalgebraic logics as developed in [2, 11] together with some recent results from [15]; the combination of the two properties will help in answering a very natural question in the present context.

Section 2 has described a situation where two logics arise naturally, one being a strengthening of the other but both sharing the same theorems. There is a more abstract situation where a similar phenomenon is observed, namely the notion of the "strong version" of a protoalgebraic logic introduced and studied in [15].

Let $\mathcal{S}$ be a protoalgebraic logic with theorems. An $\mathcal{S}$-filter F is ***Leibniz*** when $F \subseteq G$ for all $G \in \mathcal{F}i_{\mathcal{S}}\boldsymbol{A}$ with $\boldsymbol{\Omega}_{\boldsymbol{A}}(F) = \boldsymbol{\Omega}_{\boldsymbol{A}}(G)$, that is, when it is the least among all $\mathcal{S}$-filters on the same algebra having the same Leibniz congruence. The definition can be given in general, but for protoalgebraic logics, thanks to the monotonicity of the operator $\boldsymbol{\Omega}_{\boldsymbol{A}}$ on the set $\mathcal{F}i_{\mathcal{S}}\boldsymbol{A}$ for each algebra $\boldsymbol{A}$, one can show that Leibniz filters exist for every value of $\boldsymbol{\Omega}_{\boldsymbol{A}}$; more precisely, for every $\mathcal{S}$-filter F there is a (unique) Leibniz filter F^{+} with $\boldsymbol{\Omega}_{\boldsymbol{A}}(F^{+}) = \boldsymbol{\Omega}_{\boldsymbol{A}}(F)$; actually F^{+} can be obtained as the intersection of all filters with the same Leibniz congruence as F. A matrix is Leibniz when its filter is. Then with each logic $\mathcal{S}$ one can associate the logic $\mathcal{S}^{+}$ defined by the class of all Leibniz matrices of the original logic $\mathcal{S}$; this logic is called ***the strong version of*** $\mathcal{S}$, and, under certain conditions, there are strong relations between $\mathcal{S}^{+}$ and $\mathcal{S}$, as described in [15]; the first to be immediately seen are that $\mathcal{S}^{+}$ *is a strengthening of* $\mathcal{S}$ *and that these two logics have the same theorems* (because the least filter on each algebra is always Leibniz). Recall that a logic $\mathcal{S}'$ is a ***strengthening*** of a logic $\mathcal{S}$ if and only if $\vdash_{\mathcal{S}} \subseteq \vdash_{\mathcal{S}'}$ as binary relations, that is, if and only if $\Gamma \vdash_{\mathcal{S}} \psi$ implies $\Gamma \vdash_{\mathcal{S}'} \psi$ for all $\Gamma \subseteq Fm$ and all $\psi \in Fm$.

Notice that for algebraizable logics (in any of the degrees of this notion: weakly, finitely, strongly, etc.) the Leibniz operator is injective (see Section 1) therefore every $\mathcal{S}$-filter is Leibniz and $\mathcal{S}^{+} = \mathcal{S}$. Hence this issue is of interest only for protoalgebraic but non-algebraizable logics. In the case where the starting logic is semilattice-based, in [16] the following facts are proved:

Proposition 5 *Let $\mathcal{S}$ be a protoalgebraic logic that is semilattice-based (with respect to some class of algebras) but not weakly algebraizable. Then $\mathcal{S}^+$ is not selfextensional.* □

Theorem 6 *Let $\mathcal{S}$ be a protoalgebraic logic that is semilattice-based with respect to K. Then its strong version $\mathcal{S}^+$ is strongly, finitely and regularly algebraizable and its equivalent algebraic semantics is $\mathsf{V}(\mathsf{K})$. As a consequence, $\mathcal{S}^+$ is the logic defined by the class of matrices $\{\langle \boldsymbol{A}, \{1\}\rangle : \boldsymbol{A} \in \mathsf{V}(\mathsf{K})\}$.* □

The abstract setting can be intuitively read as follows: In the conditions of the theorem,

if $\mathcal{S}$ is defined from K by preserving degrees of truth

then its strong version $\mathcal{S}^+$ is defined from $\mathsf{V}(\mathsf{K})$ by preserving truth.

Obviously if K is already a variety then $\mathsf{K} = \mathsf{V}(\mathsf{K})$ and $\mathcal{S}^+$ coincides with the logic defined from K by preserving truth; however this condition is a very strong one, since often one wants to start with a very small class K (*very* often, with a single algebra!) and still have some connection between the logics defined from K by the two preservation schemes. There is another condition which will better fit in the particular situation of later sections; here $\mathsf{Q}(\mathsf{K})$ denotes the quasivariety generated by K. Then:

Theorem 7 *Let $\mathcal{S}$ be an equivalential logic that is semilattice-based with respect to K. Then its strong version $\mathcal{S}^+$ coincides with the logic defined by the class of matrices $\{\langle \boldsymbol{A}, \{1\}\rangle : \boldsymbol{A} \in \mathsf{K}\}$ if and only if $\mathsf{Q}(\mathsf{K}) = \mathsf{V}(\mathsf{K})$.* □

So here there are some conditions under which the relationship between the two logics defined from K by the two multiple-valued schemes analysed in Section 2 can be described in completely abstract terms, that is, without having to refer to $\leq, \wedge, 1$ or K. Moreover, in the equivalential case the stronger version can be syntactically reduced to the weak version:

Theorem 8 *Let $\mathcal{S}$ be an equivalential logic that is semilattice-based, let $\Delta(p,q)$ be its set of equivalence formulas, and put $X(p) = \Delta(p, \top)$. Then*

$$\Gamma \vdash_{\mathcal{S}^+} \varphi \iff X(\Gamma) \vdash_{\mathcal{S}} \varphi \tag{13}$$

for all $\Gamma \subseteq Fm$ and all $\varphi \in Fm$, where $X(\Gamma) = \bigcup\{X(\beta) : \beta \in \Gamma\}$. □

One can find a variety of particular cases where the situation is naturally found. The examples of normal modal logics are paradigmatic (there the weak and the strong versions correspond to the local and the global consequences generated by a class of Kripke frames) and have been dealt with at length in [15, Section 2B]; those of quantum logics, analysed in [15, Section 2A], constitute another typical group of examples.

The remaining sections are devoted to the multiple-valued case, where similar situations occur, although with some interesting particularities.

4 Logics defined from finite subalgebras of the real unit interval

Let **[0,1]** be the algebra on the real unit interval, with Łukasiewicz's well-known operations:

$$
\begin{aligned}
\neg x &= 1 - x \\
x \rightarrow y &= \min\{1, 1 - x + y\} \\
x \vee y &= \max\{x, y\} \\
x \wedge y &= \min\{x, y\} \\
x * y &= \max\{0, x + y - 1\} \\
x \oplus y &= \min\{1, x + y\} \\
x \leftrightarrow y &= \min\{1 - x + y, 1 - y + x\}
\end{aligned}
$$

where $+$ and $-$ are the ordinary arithmetical operations; as is well-known, one can take just a small subset of them as primitive and define the remaining ones by suitable equations; see [8, Chapter 4] for instance, but this issue is not relevant here. I use the customary abbreviations p^n and $n\,p$ to denote the iterated "star" and "plus" operations respectively; that is, $p^{n+1} = p^n * p$ and $(n+1)\,p = (n\,p) \oplus p$ for $n \geq 1$, and $p^1 = 1\,p = p$.

The algebras in the variety $\mathbf{MV} = \mathbf{V}(\mathbf{[0,1]})$ generated by the algebra **[0,1]** have received several names in the literature; the best-known two are ***MV-algebras*** and ***Wajsberg algebras***, the latter being used mostly when it is presented with the operations $\neg$ and $\rightarrow$ as the primitive ones. I assume that the language has a constant connective $\top$ such that $\top^{\boldsymbol{A}} = 1$ in any $\boldsymbol{A} \in \mathbf{MV}$; I also write 0 for $\neg 1$; the elements 0 and 1 are, respectively, the lower and upper bounds of their lattice structure. For their logical, algebraic and lattice-theoretical properties, and those of special subvarieties and subquasivarieties, one can read [8, 18, 23] and other papers therein referenced.

Let $\boldsymbol{S}$ be any subalgebra of **[0,1]**; note that $0, 1 \in S$ because I have included $\top$ in the language, but this would be the case even without this, since $a \rightarrow a = 1$ and $\neg(a \rightarrow a) = 0$ for any $a \in [0,1]$. Moreover, the set S with the natural order of real numbers $\leq$ is a bounded lattice, its operations being $\wedge$ and $\vee$, and 1 its maximum. Therefore, with each such subalgebra one can associate two sentential logics following the general schemes previously discussed: the first one, denoted by $\boldsymbol{L}_{\boldsymbol{S}}^{\leq}$, is defined by the preservation of degrees of truth schemes (11) and (12), and the second one, denoted by $\boldsymbol{L}_{\boldsymbol{S}}$, is defined by the preservation of truth scheme (3); in both cases $\mathbf{K} = \{\boldsymbol{S}\}$. Therefore:

Definition 9 *For each subalgebra $\boldsymbol{S}$ of* $[0,1]$ *the logics* $\mathrm{L}_{\boldsymbol{S}}^{\leq} = \langle \boldsymbol{Fm}, \vdash_{\boldsymbol{S}}^{\leq} \rangle$ *and* $\mathrm{L}_{\boldsymbol{S}} = \langle \boldsymbol{Fm}, \vdash_{\boldsymbol{S}}^{1} \rangle$ *are the logics defined by the following specifications:*

$$\varphi_0, \ldots, \varphi_{n-1} \vdash_{\boldsymbol{S}}^{\leq} \psi \iff v(\varphi_0) \wedge \cdots \wedge v(\varphi_{n-1}) \leq v(\psi) \quad \text{for all } v \in \mathrm{Hom}(\boldsymbol{Fm}, \boldsymbol{S}), \tag{14}$$

$$\vdash_{\boldsymbol{S}}^{\leq} \psi \iff v(\psi) = 1 \quad \text{for all } v \in \mathrm{Hom}(\boldsymbol{Fm}, \boldsymbol{S}), \tag{15}$$

and

$$\varphi_0, \ldots, \varphi_{n-1} \vdash_{\boldsymbol{S}}^{1} \psi \iff v(\psi) = 1 \text{ whenever } v(\varphi_0) = \cdots = v(\varphi_{n-1}) = 1, \quad \text{for all } v \in \mathrm{Hom}(\boldsymbol{Fm}, \boldsymbol{S}). \tag{16}$$

Some elementary properties, independent of the particular $\boldsymbol{S}$, can be immediately derived from the definitions; it is illustrative to make explicit the Hilbert-style and Gentzen-style rules common to all these logics which will be used later on. They are formulated with sentential variables, since they are understood as *rule schemes*, so that satisfying one of them means satisfying all its substitution instances (this proviso is not necessary for Hilbert-style rules, but it makes a difference for the Gentzen-style ones).

Proposition 10 *For each $\boldsymbol{S}$, the logic $\mathrm{L}_{\boldsymbol{S}}^{\leq}$ satisfies the following rules:*

1. $\dfrac{p_0, \ldots p_{n-1} \vdash q}{p_0^k, \ldots p_{n-1}^k \vdash q^k}$ *for all* $k \geq 1$.
2. $\dfrac{p \dashv\vdash q}{p^k \dashv\vdash q^k}$ *for all* $k \geq 1$.
3. $\dfrac{\vdash p}{\vdash p^k}$ *for all* $k \geq 1$.
4. $\vdash p \to p$.
5. $p \dashv\vdash p \leftrightarrow \top$.
6. $\vdash p \to (q \to p * q)$.
7. $p^s \vdash p^t$ *for all* s, t *with* $s \geq t \geq 1$.

Proof. 1: Let $a_0, \ldots, a_{n-1}, b \in [0,1]$ be such that $a_0 \wedge \ldots \wedge a_{n-1} \leq b$. Since the operation $*$ is monotonic and continuous, for each $k \geq 1$, $a_0^k \wedge \ldots \wedge a_{n-1}^k = (a_0 \wedge \ldots \wedge a_{n-1})^k \leq b^k$. Using this it is straightforward to show 1. 2 is a consequence of a particular case of 1. 3 holds because $1^k = 1$ for all $k \geq 1$, 4 because $a \leftrightarrow a = 1$ for all $a \in [0,1]$, 5 because $a \leftrightarrow 1 = a$ for all $a \in [0,1]$, and 6 because $a \to (b \to a * b) = 1$ for all $a, b \in [0,1]$. Finally, 7 holds because for all $a \in [0,1]$, if $s \geq t \geq 1$ then $a^s \leq a^t$. □

In the present context it makes sense to call ***non-trivial*** the subalgebras with more than two elements, that is, with at least one element different from 0 and from 1. The excluded case corresponds to $S = S_2 = \{0,1\}$, the Boolean algebra associated with classical propositional logic, and indeed $\mathcal{L}_{S_2} = \mathcal{L}^{\leq}_{S_2} = \mathcal{CPL}$. Then:

Theorem 11 *For each non-trivial $\boldsymbol{S}$, the logic $\mathcal{L}_{\boldsymbol{S}}$ is a proper strengthening of the logic $\mathcal{L}^{\leq}_{\boldsymbol{S}}$ and these two logics have the same theorems. In other words, $\mathcal{L}_{\boldsymbol{S}}$ is a proper, purely inferential strengthening of $\mathcal{L}^{\leq}_{\boldsymbol{S}}$. Moreover, for all $\varphi_0, \dots, \varphi_{n-1}, \psi \in Fm$,*

$$\varphi_0, \dots, \varphi_{n-1} \vdash^{\leq}_{\boldsymbol{S}} \psi \iff \vdash^{\leq}_{\boldsymbol{S}} \varphi_0 \wedge \cdots \wedge \varphi_{n-1} \to \psi \tag{17}$$

$$\iff \vdash^{1}_{\boldsymbol{S}} \varphi_0 \wedge \cdots \wedge \varphi_{n-1} \to \psi. \tag{18}$$

Proof. That $\mathcal{L}_{\boldsymbol{S}}$ is a strengthening of $\mathcal{L}^{\leq}_{\boldsymbol{S}}$ with the same theorems is a consequence of the general theory of Section 3, or it directly follows from Definition 9. To see that it is a proper one, one can show for instance that

$$p \vdash p * p \tag{19}$$

is a rule of $\mathcal{L}_{\boldsymbol{S}}$ that is not a rule of $\mathcal{L}^{\leq}_{\boldsymbol{S}}$: Since $1 * 1 = 1$ in **[0,1]**, it is clear that (19) is a rule of $\mathcal{L}_{\boldsymbol{S}}$. But $a \leq a * a$ is only true when $a = 0, 1$, while if $a \neq 0, 1$ then $a > a*a = \max\{0, 2a-1\}$. Since by assumption $\boldsymbol{S}$ is non-trivial, there are such a in S, therefore (19) is not a rule of $\mathcal{L}^{\leq}_{\boldsymbol{S}}$. Finally, the last part of the theorem follows directly from Definition 9 and the fact that on any MV-algebra, $a \leq b \iff a \to b = 1$. □

The equivalence (17) may be regarded as a kind of ***Weak Deduction Theorem***. Together with (18), these equivalences might suggest to some that there is no particular interest in $\mathcal{L}^{\leq}_{\boldsymbol{S}}$, as it can be reduced to $\mathcal{L}_{\boldsymbol{S}}$. However, the ordering structure of the real line is so natural that often it is simpler to work with $\mathcal{L}^{\leq}_{\boldsymbol{S}}$ than with $\mathcal{L}_{\boldsymbol{S}}$. Only the traditionally more accepted scheme of preserving truth has come to make $\mathcal{L}_{\boldsymbol{S}}$ appear as a more natural logic than $\mathcal{L}^{\leq}_{\boldsymbol{S}}$. To a certain extent, I would agree with the reverse judgement, and moreover in Theorem 15 below I show that for a finite $\boldsymbol{S}$ the logic $\mathcal{L}_{\boldsymbol{S}}$ can be reduced to $\mathcal{L}^{\leq}_{\boldsymbol{S}}$ in a similar way.

The general theory summarized in Section 3 already explains the main algebraic properties of the logics. By their very definition the logics $\mathcal{L}^{\leq}_{\boldsymbol{S}}$ are semilattice-based with respect to the single algebra $\boldsymbol{S}$, and hence with respect to the variety it generates; then Theorems 2, 3 and 4 automatically yield:

Proposition 12 *For each subalgebra $\boldsymbol{S}$ of* **[0,1]**, *the logic $\mathcal{L}^{\leq}_{\boldsymbol{S}}$ is selfextensional and has conjunction. Its algebraic counterpart is $\mathsf{Alg}\mathcal{L}^{\leq}_{\boldsymbol{S}} = \mathbf{V}(\boldsymbol{S})$, and* *on each algebra of this class the $L_S^{\leq}$-filters coincide with the lattice filters. A generalized matrix $\langle \boldsymbol{A}, C\rangle$ is a full model of $\mathcal{S}$ if and only if there is a strict surjective homomorphism from $\langle \boldsymbol{A}, C\rangle$ onto a generalized matrix of the form $\langle \boldsymbol{B}, \mathcal{F}ilt(\boldsymbol{B})\rangle$ with $\boldsymbol{B} \in \mathbf{V}(\boldsymbol{S})$.* □

Recall from Section 3 that $\mathcal{F}ilt(\boldsymbol{B})$ is the set of all lattice-filters of $\boldsymbol{B}$. Hence in case $\boldsymbol{B} \in \mathbf{V}(\boldsymbol{S})$ then $\mathcal{F}ilt(\boldsymbol{B}) = \mathcal{F}i_{L_S^{\leq}}\boldsymbol{B}$.

Proposition 13 *For each non-trivial $\boldsymbol{S}$, the logic L_S has conjunction, and is not selfextensional.*

Proof. Since having conjunction is expressed by Hilbert-style rules, it is a property inherited by strengthenings of any kind, so L_S has it because $L_S^{\leq}$ has it, by Proposition 12. Now to show that L_S is not selfextensional, let p be any variable, and consider the formulas $\varphi = p$ and $\psi = p * p$. Since for all $a \in [0,1]$, $a = 1 \iff a * a = 1$, $\varphi \dashv\vdash_S^1 \psi$ for any $\boldsymbol{S}$. Now let $a \in S$ be such that $0 < a \leq 1/2$ (it exists because $\boldsymbol{S}$ is nontrivial and negation makes it symmetric with respect to 1/2). Then $\neg a \neq 1$, $a * a = 0$ and $\neg(a * a) = 1$. This implies that $\neg\psi \not\dashv\vdash_S^1 \neg\varphi$, hence the interderivability relation of L_S is not a congruence with respect to negation, and this logic is not selfextensional. □

An alternative way of proving that the L_S are not selfextensional would be to use Proposition 5 and Theorem 11; but this would only work in case $L_S^{\leq}$ is protoalgebraic, something we do not know by now (and, as is shown in the next section, is not always the case), so I had to give a direct, general argument.

Proposition 12 determines the algebraic counterparts of all $L_S^{\leq}$ in both the traditional and the more abstract senses. Another typical task of Abstract Algebraic Logic is to *classify* the logics according to several criteria, notably with respect the so-called hierarchy outlined in Section 1. However, to go further in this direction it is useful (or perhaps indispensable) to treat separately the cases where $\boldsymbol{S}$ is finite, which have a perfectly standard behaviour, from the cases where $\boldsymbol{S}$ is infinite.

Therefore *I am going to assume in the rest of this section* that for some $m \geq 2$, $\boldsymbol{S} = \boldsymbol{S}_m$, the subalgebra of $[0,1]$ with m elements, that is, with universe $S_m = \{0, \frac{1}{m-1}, \ldots, \frac{m-2}{m-1}, 1\}$; here it is more practical to write $L_m^{\leq} = \langle \boldsymbol{Fm}, \vdash_m^{\leq}\rangle$ and $L_m = \langle \boldsymbol{Fm}, \vdash_m\rangle$ instead of $L_{S_m}^{\leq}$ and L_{S_m}, respectively. Wójcicki [44], see also [45, Theorem 4.3.3], showed that the logic here denoted by L_m fully coincides with the one axiomatized by the tautologies of what is usually called ***the m-valued Łukasiewicz logic*** plus the rule of Modus Ponens. In [45, Theorem 4.3.8] it is shown that L_m is an "implicative logic" in the sense of [34], therefore by the observation in page 41 of [3] it is a strongly, regularly and finitely algebraizable logic, and by the same result in [45] and Corollary 5.3 of [3] it follows that its equivalent algebraic semantics is the variety $\mathbf{Alg}L_m = \mathbf{V}(\boldsymbol{S}_m)$, often called ***the variety of m-valued***

MV-algebras, see [8, Definition 8.5.2]. For $m > 2$ this variety has been axiomatized in different ways, see [8, Theorem 8.5.1]. In the case $m = 2$ one gets $S_2 = \{0,1\}$, the two-element Boolean algebra, $Ł_2 = Ł_2^{\leq} = \mathcal{CPL}$, and $\mathbf{Alg} Ł_2 = \mathbf{V}(S_2)$ is the variety of all Boolean algebras.

To see that the pairs $(Ł_m^{\leq}, Ł_m)$ fit into the general framework of the preceding sections, one has to obtain directly a few properties of the less-known logics $Ł_m^{\leq}$. That they are semilattice-based with respect to $\boldsymbol{S}_m$ was already observed in [45, Section 4.3.14] and in [33]. As logics preserving degrees of truth they were briefly studied in GIL's unpublished Ph. D. Thesis [21], in the context of many-sided sequent calculi; one of his results adds to the general properties of Theorem 11 and Proposition 12, allowing to be more precise about their classification in the finite case:

Theorem 14 (Gil) *For each $m \geq 2$ the logic $Ł_m^{\leq}$ is finitely equivalential, the formula $(p \leftrightarrow q)^{m-1}$ being its single equivalence formula.*

Proof. The simplest way to show this is to check that the proposed equivalence formula satisfies the six syntactical conditions from [10, Definition I.10]; see also [11, Chapter 3]:

(E1) $\vdash_m^{\leq} (p \leftrightarrow p)^{m-1}$

(E2) $(p \leftrightarrow q)^{m-1} \vdash_m^{\leq} (q \leftrightarrow p)^{m-1}$

(E3) $(p \leftrightarrow q)^{m-1}, (q \leftrightarrow r)^{m-1} \vdash_m^{\leq} (p \leftrightarrow r)^{m-1}$

(E4) $(p \leftrightarrow q)^{m-1} \vdash_m^{\leq} (\neg p \leftrightarrow \neg q)^{m-1}$

(E5) $(p \leftrightarrow q)^{m-1}, (p' \leftrightarrow q')^{m-1} \vdash_m^{\leq} \big((p \to p') \leftrightarrow (q \to q')\big)^{m-1}$

(E6) $p, (p \leftrightarrow q)^{m-1} \vdash_m^{\leq} q$

(E1) is a consequence of 10.4 and 10.2. (E2) by the symmetry in the truth-valued function of $\leftrightarrow$. (E3) is proved by taking into account that if $a \in S_m$ and $a \neq 1$ then $a^{m-1} = 0$ (actually, $a^n = 0$ for all $n \geq m-1$) while $1^m = 1$, and that $a \to b = 1$ if and only if $a \leq b$; hence $a \leftrightarrow b = 1$ if and only if $a = b$. Then take any $v \in \mathrm{Hom}(\boldsymbol{Fm}, \boldsymbol{S}_m)$. If $v(p) \neq v(q)$ or $v(q) \neq v(r)$ then $v\big((p \leftrightarrow q)^{m-1}\big) \wedge v\big((q \leftrightarrow r)^{m-1}\big) = 0 \leq v\big((p \leftrightarrow r)^{m-1}\big)$. If $v(p) = v(q) = v(r)$ then $v\big((p \leftrightarrow q)^{m-1}\big) \wedge v\big((q \leftrightarrow r)^{m-1}\big) = 1 = v\big((p \leftrightarrow r)^{m-1}\big)$. This shows that (E3) holds. Similar reasonings prove (E4) and (E5). Finally, to show (E6), if $v(p) \neq v(q)$ then $v(p) \wedge v\big((p \leftrightarrow q)^{m-1}\big) = 0 \leq v(q)$, while if $v(p) = v(q)$ then $v(p) \wedge v\big((p \leftrightarrow q)^{m-1}\big) = v(p) \wedge 1 = v(p) = v(q)$. □

Every equivalential logic is a fortiori protoalgebraic [11, page 185], hence Proposition 5 and Theorems 6 and 7 apply. As one of the applications of these results one obtains:

Theorem 15 *For every $m \geq 2$, the logic L_m is the "strong version" of $L_m^{\leq}$ in the sense of Section 3, that is, $L_m = (L_m^{\leq})^+$, and for all $\varphi_0, \ldots, \varphi_{n-1}, \psi \in Fm$,*

$$\varphi_0, \ldots, \varphi_{n-1} \vdash_m \psi \iff \varphi_0^{m-1}, \ldots, \varphi_{n-1}^{m-1} \vdash_m^{\leq} \psi.$$

Proof. The first part follows from Theorem 7, because by the preceding result the logic $L_m^{\leq}$ is equivalential, and it has been proved in Theorem 3.8 of [23] that for every $m \geq 2$, $\mathbf{Q}(\boldsymbol{S}_m) = \mathbf{V}(\boldsymbol{S}_m)$; hence the equivalent condition in Theorem 7 holds in this case, and therefore the strong version of $L_m^{\leq}$ coincides with L_m. The second part follows from the first one plus Theorem 8, after taking into account that the set $X(p)$ mentioned in this last result has the form $X(p) = \Delta(p, \top)$ for any set Δ of equivalence formulas for the weak logic, here $L_m^{\leq}$. Hence here $X(p) = \{(p \leftrightarrow \top)^{m-1}\}$. However, expression (13) clearly shows that any $L_m^{\leq}$-equivalent set can replace $X(p)$. By 10.5 and 10.2, $(p \leftrightarrow \top)^{m-1} \dashv\vdash_m^{\leq} p^{m-1}$, therefore one can equally use $X(p) = \{p^{m-1}\}$. □

Corollary 16 *For every $m > 2$, the logic $L_m^{\leq}$ is not weakly algebraizable, hence a fortiori it is not algebraizable in any sense, while $L_2^{\leq} = L_2 = \mathcal{CPL}$ (classical propositional logic) is.*

Proof. As was said in Section 3, weakly algebraizable logics coincide with their strong version. Hence, by Theorem 15, if $L_m^{\leq}$ is so, then $L_m^{\leq} = (L_m^{\leq})^+ = L_m$. But by Theorem 11 L_m is a proper strengthening of $L_m^{\leq}$, hence different from it. Therefore $L_m^{\leq}$ cannot be weakly algebraizable. All this concerns the case $m > 2$, while for $m = 2$ the properties of $\mathcal{CPL}$ are well-known. □

This completes the classification of the logics $L_m^{\leq}$ in the hierarchy. Moreover, L_m *is the "strong version" of* $L_m^{\leq}$ *in two very different senses*: On one side, L_m is the logic defined by preserving truth from the same structure (the m-valued algebra $\boldsymbol{S}_m$) with respect to which $L_m^{\leq}$ is defined by preserving degrees of truth. On the other side, the logic L_m is the strong version, in the general sense of Abstract Algebraic Logic, of the logic $L_m^{\leq}$, that is, L_m is determined by the Leibniz filters of $L_m^{\leq}$. I now show that there is more: Leibniz filters of $L_m^{\leq}$ not only constitute a defining matrix semantics for L_m but they are exactly *all* its filters; moreover, they can be nicely characterised on arbitrary algebras of the signature of MV-algebras (i.e., not only on an MV-algebra):

Theorem 17 *Let F be any $L_m^{\leq}$-filter on any algebra $\boldsymbol{A}$ of the signature of MV-algebras. Then the following conditions are equivalent:*

(i) F is a Leibniz filter of $L_m^{\leq}$.

(ii) *F is an $Ł_m$-filter.*
(iii) *For all $a, b \in A$, if $a \in F$ and $a \to b \in F$ then $b \in F$.*
(iv) *For all $a, b \in A$, if $a \in F$ and $b \in F$ then $a * b \in F$.*
(v) *For all $a \in A$, if $a \in F$ then $a * a \in F$.*
(vi) *For all $a \in A$, if $a \in F$ then $a^{m-1} \in F$.*

Proof. (i)⇒(ii) by Theorem 15 and the general definition of $(Ł_m^{\leq})^+$ as the logic determined by the Leibniz filters of $Ł_m^{\leq}$. (ii)⇒(iii) as it is easy to check that $p, p \to q \vdash q$ is a rule of $Ł_m$. (iii)⇒(iv) because the formula $p \to (q \to p * q)$ is a theorem of $Ł_m^{\leq}$, by 10.6, and F is a filter of this logic. (v) is a particular case of (iv). From (v) one gets, by iteration, that if $a \in F$ then $a^{2^k} \in F$ for all $k \neq 0$, in particular for $2^k \geq m-1$; but then by 10.7 the rule $p^{2^k} \vdash p^{m-1}$ holds for $Ł_m^{\leq}$, so $a^{m-1} \in F$ because F is by assumption an $Ł_m^{\leq}$-filter. Finally only the proof of (vi)⇒(i) remains. By Theorem 14 we can use the characterization (2) of the Leibniz congruence of any $Ł_m^{\leq}$-filter F on an arbitrary algebra $\boldsymbol{A}$; this means that for all $a, b \in A$, $\langle a, b \rangle \in \boldsymbol{\Omega}_{\boldsymbol{A}}(F) \iff (a \leftrightarrow b)^{m-1} \in F$ for any $Ł_m^{\leq}$-filter F. So assume now that F is an $Ł_m^{\leq}$-filter on $\boldsymbol{A}$ satisfying condition (vi), and let G be any $Ł_m^{\leq}$-filter on the same $\boldsymbol{A}$ such that $\boldsymbol{\Omega}_{\boldsymbol{A}}(G) = \boldsymbol{\Omega}_{\boldsymbol{A}}(F)$. Take $a \in F$; by assumption $a^{m-1} \in F$. But as observed during the proof of Theorem 15, $(p \leftrightarrow \top)^{m-1} \dashv\vdash_m^{\leq} p^{m-1}$, and F is closed under all rules of $Ł_m^{\leq}$, therefore $(a \leftrightarrow 1)^{m-1} \in F$, that is, $\langle a, 1 \rangle \in \boldsymbol{\Omega}_{\boldsymbol{A}}(F) = \boldsymbol{\Omega}_{\boldsymbol{A}}(G)$. Since $1 \in G$, by compatibility also $a \in G$. This shows that $F \subseteq G$ and completes the proof that F is Leibniz. □

Corollary 18 *The logic $Ł_m$ is the inferential strengthening of the logic $Ł_m^{\leq}$ by any of the following proper rules:*

$$p, p \to q \vdash q \qquad \textit{(i.e., the rule of Modus Ponens)}$$
$$p, q \vdash p * q$$
$$p \vdash p * p$$
$$p \vdash p^{m-1}$$

□

One sometimes thinks of $Ł_m^{\leq}$ as "$Ł_m$ minus Modus Ponens". It is an open problem to find a Hilbert-style presentation of $Ł_m^{\leq}$; if one is found then adding to it any of the rules in Corollary 18 will yield one of $Ł_m$. Unfortunately, the existing presentations of $Ł_m$ do not help in this problem because they have only one rule of inference, namely Modus Ponens. A Gentzen-style presentation of $Ł_m^{\leq}$ has been proposed, without proof, in [22]. There are other syntactical relations between the two logics; besides that of Theorem 15 there is the following particular case of Theorem 11: For all $\varphi_0, \ldots, \varphi_{n-1}, \psi \in Fm$,

$$\varphi_0, \ldots, \varphi_{n-1} \vdash_m^{\leq} \psi \iff \vdash_m \varphi_0 \wedge \cdots \wedge \varphi_{n-1} \to \psi. \qquad (20)$$

A consequence of these relations is:

Theorem 19 *The sets of logics $\{L_m^{\leq} : m \geq 2\}$ and $\{L_m : m \geq 2\}$ are isomorphic ordered sets when ordered under the "strengthening" relation, and have $\mathcal{CPL}$ (classical logic) as their common upper bound. These sets are lattices where the infimum and supremum operations are given by the following arithmetical operations on subindexes:*

$$m \wedge k = \operatorname{lcm}(m-1, k-1) + 1 \tag{21}$$

$$m \vee k = \gcd(m-1, k-1) + 1 \tag{22}$$

Proof. The "strengthening" relation is clearly an ordering relation between logics, that is, between the consequence relations considered set-theoretically. From (20) it follows that for all $m, k \geq 2$, $\vdash_m^{\leq} \subseteq \vdash_k^{\leq}$ if and only if the set of theorems of $\vdash_m$ is included in the set of theorems of $\vdash_k$. It is well-known [24, Theorem 9.1.2] that this happens if and only if $k-1$ divides $m-1$, which is equivalent to saying that $\boldsymbol{S}_k$ is a subalgebra of $\boldsymbol{S}_m$. By the way the logics are defined from the algebras, this obviously implies that $\vdash_m \subseteq \vdash_k$. To show that this in turn implies that $\vdash_m^{\leq} \subseteq \vdash_k^{\leq}$ one needs just to apply the property (20), which reduces derivability in $L_m^{\leq}$ to theoremhood in L_m, and the same for k. This completes the proof that the two ordered sets are isomorphic. Using the same fundamental properties it is easy to show that both sets are lattices with the lattice operations corresponding to the operations on indexes expressed by (21) and (22), that is, $L_s^{\leq} = L_m^{\leq} \wedge L_k^{\leq}$ if and only if $L_s = L_m \wedge L_k$ if and only if $s = \operatorname{lcm}(m-1, k-1)+1$, and $L_s^{\leq} = L_m^{\leq} \vee L_k^{\leq}$ if and only if $L_s = L_m \vee L_k$ if and only if $s = \gcd(m-1, k-1)+1$. Finally, clearly $L_2^{\leq} = L_2 = \mathcal{CPL}$, and since 1 divides $m-1$ for all $m \geq 2$, it is clear that $\mathcal{CPL}$ is the common upper bound of both sets, in fact their maximum. □

In the next section the lower bounds of each of the two sets will be determined.

5 Logics defined from infinite subalgebras of the real unit interval

Logics of this kind have been analysed in [15, Section 3C] as examples of the notion of "strong version" of a protoalgebraic logic, in a context where the logics need not necessarily be finitary, that is, they are what in Section 1 was called "consequence relations"; they were also denoted by $L_S^{\leq}$ and L_S, as in the present paper, but the reader should not be confused: the "logics" of [15] are slightly different from those of the present paper. More precisely:

Theorem 20 *Let $\boldsymbol{S}$ be a non-trivial subalgebra of* $[\mathbf{0,1}]$ *and consider the two consequence relations defined by using the non-finitized schemes* (3) *and* (5) *by taking $\boldsymbol{A} = \boldsymbol{S}$ and $Val = \operatorname{Hom}(\boldsymbol{Fm}, \boldsymbol{S})$. Then each of these consequence relations is finitary if and only if the algebra $\boldsymbol{S}$ is finite.*

Proof. I have already observed in Section 2 that these consequences are those defined by a single matrix (6) and by a set of matrices (7), respectively. Now, if $\boldsymbol{S}$ is finite then the matrix (6) is finite, while the set of matrices (7) is a finite set of finite matrices. So both consequence relations are defined by a finite set of finite matrices, and it is well-known (see [45, Theorem 4.1.7] for instance) that such a consequence is finitary. Now assume that the algebra $\boldsymbol{S}$ is infinite. By McNaughton's Theorem [24, Theorem 9.1.8], it is easy to see that for each $k \in \omega$ there exists a formula $\varphi_k(p)$ in one variable p such that for all $a \in [0,1]$, $\varphi_k^{\mathbf{[0,1]}}(a) = 0$ iff $a \leq \frac{k+1}{k+3}$ and $\varphi_k^{\mathbf{[0,1]}}(a) = 1$ iff $a \geq \frac{k+2}{k+3}$. Now take $\Sigma = \{\varphi_k(p) : k \in \omega\}$ and let $\boldsymbol{S} \subseteq \mathbf{[0,1]}$ be an infinite subalgebra. The consequence relations defined from $\boldsymbol{S}$ by (3) and (5) will be represented inside this proof by $\vdash^1$ and $\vdash^{\leq}$ respectively, as in Section 2. Clearly, $\vdash^{\leq} \subseteq \vdash^1$. It is straightforward to check that $\Sigma \vdash^1 p$ and $\Sigma \vdash^{\leq} p$. However, if $\Sigma_0 \subseteq \Sigma$ is finite, then $\Sigma_0 \nvdash^1 p$: This follows from the fact, proved in Proposition 3.5.3 of [8], that S, being infinite, must be a dense subset of $[0,1]$; therefore for each $n \in \omega$ there is some $a_n \in S$ with $\frac{n+2}{n+3} < a_n < 1$, hence by the construction of the formulas $\varphi_k^{\boldsymbol{S}}(a_n) = \varphi_k^{\mathbf{[0,1]}}(a_n) = 1$ for all $k \leq n$ while $a_n \neq 1$. This shows that $\Sigma_0 \nvdash^1 p$ and a fortiori $\Sigma_0 \nvdash^{\leq} p$. Thus, neither of the two consequence relations is finitary. □

By Definition 9, the logics $Ł_{\boldsymbol{S}}^{\leq}$ and $Ł_{\boldsymbol{S}}$ in the present paper are finitary, hence they do not coincide with the consequence relations denoted by the same symbols in [15]. Moreover, since the general framework of semilattice-based logics summarized in Section 3 relies heavily on the logics being finitary, this also says that whatever results one can obtain from it cannot be based on those of [15, Section 3C].

The logics defined from the whole algebra $\mathbf{[0,1]}$ will play a special role from now on, so for ease of notation I will denote them by $Ł_{\infty}^{\leq}$ and $Ł_{\infty}$ instead of $Ł_{\mathbf{[0,1]}}^{\leq}$ and $Ł_{\mathbf{[0,1]}}$ respectively. Notice that $Ł_{\infty}$ is the *finitary* logic defined by the matrix $\langle \mathbf{[0,1]}, \{1\} \rangle$; that the consequence relation defined from this matrix is non-finitary (as Theorem 20 states for all infinite $\boldsymbol{S}$) was first noticed by Wójcicki in [44]. The tautologies of this matrix form what is usually called ***the infinite-valued Łukasiewicz logic***. After their axiomatization by Rose and Rosser, and independently by Chang, it was proved by Hay [26] that $Ł_{\infty}$ coincides with the logic axiomatized by those tautologies and the rule of Modus Ponens; see also [8, p. 101] and [25, Theorem 3.2.13]. Rodríguez, Torrens and Verdú proved in [37] that it is finitely, strongly and regularly algebraizable, that its equivalent algebraic semantics is the class MV of all MV-algebras and that on each $\boldsymbol{A} \in \mathsf{MV}$ the $Ł_{\infty}$-filters coincide with the ***implicative filters***, i.e., those $F \subseteq A$ such that $1 \in F$ and F is closed under Modus Ponens (if $a, a \to b \in F$ then b$\in F$). As a consequence, $Ł_{\infty}$ is the logic defined by the class of matrices

$$\{\langle \boldsymbol{A}, F \rangle : \boldsymbol{A} \in \mathsf{MV}, F \subseteq A \text{ is an implicative filter}\}. \tag{23}$$

The proofs in [37] show that the set of equivalence formulas for L_∞ is the set $\{p \to q, q \to p\}$, which can be replaced in this function by the single formula $p \leftrightarrow q$.

Recall that, as a particular case of Theorem 11, $L_\infty^{\leq}$ and L_∞ are linked by the relation

$$\varphi_0, \ldots, \varphi_{n-1} \vdash_\infty^{\leq} \psi \iff \vdash_\infty \varphi_0 \wedge \cdots \wedge \varphi_{n-1} \to \psi. \tag{24}$$

Concerning the other logics defined by preservation of truth from infinite subalgebras of **[0,1]**, one finds:

Theorem 21 *The logics L_S for an infinite $S \subseteq$ **[0,1]** depend only on the rationals contained in S. That is, if S and T are two infinite subalgebras of **[0,1]** then $L_S = L_T$ if and only if $S \cap \mathbb{Q} = T \cap \mathbb{Q}$. Moreover, all the logics L_S have the same theorems for all infinite subalgebras S of **[0,1]**.*

Proof. It is obvious that if $S \subseteq$ **[0,1]** then the logic L_S is a strengthening of L_∞. Hence all these L_S are algebraizable with equivalent algebraic semantics the quasivariety $\mathbf{Q}(S)$, which is a subclass of the variety $\mathbf{MV}$, and with the same equivalence formula $p \leftrightarrow q$. Now, in Theorem 2.9 of [23] it is shown that $\mathbf{Q}(S) = \mathbf{Q}(T)$ if and only if $S \cap \mathbb{Q} = T \cap \mathbb{Q}$. Since an algebraizable logic is completely determined by its equivalent algebraic semantics and its defining equations, this establishes the first part of the statement. Now, again by algebraizability, the theorems of L_S are translated into the equations holding in the quasivariety $\mathbf{Q}(S)$, which are those holding in the variety generated by S. But it is a straightforward consequence of a well-known theorem of LINDENBAUM [29, Theorem 16] that this variety is always (for an infinite S) the whole variety $\mathbf{MV}$, hence all these L_S have the same theorems. This establishes the second part of the theorem. □

Thus, only the logics defined by preservation of truth inside a *finite* truth-value-algebra have particular theorems. This constitutes another proof that the topic of the present section can only be of interest for those interested in the *inferential* side of logic: for infinite S, the logics L_S are purely inferential strengthenings of L_∞. Actually, the general theory of algebraizability shows that any set of quasiequations defining $\mathbf{Q}(S)$ relatively to $\mathbf{MV}$ yields, when translated through the equivalence formula, an axiomatization of the additional rules of L_S with respect to L_∞. In particular, in the cases where S contains all rationals in $[0,1]$ the L_S is equal to L_∞.

For logics defined by preservation of truth, there is not a one-to-one correspondence between logics and infinite subalgebras of **[0,1]**. The case of the logics defined by preservation of degrees of truth the situation is even worse:

Theorem 22 *If S is an infinite subalgebra of **[0,1]** then $L_S^{\leq} = L_\infty^{\leq}$.*

Proof. This is due to the fact that by definition all these logics are semilattice-based with respect to $\boldsymbol{S}$ and hence, by the properties summarized in Section 3, with respect to $\mathbf{V}(\boldsymbol{S})$. As I have already recalled in the previous proof, by LINDENBAUM's result [29, Theorem 16], for an infinite $\boldsymbol{S}$ all these varieties are equal and coincide with $\mathbf{MV}$. Therefore all these logics coincide with the logic that is semilattice-based with respect to $\mathbf{MV}$, that is, with $L_\infty^{\leq}$. □

Thus *there is only one logic defined by preservation of an infinite number of degrees of truth in* $[\mathbf{0,1}]$, namely $L_\infty^{\leq}$. Some of its properties follow from the general theory of Section 3: it is selfextensional and has conjunction, and $\mathbf{Alg}\, L_\infty^{\leq} = \mathbf{MV}$; moreover, on each of these algebras its filters are the lattice filters. As a consequence, in parallel with (23), $L_\infty^{\leq}$ is the logic defined by the class of matrices

$$\big\{\langle \boldsymbol{A}, F\rangle : \boldsymbol{A} \in \mathbf{MV}, F \subseteq A \text{ is a lattice filter}\big\}. \tag{25}$$

From this one can also infer that $L_\infty^{\leq}$ it coincides with the logic denoted as C_V in [22]. In this paper a Gentzen calculus is presented that is claimed (without proof) to be a Gentzen-style presentation of this logic; a reasonable conjecture is that the Gentzen system defined by this calculus is the (unique) Gentzen system fully adequate for $L_\infty^{\leq}$, which exists by the general theory of [16].

By Theorem 11, each logic $L_{\boldsymbol{S}}$ defined by the truth-preserving scheme is different from the corresponding logic $L_{\boldsymbol{S}}^{\leq}$ defined by the preservation of degrees of truth scheme. Corollary 18 provides an exact, Hilbert-style relationship between each two them in the case $\boldsymbol{S}$ is finite, and in Theorem 19 a global relationship between the two families has been established. By Theorems 21 and 22, there is no hope that such results can be replicated in the infinite case, for most of the logics of the first kind coincide, as do *all* those of the second. It is only reasonable to expect that something similar holds for the weakest cases of L_∞ and $L_\infty^{\leq}$. However:

Theorem 23 (Gil, Torrens, Verdú) *The logic* $L_\infty^{\leq}$ *is not protoalgebraic.*

Proof. This result is in fact a corollary to Theorem 5 of [22] that characterizes the protoalgebraic strengthenings of the logic denoted there as C_V, which as observed before should be equal to the logic here denoted as $L_\infty^{\leq}$. Since [22] appears in a proceedings volume very difficult to find, and moreover contains no proofs, it may be of interest to give here a direct proof.

An element a of an MV-algebra $\boldsymbol{A}$ is *archimedean*, according to [8, Corollary 6.2.4], when there is an integer $n \geq 1$ such that $\neg a \vee n\, a = 1$; an MV-algebra all whose elements are archimedean is called *hyperarchimedean* [8, Definition 6.3]. Let $\boldsymbol{A}$ be any non-archimedean MV-algebra, for instance the direct product of the algebras $\boldsymbol{S}_m$ for all $m \geq 2$ (see [8, Chapter 6] or

[41, Section 3.A]) and let $a \in A$ be any non-archimedean element of $\boldsymbol{A}$. Let $F = \mathrm{Fi}(\neg a)$ be the implicative filter generated by $\neg a$; this set is both an $L_\infty^{\leq}$-filter (thus, a lattice filter) and an L_∞-filter. This last logic is algebraizable, hence equivalential, therefore (2) can be used with $\Delta(p,q) = \{p \to q, q \to p\}$. Since $a \to 0 = \neg a \in F$ by construction, and $0 \to a = 1 \in F$ because all filters contain the maximum, by (2) $\langle a, 0\rangle \in \boldsymbol{\Omega_A}(F)$. If by contradiction one assumes that $L_\infty^{\leq}$ is protoalgebraic, then by [11, Corollary 1.1.11] from the fact that $\langle a, 0\rangle \in \boldsymbol{\Omega_A}(F)$ it follows that a and 0 must belong to the same $L_\infty^{\leq}$-filters on $\boldsymbol{A}$ that contain F; in particular this implies that $0 \in \mathcal{F}ilt(F, a)$, the lattice filter generated by $F \cup \{a\}$. Since F is itself a lattice filter, by standard lattice-filter generation this means that $0 = b \wedge a$ for some $b \in F = \mathrm{Fi}(\neg a)$. Now the characterization of implicative filters in MV-algebras found in [8, Proposition 4.2.9] or [41, 1.19] can be used, which is an algebraic version of the Local Deduction Theorem for L_∞, that is, that $\mathrm{Fi}(\neg a) = \{c \in A : (\neg a)^n \leq c \text{ for some } n \geq 1\}$. Hence $(\neg a)^n \leq b$ for some $n \geq 1$ and thus $(\neg a)^n \wedge a = 0$, which by the De Morgan Laws for both the pairs $(\wedge, \vee)$ and $(\oplus, *)$ is equivalent to $n\,a \vee \neg a = 1$, which would mean that a is archimedean, against the assumption. Therefore $L_\infty^{\leq}$ cannot be protoalgebraic. □

Thus, the general theory developed in [15] for protoalgebraic logics cannot be applied to find out an abstract relationship between $L_\infty^{\leq}$ and L_∞. We can instead consider the following, more general definition of a "strong version" of a logic: Given an arbitrary logic $\mathcal{S}$, one can always consider the logic $\mathcal{S}^{\min}$ defined by the class of matrices

$$\{\langle \boldsymbol{A}, F\rangle : \boldsymbol{A} \in \mathbf{Alg}\mathcal{S}, F = \bigcap \mathcal{F}i_{\mathcal{S}}\boldsymbol{A}\},$$

that is, the logic defined by taking only the smallest matrix on each $\mathcal{S}$-algebra. Since each of these matrices is of course an $\mathcal{S}$-matrix, the logic $\mathcal{S}^{\min}$ is a strengthening of $\mathcal{S}$. And in the present case one obtains:

Theorem 24 $(L_\infty^{\leq})^{\min} = L_\infty$.

Proof. Since $\mathbf{Alg}L_\infty^{\leq} = \mathbf{MV}$ and on any MV-algebra the least $L_\infty^{\leq}$-filter is $\{1\}$, the logic $(L_\infty^{\leq})^{\min}$ is the logic defined by the class of matrices $\{\langle \boldsymbol{A}, \{1\}\rangle : \boldsymbol{A} \in \mathbf{MV}\}$. But $\mathbf{MV}$ is the equivalent algebraic semantics for L_∞ and $\{1\}$ is the designated filter on each $\boldsymbol{A} \in \mathbf{MV}$, therefore $(L_\infty^{\leq})^{\min}$ coincides with the logic L_∞. □

Thus the relation between $L_\infty^{\leq}$ and L_∞ is parallel to that found between $L_m^{\leq}$ and L_m for each finite m: On one hand L_∞ is the logic defined by preserving truth from the same structure (the real unit interval) with respect to which $L_\infty^{\leq}$ is defined by preserving degrees of truth. On the other side, the logic L_∞ is the strong version, in a different abstract sense, of the logic $L_\infty^{\leq}$. There is also a version of Corollary 15:

Theorem 25 *The logic L_∞ is the inferential strengthening of the logic $L_\infty^{\leq}$ by any of the following proper rules*

$$p\,,p \to q \vdash q \qquad \textit{(i.e., the rule of Modus Ponens)}$$
$$p\,,q \vdash p * q$$
$$p \vdash p * p$$

or by the infinite set of proper rules

$$p \vdash p^n \quad \textit{for all } n \geq 2\,.$$

Proof. $\mathbf{Alg}L_\infty^{\leq} = \mathbf{Alg}L_\infty = \mathbf{MV}$ and by Theorem 2.22 of [14], each logic $\mathcal{S}$ is complete with respect to the class of matrices $\{\langle \boldsymbol{A}\,, F\rangle : \boldsymbol{A} \in \mathbf{Alg}\mathcal{S}\ ,\ F \in \mathcal{F}i_{\mathcal{S}}\boldsymbol{A}\}$; hence it is enough to compare the filters of the two logics in the class of MV-algebras. It is true that an $L_\infty^{\leq}$-filter on an MV-algebra is just a lattice filter, and that an L_∞-filter on an MV-algebra is just an implicative filter. And as a consequence of Proposition 9 of [18], see also [8, Lemma 4.2.7], a lattice filter on an MV-algebra is an implicative filter if and only if it is closed under Modus Ponens, and also if and only if it is closed under the rule $p, q \vdash p * q$. This shows that the first two rules do axiomatize L_∞ relatively to $L_\infty^{\leq}$. In order to show that the apparently weaker rule $p \vdash p * p$ also defines L_∞ from $L_\infty^{\leq}$, I show that an $L_\infty^{\leq}$-theory that is closed under it is also closed under the rule $p, q \vdash p * q$: Let Γ be any such theory, and $\varphi, \psi \in \Gamma$. Take any $t \in [0,1]$ and any $v \in \mathrm{Hom}(\boldsymbol{Fm}, \mathbf{[0,1]})$ such that $v(\gamma) \geq t$ for all $\gamma \in \Gamma$. In particular $v(\varphi)\,, v(\psi) \geq t$. Since $[0,1]$ is linearly ordered, let's say $v(\varphi) \leq v(\psi)$. By monotonicity of $*$, $v(\varphi * \varphi) = v(\varphi) * v(\varphi) \leq v(\varphi) * v(\psi) = v(\varphi * \psi)$. But by assumption $\varphi \in \Gamma$ implies $\varphi * \varphi \in \Gamma$, and one concludes that $v(\varphi * \psi) \geq t$. Since t and v are arbitrary, this shows that $\Gamma \vdash_\infty^{\leq} \varphi * \psi$, which implies $\varphi * \psi \in \Gamma$ because Γ is assumed to be an $L_\infty^{\leq}$-theory. Finally, an $L_\infty^{\leq}$-theory closed under the rule $p \vdash p * p$ is also closed under all rules $p \vdash p^{2^k}$ for all $k \geq 1$, by iteration. Since by $p^{2^k} \vdash p^n$ when $n \geq 2^k$ is also a rule of $L_\infty^{\leq}$, such theory is also closed under all rules $p \vdash p^n$ for all $n \geq 2$. That all these rules are proper follows from Theorem 11 and their mutual equivalence just established. □

Again, as in the finite case, I do not know of a Hilbert-style presentation of $L_\infty^{\leq}$, while a Gentzen-style one is proposed, without proof, in [22].

While the last results have been logical applications of the main results, I offer below an algebraic application of the logical results:

Proposition 26 *Let $\boldsymbol{A}$ be an MV-algebra, and let F be a lattice filter on $\boldsymbol{A}$. Then the largest implicative filter on $\boldsymbol{A}$ contained in F is the set $G = \{a \in A : a^k \in F \text{ for all } k \geq 1\}$.*

Proof. It is clear that $G \subseteq F$, because $a^1 = a$. In order to see that G is an $Ł_\infty^{\leq}$-filter, one assumes that $\varphi_0, \ldots, \varphi_{n-1} \vdash_\infty^{\leq} \psi$ and that $h(\varphi_0), \ldots, h(\varphi_{n-1}) \in G$ for some $h \in \mathrm{Hom}(\boldsymbol{Fm}, \boldsymbol{A})$; by definition of G this means that for all $k \geq 1$, $(h(\varphi_0))^k, \ldots, (h(\varphi_{n-1}))^k \in F$. Now by Proposition 10.1 also $\varphi_0^k, \ldots, \varphi_{n-1}^k \vdash_\infty^{\leq} \psi^k$, and since F is an $Ł_\infty^{\leq}$-filter, one concludes that $h(\psi^k) \in F$ for all $k \geq 1$, that is, $h(\psi) \in G$. This shows that G is an $Ł_\infty^{\leq}$-filter. By its own definition G is trivially a model of the rule $p \vdash p * p$, hence by Theorem 25 it is an $Ł_\infty$-filter, that is, an implicative filter. Finally, let $H \subseteq F$ be another implicative filter, and $a \in H$. Again by Theorem 25 $a^k \in H$ for all $k \geq 1$, so $a^k \in F$ for all $k \geq 1$, which implies $a \in G$, that is, $H \subseteq G$. Thus G is the largest such implicative filter. □

The final observation on the logics examined in this paper is a relation between the finite and the infinite cases. At the end of Section 4 I promised to determine the lower bounds of the families of logics generated by the two schemes from finite subalgebras of $[\mathbf{0},\mathbf{1}]$, that is, of the families of all $Ł_m^{\leq}$ and of all the $Ł_m$. In doing this one finds an interesting fact: the desired infima are not the same if computed in the (natural) lattice of all logics over the same language, or if computed in the larger lattice of all consequence relations over the same language; such sets are complete lattices, see [45, Section 1.5].

Theorem 27 *The infimum of the family $\{Ł_m^{\leq} : m \geq 2\}$ in the lattice of all (finitary) logics is the logic $Ł_\infty^{\leq}$, while the infimum of the same family in the larger lattice of all consequence relations is a non-finitary logic different from $Ł_\infty^{\leq}$.*

Proof. This uses TARSKI's theorem [29, Theorem 20] that $\mathbf{V}(\{\boldsymbol{S}_m : m \geq 2\}) = \mathbf{V}([\mathbf{0},\mathbf{1}]) = \mathbf{MV}$; see also [8, Proposition 8.1.2]. Then the following chain of equivalences should be obvious:

$$\begin{array}{lll}
\varphi_0, \ldots, \varphi_{n-1} \vdash_m^{\leq} \psi \quad \text{for all } m \geq 2 & & \\
\iff \vdash_m \varphi_0 \wedge \cdots \wedge \varphi_{n-1} \to \psi \quad \text{for all } m \geq 2 & & \text{by (20)} \\
\iff \varphi_0 \wedge \cdots \wedge \varphi_{n-1} \to \psi \approx \top \ \text{ holds in } \boldsymbol{S}_m, \text{ for all } m \geq 2 & & \\
\iff \varphi_0 \wedge \cdots \wedge \varphi_{n-1} \to \psi \approx \top \ \text{ holds in } \mathbf{MV} & & \text{by TARSKI's Theorem} \\
\iff \vdash_\infty \varphi_0 \wedge \cdots \wedge \varphi_{n-1} \to \psi & & \text{by algebraizability} \\
\iff \varphi_0, \ldots, \varphi_{n-1} \vdash_\infty^{\leq} \psi & & \text{by (24).}
\end{array}$$

So $Ł_\infty^{\leq} = \inf\{Ł_m^{\leq} : m \geq 2\}$ in the lattice of (finitary) logics (over the same language, of course). Now put $\mathcal{S}_\infty = \inf_c\{Ł_m^{\leq} : m \geq 2\}$, the infimum of the same family in the bigger lattice of all consequence relations. That is, for all $\Gamma \subseteq Fm$ and all $\psi \in Fm$, $\Gamma \vdash_{\mathcal{S}_\infty} \psi$ if and only if for all $m \geq 2$, $\Gamma \vdash^{\leq}_{m} \psi$. Therefore, $L^{\leq}_{\infty}$ is the finitary part of $\mathcal{S}_\infty$, and so seeing that $\mathcal{S}_\infty$ is non-finitary and seeing that $\mathcal{S}_\infty \neq L^{\leq}_{\infty}$ amount to the same thing. Consider the set of formulas in two variables $\Delta(p,q) = \{(p \leftrightarrow q)^{m-1} : m \geq 2\}$. By Theorem 14, $p, (p \leftrightarrow q)^{m-1} \vdash^{\leq}_{m} q$ for each $m \geq 2$, hence a fortiori $p, \Delta(p,q) \vdash^{\leq}_{m} q$ for all $m \geq 2$, and thus $p, \Delta(p,q) \vdash_{\mathcal{S}_\infty} q$. However, $p, \Delta(p,q) \nvdash^{\leq}_{\infty} q$. In order to see this, since $L^{\leq}_{\infty}$ is finitary and $p^s \vdash p^t$ whenever $s \geq t \geq 1$, it is enough to see that $p, (p \leftrightarrow q)^{m-1} \nvdash^{\leq}_{\infty} q$ for each $m \geq 2$. For $m = 2$ it is enough to take $v(q) < v(p)$ while for $m > 2$ take $v(p) = 1/2$ and $\frac{m-3}{2m-4} < v(q) < 1/2$; a straightforward computation shows that in both cases $v(q) < v(p) \wedge v\big((p \leftrightarrow q)^{m-1}\big)$, which is against $p, (p \leftrightarrow q)^{m-1} \nvdash^{\leq}_{\infty} q$. □

Thus, this yields a beautiful, natural example that the lattice of all logics on a fixed language is not a *complete* sublattice of the lattice of all consequence relations over the same language; notice that by [45, Theorem 1.5.6] the former is indeed a sublattice of the latter. At the same time, since by Theorem 14 all $L^{\leq}_{m}$ are protoalgebraic while by Theorem 23 $L^{\leq}_{\infty}$ is not protoalgebraic, this also shows that the set of all protoalgebraic logics (over a fixed language) is not a complete sublattice of the lattice of all logics either. As the logics $L^{\leq}_{m}$ are all finitely equivalential, the same applies to the sets of all equivalential logics and of all finitely equivalential logics (always on the same language). However, the issue of the order structure of all these particular sets has not been touched in depth in the literature; it is however known that each of these sets is closed under strengthenings, see [11, pages 71,187,316].

Finally, almost the same situation is found for the strong versions:

Theorem 28 *The infimum of the family $\{L_m : m \geq 2\}$ in the lattice of all (finitary) logics is the logic L_∞, while the infimum of the same family in the larger lattice of all consequence relations is a non-finitary logic different from L_∞.*

Proof. The proof of the first part is similar to the first part of the proof of Theorem 27, but using a much more recent and sophisticated algebraic result. By definition of the logics L_m, the entailment $\varphi_0, \ldots, \varphi_{n-1} \vdash_m \psi$ holds for all $m \geq 2$ if and only if the quasiequation

$$\varphi_0 \approx \top \ \ \& \ \ \ldots \ \ \& \ \ \varphi_{n-1} \approx \top \ \Rightarrow \ \psi \approx \top \tag{26}$$

holds in $\boldsymbol{S}_m$ for all $m \geq 2$, that is, iff (26) holds in the quasivariety generated by the family of algebras $\{\boldsymbol{S}_m : m \geq 2\}$. Clearly, the union of all their universes contains all the rational points in $[0,1]$, hence as a consequence of Theorem 3.8 of [23] this quasivariety coincides with the variety generated by the same family, which by TARSKI's Theorem used in the preceding proof, is the whole class **MV**. Then, by the algebraizability of $\mathcal{L}_\infty$ with **MV** as equivalent algebraic semantics, (26) holds in **MV** if and only if $\varphi_0, \ldots, \varphi_{n-1} \vdash_\infty \psi$. This proves that $\mathcal{L}_\infty = \inf\{\mathcal{L}_m : m \geq 2\}$. In order to prove the second part of the theorem one considers $\mathcal{S}_\infty = \inf_c\{\mathcal{L}_m : m \geq 2\}$, the infimum of the same family in the bigger lattice of all consequence relations. To reach a contradiction let's assume that this consequence is finitary; then it belongs to the smaller lattice and hence $\mathcal{S}_\infty = \mathcal{L}_\infty$. By definition $\mathcal{L}_\infty$ is the finitary part of the consequence relation defined by the infinite matrix $\langle [0,1], \{1\} \rangle$, which I denote by $\mathcal{S}^\infty$. As I recalled before and is well-known, $\mathcal{S}^\infty$ is not finitary, hence it is strictly stronger than its finitary part $\mathcal{L}_\infty$. But for each $m \geq 2$, the matrix $\langle \S_m, \{1\} \rangle$ is a submatrix of $\langle [0,1], \{1\} \rangle$, and by [11, Proposition 0.3.3] the consequence it defines is stronger than that defined by the larger matrix: Each $\mathcal{L}_m$ is stronger than $\mathcal{S}^\infty$, hence so is their infimum in the bigger lattice, that is, $\mathcal{S}_\infty = \mathcal{L}_\infty$ is stronger than $\mathcal{S}^\infty$, which contradicts the fact proved before that $\mathcal{S}^\infty$ is strictly stronger than $\mathcal{L}_\infty$. This shows that $\mathcal{S}_\infty$ is non-finitary and hence different from $\mathcal{L}_\infty$. □

The distinction between the greatest lower bounds of each family in the two different complete lattices does not carry over to their lowest upper bound: By Theorem 19 they coincide with $\mathcal{CPL} = \mathcal{L}_2 = \mathcal{L}_2^{\leq}$, which is a member of both families, hence it is actually their maximum inside the two lattices.

6 Conclusions

In this paper I have presented the theory of semilattice-based logics as a formalization, in the context of Abstract Algebraic Logic, of the (already known) idea of *preserving degrees of truth*, a scheme for semantic definition of logics particularly suited for multiple-valued logics opposed to the (more standard) scheme of *preservation of truth*. After summarizing the main notions and results in this theory, I have applied it, together with other more general parts of Abstract Algebraic Logic, to the study of the logics obtained in both ways from a single subalgebra of the real unit interval, endowed with Łukasiewicz's operations. The algebraic counterparts of these logics have been determined, and they have been classified according to the so-called protoalgebraic hierarchy. Some further properties have been obtained, either of an algebraic or of a logical character. The behaviour of the logics resulting from finite subalgebras has been seen as much more standard and regular than that of those arising from infinite subalgebras; in particular, the scheme of preservation of degrees of truth obtains only one logic in the latter case, while the scheme of preservation of truth obtains a different logic for each infinite subalgebra containing different sets of rationals from the real unit interval.

An interesting observation arises, that preservation of truth seems to depend on the rational points belonging to the defining truth-value algebra, while preservation of degrees of truth seems to do so only in case there are only a finite number of degrees of truth to be preserved. No interpretation of this fact has been attempted at.

The development of the last two sections of the paper shows many examples of how the interplay logic-algebra works, and in both directions. General theories of logic and of its algebraization have been used, as well as central notions of universal algebra. More importantly, several deep facts concerning the algebraic properties of the particular classes of algebras involved have been used in an essential way in order to obtain some of the main logical results.

I have given just a sample of how these methods work when dealing with logics defined from a subalgebra of the real unit interval. Clearly this does not exhaust the family of multiple-valued logics defined by preservation of degrees of truth in this context; actually, there is one such logic for each proper subvariety of the variety **MV** of all MV-algebras. By LINDENBAUM's and TARSKI's Theorems quoted before, such a subvariety can contain only a finite number of the subalgebras S_m and can contain no infinite subalgebra of [0,1]; KOMORI [27] described all such subvarieties. The situation concerning subquasivarieties of **MV** (which are related with the logics obtained by preservation of truth) is far more complicated; those generated by families of subalgebras of [0,1] have been classified and described in [23].

The paper also shows the importance of having a general framework, as a way of placing the investigations on particular logics and/or particular classes of algebras in the context of a more far-reaching research program. I think that the contents of the paper shows that Abstract Algebraic Logic is a good candidate for playing such a role, and offers some powerful tools for these tasks.

Acknowledgements

The writing of this paper was part of my ordinary research work, partially funded by grants PB97-0888 of the Spanish government and 2000SGR-00007 of the Catalan government. I want to thank my colleague RAMON JANSANA and an anonymous referee for their observations on the first version of this paper.

References

1. Ardeshir, M., and Ruitenburg, W.: Basic propositional calculus I. Mathematical Logic Quarterly 44 (1998), 317–343
2. Blok, W., and Pigozzi, D.: Protoalgebraic logics. Studia Logica 45 (1986), 337–369
3. Blok, W., and Pigozzi, D.: Algebraizable logics, vol. 396 of Mem. Amer. Math. Soc. A.M.S., Providence, January (1989)

4. Blok, W., and Pigozzi, D.: Algebraic semantics for universal Horn logic without equality. In Universal Algebra and Quasigroup Theory, Romanowska, A. and Smith, J.D.H. Eds. Heldermann, Berlin (1992), 1–56
5. Blok, W., and Pigozzi, D.: Abstract algebraic logic and the deduction theorem. Bulletin of Symbolic Logic (200x). To appear.
6. Brown, D. J., and Suszko, R.: Abstract logics. Dissertationes Math. (Rozprawy Mat.) 102 (1973), 9–42
7. Celani, S., and Jansana, R.: A closer look at some subintuitionistic logics. Manuscript (2000)
8. Cignoli, R., Mundici, D., and D'Ottaviano, I.: Algebraic foundations of many-valued reasoning, vol. 7 of Trends in Logic, Studia Logica Library. Kluwer, Dordrecht (1999)
9. Czelakowski, J.: Model-theoretic methods in methodology of propositional calculi. Polish Academy of Sciences, Warszawa (1980)
10. Czelakowski, J.: Equivalential logics, I, II. Studia Logica 40 (1981), 227–236 and 355–372
11. Czelakowski, J.: Protoalgebraic Logics, vol. 10 of Trends in Logic, Studia Logica Library. Kluwer Academic Publishers, Dordrecht (2001)
12. Czelakowski, J., and Jansana, R.: Weakly algebraizable logics. The Journal of Symbolic Logic 65, 2 (2000), 641–668
13. Font, J.M.: Belnap's four-valued logic and De Morgan lattices. Logic Journal of the I.G.P.L. 5, 3 (1997), 413–440
14. Font, J. M., and Jansana, R.: A general algebraic semantics for sentential logics, vol. 7 of Lecture Notes in Logic. Springer-Verlag, (1996) 135 pp. Presently distributed by the Association for Symbolic Logic.
15. Font, J. M., and Jansana, R.: Leibniz filters and the strong version of a protoalgebraic logic. Archive for Mathematical Logic 40 (2001), 437–465
16. Font, J. M., and Jansana, R.: On semilattice-based deductive systems. Manuscript (2001)
17. Font, J. M., Jansana, R., and Pigozzi, D.: A survey on abstract algebraic logic. Studia Logica, Special Issue on Abstract Algebraic Logic, Part II (200x). To appear
18. Font, J. M., Rodríguez, A. J., and Torrens, A.: Wajsberg algebras. Stochastica 8 (1984), 5–31
19. Font, J. M., and Rodríguez, G. Algebraic study of two deductive systems of relevance logic. Notre Dame Journal of Formal Logic 35, 3 (1994), 369–397
20. Font, J. M., and Verdú, V.: Algebraic logic for classical conjunction and disjunction. Studia Logica, Special Issue on Algebraic Logic 50 (1991), 391–419
21. Gil, A.J.: Sistemes de Gentzen multidimensionals i lògiques finitament valorades. Teoria i aplicacions. Ph. D. Thesis, University of Barcelona (1996)
22. Gil, A. J., Torrens, A., and Verdú, V.: Lógicas de Łukasiewicz congruenciales. Teorema de la deducción. In Actas del I Congreso de la Sociedad de Lógica, Metodología y Filosofía de la Ciencia en España (Madrid, 1993), Bustos, E., Echeverría, J., Pérez Sedeño, E. and Sánchez Balmaseda, M.I., Eds., 71–74
23. Gispert, J., and Torrens, A.: Quasivarieties generated by simple MV-algebras. Studia Logica 61 (1998), 79–99
24. Gottwald, S.: A treatise on many-valued logics, vol. 9 of Studies in Logic and Computation. Research Studies Press, Baldock (2001)

25. Hájek, P.: Metamathematics of fuzzy logic, vol. 4 of Trends in Logic, Studia Logica Library. Kluwer, Dordrecht (1998)
26. Hay, L.: Axiomatization of the infinite-valued predicate calculus. The Journal of Symbolic Logic 28 (1963), 77–86
27. Komori, Y.: Super-Łukasiewicz propositional logic. Nagoya Mathematical Journal 84 (1981), 119–133
28. Łukasiewicz, J. Selected Works, edited by L. Borkowski. Studies in Logic and the Foundations of Mathematics. North-Holland, Amsterdam (1970)
29. Łukasiewicz, J., and Tarski, A.: Untersuchungen über den Aussagenkalkül. Comptes Rendus des Séances de la Société des ciences et des Lettres de Varsovie, Cl. III 23 (1930), 30–50. Reprinted in [28], 131–152
30. Miyakawa, M., Nakamura, K., Ramik, J., and Rosenberg, I.: Joint canonical fuzzy numbers. Fuzzy Sets and Systems 53 (1993), 39–47
31. Murali, V.: Lattice of fuzzy subalgebras and closure systems in I^X. Fuzzy Sets and Systems 41 (1991), 101–111
32. Nakamura, K.: Canonical fuzzy number of dimension two and fuzzy utility difference for understanding preferential judgements. Information Science 50 (1990), 1–22
33. Nowak, M.: Logics preserving degrees of truth. Studia Logica 49 (1990), 483–499
34. Rasiowa, H.: An algebraic approach to non-classical logics, vol. 78 of Studies in Logic and the Foundations of Mathematics. North-Holland, Amsterdam (1974)
35. Rebagliato, J., and Verdú, V.: A finite Hilbert-style axiomatization of the implication-less fragment of the intuitionistic propositional calculus. Mathematical Logic Quarterly 40 (1994), 61–68
36. Restall, G.: Subintuitionistic logics. Notre Dame Journal of Formal Logic 35 (1994), 116–129
37. Rodríguez, A. J., Torrens, A., and Verdú, V.: Łukasiewicz logic and Wajsberg algebras. Bulletin of the Section of Logic 19 (1990), 51–55
38. Sofronie-Stokkermans, V.: Representation theorems and the semantics of (semi)lattice-based logics. In Proceedings of the 31st ISMVL Warsaw (2001), Konikowska, B., Ed., The IEEE Computer Society Press, 125–134
39. Sofronie-Stokkermans, V.: Representation theorems and the semantics of non-classical logics, and applications to automated theorem proving. This volume, 57–98
40. Suzuki, Y., Wolter, F., and Zacharyaschev, M.: Speaking about transitive frames in propositional languages. Journal of Logic, Language and Information 7 (1998), 317–339
41. Torrens, A.: W-algebras which are Boolean products of members of SR[1] and CW-algebras. Studia Logica 47 (1987), 265–274
42. Wansing, H.: Displaying as temporalizing. Sequent systems for subintuitionistic logic. In Logic, Language and Computation, Akama, S., Ed. Kluwer (1997) 159–178
43. Wójcicki, R.: Matrix approach in the methodology of sentential calculi. Studia Logica 32 (1973), 7–37
44. Wójcicki, R.: On matrix representation of consequence operations on Łukasiewicz's sentential calculi. Zeitschrift für Mathematische Logik und Grundlagen der Mathematik 19 (1976), 239–247
45. Wójcicki, R.: Theory of logical calculi. Basic theory of consequence operations, vol. 199 of Synthese Library. Reidel, Dordrecht (1988)

Chapter 3
Representation Theorems and the Semantics of Non-classical Logics, and Applications to Automated Theorem Proving

Viorica Sofronie-Stokkermans

Max-Planck-Institut für Informatik, Stuhlsatzenhausweg 85, Saarbrücken, Germany
sofronie@mpi-sb.mpg.de

Abstract. We give a uniform presentation of representation and decidability results related to the Kripke-style semantics of several non-classical logics. We show that a general representation theorem (which has as particular instances the representation theorems as algebras of sets for Boolean algebras, distributive lattices and semilattices) extends in a natural way to several classes of operators and allows to establish a relationship between algebraic and Kripke-style models. We illustrate the ideas on several examples. We conclude by showing how the Kripke-style models thus obtained can be used (if first-order axiomatizable) for automated theorem proving by resolution for some non-classical logics.

1 Introduction

The goal of this paper is to give a uniform presentation of representation and decidability results related to a Kripke-style semantics of several non-classical logics. The logics we consider have a natural algebraic semantics. According to the properties of premise combination, their algebraic models can for instance be bounded lattices or semilattices with additional operators, or just partially ordered semigroups or monoids. There are strong links between the structural rules the logics satisfy and the properties of their algebraic models. Besides algebraic models also Kripke-style or relational models exist. Kripke-style models have usually a much simpler structure than the algebraic models.

The main contribution of this paper is a general way of establishing links between algebraic and Kripke-style semantics. We show that a general representation theorem (which has as particular instances the representation theorems as algebras of sets for Boolean algebras, distributive lattices and semilattices) extends in a uniform way to certain classes of operators. This extension allows in many cases to establish a direct relationship between algebraic and Kripke-style models of non-classical logics. We illustrate the ideas on several examples. For instance, we show that representation theorems allow us to explain, in a uniform way, the choice of Kripke-style models for modal logics and intuitionistic modal logics; for relevant logics; and for other substructural logics such as *BCC* and related logics. Thus, we obtain a simple and uniform explanation for the choice of a multitude of Kripke-style models for non-classical logics, as well as a way of constructing such models in more general cases. The method we use explains why in some cases it is sufficient if we choose as Kripke-style models sets of possible worlds endowed with relations, and when it is necessary to additionally equip them with a preorder or a semilattice structure. It also explains, in a uniform way, which is the appropriate way of defining meaning functions in such models, and the properties these meaning functions should have. Based on the Kripke-style models obtained this way, we present a method for automated theorem proving by resolution in various classes of non-classical logics.

The paper is structured as follows. Section 2 contains generalities on non-classical logics and their semantics. In Section 3 one of the main results of the paper is presented: a general criterion that explains the link between algebraic and Kripke-style semantics. Section 4 contains classical representation theorems for Boolean algebras, distributive lattices and (semi)lattices, and shows how they arise as particular instances of a more general representation theorem. Section 5 contains another of the main results of this paper. We show that the representation theorems described above extend according to a common pattern when we consider certain classes of operators. We also discuss the way these representations can be used for defining Kripke-style semantics. In Sections 6 and 7 the ideas above are illustrated on concrete examples, namely logics based on distributive lattices with operators and logics based on (semi)lattices with residuation. In Section 8 we present a resolution-based method that exploits the Kripke-style models obtained by using the methods described in Section 3–5.

2 Propositional non-classical logics

A propositional logic $\mathcal{L}$ can usually be described by a set of formulae, $\mathsf{Fma}(\mathcal{L})$, and an entailment relation $\vdash$ between sets or sequences of formulae and formulae. A formula $\phi \in \mathsf{Fma}(\mathcal{L})$ is a *theorem* of $\mathcal{L}$ (denoted by $\mathcal{L} \vdash \phi$) if $\emptyset \vdash \phi$. Usually, $\mathsf{Fma}(\mathcal{L})$ is constructed starting from a set Var of propositional variables and a set O of logical connectives. In this case the logic is described as $\mathcal{L} = (\mathsf{Var}, O, \vdash)$. The properties of entailment can be expressed by Gentzen-style calculi, Hilbert-style calculi, natural deduction systems, etc.

2.1 Gentzen-style calculi

The fundamental objects in Gentzen-style calculi are *sequents*. A sequent is, as in [25], an expression of the form $X \vdash Y$, where X and Y are sequences of formulae, asserting that there is an inference from the premises X to the conclusion Y. If Y is empty or consists of a single formula, $X \vdash Y$ is a single conclusion sequent.

Logical consequence can be expressed either within the language of a logic, by means of a special operator for logical implication, or by using the entailment relation $\vdash$. If the language of $\mathcal{L}$ contains a symbol $\rightarrow$ denoting logical implication and ',' is used for premise combination, the link between implication and logical consequence can usually be expressed by a *residuation condition* (also known as the *deduction theorem* for $\vdash$):

$$p, q \vdash r \text{ iff } p \vdash q \rightarrow r. \tag{1}$$

The properties of the premise combination in the left-hand side of the $\vdash$ sign reflect on the properties of $\rightarrow$. Some of the rules expressing properties of the premise combination, known in literature as "structural rules" (appropriate e.g. for single conclusion sequent systems) are *weakening* (W), *exchange* (E), *associativity* (A), and *contraction* (C).

$$(W)\ \frac{\Gamma, \Delta \vdash A}{\Gamma, Y, \Delta \vdash A} \qquad (E)\ \frac{\Gamma, X, Y, \Delta \vdash A}{\Gamma, Y, X. \Delta \vdash A}$$

$$(A)\ \frac{(\Gamma, \Delta), \Sigma \vdash A}{\Gamma, (\Delta, \Sigma) \vdash A} \qquad (C)\ \frac{\Gamma, X, X, \Delta \vdash A}{\Gamma, X, \Delta \vdash A}$$

2.2 Substructural logics

Substructural logics are non-classical logics in which not all the structural rules hold. They are motivated by considerations from philosophy (relevant logics), linguistics (the Lambek calculus) or computing (linear logic). In *relevant logic* [61] for instance, in order for $X \vdash A$ to hold, X must be relevant to A. It may happen that $X \vdash A$ holds but $X, Y \vdash A$ does not hold, namely when Y is not relevant for A. Thus, weakening may not hold. *Linear logic* [26] is based on another way of restricting premise combination: premises are seen as resources which must be used, cannot be reused, and do not extend indefinitely. Thus, in this case weakening and the contraction rule do not hold. There are other logics in which the contraction rule is absent, such as, for instance, Łukasiewicz's logics, as well as other fuzzy logics. The *Lambek calculus* was introduced by Lambek [44, 45] to mathematically model notions such as language and syntax. Since in this case premise combination corresponds to composition of linguistic units (e.g. strings) both the *number* and the *order* of the premises are important, hence neither the contraction nor the exchange rule hold. Variants of the Lambek calculus, like the non-associative Lambek calculus, do not even accept the associativity rule.

Note: Many papers on substructural logics define their theorems with a *Hilbert-style calculus*, and hence, with a provability relation which does satisfy all structural rules, but does not satisfy a deduction theorem.

2.3 Algebraic semantics

Non-classical logics usually have a natural algebraic model, namely their Lindenbaum algebra. Let $\mathcal{L} = (\mathsf{Var}, O, \vdash)$. The set of all formulae $\mathsf{Fma}(\mathcal{L})$ has an O-algebra structure. The *Lindenbaum algebra* of $\mathcal{L}$ can usually be constructed by identifying provably equivalent formulae. Formally, the Lindenbaum algebra of $\mathcal{L}$ is the quotient $\mathsf{Fma}(\mathcal{L})/\equiv$, where $\equiv$ is defined by $\phi \equiv \psi$ iff $\phi \vdash \psi$ and $\psi \vdash \phi$. ($\equiv$ is a congruence relation under certain assumptions about the properties of $\vdash$.) A relation $\leq$ can be defined on $\mathsf{Fma}(\mathcal{L})/\equiv$ by $[\phi] \leq [\psi]$ iff $\phi \vdash \psi$. The structural rules valid in $\mathcal{L}$ reflect into properties of operators in O and of $\leq$ in $\mathsf{Fma}(\mathcal{L})/\equiv$. For instance, the following correspondences exist (for a more detailed presentation of the links between structural rules and properties of $\mathsf{Fma}(\mathcal{L})/\equiv$ see [12]):

LOGIC	LINDENBAUM ALGEBRA	
cut rule	$\leq$ is a partial order and $\circ$ is monotone	
weakening (W)	$[\phi] \circ [\psi] \leq [\psi]$	for all $\phi, \psi \in \mathsf{Fma}(\mathcal{L})$
exchange (E)	$[\phi] \circ [\psi] = [\psi] \circ [\phi]$	for all $\phi, \psi \in \mathsf{Fma}(\mathcal{L})$
associativity (A)	$[\phi] \circ ([\psi] \circ [\gamma]) = ([\psi] \circ [\phi]) \circ [\gamma]$	for all $\phi, \psi, \gamma \in \mathsf{Fma}(\mathcal{L})$
contraction (C)	$[\phi] \leq [\phi] \circ [\phi]$	for all $\phi \in \mathsf{Fma}(\mathcal{L})$
residuation condition	$[\phi] \circ [\psi] \leq [\gamma]$ iff $[\phi] \leq [\psi] \to [\gamma]$	for all $\phi, \psi, \gamma \in \mathsf{Fma}(\mathcal{L})$

The equivalence classes of the theorems of $\mathcal{L}$ can be regarded as *distinguished elements*. In what follows we will consider algebraic models (called O-matrices) of the form $(\mathbf{A}, D)$, where $\mathbf{A} = (A, \{o_A\}_{o \in O})$ is an algebra of similarity type O and $D \subseteq A$ is a subset of distinguished elements. Let $\mathcal{L} = (\mathsf{Var}, O, \vdash)$ be a logic, $\phi \in \mathsf{Fma}(\mathcal{L})$, and $(\mathbf{A}, D)$ a O-matrix. Let $f : \mathsf{Var} \to A$ be a map and $\overline{f} : \mathsf{Fma}(\mathcal{L}) \to \mathbf{A}$ be its unique extension to a homomorphism of O-algebras. Let $\mathcal{M}$ be a class of O-matrices. We define the relation $\overset{a}{\models}$ as follows:

- $(\mathbf{A}, D) \overset{a}{\models}_f \phi$ iff $\overline{f}(\phi) \in D$;
- $(\mathbf{A}, D) \overset{a}{\models} \phi$ iff $\overline{f}(\phi) \in D$ for every $f : \mathsf{Var} \to A$;
- $\mathcal{M} \overset{a}{\models} \phi$ iff $(\mathbf{A}, D) \overset{a}{\models} \phi$ for every $(\mathbf{A}, D) \in \mathcal{M}$.

The logic $\mathcal{L}$ is *sound* with respect to $\mathcal{M}$ iff for every $\phi \in \mathsf{Fma}(\mathcal{L})$, if $\mathcal{L} \vdash \phi$ then $\mathcal{M} \overset{a}{\models} \phi$; $\mathcal{L}$ is *complete* with respect to $\mathcal{M}$ iff for every $\phi \in \mathsf{Fma}(\mathcal{L})$, if $\mathcal{M} \overset{a}{\models} \phi$ then $\mathcal{L} \vdash \phi$.

Examples. Classical, intuitionistic, and modal logic, and several other non-classical logics are sound and complete with respect to matrices of the form $(A, \{1_A\})$, where A is a partially-ordered set (e.g. Boolean algebra or Heyting algebra, possibly with operators) with greatest element 1_A. In many non-classical logics one has to consider larger sets of distinguished elements. The standard system of relevance logic is one example: one has to take as a set of distinguished elements the set $D = \{a \in A \mid a \geq e\}$, where e is the neutral element of the De Morgan monoid A (see Anderson and Belnap [1] p.364); the full Lambek calculus is another (see [55], p.272).

In what follows we will consider the following classes of logics:

(i) positive logics (logics which do not have the implication symbol as a logical connective[1]);
(ii) logics having as algebraic models Heyting algebras with operators (e.g. classical and intuitionistic logic, modal logics, Post logics, and generalizations thereof);
(iii) logics having as algebraic models residuated (semi)lattices with operators (e.g. the substructural logics mentioned in Section 2.2, as well as many fuzzy logics).

Since the difference between the classes (ii) and (iii) is similar to the difference between Post and Łukasiewicz logics, we will sometimes refer to the logics in class (ii) as *Post-style* and those in class (iii) as *Łukasiewicz-style* (terminology due to Ewa Orłowska [57]).

2.4 Kripke-style semantics

A Kripke model consists of a set of *possible worlds*, with additional accessibility relations between possible worlds associated with the logical operators, and a *meaning function* m which assigns to every propositional variable p the set $m(p)$ of all possible worlds where p is true. A satisfiability relation $\overset{r}{\models}$ can usually be defined for every class $\mathcal{K}$ of Kripke-style models by induction on the structure of the formulae. The distinctive feature of Kripke-style semantics is that the accessibility relations are used in the definition of satisfiability, which is not just a mechanical truth-functional translation of the formula structure into the model. (Several examples of such inductive definitions are presented in Section 6 and Section 7.)

A logic $\mathcal{L}$ is *sound* with respect to a class $\mathcal{K}$ of Kripke-style models iff for every $\phi \in \mathsf{Fma}(\mathcal{L})$, if $\mathcal{L} \vdash \phi$ then $\mathcal{K} \overset{r}{\models} \phi$; it is *complete* with respect to $\mathcal{K}$ iff for every $\phi \in \mathsf{Fma}(\mathcal{L})$, if $\mathcal{K} \overset{r}{\models} \phi$ then $\mathcal{L} \vdash \phi$. The relationship between algebraic and Kripke-style models will be analyzed in Section 3. Several examples will be discussed in Section 6 and Section 7. (Note that Kripke modeling has its limitations. Fine and Thomasson constructed modal logics which have no characteristic set of Kripke-style models (see also [31], p.11).)

[1] In positive logics, logical consequence can only be expressed by using the provability relation $\vdash$ (cf. also the so-called *binary logics* [31] Ch.2, or the similar concept in [13]).

3 A link between algebraic and Kripke-style models

In this section we show that if a logic $\mathcal{L}$ is sound and complete with respect to a class $\mathcal{M}$ of matrices, and if the underlying algebras of the matrices in $\mathcal{M}$ can be represented as algebras of subsets of relational structures in a class $\mathcal{R}$, then the relational structures in $\mathcal{R}$ are good candidates for defining a class $\mathcal{K}_{\mathcal{M},\mathcal{R}}$ of Kripke-style models for $\mathcal{L}$. In particular we show that, under certain conditions, soundness and completeness of $\mathcal{L}$ with respect to the class $\mathcal{K}_{\mathcal{M},\mathcal{R}}$ is a consequence of the soundness and completeness of $\mathcal{L}$ with respect to $\mathcal{M}$.

Let $\mathcal{M}$ be a class of O-matrices. Assume that there exists a class $\mathcal{R}$ of relational structures such that the following condition holds:

(C) there exist $\mathbb{D} : \mathcal{M} \to \mathcal{R}$, $\mathbb{E} : \mathcal{R} \to \mathcal{M}$ such that:
  (i) for every $K \in \mathcal{R}$, $\mathbb{E}(K) = (\mathbf{A}_K, D_K) \in \mathcal{M}$, where $\mathbf{A}_K$ is an algebra of *subsets* of the support of K; and
  (ii) for every $M = (\mathbf{A}, D) \in \mathcal{M}$, if $\mathbb{E}(\mathbb{D}(M)) = (\mathbf{A}_{\mathbb{D}(M)}, D_{\mathbb{D}(M)})$ then there is an injective homomorphism $i : \mathbf{A} \to \mathbf{A}_{\mathbb{D}(M)}$ with $i^{-1}(D_{\mathbb{D}(M)}) \subseteq D$.

Let $\mathcal{K}_{\mathcal{M},\mathcal{R}} = \{(K, m) \mid K \in \mathcal{R}, m : \mathsf{Var} \to A_K, \text{ where } \mathbb{E}(K) = (\mathbf{A}_K, D_K)\}$ be the class of all pairs (K, m) consisting of a frame K and a meaning function m associating with every propositional variable p a subset $m(p) \in A_K$ of the support of K (where A_K is in the support of $\mathbf{A}_K$). A relation $\overset{r}{\models}$ can be defined as follows:

Definition 1. *Assume that $\mathcal{M}$ and $\mathcal{R}$ satisfy condition (C)(i). Let $K \in \mathcal{R}$, $m : \mathsf{Var} \to A_K$, and $\overline{m} : \mathsf{Fma}(\mathcal{L}) \to \mathbf{A}_K$ be the unique homomorphism of O-algebras that extends m. Let x be an element in the support of K. Then:*

- $K \overset{r}{\models}_{m,x} \phi$ *iff* $x \in \overline{m}(\phi)$;
- $K \overset{r}{\models}_{m} \phi$ *iff* $\overline{m}(\phi) \in D_K$;
- $K \overset{r}{\models} \phi$ *iff for every meaning function* m, $K \overset{r}{\models}_{m} \phi$.

Theorem 1. *Assume that $\mathcal{M}$ and $\mathcal{R}$ satisfy condition (C)(i). Then, for every $\phi \in \mathsf{Fma}(\mathcal{L})$, if $\mathcal{M} \overset{a}{\models} \phi$ then for every $K \in \mathcal{R}$, $K \overset{r}{\models} \phi$. If in addition $\mathcal{M}$ and $\mathcal{R}$ satisfy condition (C)(ii) the converse also holds.*

Proof: Assume that $\mathcal{M}$ and $\mathcal{R}$ satisfy condition (C)(i) and let $\phi \in \mathsf{Fma}(\mathcal{L})$ such that $\mathcal{M} \overset{a}{\models} \phi$. Let $K \in \mathcal{R}$, $m : \mathsf{Var} \to A_K$, and $\overline{m} : \mathsf{Fma}(\mathcal{L}) \to \mathbf{A}_K$ be the unique homomorphism of O-algebras that extends m. By (C)(i), $\mathbb{E}(K) = (\mathbf{A}_K, D_K) \in \mathcal{M}$. Hence, as $\mathcal{M} \overset{a}{\models} \phi$, $\overline{m}(\phi) \in D_K$, i.e. $K \overset{r}{\models}_{m} \phi$. Assume that, in addition, also (C)(ii) holds. Let $\phi \in \mathsf{Fma}(\mathcal{L})$, with $\mathcal{R} \overset{r}{\models} \phi$. Let $M = (\mathbf{A}, D) \in \mathcal{M}$, and $f : \mathsf{Var} \to A$. By (C)(ii), there is an injective homomorphism $i : \mathbf{A} \to \mathbf{A}_{\mathbb{D}(M)}$ with $i^{-1}(D_{\mathbb{D}(M)}) \subseteq D$. Let $m = i \circ f : \mathsf{Var} \to \mathbf{A}_{\mathbb{D}(M)}$. Since $\mathbb{D}(M) \in \mathcal{R}$, we know that $\mathbb{D}(M) \overset{r}{\underset{a}{\models}}_m \phi$, i.e. $\overline{m}(\phi) = i(\overline{f}(\phi)) \in D_{\mathbb{D}(M)}$. Then $\overline{f}(\phi) \in i^{-1}(D_{\mathbb{D}(M)}) \subseteq D$, hence $(\mathbf{A}, D) \models_f \phi$. □

Corollary 1. *Let $\mathcal{L} = (\mathsf{Var}, O, \vdash)$ be sound and complete with respect to a class $\mathcal{M}$ of O-matrices. Assume that there exists a class $\mathcal{R}$ such that condition (C) holds. Then $\mathcal{L}$ is sound and complete with respect to the class $\mathcal{K}_{\mathcal{M},\mathcal{R}}$ defined above, which can be regarded as a class of Kripke-style models.*

The choice of $\mathcal{R}$, $\mathbb{D}$ and $\mathbb{E}$ is often inspired by representation theorems for the algebras in $\mathcal{M}$. In Section 6 and Section 7 we will show how the ideas above can be used in order to establish Kripke-style semantics, or in order to explain the link between algebraic and Kripke-style semantics, for several logics.

In what follows we will only consider classes $\mathcal{M}$ of matrices of the form $(\mathbf{A}, D_A)$, where the support A of $\mathbf{A}$ is a partially-ordered set, and $D_A = \{a \in A \mid a \geq c_A\}$, where c is a constant symbol. If c_A is the largest element 1_A of A, then $D_A = \{1_A\}$. The class of logics whose matrices have this form is sufficiently large. Classical, intuitionistic and modal logic have matrices of the form $(\mathbf{A}, \{1_A\})$. Many substructural logics (e.g. the standard system of relevance logic [1] and full Lambek calculus (see [55], p.272)) have matrices of the form $(\mathbf{A}, \{a \in A \mid a \geq c_A\})$.

4 Representation theorems for finitely-generated (quasi)varieties

Representation theorems provide a better understanding of the properties of algebraic structures. They state that algebras in a certain class are isomorphic to (i.e. *can be represented as*) algebras whose structure is simpler to describe, and usually much better understood, such as algebras of sets, functions, or relations.

We start this section by presenting the main idea of a class of representation theorems (duality theorems, in fact) for quasi-varieties generated by a finite algebra. Stone's duality for Boolean algebras, Priestley's duality for bounded distributive lattices and known duality theorems for semilattices are all based on this idea. Then we briefly discuss other ways in which Priestley's duality for bounded distributive lattices can be extended to general lattices.

In Section 5 we then show that the representation theorems for Boolean algebras, distributive lattices and semilattices can all be extended according to a uniform pattern to such algebras *with additional operators*. (Note that, whereas the class of Boolean algebras, the class of distributive lattices and that of semilattices are quasi-varieties generated by a two-element algebra, the classes of Boolean algebras (resp. distributive lattices or semilattices) with operators are in general not finitely-generated.)

4.1 Representations as algebras of homomorphisms

Given a class $\mathcal{V} = ISP(\underline{P})$ of algebras, the main idea of so-called *natural duality theorems* [7], is to seek an "alter ego" for the generating algebra $\underline{P}$, which we require to be a topological structure $\underset{\approx}{P} = (P, \tau, R)$ on the underlying set P of $\underline{P}$, so chosen that there is a dual equivalence between $\mathcal{V}$ and a suitable category of topological relational structures of the same type as $\underset{\approx}{P}$. The goal is to obtain a concrete representation of any $\mathbf{A} \in \mathcal{V}$ as the algebra of continuous R-preserving maps from $D(\mathbf{A}) = \mathsf{Hom}_{\mathcal{V}}(\mathbf{A}, \underline{P})$ into $\underset{\approx}{P}$, this algebra inheriting its operations pointwise from $\underline{P}^{D(\mathbf{A})}$. Several classes of algebras for which such duality theorems exist are presented in [7]. Such theorems can be established in particular for the classes $\mathsf{B} = ISP(B_2)$ of Boolean algebras, $\mathsf{D}_{01} = ISP(L_2)$ of bounded distributive lattices, and $\mathsf{SL} = ISP(S_2)$ of meet semilattices (where B_2 is the 2-element Boolean algebra, L_2 the 2-element bounded lattice, and S_2 the 2-element meet semilattice $(\{0,1\}, \min)$).

4.2 Representations as algebras of sets for Boolean algebras, distributive lattices, semilattices and lattices

For $\mathcal{V} = \mathsf{B}$, we obtain Stone's representation theorem by choosing as an alter-ego for B_2 the 2-element set with the discrete topology and noting that, for every $\mathbf{B} \in \mathsf{B}$, $\mathsf{Hom}_{\mathsf{B}}(\mathbf{B}, B_2)$ is homeomorphic to $D(\mathbf{B}) = (\mathcal{F}_{\mathsf{max}}(\mathbf{B}), \tau)$, where $\mathcal{F}_{\mathsf{max}}(\mathbf{B})$ is the set of maximal filters of $\mathbf{B}$, and τ the topology generated by the sets $X_a = \{F \in \mathcal{F}_{\mathsf{max}}(\mathbf{B}) \mid a \in F\}$.

Theorem 2 (Stone's representation theorem [73]). *Every Boolean algebra* $\mathbf{B}$ *is isomorphic to the Boolean algebra* $\mathsf{Clopen}(D(\mathbf{B}))$ *of all closed and open subsets of* $D(\mathbf{B})$. *The isomorphism* $\eta_{\mathbf{B}} : \mathbf{B} \to \mathsf{Clopen}(D(\mathbf{B}))$ *is defined by* $\eta_{\mathbf{B}}(x) = \{F \in \mathcal{F}_{\mathsf{max}}(\mathbf{B}) \mid x \in F\}$.

If $\mathcal{V} = \mathsf{D}_{01}$ we obtain the Priestley representation theorem by choosing as an alter-ego for L_2 the two-element partially-ordered set $\{0,1\}$ with $0 \leq 1$ and the discrete topology, and noting that for every $\mathbf{L} \in \mathsf{D}_{01}$, $\mathsf{Hom}_{\mathsf{D}_{01}}(\mathbf{L}, L_2)$ is homeomorphic to $D(\mathbf{L}) = (\mathcal{F}_p(\mathbf{L}), \subseteq, \tau)$, where $\mathcal{F}_{\mathsf{p}}(\mathbf{L})$ is the set of prime filters of $\mathbf{L}$ and τ is the topology generated by the sets of the form $X_a = \{F \in \mathcal{F}_p(\mathbf{L}) \mid a \in F\}$ and their complements as a subbasis.

Theorem 3 (Priestley representation theorem [59]). *Every bounded distributive lattice* $\mathbf{L}$ *is isomorphic to the lattice* $\mathsf{ClopenOF}(D(\mathbf{L}))$ *of closed and open order filters of* $D(\mathbf{L})$. *The isomorphism* $\eta_{\mathbf{L}} : \mathbf{L} \to \mathsf{ClopenOF}(D(\mathbf{L}))$ *is defined by* $\eta_{\mathbf{L}}(x) = \{F \in \mathcal{F}_p(\mathbf{L}) \mid x \in F\}$.

If $\mathcal{V} = \mathsf{SL}$ we can choose as an alter-ego $\underset{\approx}{S_2}$ of S_2 the semilattice S_2 with the discrete topology. Then every $\mathbf{S} \in \mathsf{SL}$ is isomorphic to the algebra of continuous bounded semilattice homomorphisms from $D(\mathbf{S}) = \mathsf{Hom}_{\mathsf{SL}}(\mathbf{S}, S_2)$ (a topological bounded semilattice) into $\underset{\approx}{S_2}$. $\mathsf{Hom}_{\mathsf{SL}}(\mathbf{S}, S_2)$ is isomorphic to the topological bounded semilattice $\mathcal{SF}(\mathbf{S})$ of semilattice-filters of $\mathbf{S}$.

Theorem 4 (Representation theorem for semilattices [8] p.99). *Every meet semilattice* $\mathbf{S}$ *is isomorphic to the semilattice* $\mathsf{Clopen}\mathcal{SF}(D(\mathbf{S}))$ *of all closed and open semilattice-filters of* $D(\mathbf{S})$. *The isomorphism* $\eta_{\mathbf{S}} : \mathbf{S} \to \mathsf{Clopen}\mathcal{SF}(D(\mathbf{S}))$ *is defined by* $\eta_{\mathbf{S}}(x) = \{F \in \mathcal{SF}(\mathbf{S}) \mid x \in F\}$.

For every semilattice $\mathbf{S}$, $(\mathcal{SF}(D(\mathbf{S})), \cap, \vee)$ is a lattice, where if $F_1, F_2 \in \mathcal{SF}(D(\mathbf{S}))$ then $F_1 \vee F_2$ is the semilattice filter generated by $F_1 \cup F_2$. We now prove that the representation theorem for semilattices can be extended to one for lattices.

Theorem 5. *For every lattice* $\mathbf{L}$, *the map* $\eta_{\mathbf{L}} : \mathbf{L} \to (\mathcal{SF}(D(\mathbf{L})), \cap, \vee)$ *defined by* $\eta_{\mathbf{L}}(x) = \{F \in \mathcal{SF}(\mathbf{L}) \mid x \in F\}$ *is an injective homomorphism of lattices.*

Proof: It is clear that $\eta_{\mathbf{L}}(x)$ is an injective $\wedge$-homomorphism. The fact that it is also a $\vee$-homomorphism can be shown as follows. Let $i_1 : L \to \mathcal{SF}(L)$ be defined for every $x \in L$ by $i_1(x) = \uparrow x$. It can be seen that $i_1(x \wedge y) = \uparrow(x \wedge y) = \uparrow x \vee \uparrow y = i_1(x) \vee i_1(y)$, and $i_1(x \vee y) = \uparrow(x \vee y) = \uparrow x \cap \uparrow y$. Let $i_2 : \mathcal{SF}(L) \to \mathcal{SF}(\mathcal{SF}(L))$ be defined, for every $F \in \mathcal{SF}(L)$, by $i_2(F) = \uparrow F = \{G \in \mathcal{SF}(L) \mid F \subseteq G\}$. Then $\eta_{\mathbf{L}} = i_2 \circ i_1$, and $\eta_{\mathbf{L}}(x \vee y) = i_2(i_1(x \vee y)) = i_2(i_1(x) \cap i_1(x)) = i_2(i_1(x)) \vee i_2(i_1(x)) = \eta_{\mathbf{L}}(x) \vee \eta_{\mathbf{L}}(y)$. □

4.3 Other representation theorems for lattices

Other representation theorems for lattices generalize the Stone and Priestley representation theorem in different ways. Nevertheless, in establishing them the same pattern as for the Stone and Priestley representation theorems is followed: First, for every lattice $\mathbf{L}$ a description of a so-called "dual space", $D(\mathbf{L})$, is given. $D(\mathbf{L})$ usually is a topological space with additional structure. The lattice $\mathbf{L}$ is then proved to be isomorphic with the lattice of all those subsets of $D(\mathbf{L})$ which are closed, both topologically and with respect to a closure operator Γ induced by the non-topological structure of $D(\mathbf{L})$.

In [77], Urquhart proves the following representation theorem. Let $\mathbf{L} = (L, \vee, \wedge, 0, 1)$ be a bounded lattice. The dual space $D(\mathbf{L}) = (X, \leq_1, \leq_2, \tau)$ of $\mathbf{L}$ is a doubly-ordered topological space, where (i) X is the set of all filter-ideal pairs (F, I) with the property that F is maximal among the filters of $\mathbf{L}$ with $F \cap I = \emptyset$ and I is maximal among all ideals of $\mathbf{L}$ with $F \cap I = \emptyset$; (ii) for $(F_1, I_1), (F_2, I_2) \in X$, $(F_1, I_1) \leq_1 (F_2, I_2)$ iff $F_1 \subseteq F_2$ and $(F_1, I_1) \leq_2 (F_2, I_2)$ iff $I_1 \subseteq I_2$; (iii) the topology τ is defined by taking the family $\{X \backslash X^1_a \mid a \in L\} \cup \{X \backslash \rho(X^1(a)) \mid a \in L\}$ as a subbasis, where $X^1_a = \{(F, I) \in X \mid a \in F\}$, and

$$\rho(Y) := \{x \in X \mid Y \subseteq X \backslash \uparrow_{\leq_2} x\}; \quad \lambda(Z) := \{x \in X \mid Z \subseteq X \backslash \uparrow_{\leq_1} x\}.$$

Let $\Gamma = \lambda \circ \rho$. With the notations above, the following holds.

Theorem 6 ([77]). *Every bounded lattice* **L** *is isomorphic to the lattice of all* Γ*-closed subsets* Y *of* X *with the property that* Y *and* $\rho(Y)$ *are closed in* τ.

In [36], Hartonas and Dunn combine filters and ideals in a different way. The dual space of a lattice **L** is in this case a triple of the form $(X, \perp, Y)$, where X and Y are the filter space resp. the ideal space of **L** (endowed with a natural topology) and $\perp \subseteq X \times Y$ is defined by $F \perp I$ iff $F \cap I \neq \emptyset$. Let

$$\lambda(U) = \{y \in Y \mid U \perp y\}; \quad \rho(V) = \{x \in X \mid x \perp V\}.$$

and let $\Gamma = \lambda \circ \rho$. (A similar idea is used in [28, 39] for ortholattices.) With the notations above the following holds.

Theorem 7 ([36]). *Every lattice* **L** *is isomorphic to the lattice of all clopen,* Γ*-stable subsets of the filter space of* **L**.

The Stone representation theorem for lattices in [36] is further simplified and extended to certain classes of lattice-ordered algebras by Hartonas in [35]. Here, the dual space $D(\mathbf{L}) = (\mathcal{F}(\mathbf{L}), \Gamma, \tau)$ of a lattice **L** (with upper bound) is a closure system, where (i) $\mathcal{F}(\mathbf{L})$ is the set of all filters of **L**, (ii) $\Gamma = \lambda \circ \rho$ is a closure operator, where λ, ρ are defined by:

$$\lambda(U) = \{x \in \mathcal{F}(\mathbf{L}) \mid U \subseteq x\}; \quad \rho(V) = \{x \in \mathcal{F}(\mathbf{L}) \mid x \subseteq V\};$$

(iii) the topology τ is generated by the family consisting of the sets $X_a = \{x \in \mathcal{F}(\mathbf{L}) \mid \uparrow a \subseteq x\}$, $a \in L$ and complements thereof as a subbasis.

Theorem 8 ([35]). *Every lattice* **L** *is isomorphic to the lattice of all clopen,* Γ*-closed subsets of* $D(\mathbf{L})$.

5 Representation of (semi)lattices with operators

The representation theorems for Boolean algebras, distributive lattices and (semi)lattices presented in Section 4.2 can be extended to Boolean algebras, distributive lattices and (semi)lattices with certain types of operators. Such extensions have been studied on a considerable number of concrete examples, which lack of space prevents us from mentioning here exhaustively, as well as in more general settings. For instance, the Stone duality has been extended to Boolean algebras with additive operators by Jónsson and Tarski [42, 43]; the Priestley duality has been extended to distributive lattices with operators that are join and meet hemimorphisms by Goldblatt [30], and to operators that are hemimorphisms in some arguments and "anti"hemimorphisms in the others in [67, 71]. Extensions to Heyting algebras are also known [9, 30]. Representation theorems exist also for lattices with operators [12, 35], and even weaker structures, such as partial orders [12]. Very similar ideas are used in all cases: we will show that such extensions follow a common pattern.

5.1 Operators on lattices and semilattices

Bounded lattices or semilattices with additional operators occur often as algebraic models of non-classical logics. The operations $\vee$ and $\wedge$ model logical disjunction and conjunction; the additional operations are usually interpretations of other logical connectives such as the modal connectives for necessity ($\Box$) or possibility ($\diamond$), or various types of implication. The operators that correspond to the modal connectives commute with part of the lattice structure, e.g.

$$\Box(1) = 1, \ \Box(x \wedge y) = \Box(x) \wedge \Box(y); \tag{2}$$
$$\diamond(0) = 0, \ \diamond(x \vee y) = \diamond(x) \vee \diamond(y). \tag{3}$$

Definition 2. *A* join hemimorphism[2] *on a bounded lattice* $\mathbf{L} = (L, \vee, \wedge, 0, 1)$ *is a function* $f : L^n \to L$ *such that for every* $i, 1 \leq i \leq n$,

(1) $f(a_1, \dots, a_{i-1}, 0, a_{i+1}, \dots, a_n) = 0$,
(2) $f(a_1, \dots, a_{i-1}, b_1 \vee b_2, a_{i+1}, \dots, a_n) =$
$= f(a_1, \dots, a_{i-1}, b_1, a_{i+1}, \dots, a_n) \vee f(a_1, \dots, a_{i-1}, b_2, a_{i+1}, \dots, a_n)$.

Hemimorphisms on join semilattices with 0 can be defined in a similar way.

We want to make the classes of distributive lattices with operators we consider broad enough to also encompass those obtained by considering weakened negations and implications, which satisfy identities such as:

$$\sim 0 = 1, \ \sim(x \vee y) = \sim x \wedge \sim y, \tag{4}$$
$$\sim 1 = 0, \ \sim(x \wedge y) = \sim x \vee \sim y; \tag{5}$$
$$0 \to z = 1, \ (x \vee y) \to z = (x \to z) \wedge (y \to z), \tag{6}$$
$$x \to 1 = 1, \ x \to (y \wedge z) = (x \to y) \wedge (x \to z). \tag{7}$$

This means that we need to allow the operators to be hemimorphisms in some arguments and anti-hemimorphisms in other arguments. We now formally define operators that have such properties. Similar definitions appear in [12].

If $\mathbf{L}$ is a (semi)lattice we denote $\mathbf{L}^{+1} = \mathbf{L}$, $\mathbf{L}^{-1} = \mathbf{L}^d$ (the order-dual of $\mathbf{L}$).

Definition 3. *Let* $\mathbf{L} = (L, \vee, \wedge, 0, 1)$ *be a bounded lattice and let* $\varepsilon_1, \dots, \varepsilon_n, \varepsilon \in \{-1, 1\}$. *A function* $f : L^n \to L$ *has type* $\varepsilon_1 \dots \varepsilon_n \to \varepsilon$ *if* $f : \mathbf{L}^{\varepsilon_1} \times \dots \times \mathbf{L}^{\varepsilon_n} \to \mathbf{L}^{\varepsilon}$ *is a join hemimorphism.*

Let $(L, \vee, \wedge)$ be a lattice. For every $a, b \in L$, the *relative pseudocomplement of a with respect to b* (denoted by $a \Rightarrow b$) is the largest element of $\{c \in L \mid a \wedge c \leq b\}$ (if any).

[2] The term "hemimorphism" was introduced by Halmos in [34]; the concept was used for the representation of the necessity operator of modal logic.

Definition 4. *A* Heyting algebra *is an algebra* $(A, \vee, \wedge, \Rightarrow, \neg, 0, 1)$ *such that* $(A, \vee, \wedge, 0, 1)$ *is a bounded distributive lattice, for all* $a, b \in A$, $a \Rightarrow b$ *is the relative pseudocomplement of* a *with respect to* b, *and for all* $a \in A$, $\neg a = a \Rightarrow 0$.

The condition that for all $a, b \in A$, $a \Rightarrow b$ is the relative pseudocomplement of a with respect to b can be also expressed by the condition

(H) $\quad a \wedge b \leq c$ if and only if $a \leq b \Rightarrow c$.

5.2 A common pattern for representation theorems

Assume that the algebra $\mathbf{A}$ has additional operators indexed by a set Σ, $\{\sigma_f\}_{f \in \Sigma}$. The duality theorems in Section 4.2 can be extended to representation theorems that also encompass operators in Σ according to the following pattern:

(a) The duality theorems are relaxed by ignoring the topology on the dual spaces. In particular the following representation theorems can be obtained this way:

1. For every Boolean algebra $\mathbf{B}$, $\eta_{\mathbf{B}} : \mathbf{B} \rightarrow \mathcal{P}(D(\mathbf{B}))$ is an injective homomorphism of Boolean algebras into the algebra of all subsets of $D(\mathbf{B})$.

2. For every bounded distributive lattice $\mathbf{L}$, $\eta_{\mathbf{L}} : \mathbf{L} \rightarrow (\mathcal{O}(D(\mathbf{L})), \cap, \cup, \emptyset, D(\mathbf{L}))$ is an injective homomorphism of bounded distributive lattices into the lattice of order-filters of $D(\mathbf{L})$.

3. For every semilattice $\mathbf{S}$, $\eta_{\mathbf{S}} : \mathbf{S} \rightarrow (\mathcal{SF}(D(\mathbf{S})), \cap)$ is an injective homomorphism of semilattices into the semilattice of semilattice filters of $D(\mathbf{S})$.

4. For every lattice $\mathbf{L}$, $\eta_{\mathbf{L}} : \mathbf{L} \rightarrow (\mathcal{SF}(D(\mathbf{L})), \cap, \vee)$ is an injective homomorphism of lattices into the lattice of semilattice filters of $D(\mathbf{L})$.

The maps $\eta_{\mathbf{A}} : \mathbf{A} \rightarrow E(D(\mathbf{A}))$ (where $E(D(\mathbf{A}))$ stands for $\mathcal{P}(D(\mathbf{A}))$ if $\mathbf{A}$ is a Boolean algebra, for $\mathcal{O}(D(\mathbf{A}))$ is $\mathbf{A}$ is a bounded distributive lattice, and for $\mathcal{SF}(D(\mathbf{A}))$ if $\mathbf{A}$ is a (semi)lattice) are in all cases defined by

$$\eta_{\mathbf{A}}(x) = \{F \in D(\mathbf{A}) \mid a \in F\}.$$

(b) Relations and maps on $D(\mathbf{A})$ are associated with the operators of $\mathbf{A}$.

(c) Operators on $E(D(\mathbf{A}))$ are associated with the relations and maps on $D(\mathbf{A})$.

(d) It is proved that the maps $\eta_{\mathbf{A}} : \mathbf{A} \rightarrow E(D(\mathbf{A}))$ are homomorphisms also with respect to the additional operators.

Remark: It may be possible that a function $\sigma_f : A^n \to A$ has several types. (For instance, a unary lattice homomorphism is both of type $1 \to 1$ and of type $-1 \to -1$.) This needs to be taken into accound in the steps (b) and (c) above. There are several ways to deal with this situation. One possibility is to "duplicate" the function f, i.e. replace f with a family $\{f_t \mid f \text{ of type } t\}$, with the additional condition that $f_t = f_{t'}$ whenever f is both of type t and t'. Sometimes it is possible to establish relationships between the relations associated with these copies of f. However, in general complications may arise due to this duplication, since the conditions $f_t = f_{t'}$ may not be preserved in $E(D(A))$ after step (c). Another solution, which we adopted in [67], is to consider separate classes of operators with "composite types" (such as Lh (lattice homomorphisms) or La (lattice antimorphisms)) for which special constructions are used.

In what follows we will discuss the steps (b) and (c) in more detail.

1. Relations on $D(\mathbf{A})$ associated with the operators of A. The operators on an algebra $\mathbf{A}$ induce in a canonical way relations on $D(\mathbf{A})$. For instance, if $\sigma_f : A^n \to A$ and $\varepsilon_1 \dots \varepsilon_n, \varepsilon \in \{-1, 1\}$ then we can define the relation $R_f^{\varepsilon_1 \dots \varepsilon_n \to \varepsilon} \subseteq D(\mathbf{A})^{n+1}$ by:

$$R_f^{\varepsilon_1 \dots \varepsilon_n \to \varepsilon}(F_1, \dots, F_n, F) \text{ iff } \sigma_f(F_1^{\varepsilon_1}, \dots, F_n^{\varepsilon_n}) \subseteq F^{\varepsilon} \tag{8}$$

where $F^{+1} := F$ and $F^{-1} := L \backslash F$. If $\mathbf{A} = (A, \{\sigma_f\}_{f \in \Sigma})$ has the property that each $f \in \Sigma$ has a unique type then, for every $f \in \Sigma$ of (unique) type $\varepsilon_1 \dots \varepsilon_n \to \varepsilon$, we denote the relation $R_f^{\varepsilon_1 \dots \varepsilon_n \to \varepsilon}$ by R_f.

In what follows we discuss two situations in which the relation R_f can be replaced by a *function* φ_f.

Distributive lattices and Boolean algebras. If $\mathbf{A}$ is a distributive lattice or a Boolean algebra, R_f may be replaced by a function if $f \in \Sigma$ is both of type $\varepsilon_1 \dots \varepsilon_n \to \varepsilon$ and $-\varepsilon_1 \cdots - \varepsilon_n \to -\varepsilon$.

Theorem 9. *Let $f \in \Sigma$ be both of type $\varepsilon_1 \dots \varepsilon_n \to \varepsilon$ and $-\varepsilon_1 \cdots - \varepsilon_n \to -\varepsilon$ (i.e. σ_f is a lattice homomorphism or a lattice antimorphism in each argument). Then, for every $1 \leq i \leq n$*

$$R_f^{-\varepsilon_1 \cdots - \varepsilon_n \to -\varepsilon}(F_1, \dots, F_n, F) \;\; \textit{iff} \;\; \varphi(F_1, \dots, F_{i-1}, F, F_{i+1}, \dots, F_n) \subseteq F_i^{\varepsilon_i},$$

where $\varphi(F_1, \dots, F_{i-1}, F, F_{i+1}, \dots, F_n) = \{y_i \mid \exists x_1, \dots, x_{i-1}, x_{i+1}, \dots, x_n : (\bigwedge_{j \neq i} x_j \notin F_j^{\varepsilon_j} \wedge \sigma_f(x_1, \dots, x_{i-1}, y_i, x_{i+1}, \dots, x_n) \in F^{\varepsilon})\}$.

Proof: By the definition of $R_f^{-\varepsilon_1\cdots-\varepsilon_n\to-\varepsilon}$ we have:
$\neg R_f^{-\varepsilon_1\cdots-\varepsilon_n\to-\varepsilon}(F_1,\dots,F_n,F)$

$$\begin{aligned}
&\text{iff } \sigma_f(F_1^{-\varepsilon_1},\dots,F_n^{-\varepsilon_n}) \not\subseteq F^{-\varepsilon}\\
&\text{iff } \exists x_1\dots x_n(\bigwedge_{i=1}^{n} x_i \in F_i^{-\varepsilon_i} \wedge \sigma_f(x_1,\dots,x_n) \in F^{\varepsilon})\\
&\text{iff } \exists x_i \notin F_i^{\varepsilon_i} \wedge (\exists x_1\dots x_{i-1},x_{i+1},\dots,x_n\\
&\qquad (\bigwedge_{j\neq i} x_j \notin F_j^{\varepsilon_j} \wedge \sigma_f(x_1,\dots,x_n) \in F^{\varepsilon}))\\
&\text{iff } \exists y_i \notin F_i^{\varepsilon_i} \wedge y_i \in \{x_i \mid \exists x_1\dots x_{i-1},x_{i+1},\dots,x_n\\
&\qquad (\bigwedge_{j\neq i} x_j \notin F_j^{\varepsilon_j} \wedge \sigma_f(x_1,\dots,x_n) \in F^{\varepsilon})\}\\
&\text{iff } \exists y_i \notin F_i^{\varepsilon_i} \text{ with } y_i \in \phi(F_1,\dots,F_{i-1},F,F_{i+1},\dots,F_n)\\
&\text{iff } \varphi(F_1,\dots,F_{i-1},F,F_{i+1},\dots,F_n) \not\subseteq F_i^{\varepsilon_i}
\end{aligned}$$

We show that $\varphi(F_1,\dots,F_{i-1},F,F_{i+1},\dots,F_n)$ is a prime filter of A^{ε_i}. To prove that it is upwards-closed in A^{ε_i}, let $y_i \in \varphi(F_1,\dots,F_{i-1},F,F_{i+1},\dots,F_n)$, and let $y_i' \geq_{\varepsilon_i} y_i$. Then there exist $x_1,\dots,x_{i-1},x_{i+1},\dots,x_n$ such that for all $j \neq i$, $x_j \notin F_j^{\varepsilon_j}$ and $\sigma_f(x_1,\dots,x_{i-1},y_i,x_{i+1},\dots,x_n) \in F^{\varepsilon}$. By the monotonicity of σ_f, $\sigma_f(x_1,\dots,x_{i-1},y_i',x_{i+1},\dots,x_n) \geq_{\varepsilon} \sigma_f(x_1,\dots,x_{i-1},y_i,x_{i+1},\dots,x_n)$, hence $\sigma_f(x_1,\dots,x_{i-1},y_i',x_{i+1},\dots,x_n) \in F^{\varepsilon}$, and so $y_i' \in \varphi(F_1,\dots,F_{i-1},F,F_{i+1},\dots,F_n)$.

We prove that $\varphi(F_1,\dots,F_{i-1},F,F_{i+1},\dots,F_n)$ is closed under $\wedge_{\varepsilon_i}$. Let y_i^1, $y_i^2 \in \varphi(F_1,\dots,F_{i-1},F,F_{i+1},\dots,F_n)$. Then there exist $x_1^k,\dots,x_{i-1}^k,x_{i+1}^k,\dots,x_n^k$ such that $x_j^k \notin F_j^{\varepsilon_j}$ and $\sigma_f(x_1^k,\dots,x_{i-1}^k,y_i^k,x_{i+1}^k,\dots,x_n^k) \in F^{\varepsilon}$, for $k=1,2$. For all $j=1,\dots,i-1,i+1,\dots,n$ let $x_j := x_j^1 \vee_{\varepsilon_j} x_j^2$. As F_j are prime filters, $x_j \notin F_j^{\varepsilon_j}$. By the monotonicity of σ_f, $\sigma_f(x_1,\dots,y_i^k,\dots,x_n) \geq_{\varepsilon} \sigma_f(x_1^k,\dots,x_{i-1}^k,y_i^k,x_{i+1}^k,\dots,x_n^k)$ for $k=1,2$. Hence, $\sigma_f(x_1,\dots,y_i^1 \wedge_{\varepsilon_i} y_i^2,\dots,x_n) = \sigma_f(x_1,\dots,y_i^1,\dots,x_n) \wedge_{\varepsilon} \sigma_f(x_1,\dots,y_i^2,\dots,x_n) \in F^{\varepsilon}$. Therefore, $y_i^1 \wedge_{\varepsilon_i} y_i^2 \in \varphi(F_1,\dots,F_{i-1},F,F_{i+1},\dots,F_n)$.

To show that $\varphi(F_1,\dots,F_{i-1},F,F_{i+1},\dots,F_n)$ is a prime filter of A^{ε_i}, assume that $y_i^1 \vee_{\varepsilon_i} y_i^2 \in \varphi(F_1,\dots,F_{i-1},F,F_{i+1},\dots,F_n)$. Then there exist $x_1,\dots,x_{i-1},x_{i+1},\dots,x_n$ such that for all $j \neq i$, $x_j \notin F_j^{\varepsilon_j}$ and $\sigma_f(x_1,\dots,x_{i-1},y_i^1 \vee_{\varepsilon_i} y_i^2,x_{i+1},\dots,x_n) \in F^{\varepsilon}$. As σ_f is of type $\varepsilon_1\dots\varepsilon_n \to \varepsilon$, $\sigma_f(x_1,\dots,y_i^1 \vee_{\varepsilon_i} y_i^2,\dots,x_n) = \sigma_f(x_1,\dots,y_i^1,\dots,x_n) \vee_{\varepsilon} \sigma_f(x_1,\dots,y_i^2,\dots,x_n)$. Hence, $\sigma_f(x_1,\dots,y_i^1,\dots,x_n) \in F^{\varepsilon}$ or $\sigma_f(x_1,\dots,y_i^2,\dots,x_n) \in F^{\varepsilon}$, so $y_i^1 \in \varphi(F_1,\dots,F_{i-1},F,F_{i+1},\dots,F_n)$ or $y_i^2 \in \varphi(F_1,\dots,F_{i-1},F,F_{i+1},\dots,F_n)$.

It is easy to see that φ completely describes the relation $R_f^{-\varepsilon_1\cdots-\varepsilon_n\to-\varepsilon}$. □

Example 1. Let $h : L \to L$ be a unary lattice homomorphism, i.e. an operator of type $-1 \to -1$ and $1 \to 1$. Then $R_h(F_1, F)$ iff $\{x \mid \sigma_h(x) \in F\} \subseteq F_1$. The considerations above show that R_h is completely described by the ($\leq$-preserving) function H_h defined by $H_h(F) = \sigma_h^{-1}(F)$. Similarly, the function corresponding to a unary lattice antimorphism k is the ($\leq$-reversing) function K_k defined by $K_k(F) = L \backslash \sigma_k^{-1}(F)$. In Section 6.3 we illustrate the construction above for a binary operator $\rightarrow$ which is of type $-1, 1 \rightarrow 1$ and $1, -1 \rightarrow -1$.

Semilattices. We show that for any monotone operator on a meet semilattice A, the relation $R_f^{1 \dots 1 \rightarrow 1}$ can be replaced by a function. The idea is illustrated on an example in Section 7.2.

Theorem 10. *If* $\mathbf{A}$ *is a meet semilattice and* $\sigma_f : \mathbf{A}^n \rightarrow \mathbf{A}$ *is monotone in each argument then the function* $\varphi_f : \mathcal{SF}(\mathbf{A})^n \rightarrow \mathcal{SF}(\mathbf{A})$, *defined for every* $F_1, \dots, F_n \in \mathcal{SF}(\mathbf{A})$ *by* $\varphi_f(F_1, \dots, F_n) = \uparrow \sigma_f(F_1, \dots, F_n)$, *completely describes* $R_f^{1 \dots 1 \rightarrow 1}$.

Proof: We show that for every $F_1, \dots, F_n \in \mathcal{SF}(\mathbf{A})$, $\uparrow \sigma_f(F_1, \dots, F_n) \in \mathcal{SF}(\mathbf{A})$. It is clear that $\uparrow \sigma_f(F_1, \dots, F_n)$ is upwards-closed. We show that it is closed under meets in A. Let $y, z \in \uparrow \sigma_f(F_1, \dots, F_n)$. Then there exist $y_i, z_i \in F_i$, $1 \leq i \leq n$, such that $y \geq \sigma_f(y_1, \dots, y_n)$ and $z \geq \sigma_f(z_1, \dots, z_n)$. By the monotonicity of σ_f, $\sigma_f(y_1, \dots, y_n) \geq \sigma_f(y_1 \wedge z_1, \dots, y_n \wedge z_n)$ and $\sigma_f(z_1, \dots, z_n) \geq \sigma_f(y_1 \wedge z_1, \dots, y_n \wedge z_n)$. Hence, $y \wedge z \geq \sigma_f(y_1 \wedge z_1, \dots, y_n \wedge z_n)$, where $y_i \wedge z_i \in F_i$, i.e. $y \wedge z \in \uparrow \sigma_f(F_1, \dots, F_n)$. As $R_f^{1 \dots 1 \rightarrow 1}(F_1, \dots, F_n)$ iff $\varphi_f(F_1, \dots, F_n) \subseteq F$, φ_f completely describes $R_f^{1 \dots 1 \rightarrow 1}$. □

2. Operators on $E(D(\mathbf{A}))$ induced by the relations on $D(\mathbf{A})$. The maps and relations on $X = D(\mathbf{A})$ induce in their turn, in a canonical way, operations on $E(D(\mathbf{A}))$. For instance, if $\varepsilon_1, \dots, \varepsilon_n \in \{-1, +1\}$ and $R \subseteq X^{n+1}$ is an increasing[3] relation then an operator $f_R^{\varepsilon_1 \dots \varepsilon_n} : E(X)^n \rightarrow E(X)$ can be defined by

$$f_R^{\varepsilon_1 \dots \varepsilon_n}(U_1, \dots, U_n) = R^{-1}(U_1^{\varepsilon_1}, \dots, U_n^{\varepsilon_n}); \tag{9}$$

if $R \subseteq X^{n+1}$ is decreasing then $f_R^{\varepsilon_1 \dots \varepsilon_n} : E(X)^n \rightarrow E(X)$ can be defined by

$$f_R^{\varepsilon_1 \dots \varepsilon_n}(U_1, \dots, U_n) = X \backslash R^{-1}(U_1^{\varepsilon_1}, \dots, U_n^{\varepsilon_n}), \tag{10}$$

where $R^{-1}(U_1, \dots, U_n) = \{x \mid \exists x_1 \dots x_n (x_1 \in U_1, \dots, x_n \in U_n, R(x_1, \dots, x_n, x))\}$ and, for every $U \subseteq X$, $U^{+1} := U$ and $U^{-1} := X \backslash U$.

With these definitions the following hold:

Theorem 11. *(1) Let A be a bounded distributive lattice with operators in Σ. If $\sigma_f : A^n \rightarrow A$ is an operator of type $\varepsilon_1 \dots \varepsilon_n \rightarrow \varepsilon$ then $R_f \subseteq D(A)^{n+1}$ is an* increasing *relation if $\varepsilon = 1$ and a* decreasing *relation if $\varepsilon = -1$.*

[3] A relation $R \subseteq X^{n+1}$ is increasing if for every $\overline{x} \in X^n$ (if $R(\overline{x}, y)$ and $y \leq y'$ then $R(\overline{x}, y')$). R is decreasing if for every $\overline{x} \in X^n$ (if $R(\overline{x}, y)$ and $y' \leq y$ then $R(\overline{x}, y')$).

(2) *Let $(X, \leq)$ be a partially-ordered set, let $E(X)$ be the set of all order-filters of X, and let $R \subseteq X^{n+1}$. It R is an increasing relation then $f_R^{\varepsilon_1 \dots \varepsilon_n}$ as defined in (9) is an operator of type $\varepsilon_1 \dots \varepsilon_n \to 1$ on $E(X)$. If R is decreasing then $f_R^{\varepsilon_1 \dots \varepsilon_n}$ as defined in (10) is an operator of type $\varepsilon_1 \dots \varepsilon_n \to -1$ on $E(X)$.*

Theorem 12. *(1) Let A be a semilattice and $\sigma_f : A^n \to A$ a monotone map. Then $R_f^{1 \dots 1 \to 1} \subseteq D(A)^{n+1}$ is an increasing relation.*

(2) If X is a semilattice and $R \subseteq X^{n+1}$ is an increasing relation then the map $f_R^{1 \dots 1} : D(A)^n \to D(A)$ defined as in (9) is monotone.

3. Representation theorems. Finally, we have to show that, for every A in the class $\mathcal{K}$ of algebras for which the representation theorem is sought, the following hold:

(i) the algebra $E(D(\mathbf{A}))$ is an algebra in $\mathcal{K}$;

(ii) the map $\eta_A : \mathbf{A} \to E(D(\mathbf{A}))$ is a homomorphism with respect to the whole signature, i.e. for every $f \in \Sigma$ of arity n, $\eta_A(\sigma_f(x_1, \dots, x_n)) = f_{R_f}(\eta_A(x_1), \dots, \eta_A(x_n))$.

Assume that the class $\mathcal{K}$ has the property that every operator $f \in \Sigma$ is of exactly one type $\varepsilon_1 \dots \varepsilon_i \to \varepsilon$. Let $D(\mathbf{A})$ be the relational structure $(D(A), \{R_f\}_{f \in \Sigma})$, where if f is of type $\varepsilon_1 \dots \varepsilon_i \to \varepsilon$ then $R_f = R_f^{\varepsilon_1 \dots \varepsilon_i \to \varepsilon}$, as defined in (8). Let $E(D(\mathbf{A}))$ be the algebra $(E(D(A)), \{f_{R_f}\}_{f \in \Sigma})$, where if f is of type $\varepsilon_1 \dots \varepsilon_i \to \varepsilon$ then $f_{R_f}(U_1, \dots, U_n) = \left(R_f^{-1}(U_1^{\varepsilon_1}, \dots, U_n^{\varepsilon_n})\right)^{\varepsilon}$. Then the following theorems hold.

Theorem 13. *For every semilattice A with monotone operators in Σ, the map $\eta_A : \mathbf{A} \to E(D(\mathbf{A}))$ is a homomorphism with respect to Σ.*

Proof: Let $f \in \Sigma$. To prove that $\eta_A(\sigma_f(x_1, \dots, x_n)) = f_{R_f}(\eta_A(x_1), \dots, \eta_A(x_n))$ it suffices to show that $F \in \eta_A(\sigma_f(x_1, \dots, x_n))$ iff $F \in f_{R_f}(\eta_A(x_1), \dots, \eta_A(x_n)) = R_f^{-1}(\eta_A(x_1), \dots, \eta_A(x_n))$. This reduces to proving that, for every $x_1, \dots, x_n \in A$,

$$\sigma_f(x_1, \dots, x_n) \in F \text{ iff } \exists F_1, \dots, F_n (x_i \in F_i \text{ and } \sigma_f(F_1, \dots, F_n) \subseteq F).$$

The direct implication is a direct consequence of the fact that for every $x \in A$ there exists a semilattice-filter $F = \uparrow x$ such that $x \in F$. The inverse implication is immediate. □

Theorem 14. *For every bounded distributive lattice (or Boolean algebra) A with operators in Σ, the map $\eta_A : \mathbf{A} \to E(D(\mathbf{A}))$ is a homomorphism with respect to Σ.*

Proof: We give the proof for bounded distributive lattices. The case of Boolean algebras follows as a corollary by ignoring the order on the dual space.

Let $\eta_A : A \to \mathcal{O}(D(A))$ defined by $\eta_A(a) = X_a = \{F \in \mathcal{F}_p(A) \mid a \in F\}$. By the Priestley representation theorem we know that η_A is an injective homomorphism of bounded lattices. We have to prove that for every $f \in \Sigma$ of type $\varepsilon_1 \dots \varepsilon_n \to \varepsilon$, $\eta_A(\sigma_f(a_1, \dots, a_n)) = f_{R_f}(\eta_A(a_1), \dots, \eta_A(a_n))$. Taking into account the definition of η_A and of f_{R_f}, it is sufficient to show that the following conditions are equivalent for every $F \in D(\mathbf{A})$:

1. $\sigma_f(a_1, \dots, a_n) \in F^\varepsilon$;
2. $\exists G_1, \dots, G_n \in D(A)$ such that $a_1 \in G_1^{\varepsilon_1}$, $\dots$, $a_n \in G_n^{\varepsilon_n}$ and $R_f(G_1, \dots, G_n, F))$.

That 2. implies 1. is obvious: If $G_1, \dots, G_n \in D(A)$ are such that $(a_1 \in G_1^{\varepsilon_1}, \dots, a_n \in G_n^{\varepsilon_n}$ and $R_f(G_1, \dots, G_n, F))$ then for all $b_1, \dots, b_n$, if $b_i \in G_i^{\varepsilon_i}$ for $1 \leq i \leq n$ then $\sigma_f(b_1, \dots, b_n) \in F^\varepsilon$. In particular, taking $b_1 = a_1, \dots, b_n = a_n$, it follows that $f(a_1, \dots, a_n) \in F^\varepsilon$.

In order to prove that 1. implies 2., we use an argument similar to the one used by Goldblatt in Theorem 2.2.1 in [30]. We inductively construct a sequence of prime filters $G_1, \dots, G_n$ such that for every $i \leq n$ the following hold:

(i) $a_i \in G_i^{\varepsilon_i}$,
(ii) If $y_k \in G_k^{\varepsilon_k}$ for all $k \leq i$, then $\sigma_f(y_1, \dots, y_i, a_{i+1}, \dots, a_n) \in F^\varepsilon$.

Let $j \leq n$ be fixed and assume that for every $i < j$, G_i has been defined such that the conditions (i) and (ii) are satisfied. Then G_j will be defined such that (i) and (ii) are satisfied for j.
Let $H_j = \{z \mid \exists y_1 {\in} G_1^\varepsilon \dots \exists y_{j-1} {\in} G_{j-1}^\varepsilon {:} \sigma_f(y_1, \dots, y_{j-1}, z, a_{j+1}, \dots, a_n) {\notin} F^\varepsilon\}$.

Claim 1. H_j is closed under all finite joins[4] in A^{ε_j}.

Proof: Let $z_1, z_2 \in H_j$. Then for all $i < j$ there exist $y_i^1, y_i^2 \in G_i^{\varepsilon_i}$ such that $f_A(y_1^k, \dots, y_{j-1}^k, z_k, a_{j+1}, \dots, a_n) \notin F^\varepsilon$ for $k = 1, 2$. Let $y_i = y_i^1 \wedge^{\varepsilon_i} y_i^2$ for $i = 1, \dots, j-1$. Since $\sigma_f : A^{\varepsilon_1} \times \dots \times A^{\varepsilon_n} \to A^\varepsilon$ is a join-hemimorphism, it is order-preserving in every argument with respect to the orderings in $A^{\varepsilon_1}, \dots, A^{\varepsilon_n}$ and, respectively, A^ε. Hence, $\sigma_f(y_1, \dots, y_{j-1}, z_k, a_{j+1}, \dots, a_n) \notin F^\varepsilon$ for $k = 1, 2$. It follows therefore that $\sigma_f(y_1, \dots, y_{j-1}, z_1 \vee^{\varepsilon_j} z_2, a_{j+1}, \dots, a_n) = \sigma_f(y_1, \dots, y_{j-1}, z_1, a_{j+1}, \dots, a_n) \vee^\varepsilon \sigma_f(y_1, \dots, y_{j-1}, z_2, a_{j+1}, \dots, a_n) \notin F^\varepsilon$ and thus, $z_1 \vee^{\varepsilon_j} z_2 \in H_j$. Note also that $0^{\varepsilon_j} \in H_j$. It follows that $\bigvee^{\varepsilon_j} I \in H_j$ for every finite (possibly empty) subset $I \subseteq H_j$.

[4] Due to the definition of the lattice A^{ε_j} (equal to A if $\varepsilon_j = 1$, and to the dual of A if $\varepsilon_j = -1$) this means: H_j is closed under all finite joins if $\varepsilon_j = 1$ and under all finite meets if $\varepsilon_j = -1$.

Claim 2. H_j is separated from a_j in A^{ε_j}, i.e. for every finite subset $I \subseteq H_j$, $\bigvee^{\varepsilon_j} I \not\geq a_j$.
Proof: Assume that there exists a finite subset $I \subseteq H_j$, such that $\bigvee^{\varepsilon_j} I \geq a_j$. Since H_j is closed under all finite joins in A^{ε_j}, $z = \bigvee^{\varepsilon_j} I \in H_j$. Hence, there exist $y_1 \in G_1^{\varepsilon_1}, \dots, y_{j-1} \in G_{j-1}^{\varepsilon_{j-1}}$ such that $\sigma_f(y_1, \dots, y_{j-1}, z, a_{j+1}, \dots, a_n) \notin F^{\varepsilon}$. On the other hand, since $\sigma_f : A^{\varepsilon_1} \times \dots \times A^{\varepsilon_n} \to A^{\varepsilon}$ is order-preserving in every argument and $\bigvee^{\varepsilon_j} I \geq_{\varepsilon_j} a_j$, $\sigma_f(y_1, \dots, y_{j-1}, \bigvee^{\varepsilon_j} I, a_{j+1}, \dots, a_n) \geq_{\varepsilon} \sigma_f(y_1, \dots, y_{j-1}, a_j, a_{j+1}, \dots, a_n)$. By the induction hypothesis we know that $\sigma_f(y_1, \dots, y_{j-1}, a_j, a_{j+1}, \dots, a_n) \in F^{\varepsilon}$, so $\sigma_f(y_1, \dots, y_{j-1}, \bigvee^{\varepsilon_j} I, a_{j+1}, \dots, a_n) \in F^{\varepsilon}$, which is a contradiction.

Since H_j is separated from a_j in A^{ε_j}, there exists a prime filter $\overline{G}_j$ of A^{ε_j} such that $H_j \subseteq A \backslash \overline{G}_j$ and $a_j \in \overline{G}_j$. This actually means that there exists a prime filter G_j of A such that $\overline{G}_j = G_j^{\varepsilon_j}$, i.e. $H_j \subseteq A \backslash G_j^{\varepsilon_j}$ and $a_j \in G_j^{\varepsilon_j}$. Thus, (i) is fulfilled by G_j. In order to prove that (ii) also holds, let $y_1, \dots, y_j$ such that $y_k \in G_k^{\varepsilon_k}$ for every $k \leq j$. Thus, $y_j \in G_j^{\varepsilon_j}$ and therefore $y_j \notin H_j$. It follows that for all $\overline{y}_1 \in G_1^{\varepsilon_1}, \dots, \overline{y}_{j-1} \in G_{j-1}^{\varepsilon_{j-1}}, f(\overline{y}_1, \dots, \overline{y}_{j-1}, y_j, a_{j+1}, \dots, a_n) \in F^{\varepsilon}$. Hence, for $\overline{y}_1 = y_1, \dots, \overline{y}_{j-1} = y_{j-1}$ we obtain $f(y_1, \dots, y_j, a_{j+1}, \dots, a_n) \in F^{\varepsilon}$. □

The problem of checking whether $E(D(\mathbf{A}))$ is an algebra in $\mathcal{K}$ is more difficult. If $\mathcal{K}$ is a class of Boolean algebras or bounded distributive lattices with operators, this reduces to showing that the class $\mathcal{K}$ is closed under so-called *canonical extensions.* The *canonical extension* of a Boolean algebra B is $\mathcal{P}(D(B))$; that of a bounded distributive lattice L is $\mathcal{O}(D(L))$, the lattice of order-filters of $D(L)$. Operators on Boolean algebras or bounded distributive lattices (in fact, more generally, all finitary operations that are isotone or antitone in every argument) can be extended to operators on the canonical extension of such algebras. Criteria that guarantee that a class of bounded distributive lattices (or Boolean algebras) with operators is closed under canonical extensions are given e.g. in [30, 23, 67, 32, 24]. For example, all finitely-generated varieties of Boolean algebras or distributive lattices with operators are canonical; in fact, so are all varieties of such algebras generated by a class $\mathcal{A}_{\text{gen}}$ of algebras closed under ultraproducts and under canonical extensions [24, 32]. However, there exist classes of distributive lattices and of Boolean algebras with operators which are not closed under canonical extensions. For instance, Gehrke and Priestley [21] proved that the variety $\mathcal{MV}$ of MV-algebras is as non-canonical as possible: every finitely-generated subvariety is canonical , but no non-finitely generated subvariety is canonical.

The notion of canonical extension has been extended to lattices in [22]. The canonical extensions for lattices defined in [22] are close to Urquhart's representation theorem for lattices [77]. However, the results on canonical extensions for lattices studied in [22] are not relevant for the considerations we make here for non-distributive lattices and semilattices because the lattice $\mathcal{SF}(D(L))$ is in general larger than the canonical extension of L [20].

5.3 Kripke-style models

Representation theorems such as those described above can be used in order to define classes of Kripke-style models naturally associated with classes of algebras, or with classes of matrices $(\mathbf{A}, D_A)$, where the sets of designated elements are of the form $D_A = \{a \in A \mid a \geq c_A\}$, where c is a constant.

Let $\mathcal{A}$ be a class of algebras which are semilattices or lattices with operators. The results in Section 3 show that in order to find a corresponding class of Kripke-style models, it is sufficient to find a class $\mathcal{R}$ of relational structures that satisfies the following conditions:

(C) there exist $\mathbb{D} : \mathcal{A} \to \mathcal{R}$, $\mathbb{E} : \mathcal{R} \to \mathcal{A}$ such that for every $\mathbf{A} \in \mathcal{A}$ there exists an injective homomorphism $i : \mathbf{A} \to \mathbb{E}(\mathbb{D}(\mathbf{A}))$ with the property that the inverse image of the set of all designated elements in its codomain is a set of designated elements in its domain.

In the light of the previous remarks on representation theorems, one of the most natural choices for a class $\mathcal{R}$ of relational structures such that condition (C) holds would be the family of all duals of algebras in $\mathcal{A}$, $\mathcal{R}_D = \{D(\mathbf{A}) \mid \mathbf{A} \in \mathcal{A}\}$, with $\mathbb{D}(\mathbf{A}) := D(\mathbf{A})$ for every $\mathbf{A} \in \mathcal{A}$, and $\mathbb{E}(D(\mathbf{A})) := E(D(\mathbf{A}))$ for every $D(\mathbf{A}) \in \mathcal{R}_D$. However, the class $\mathcal{R}_D$ may be difficult to describe in a simple logical language.

Finding a class $\mathcal{R}$ with a *simple description* is very important, for instance if we want to use the Kripke-style semantics for devising efficient algorithms for automated theorem proving. In particular, we would like to determine (if at all possible) classes of relational models that have a *first-order description*. This is not always possible[5]. In many cases, however, first-order axiomatizable classes $\mathcal{R}$ of Kripke-style models can be obtained by abstracting non-topological properties of the duals of algebras in $\mathcal{A}$. As before, for every $\mathbf{A} \in \mathcal{A}$, $\mathbb{D}(\mathbf{A}) = D(\mathbf{A})$. For every $K \in \mathcal{R}$, $\mathbb{E}(K)$ can be constructed by associating operators to the relations on K. The construction is analogous to that presented in Section 5.2.2.

- The fact that $\mathbb{D} : \mathcal{A} \to \mathcal{R}$ obviously holds; the fact that, for every $K \in \mathcal{R}$, $\mathbb{E}(K) \in \mathcal{A}$, in general needs to be proved for each special class of algebras, in order to ensure that the class $\mathcal{R}$ is not too large.
- The representation theorems for algebras in $\mathcal{A}$ amount to the fact that for every $\mathbf{A} \in \mathcal{A}$ there exists an injective homomorphism $\eta_A : \mathbf{A} \to \mathbb{E}(\mathbb{D}(\mathbf{A}))$.
- We assume that the sets of distinguished elements are all of the form $D_A = \{a \in A \mid a \geq c_A\}$ for some constant symbol c.
  If $\eta_A(a) \in D_{\mathbb{E}(\mathbb{D}(\mathbf{A}))}$ then $\eta_A(a) \geq c_{\mathbb{E}(\mathbb{D}(\mathbf{A}))} = \eta_A(c_A)$, i.e. (under the assumption that all structures in $\mathcal{A}$ have at least an underlying semilattice

[5] For instance, the McKinsey axiom $\Box \Diamond \phi \to \Diamond \Box \phi$ is not determined by any elementary (i.e. first-order definable) class of Kripke frames (cf. [31] p. 5 and 65). Goldblatt also proved that no class of first-order definable frames corresponds to the class of orthomodular lattices [29] (cf. also [31] pp.99ff).

structure that induces the ordering $\leq$), $\eta_A(c_A) = \eta_A(a) \wedge \eta_A(c_A) = \eta_A(a \wedge c_A)$, hence, by the injectivity of η_A, $a \geq c_A$.

In Section 6 we illustrate our ideas for the following classes of logics:

(i) positive logics (which do not have the implication as a logical connective);
(ii) logics having as algebraic models Heyting algebras with operators (e.g. classical and intuitionistic logic, modal logics, Post logics, and generalizations thereof);

In Section 7 we illustrate our ideas for:

(iii) logics having as algebraic models residuated (semi)lattices with operators.

6 Example 1: Logics based on distributive lattices and Heyting algebras with operators

Let $\mathcal{V}$ be one of the varieties D_{01} (of bounded distributive lattices) or H (of Heyting algebras). Let $\mathcal{V}\mathsf{O}_\Sigma$ be the class of all structures $(A, \{\sigma_f\}_{f \in \Sigma})$, where A is an algebra in $\mathcal{V}$ with additional operators in Σ which are either lattice homomorphisms, or lattice antimorphisms, or, more general, of the (unique) type $\varepsilon_1 \dots \varepsilon_n \to \varepsilon$, for some $\varepsilon_1, \dots, \varepsilon_n, \varepsilon \in \{-1, +1\}$. In Section 5.2.1 we showed that for every $f \in \Sigma$ of type $\varepsilon_1 \dots \varepsilon_n \to \varepsilon$ we can define a relation $R_f \subseteq D(A)^{n+1}$ by

$$R_f(F_1, \dots, F_n, F) \text{ iff } \sigma_f(F_1^{\varepsilon_1}, \dots, F_n^{\varepsilon_n}) \subseteq F^\varepsilon.$$

For every unary lattice homomorphism $h : A \to A$ or antimorphism $k : A \to A$, a unary function R_h resp. R_k on $D(A)$ can be defined, as pointed out in Example 1, by

$$R_h(F) = h^{-1}(F) \qquad R_k(F) = A \backslash k^{-1}(F).$$

The non-topological properties of the duals of algebras in $\mathcal{V}\mathsf{O}_\Sigma$ justify the definition of the class $\mathcal{R}p_\Sigma$ of all structures $(X, \leq, \{R_f\}_{f \in \Sigma})$, where $\leq$ is a preorder on X and

- if $f \in Lh$, $R_f : X \to X$ preserves $\leq$; if $f \in La$, $R_f : X \to X$ reverses $\leq$;
- if f is of type $\varepsilon_1 \dots \varepsilon_n \to +1$ then $R_f \subseteq X^{n+1}$ is increasing; if f is of type $\varepsilon_1 \dots \varepsilon_n \to -1$ then $R_f \subseteq X^{n+1}$ is decreasing.

For every $\mathbf{A} \in \mathcal{V}\mathsf{O}_\Sigma$, the Priestley dual $D(\mathbf{A})$ of $\mathbf{A}$ (ignoring the topology), on which maps and relations are defined as explained above, is in this way transformed into a structure in $\mathcal{R}p_\Sigma$. Conversely, for every $\mathbf{X} = (X, \leq, \{R_f\}_{f \in \Sigma}) \in \mathcal{R}p_\Sigma$ let $\mathbb{E}(\mathbf{X})$ be the algebra $(\mathcal{O}(\mathbf{X}), \cup, \cap, \emptyset, X, \{\sigma_f\}_{f \in \Sigma})$ if $\mathcal{V} = \mathsf{D}_{01}$, resp. $(\mathcal{O}(\mathbf{X}), \cup, \cap, \Rightarrow, \neg, \emptyset, X, \{\sigma_f\}_{f \in \Sigma})$ if $\mathcal{V} = \mathsf{H}$. Here $\mathcal{O}(\mathbf{X})$ is the set of all upwards closed subsets of $\mathbf{X}$ with respect to $\leq$; $\cup$ and $\cap$ are the set-theoretical union and intersection;

- $U \Rightarrow V = \{x \mid \forall y((x \leq y, y \in U) \rightarrow y \in V)\}$;
- $\neg U = \{x \mid \forall y(x \leq y \rightarrow y \notin V)\}$;
- $\sigma_f(U_1, \ldots, U_n) = \left(R_f^{-1}(U_1^{\varepsilon_1}, \ldots, U_n^{\varepsilon_n})\right)^{\varepsilon}$ for every $f \in \Sigma$ of type $\varepsilon_1 \ldots \varepsilon_n \rightarrow \varepsilon$, where $U^1 = U$, $U^{-1} = X \backslash U$ and $R^{-1}(U_1, \ldots, U_n) = \{x \in X \mid \exists x_1 \ldots x_n$ $(x_1 \in U_1, \ldots, x_n \in U_n,$ and $R(x_1, \ldots, x_n, x))\}$;
- if $f \in Lh \cup La$ then $\sigma_f : \mathcal{O}(\mathbf{X}) \rightarrow \mathcal{O}(\mathbf{X})$ is defined by $\sigma_f(U) = R_f^{-1}(U)$ if $f \in Lh$, and by $\sigma_f(U) = X \backslash R_f^{-1}(U)$ if $f \in La$.

Theorem 15. *$\mathcal{V}\mathsf{O}_\Sigma$ and $\mathcal{R}p_\Sigma$ fulfill condition (C), where the map $\mathbb{D}$ associates with every algebra $\mathbf{A}$ in $\mathcal{V}\mathsf{O}_\Sigma$ the ordered relational space obtained from the Priestley dual of $\mathbf{A}$ by ignoring the topology; the map $\mathbb{E}$ associates with every space $\mathbf{X} = (X, \leq, \{R\}_{R \in \Sigma}) \in \mathcal{R}p_\Sigma$ the algebra $\mathbb{E}(\mathbf{X})$ defined above.*

Proof: This follows from Theorem 11 and Theorem 14 using arguments similar to those in [30, 67, 65]. □

Corollary 2. *Let $\mathcal{L} = (\mathsf{Var}, O, \vdash)$ be sound and complete with respect to $\mathcal{V}\mathsf{O}_\Sigma$, where $\mathcal{V}$ is D_{01} or H. Then $\mathcal{L}$ is sound and complete with respect to the class $\mathcal{K}p_\Sigma$ of all Kripke-style models (K, m) consisting of a frame $K = (X, \leq, \{\sigma_X\}_{\sigma \in \Sigma}) \in \mathcal{R}p_\Sigma$ and a meaning function $m : \mathsf{Var} \rightarrow \mathcal{O}(K)$.*

Similar results hold for $\mathcal{V} = \mathsf{B}$, the class of Boolean algebras: It can be shown that the class BO_Σ of all Boolean algebras with operators in Σ and the class $\mathcal{R}_\Sigma$ of all spaces in $\mathcal{R}p_\Sigma$ where $\leq$ is discrete (i.e. $x \leq y$ iff $x = y$) satisfy condition (C). In this case meaning functions have can take any value in $\mathcal{P}(K)$. Representation theorems can also be also used for logics based on subclasses of $\mathcal{V}\mathsf{O}_\Sigma$ (where $\mathcal{V}$ is $\mathsf{D}_{01}, \mathsf{H}$ or B) [69, 67, 68]. Some examples are presented below.

6.1 Generalizations of modal logic

These logics are based on classes of Boolean algebras with operators (in a set Σ) which are join hemimorphisms. There exists a very large number of papers in which Stone's representation theorem is extended to representation theorems for various classes of Boolean algebras with operators; the references are too numerous to be mentioned here. The main ideas go back to the work of Jónsson and Tarski [42]. A survey of those aspects in the theory of varieties of Boolean algebras with operators that emphasizes connections with modal logics is [32].

The relational structures that correspond to Boolean algebras with operators in a set Σ belong to the class

$$\mathsf{R}_\Sigma = \{(X, \{R_f\}_{f \in \Sigma}) \mid (X, \leq \{R_f\}_{f \in \Sigma}) \in \mathsf{R}p_\Sigma \text{ with } \leq \text{ discrete}\}$$

consisting of those elements of Rp_Σ for which the preorder $\leq$ is discrete (i.e. $x \leq y$ iff $x = y$) and hence can be ignored. The functors $\mathbb{D}$ and $\mathbb{E}$ are defined by:

$$\mathbb{D}(A) = (\mathcal{F}_{\max}(A), \{R_f\}_{f\in\Sigma});$$
$$\mathbb{E}(X, \{R_f\}_{f\in\Sigma}) = (\mathcal{P}(X), \cup, \cap, C, \emptyset, X, \{f_{R_f}\}_{f\in\Sigma});$$

meaning functions $m : \mathsf{Var} \to \mathcal{P}(X)$ assign sets of possible worlds to propositional variables.

Example 2 (Normal modal logic (K)). Here $\mathcal{M} = \mathsf{MAlg}$, the class of modal algebras (Boolean algebras endowed with a meet hemimorphism $\Box$); $\mathcal{K}$ is the class of relational spaces $K = (X, R)$, where $R \subseteq X^2$. We illustrate the way $\overset{r}{\models}$ can be defined for formulae of the form $\Box\phi$.

$$K \overset{r}{\models}_{m,x} \Box\phi \text{ iff } x \in \overline{m}(\Box\phi) = \Box_R(\overline{m}(\phi)) \text{ iff } \forall y \in X, \text{if } R(y,x) \text{ then } y \in \overline{m}(\phi)$$
$$\text{iff } \forall y \in X, \text{if } R(y,x) \text{ then } K \overset{r}{\models}_{m,y} \phi.$$

6.2 Logics based on Heyting algebras with operators

The relational spaces that correspond to Heyting algebras with operators are preordered relational spaces; the maps $\mathbb{D}$ and $\mathbb{E}$ are as in Theorem 15. Meaning functions are of the form $m : \mathsf{Var} \to \mathcal{O}(X)$, i.e. assign order-filters of X to propositional variables (they are *hereditary* cf. e.g. [15, 41]).

Example 3 (Intuitionistic logic). In this case $\mathcal{M} = \mathsf{H}$, the class of Heyting algebras; $\mathcal{K}$ is the class of all preordered sets $(X, \leq)$. We illustrate the way $\overset{r}{\models}$ can be defined for formulae of the form $\phi_1 \Rightarrow \phi_2$:

$$K \overset{r}{\models}_{m,x} \phi_1 \Rightarrow \phi_2 \text{ iff } x \in \overline{m}(\phi_1 \Rightarrow \phi_2) = \overline{m}(\phi_1) \Rightarrow \overline{m}(\phi_2)$$
$$\text{iff } \forall y \in X, \text{if } (x \leq y \text{ and } y \in \overline{m}(\phi_1)) \text{ then } y \in \overline{m}(\phi_2)$$
$$\text{iff } \forall y \in X, \text{if } (x \leq y \text{ and } K \overset{r}{\models}_{m,y} \phi_1) \text{ then } K \overset{r}{\models}_{m,y} \phi_2.$$

Example 4 (LC or Dummet's logic [11][6]). In this case $\mathcal{M} = \mathsf{LinH}$, the class of linear Heyting algebras (Heyting algebras satisfying $a \Rightarrow b \vee b \Rightarrow a = 1$); $\mathcal{K}$ is the class of all totally ordered sets $(X, \leq)$. $\overset{r}{\models}$ is defined as for intuitionistic logic.

[6] In the fuzzy logic community this logic is known as "Gödel's logic"; this comes from the fact that in the fuzzy logic framework this logic results from using the operations for conjunction and implication defined by Gödel on finite linear Heyting algebras as the t-norm and its residual, cf. also Example 6 in Section 7.1 and [37].

Example 5 (SHn-logic.). In this case $\mathcal{M} = \mathsf{SHn}$, the class of SHn-algebras (algebras of the type $(A, \sim, s_1, \dots, s_{n-1})$ where A is a Heyting algebra, and $g \in La, s_1, \dots, s_{n-1} \in Lh$ satisfy certain additional properties [40]); $\mathcal{K} = \mathsf{SHnSp}$, the class of SHn-spaces (preordered stuctures with unary maps $(X, \leq, g, S_1, \dots, S_{n-1})$, where g is $\leq$-reversing and $S_1, \dots, S_{n-1}$ are $\leq$-preserving, and satisfy additional properties [41, 69]). $\overset{r}{\models}$ is defined as for intuitionistic logic for $\vee, \wedge, \Rightarrow$. For formulae of the form $\sim\phi$ it is defined as follows:

$K \overset{r}{\models}_{m,x} \sim\phi$ iff $x \in \overline{m}(\sim \phi) = \sim_g (\overline{m}(\phi))$ iff $g(x) \not\in \overline{m}(\phi)$ iff $K \overset{r}{\not\models}_{m,g(x)} \phi$.

Similar results can be also obtained for distributive lattices which are not necessarily bounded. In that case the corresponding relational spaces are preordered spaces with endpoints.

6.3 Logics based on implicative lattices

We illustrate the considerations in Section 5.2, Example 1, for a binary operator of type $1, -1 \to -1$ and $-1, 1 \to 1$, and show that in this case the ternary relation $R_{\to}$ can be replaced by a binary *function* φ.

Definition 5. *An* implicative lattice *is an algebraic structure* $(A, \vee, \wedge, \to)$, *such that* $(A, \vee, \wedge)$ *is a distributive lattice and* $\to: A^2 \to A$ *is a lattice homomorphism in the second argument and a lattice antimorphism in the first argument.*

Examples are (reducts of) Boolean algebras, Wajsberg algebras [60], Ockham algebras [5], and lattice-ordered groups. Since implicative lattices need not be bounded, the dual $D^*(A)$ of an implicative lattice A is equal to $D(A) \cup \{\emptyset, A\}$, where $D(A)$ is the Priestley dual of A. A relation $R_{\to}$ associated with $\to$ can be defined by:

$$R_{\to}(F_1, F_2, F_3) \text{ iff } F_1 \to (L \backslash F_2) \subseteq (L \backslash F_3).$$

As pointed out in Section 5.2.1, $R_{\to}(F_1, F_2, F_3)$ iff $\{y \mid \exists x (x \in F_1, x \to y \in F_3\} \subseteq F_2$, and $R_{\to}$ is completely described by the binary operation φ defined by

$$\varphi(F_1, F_3) = \{y \mid \exists x (x \in F_1, x \to y \in F_3)\} = \bigcup_{x \in F_1} \{y \mid x \to y \in F_3\}.$$

Moreover, an operation $\rightarrow$ can be canonically defined on the lattice of proper order-filters of $D^*(A)$, $\mathcal{O}^*(D^*(A))$ by:

$$\begin{aligned} U \rightarrow V &= D^*(A)\backslash R_{\rightarrow}(U, D^*(A)\backslash V) \\ &= D^*(A)\backslash\{F \mid \exists F_1, F_2(F_1 \in U, F_2 \notin V, R_{\rightarrow}(F_1, F_2, F))\} \\ &= D^*(A)\backslash\{F \mid \exists F_1, F_2(F_1 \in U, F_2 \notin V, \varphi(F_1, F) \subseteq F_2)\} \\ &= \{F \mid \forall F_1(F_1 \in U \Rightarrow \forall F_2(\varphi(F_1, F) \subseteq F_2 \Rightarrow F_2 \in V))\} \\ &= \{F \mid \forall F_1(F_1 \in U \Rightarrow \varphi(F_1, F) \in V)\} \\ &= \bigcap_{F_1 \in U} \{F \mid \varphi(F_1, F) \in V\}. \end{aligned}$$

Priestley representation theorems for implicative lattices have been studied by Martínez in [50]. The definition of the map φ on $D^*(A)$ as used in [50], as well as the definition of the map $\rightarrow$ on $\mathcal{O}^*(D^*(A))$ are justified by the remarks above. Actually, in [50] a dual equivalence is established between the category of implicative lattices and a category of implicative spaces.

Definition 6 (Implicative space). *An* implicative space *is a space* $(X, \leq, p_m, p_M, \varphi)$, *where* $(X, \leq, p_m, p_M)$ *is a partially-ordered set with endpoints* p_m *and* p_M *(i.e. such that for all* $x \in X, p_m \leq x \leq p_M$*),* $\varphi : X \times X \rightarrow X$ *is order-preserving in each argument,* $\varphi(x, p_M) = p_M$ *for all* $x \neq p_m$*, and* $\varphi(p_M, p_m) = \varphi(p_m, p_M) = p_m$.

Let $\mathsf{IL_t}$ be the class of all *totally ordered* implicative lattices, and $\mathsf{ISp_t}$ the class of all *totally ordered* implicative spaces.

Theorem 16. *Condition (C) is fulfilled for* $\mathcal{M} = \mathsf{IL_t}$ *and* $\mathcal{R} = \mathsf{ISp_t}$*, where for every* $\mathbf{L} \in \mathsf{IL_t}$*,* $\mathbb{D}(\mathbf{L}) = (D^*(\mathbf{L}), \subseteq, \emptyset, L, \varphi)$*, and for every* $\mathbf{X} \in \mathsf{ISp_t}$*,* $\mathbb{E}(\mathbf{X}) = (\mathcal{O}^*(\mathbf{X}), \cup, \cap, \emptyset, X, \rightarrow)$*, where* $\mathcal{O}^*(\mathbf{X})$ *is the set of all proper order-filters of* $\mathbf{X}$*, and for every* $U, V \in \mathcal{O}^*(\mathbf{X})$*,*

$$U \rightarrow V = \{x \mid \forall x_1(x_1 \in U \text{ implies } \varphi(x_1, x) \in V)\}.$$

Proof: The fact that for every $\mathbf{L} \in \mathsf{IL_t}$, $\mathbb{D}(L) \in \mathsf{ISp_t}$ is an immediate consequence of the results in [50]. To show that if $(X, \leq, p_m, p_M, \varphi) \in \mathsf{ISp_t}$ then $\mathbb{E}(X, \leq, p_m, p_M, \varphi) \in \mathsf{IL_t}$ we first show that ϕ is well-defined, i.e. if $U, V \in \mathcal{O}^*(X)$ then $U \rightarrow V \in \mathcal{O}^*(X)$. In order to prove this, note that if $U, V \in \mathcal{O}^*(X)$ then $p_m \notin U \cup V$ and $p_M \in U \cap V$. As $\varphi(x, p_M) = p_M$ for every $x \neq p_m$, using the definition of $U \rightarrow V$, it follows that $p_M \in U \rightarrow V$. Hence, $U \rightarrow V \neq \emptyset$. To show that $U \rightarrow V \neq X$, assume that $p_m \in U \rightarrow V$. Then for all $p \in U, \varphi(p, p_m) \in V$, hence in particular for $p = p_M$, $\varphi(p_M, p_m) = p_m \in V$, which is a contradiction. The fact that $U \rightarrow V$ is upwards closed follows immediately from the fact that φ is order-preserving in the second argument. We prove that $\rightarrow$ is of type $-1, 1 \rightarrow 1$ and $1, -1 \rightarrow -1$. It is obvious that it is of type $1, -1 \rightarrow -1$, i.e.

$$U \to (V \cap V') = (U \to V) \cap (U \to V')$$
$$(U \cup U') \to V = (U \to V) \cap (U' \to V).$$

We now prove that if X is linearly ordered then

$$(1) \quad U \to (V \cup V') = (U \to V) \cup (U \to V');$$

$$(2) \quad (U \cap U') \to V = (U \to V) \cup (U' \to V).$$

(1) It can easily be seen that $(U \to V) \cup (U \to V') \subseteq U \to (V \cup V')$. We prove the other inclusion. Let $q \in U \to (V \cup V')$. Then for all $p \in U$, $\varphi(p,q) \in V \cup V'$. Assume that $q \notin (U \to V) \cup (U \to V')$. Then $\varphi(p,q) \notin V$ and $\varphi(p',q) \notin V'$ for some $p, p' \in U$. Since X is linearly ordered by $\leq$, either $p \leq p'$ or $p' \leq p$. Assume first that $p \leq p'$. Since φ is order-preserving in the first argument and V' is upwards closed, it follows that $\varphi(p,q) \notin V'$. Therefore, there exists $p \in U$ such that $\varphi(p,q) \notin V \cup V'$. This contradicts the fact that $q \in U \to (V \cup V')$. The case when $p' \leq p$ can be treated similarly. This proves that $U \to (V \cup V') \subseteq (U \to V) \cup (U \to V')$.

(2) It can easily be seen that $(U \to V) \cup (U' \to V) \subseteq (U \cap U') \to V$. We prove the other inclusion. Assume that there exists $q \in (U \cap U') \to V$ and $q \notin (U \to V) \cup (U' \to V)$. Then $\varphi(p,q), \varphi(p',q) \notin V$ for some $p \in U, p' \in U'$. As X is linearly ordered either $p \leq p'$ or $p' \leq p$. If $p \leq p'$ then $p' \in U \cap U'$ has the property that $\varphi(p',q) \notin V$ which contradicts the fact that $q \in (U \cap U') \to V$. The case $p' \leq p$ can be treated similarly. This proves that $(U \cap U') \to V \subseteq (U \to V) \cup (U' \to V)$. □

Corollary 3. *Let $\mathcal{L} = (\mathsf{Var}, O, \vdash)$ be a logic having as class of algebraic models $\mathsf{IL_t}$. Then $\mathcal{L}$ is sound and complete with respect to the class of all Kripke-style models (K, m), where $K \in \mathsf{ISp_t}$ and $m : \mathsf{Var} \to \mathcal{O}^*(K)$.*

Similar results can be established for logics having as algebraic models subclasses of $\mathsf{IL_t}$. For instance, $x \leq (x \to y) \to y$ corresponds to the commutativity of φ, and $x \to (y \to z) = y \to (x \to z)$ to the associativity of φ (cf. [50]).

7 Example 2: Logics based on residuated (semi)lattices

We now consider some classes of logics having as algebraic models (semi)lattices with operators that satisfy certain residuation conditions.

Definition 7 (Residuation). *Let $(L, \leq)$ be a partially-ordered set, and let $\circ, \to$ be two binary operations on L. $\to$ is the*[7] left residuation *of $\circ$ if*

$$a \circ b \leq c \text{ iff } a \leq b \to c,$$

and the right residuation *of $\circ$ if*

$$b \circ a \leq c \text{ iff } a \leq b \to c.$$

[7] Two left (resp. right) residuations of the same operator coincide.

Definition 8 (Residuated semigroup, monoid). *A* left (right) residuated semigroup *is a structure* $(M, \leq, \circ, \rightarrow)$ *where* $(M, \leq)$ *is a partially ordered set,* $\circ : M^2 \rightarrow M$ *is associative and monotone in both arguments, and* $\rightarrow$ *is the left (right) residuation of* $\circ$. *A* left (right) residuated monoid *is a structure* $(M, \leq, \circ, \rightarrow, 1)$ *where* $(M, \circ, 1)$ *is a monoid and* $(M, \leq, \circ, \rightarrow)$ *is a left (right) residuated semigroup.*

Definition 9 (Semilattices and lattices with residuation). *A* semilattice with left (right) residuation *is an algebra* $(S, \vee, \circ, \rightarrow)$, *where* $(S, \vee)$ *is a semilattice,* $\circ$ *is a binary join hemimorphism, and* $\rightarrow$ *is the left (right) residuation of* $\circ$. *A* lattice with left (right) residuation *is an algebra* $(L, \vee, \wedge, \circ, \rightarrow)$, *where* $(L, \vee, \wedge)$ *is a lattice, and* $(L, \vee, \circ, \rightarrow)$ *is a semilattice with left (right) residuation.*

As examples, we discuss a system of relevant logic and substructural logics such as BCC (and related) logics. We also mention some variants of the Lambek calculus. Generalizations of residuation conditions to operators of type $\varepsilon_1 \dots \varepsilon_n \rightarrow \varepsilon$ have been proposed and studied in [12, 72], but we will not consider these general cases here.

7.1 Logics based on residuated distributive lattices

Residuated distributive lattices occur in a natural way as algebraic models for some fuzzy logics and relevant logics.

Example 6 (Fuzzy logics). Fuzzy logics are many-valued logics having the interval $[0, 1]$ as set of truth values; premise combination $\circ$ is modeled by t-norms. Every continuous t-norm $\circ$ on $[0, 1]$ has a unique right residuum $\rightarrow$. Many fuzzy logics are sound and complete with respect to *classes* of residuated distributive lattices. The basic fuzzy logic (BL) for instance has as algebraic models the class of all linearly ordered BL-algebras (i.e. linearly ordered bounded lattices with two binary operators $\circ$ and $\rightarrow$, $(L, \vee, \wedge, 0, 1, \circ, \rightarrow)$, where $(L, \circ, 1)$ is a commutative semigroup with 1, $\circ$ is monotone in both arguments, and where for all $x, y, z \in L$, $x \circ z \leq y$ iff $z \leq (x \rightarrow y)$, and $x \wedge y = x \circ (x \rightarrow y)$ [37]). By choosing *the Gödel t-norm*, $x \circ y = \min(x, y)$; *the Łukasiewicz t-norm*, $x \circ y = \max(0, x+y-1)$; or, respectively, *the product t-norm*, $x \circ y = x \cdot y$ (product of reals), we can define Gödel, Łukasiewicz, resp. product logics. These logics are also sound and complete with respect to *classes of residuated lattices* (cf. e.g. [37] Ch.2–4):

- the Gödel logics have as algebraic models linearly ordered Heyting algebras;
- the Łukasiewicz logics [46] have as algebraic models the class of linearly ordered MV-algebras (BL-algebras in which the identity $x = ((x \rightarrow 0) \rightarrow 0)$ holds). Since in an MV-algebra all operations can be expressed in terms of $\rightarrow$ and 0 by: $x \circ y = (x \rightarrow (y \rightarrow 0)) \rightarrow 0$; $x \vee y = (x \rightarrow y) \rightarrow y$

and $x \wedge y = x \circ (x \rightarrow y)$, one can also choose the class W of all (linearly ordered) Wajsberg algebras (of type $(A, \rightarrow, 0)$) instead (cf. e.g. [37]).

- the product logics have as algebraic models the class of all (linearly ordered) product algebras, BL-algebras that satisfy

$$(z \rightarrow 0) \rightarrow 0 \leq ((x \circ z \rightarrow y \circ z) \rightarrow (x \rightarrow y)) \quad \text{and} \quad x \cap (x \rightarrow 0) = 0.$$

Example 7 (The relevant logic RL). The relevant logic RL introduced by Urquhart in [78] has as class of algebraic models the class of relevant algebras (cf. Definition 10 below).

Let $\mathbf{L} = (L, \vee, \wedge, 0, 1, \circ, \rightarrow, \{\sigma_f\}_{f \in \Sigma})$ be a bounded distributive lattice with left residuation, and with additional operators $\{\sigma_f\}_{f \in \Sigma}$ in the classes discussed at the beginning of Section 6. The results presented in Section 6 apply to the structure $(L, \vee, \wedge, 0, 1, \circ, \{\sigma_f\}_{f \in \Sigma})$. In particular, $\circ$ induces a ternary relation $R_\circ$ on $D(\mathbf{L})$, increasing in the last argument, defined by $R_\circ(F_1, F_2, F_3)$ iff $F_1 \circ F_2 \subseteq F_3$. The join hemimorphism $\rightarrow: \mathbf{L} \times \mathbf{L}^d \rightarrow \mathbf{L}^d$, (cf. e.g. [1], p.358, [12]) induces a ternary relation $R_\rightarrow$ on $D(\mathbf{L})$, defined by:

$$R_\rightarrow(F_1, F_2, F_3) \text{ iff } \forall x, y((x \in F_1 \text{ and } y \notin F_2) \Rightarrow x \rightarrow y \notin F_3).$$

Lemma 1. *Assume that $\rightarrow$ is the left residuation of $\circ$. Then $R_\rightarrow(F_1, F_2, F_3)$ iff $R_\circ(F_3, F_1, F_2)$.*

Since $R_\rightarrow$ and $R_\circ$ coincide up to the permutation of their arguments, it is sufficient to use only one relation on $X = D(\mathbf{L})$, for instance $R_\circ$. $R_\circ$ and $R_\rightarrow$ induce binary operations $\circ$ and $\rightarrow$ on $\mathcal{O}(D(\mathbf{L}))$. By Lemma 1, these can be described in terms of $R_\circ$ as follows:

$$U \circ V = \{z \in X \mid \exists x, y(x \in U, y \in V, R_\circ(x, y, z))\} \tag{11}$$
$$U \rightarrow V = \{x \in X \mid \forall y, z(R_\circ(x, y, z), y \in U \Rightarrow z \in V)\} \tag{12}$$

Theorem 17. *Let* DR *be the class of all* distributive left residuated lattices, *and* KR *the class of all spaces $(X, \leq, R)$ where $\leq$ is a preorder and R a ternary relation such that*

(M) *if $R(x, y, z), x' \leq x, y' \leq y, z \leq z'$ then $R(x', y', z')$.*

Then condition (C) is fulfilled, where for every $\mathbf{L} \in$ DR, $\mathbb{D}(\mathbf{L}) = (\mathcal{F}_p(\mathbf{L}), \subseteq, R_\circ)$; and for every $\mathbf{X} = (X, \leq, R) \in$ KR, $\mathbb{E}(\mathbf{X}) = (\mathcal{O}(\mathbf{X}), \cup, \cap, \emptyset, X, \circ, \rightarrow)$, where $U \circ V$ and $U \rightarrow V$ are given by (11) and (12) above.

Theorem 17 extends to subclasses of the class DR_Σ of lattices in DR with operators in Σ. In fact, representation theorems have been used in various papers on relevant logics or algebras arising from them (cf. e.g. [6, 78]). We mention as an example, a system of relevant logics studied by Urquhart in [78].

Definition 10 (Relevant algebra). *A* relevant algebra *is an algebra* $(L, \vee, \wedge, \circ, \rightarrow, \neg, e, 0, 1)$, *where* $(L, \vee, \wedge, 0, 1)$ *is a bounded distributive lattice,* $e \in L$, $\circ$ *is a binary join hemimorphism with* $e \circ a = a$; $\neg$ *is a lattice antimorphism, and* $\rightarrow$ *is the left residuation of* $\circ$.

The relevant logic RL, sound and complete with respect to the class RL of relevant algebras, is studied in [78].

Let RS be the class of all structures $K = (X, \leq, R, g, t)$, where $\leq$ is a preorder, $g : X \rightarrow X$ is order reversing, R satisfies condition (M), and t is an increasing subset of X satisfying the condition $\forall y, z (y \leq z \text{ iff } (\exists x \in t : R(x, y, z)))$.

Theorem 18 (The relevant logic RL [78]). *The classes* $\mathcal{M} = \mathsf{RL}$ *and* $\mathcal{R} = \mathsf{RS}$ *satisfy condition* (C), *where, if* $\mathbf{A} \in \mathsf{RL}$, $\mathbb{D}(\mathbf{A}) = (\mathcal{F}_p(\mathbf{A}), \subseteq, R_\circ, H_\sim, R_e)$ *is the relational space obtained from the Priestley dual of* $\mathbf{A}$ *(cf. [78]) by ignoring the topology and if* $K = (X, \leq, R, g, t) \in \mathsf{RS}$, $\mathbb{E}(K) = (\mathcal{O}(K), \cup, \cap, \circ_R, \sim_g, 0, 1, t)$. *Kripke models are pairs* (K, m) *where* $K \in \mathcal{R}$ *and* m *is a meaning function,* $m : \mathsf{Var} \rightarrow \mathcal{O}(X)$. *The definition of* $\overset{r}{\models}$ *specializes for formulae of the form* $\phi_1 \circ \phi_2$ *and* $\phi_1 \rightarrow \phi_2$ *as follows:*

$$
\begin{aligned}
K \overset{r}{\models}_{m,x} \phi_1 \circ \phi_2 \ & \textit{iff}\ x \in \overline{m}(\phi_1 \circ \phi_2) = \overline{m}(\phi_1) \circ \overline{m}(\phi_2) \\
& \textit{iff}\ \exists y, z (y \in \overline{m}(\phi_1), z \in \overline{m}(\phi_2), R(y, z, x)) \\
& \textit{iff}\ \exists y, z (K \overset{r}{\models}_{m,y} \phi_1, K \overset{r}{\models}_{m,z} \phi_2, R(y, z, x)); \\
K \overset{r}{\models}_{m,x} \phi_1 \rightarrow \phi_2 \ & \textit{iff}\ x \in \overline{m}(\phi_1 \rightarrow \phi_2) = \overline{m}(\phi_1) \rightarrow \overline{m}(\phi_2) \\
& \textit{iff}\ \forall y, z ((R(x, y, z), y \in \overline{m}(\phi_1)) \Rightarrow z \in \overline{m}(\phi_2)) \\
& \textit{iff}\ \forall y, z ((R(x, y, z), K \overset{r}{\models}_{m,y} \phi_1) \Rightarrow K \overset{r}{\models}_{m,z} \phi_2).
\end{aligned}
$$

Similar considerations apply also to many fuzzy logics.

Note: Heyting algebras are residuated distributive lattices with $\circ = \wedge$. In this case $R_\circ(x, y, z)$ iff $x \leq z$ and $y \leq z$, hence $R_\circ$ is not needed in the definition of the Kripke-style frames. Thus, for the variety of Heyting algebras, Theorem 15 is a consequence of Theorem 17.

7.2 *BCC* and related logics

BCC logic is obtained from intuitionistic logic by eliminating the contraction and the exchange rules. *BCK* logic is obtained from *BCC* logic by adding the exchange rule; *DBCC* and *DBCK* logics are obtained from *BCC*, resp. *BCK* logics, by adding distributivity. The roots of *BCK* logic can be traced back to Fitch's and Tarski's work [14, 75]; the name "*BCK*" (derived from the combinators corresponding to the axioms of implicational *BCK* logic) is used for the first time in [51]. The algebraic models of (implicational) *BCK* logic, algebras with signature $\{\rightarrow, 1\}$, have been introduced by Iseki, and studied under the name of *BCK* algebras (see e.g. [38]). Also Wronski and his collaborators made significant contributions in this area. For more details cf. [10]. Since BCK algebras, as defined in [38], can also be described in terms of partially-ordered (commutative) residuated integral monoids, we do not present the usual definition of BCK algebras here, but define partially-ordered residuated integral monoids, and then a class of structures which are *lattices* with a BCK-structure, which for this reason we call *"full BCK algebras"*.

Definition 11 (Partially-ordered residuated integral monoids). *A structure* $(M, \leq, \circ, 1)$ *is a* partially-ordered monoid *if* $(M, \leq)$ *is a partially-ordered set,* $(M, \circ, 1)$ *is a monoid and* $\circ$ *is monotone in every argument. A partially-ordered monoid is* integral *if for all* $x \in M$, $x \leq 1$. *A structure* $(M, \leq, \circ, 1, \rightarrow)$ *is a* partially-ordered residuated (integral) monoid *if* $(M, \leq, \circ, 1)$ *is a partially-ordered (integral) monoid and* $\rightarrow$ *is the left residuation of* $\circ$*; it is a* partially-ordered commutative residuated (integral) monoid *(pocrim) if* $\circ$ *is commutative.*

For every pocrim $(M, \leq, \circ, 1, \rightarrow)$, $(M, \rightarrow, 1)$ is a BCK algebra. Conversely, every BCK algebra is a subreduct of a pocrim; this was shown independently in [58, 56, 16] cf. also [4]. Therefore in what follows we consider lattice-ordered pocrims (and their non-commutative versions) which we call here full BCK (resp. BCC) algebras.

Definition 12. *A* full BCC algebra *is a left residuated lattice* $(L, \vee, \wedge, \circ, \rightarrow, 0, 1)$ *with first element 0 and last element 1, and with the property that* $(L, \circ, 1)$ *is a monoid. A* full BCK algebra *is a full* BCC*-algebra with the property that* $\circ$ *is commutative.* Full $DBCC$ *(resp.* $DBCK$*)* algebras *are distributive full* BCC *(resp.* BCK*) algebras.*

We denote by BCC, BCK, DBCC, resp. DBCK the classes of full BCC, BCK, $DBCC$, resp. $DBCK$ algebras. BCC, BCK, $DBCC$ and $DBCK$ logics are sound and complete with respect to BCC, BCK, DBCC, and DBCK.

A Kripke-style semantics for BCC and related logics (in terms of semilattice-ordered monoids) is given by Ono and Komori in [56]. We show that the relationship between the algebraic and Kripke-style models is a consequence of Theorem 1 and Theorem 5.

Lemma 2. *Let* $\mathbf{S} = (S, \wedge, \circ, 1)$ *be a* $\wedge$*-semilattice with 1, with a binary operator* $\circ$ *which is monotone in both arguments. Let* $\mathcal{SF}^*(\mathbf{S})$ *be the set of all proper semilattice filters of* $\mathbf{S}$*, and let* $D(\mathbf{S}) := (\mathcal{SF}^*(\mathbf{S}), \cap, \{1\}, \varphi)$*, where*

$$\varphi(F_1, F_2) = \uparrow(F_1 \circ F_2) = \{z \mid \exists x_1 \in F_1, \exists x_2 \in F_2 : z \geq x_1 \circ x_2\}$$

(the order-filter generated by $F_1 \circ F_2 = \{x_1 \circ x_2 \mid x_1 \in F_1, x_2 \in F_2\}$*) for all* $F_1, F_2 \in \mathcal{SF}^*(\mathbf{S})$*. Then the following hold:*

(i) $(\mathcal{SF}^*(\mathbf{S}), \cap, \{1\}, \varphi)$ *is a (meet) semilattice with smallest element* $\{1\}$ *and an operator* φ *which is monotone in both arguments;*

(ii) *if* $\circ$ *is associative then* φ *is associative; if* $\circ$ *is commutative then* φ *is commutative; if* $(S, \circ, 1)$ *is a monoid then* $(\mathcal{SF}^*(\mathbf{S}), \varphi, \{1\})$ *is a monoid;*

Let $\mathbf{L} = (L, \wedge, \vee, \circ, 0, 1)$ *be a bounded lattice, with a binary operator* $\circ$ *which is a join hemimorphism in both arguments. Then*

(iii) φ *is a* $\wedge$*-hemimorphism in both arguments;*
(iv) *if* $\mathbf{L}$ *is distributive then for every* $F_1, F_2, F_3 \in \mathcal{SF}^*(\mathbf{L})$ *if* $F_1 \cap F_2 \subseteq F_3$ *then there exist*
$F_1', F_2' \in \mathcal{SF}^*(\mathbf{L})$ *such that* $F_1 \subseteq F_1'$, $F_2 \subseteq F_2'$, *and* $F_1' \cap F_2' = F_3$.

Proof: (i) It is is clear that $(\mathcal{SF}^*(\mathbf{S}), \cap, \{1\})$ is a semilattice with smallest element $\{1\}$. The map φ is well-defined: for every $F_1, F_2 \in \mathcal{SF}^*(\mathbf{S})$, $\varphi(F_1, F_2)$ is by definition upwards closed and is clearly non-empty. We show now that $\varphi(F_1, F_2)$ is also closed under $\wedge$. Let $y, y' \in \varphi(F_1, F_2)$. Then there exist $x_1, x_1' \in F_1$ and $x_2, x_2' \in F_2$ such that $y \geq x_1 \circ x_2, y' \geq x_1' \circ x_2'$. By the monotonicity of $\circ$, $y \geq (x_1 \wedge x_1') \circ (x_2 \wedge x_2')$ and $y' \geq (x_1 \wedge x_1') \circ (x_2 \wedge x_2')$, hence $y \wedge y' \geq z_1 \circ z_2$, where $z_1 = x_1 \wedge x_1' \in F_1$ and $z_2 = x_2 \wedge x_2' \in F_2$.

We now prove that φ is monotone in the first argument (monotonicity in the second argument can be proved analogously). Let $F_1, F_2, F \in \mathcal{SF}^*(\mathbf{S})$ be such that $F_1 \subseteq F_2$. If $x \in \varphi(F_1, F)$ then $x \geq x_1 \circ y$ for some $x_1 \in F_1, y \in F$. As $F_1 \subseteq F_2$, $x \geq x_1 \circ y$ for some $x_1 \in F_2, y \in F$, i.e. $\varphi(F_1, F) \subseteq \varphi(F_2, F)$.

(ii) The associativity (commutativity) of φ follows from the associativity (commutativity) of $\circ$. If $(S, \circ, 1)$ is a monoid then $\varphi(\{1\}, F) = \{1\} \circ F = F = F \circ \{1\} = \varphi(F, \{1\})$, hence $(\mathcal{SF}^*(\mathbf{S}), \varphi, \{1\})$ is a monoid.

(iii) Clearly, $\varphi(F_1 \cap F_2, F) \subseteq \varphi(F_1, F) \cap \varphi(F_2, F)$. We prove the converse inclusion. Let $x \in \varphi(F_1, F) \cap \varphi(F_2, F)$. Then there exist $x_1 \in F_1, y_1 \in F, x_2 \in F_2, y_2 \in F$ such that $x \geq x_1 \circ y_1$ and $x \geq x_2 \circ y_2$. By the monotonicity of $\circ$ and the fact that F is a lattice filter we know that $x \geq x_i \circ y$ for $i = 1, 2$, where $y = y_1 \wedge y_2 \in F$. Hence, as $\circ$ is join hemimorphism, $x \geq x_1 \circ y \vee x_2 \circ y = (x_1 \vee x_2) \circ y$, where $x_1 \vee x_2 \in F_1 \cap F_2$. This shows that $x \in \varphi(F_1 \cap F_2, F)$. The fact that $\varphi(F, F_1 \cap F_2) = \varphi(F, F_1) \cap \varphi(F, F_2)$ can be proved analogously.

(iv) Assume that $\mathbf{L}$ is distributive. Let $F_1, F_2, F_3 \in \mathcal{SF}^*(\mathbf{L})$ be such that $F_1 \cap F_2 \subseteq F_3$. Let $F_1' = F_1 \vee F_3$ and $F_2' = F_2 \vee F_3$. It is easy to see that $F_3 \subseteq F_1' \cap F_2'$. To prove the converse inclusion let $x \in F_1' \cap F_2'$. Then $x \geq x_1 \wedge y_1$ for some $x_1 \in F_1, y_1 \in F_3$ and $x \geq x_2 \wedge y_2$ for some $x_2 \in F_2, y_2 \in F_3$. In particular, $x \geq x_i \wedge y$, $i = 1, 2$, where $y = y_1 \wedge y_2 \in F_3$, i.e. $x \geq (x_1 \wedge y) \vee (x_2 \wedge y)$. Hence, by the distributivity of $\mathbf{L}$, $x \geq (x_1 \vee x_2) \wedge y$, where $x_1 \vee x_2 \in F_1 \cap F_2 \subseteq F_3$ and $y \in F_3$, i.e. $x \in F_3$. □

Theorem 19. *Condition* (*C*) *holds for:*

(i) $\mathcal{M}_{BCC} =$ BCC *and* $\mathcal{R}_{BCC}$*: the class of semilattice-ordered monoids,*
(ii) $\mathcal{M}_{BCK} =$ BCK *and* $\mathcal{R}_{BCK}$*: the class of semilattice-ordered commutative monoids,*

(iii) $\mathcal{M}_{DBCC}$ = DBCC *and* $\mathcal{R}_{DBCC}$: *the class of distributive*[8] *semilattice-ordered monoids,*

(iv) $\mathcal{M}_{DBCK}$ = DBCK *and* $\mathcal{R}_{DBCK}$: *the class of distributive semilattice-ordered commutative monoids;*

where if $\mathbf{L} = (L, \vee, \wedge, \circ, \rightarrow, 0, 1) \in \mathcal{M}_{\mathcal{L}}$, $\mathbb{D}(\mathbf{L}) = (\mathcal{SF}((L, \wedge, 1)), \cap, \{1\}, \varphi)$, *and if* $\mathbf{S} = (S, \wedge, 1, \circ) \in \mathcal{R}_{\mathcal{L}}$, $\mathbb{E}(\mathbf{S}) = (\mathcal{SF}(\mathbf{S}), \cap, \vee, \{1\}, S, \varphi, \rightarrow)$, *where* $\varphi, \vee$ *and* $\rightarrow$ *are defined for every* $F_1, F_2 \in \mathcal{SF}(\mathbf{S})$ *by:*

- $\varphi(F_1, F_2) = \uparrow(F_1 \circ F_2)$; $\quad F_1 \rightarrow F_2 = \{x \mid \forall y (y \in F_1$ *implies* $x \circ y \in F_2)\}$;
- $F_1 \vee F_2$ *is the semilattice filter generated by* $F_1 \cup F_2$.

Proof: Let $\mathbf{L} = (L, \vee, \wedge, \circ, \rightarrow, 0, 1) \in$ BCC. By Lemma 2, $\mathbb{D}(\mathbf{L})$ is a semilattice-ordered monoid; if $\circ$ is commutative then φ is also commutative; if L is distributive then $\mathbb{D}(\mathbf{L})$ is distributive. This shows that $\mathbb{D} : \mathcal{M}_{\mathcal{L}} \rightarrow \mathcal{R}_{\mathcal{L}}$ for every $\mathcal{L} \in \{BCC, BCK, DBCC, DBCK\}$.

We now show that $\mathbb{E} : \mathcal{R}_{\mathcal{L}} \rightarrow \mathcal{M}_{\mathcal{L}}$ satisfies the first item in condition (C). Let $\mathbf{S} = (S, \wedge, 1, \circ)$ be a semilattice-ordered monoid. Then $(\mathcal{SF}(\mathbf{S}), \cap, \vee, \{1\}, S)$ is a lattice of subsets of S, φ is an associative binary operation which is monotone, and $\varphi(F_1, F_2) \leq F_3$ iff $\uparrow(F_1 \circ F_2) \subseteq F_3$ iff $F_1 \circ F_2 \subseteq F_3$. It remains to show that $\rightarrow$ is the left residuation of $\circ$, i.e. that $\varphi(F_1, F_2) \subseteq F_3$ if and only if $F_1 \subseteq F_2 \rightarrow F_3$. Assume that $F_1 \circ F_2 \subseteq F_3$. Let $x \in F_1$. We show that $x \in F_2 \rightarrow F_3$. Indeed, let $y \in F_2$. Then $x \circ y \in F_1 \circ F_2 \subseteq F_3$, hence $x \circ y \in F_3$. Assume now that $F_1 \subseteq F_2 \rightarrow F_3$. Let $z \in F_1 \circ F_2$. Then there exist $x \in F_1, y \in F_2$ such that $z = x \circ y$. But then $x \in F_2 \rightarrow F_3$, hence as $y \in F_2$, it follows that $z = x \circ y \in F_3$.

As before, it is easy to see that if $\circ$ is commutative then φ is commutative, and if $\mathbf{S}$ is distributive then $\mathbb{E}(\mathbf{S})$ is distributive [56].

Finally, we prove that if $\mathbf{L} = (L, \vee, \wedge, \circ, \rightarrow, 0, 1) \in$ BCC then $\eta_L : \mathbf{L} \rightarrow \mathbb{E}(\mathbb{D}(\mathbf{L}))$ defined for every $a \in L$ by $\eta_L(a) = \{F \in D(L) \mid a \in F\}$ is an injective homomorphism of BCC algebras. By Theorem 5, η_L is an injective lattice homomorphism, and it preserves 0 and 1. Moreover,

$$\begin{aligned}
\eta_L(a) \circ \eta_L(b) &= \uparrow(\{F \in D(L) \mid a \in F\} \circ \{G \in D(L) \mid b \in G\}) \\
&= \{H \in D(L) \mid \exists F, G \in D(L), a \in F, b \in G, F \circ G \subseteq H\} \\
&= \{H \in D(L) \mid \uparrow a \circ \uparrow b \subseteq H\} \\
&= \{H \in D(L) \mid a \circ b \in H\} = \eta_L(a \circ b). \\
\eta_L(a) \rightarrow \eta_L(b) &= \{F \mid \forall G (G \in \eta_L(a) \text{ implies } F \circ G \in \eta_L(b))\} \\
&= \{F \mid \forall G (a \in G \text{ implies } b \in F \circ G)\} = \{F \mid b \in F \circ \uparrow a\} \\
&= \{F \mid \exists x \in F : b \geq x \circ a\} = \{F \mid \exists x \in F : x \leq a \rightarrow b\} \\
&= \{F \mid a \rightarrow b \in F\} = \eta_L(a \rightarrow b).
\end{aligned}$$

This shows that in all cases above $\mathbb{D}$ and $\mathbb{E}$ satisfy condition (C). □

[8] A semilattice $(S, \wedge)$ is called *distributive* if, for all $a, b, c \in S$, if $a \wedge b \leq c$ then there exist $a', b' \in S$ such that $a \leq a', b \leq b'$ and $a' \wedge b' = c$ [56].

Corollary 4. *Let $\mathcal{L}$ be one of the BCC, BCK, DBCC, DBCK logics with set of propositional variables* Var. *Then $\mathcal{L}$ is sound and complete with respect to the class of all Kripke-style models of the form (K, m), where $K = (S, \wedge, 1, \varphi) \in \mathcal{R}_{\mathcal{L}}$ and $m : \mathsf{Var} \to \mathcal{SF}(K)$. In particular:*

$$
\begin{array}{ll}
K \overset{r}{\models}_{m,x} \phi_1 \wedge \phi_2 & \textit{iff } x \in \overline{m}(\phi_1) \cap \overline{m}(\phi_2) \textit{ iff } K \overset{r}{\models}_{m,x} \phi_1 \textit{ and } K \overset{r}{\models}_{m,x} \phi_2; \\
K \overset{r}{\models}_{m,x} \phi_1 \vee \phi_2 & \textit{iff } x \in \overline{m}(\phi_1) \vee \overline{m}(\phi_2) \\
& \textit{iff } \exists y, z (y, z \in \overline{m}(\phi_1) \cup \overline{m}(\phi_2), x \geq y \wedge z) \\
& \textit{iff } \exists y, z [x \geq y \wedge z, (K \overset{r}{\models}_{m,y} \phi_1 \textit{ or } K \overset{r}{\models}_{m,y} \phi_2), \\
& \qquad\qquad (K \overset{r}{\models}_{m,z} \phi_1 \textit{ or } K \overset{r}{\models}_{m,z} \phi_2)]; \\
K \overset{r}{\models}_{m,x} \phi_1 \circ \phi_2 & \textit{iff } x \in \overline{m}(\phi_1) \circ \overline{m}(\phi_2) = \uparrow\varphi(\overline{m}(\phi_1), \overline{m}(\phi_2)) \\
& \textit{iff } \exists y, z (y \in \overline{m}(\phi_1), z \in \overline{m}(\phi_2), x \geq \varphi(y, z)) \\
& \textit{iff } \exists y, z (K \overset{r}{\models}_{m,y} \phi_1, K \overset{r}{\models}_{m,z} \phi_2, x \geq \varphi(y, z)); \\
K \overset{r}{\models}_{m,x} \phi_1 \to \phi_2 & \textit{iff } x \in \overline{m}(\phi_1) \to \overline{m}(\phi_2) \\
& \textit{iff } \forall y (y \in \overline{m}(\phi_1) \textit{ implies } \varphi(x, y) \in \overline{m}(\phi_2)) \\
& \textit{iff } \forall y (K \overset{r}{\models}_{m,y} \phi_1 \textit{ implies } K \overset{r}{\models}_{m,\varphi(x,y)} \phi_2).
\end{array}
$$

Since $\circ$ stands for premise combination, one could argue that the operation $\wedge$ has no clear logical meaning. The algebraic models of restrictions of substructural logics to the signature $\{\vee, \circ, \to\}$ are residuated *join semilattices*, i.e. algebras $(L, \vee, \circ, \to, \{\sigma\}_{\sigma \in \Sigma})$, where $(L, \vee)$ is a semilattice, $\circ$ is a binary and $\vee$-preserving operation, $\to$ is the left residuation of $\circ$, and the operations $\sigma \in \Sigma$ (if any) are $\vee$-preserving. The restriction of logics based on residuated lattices to subsignatures has been studied e.g. by Ono and Komori [56]. These results have been used by MacCaull [49] for giving Kripke-style semantics for so-called *residuated logics*, again in terms of semilattice-ordered monoids. The soundness and completeness results in [49] again follow as a consequence of Theorem 1.

Alternatively, we can start with representations for join semilattices instead of meet semilattices; the Kripke-style semantics (including also the properties of meaning functions) we would thus obtain is different from that of Ono and Komori and is therefore not presented here.

7.3 The Lambek calculus

In the Lambek calculus [44, 45] neither exchange nor contraction hold. The *original Lambek calculi* L and $L1$ have as connectives $\to$, $\leftarrow$ (logical consequence) and $\circ$ (premise combination). Lambek calculi can be seen as calculi of types: the propositional variables are *atomic types*, and if τ_1, τ_2 are types so are $\tau_1 \circ \tau_2, \tau_1 \to \tau_2$ and $\tau_1 \leftarrow \tau_2$. The algebraic models of L and $L1$ are residuated semigroups resp. monoids. The *full Lambek calculus* FL is an extension of the Lambek calculus obtained by adding connectives $\wedge$ and $\vee$ and rules for these connectives. The algebraic models related with various structural extensions of FL have been studied in [54, 55]: they are algebras $\mathbf{A}$ with sets of distinguished elements $D_A = \{a \in A \mid a \geq e\}$. In [55], Ono proved that FL is sound and complete with respect to the class of Kripke models consisting of all pairs $(\mathbf{S}, m)$, where $\mathbf{S}$ is a semilattice-ordered monoid, and $m : \mathsf{Var} \to \mathcal{SF}(\mathbf{S})$ is a meaning function; the definition of $\overset{r}{\models}$ is similar to that in Corollary 4 (with the difference that a different set of distinguished elements has to be taken into account). Considerations similar to those in Theorem 19 also explain most of these results.

8 Decidability and automated theorem proving

There exist several methods of proving decidability of a logic. One possibility is to use the inference rules that define the logic, e.g. a Gentzen-type calculus, and prove that cut elimination holds (or that the logic has the subformula property). Another possibility is to prove that the logic has the finite model property, i.e. that for every formula that is not provable in $\mathcal{L}$ there exists a finite model of $\mathcal{L}$ in which the formula is not true. This type of decidability proofs, although theoretical very interesting, are not directly usable for devising practical algorithms. An alternative way of proving decidability is by providing an embedding into a decidable fragment of (first-order) logic.

In what follows we mention some methods for proving decidability that turned out to be fruitful for automated theorem proving. We focus on methods that use Kripke-style models. Because of space limitation, neither tableau nor proof-theoretic methods are mentioned, although they often provide good time and space complexity bounds.

Embeddings into decidable fragments of first-order logic. In the attempt of understanding why so many modal logics are decidable many authors noticed that the definition of the Kripke-style semantics justifies an embedding into (decidable fragments of) classical logic. For instance, Gabbay [17] showed that many modal logics can be embedded into the 2-variable fragment of classical first-order logic, which is decidable [52] (in fact, as shown in [27], NEXPTIME-complete). More generally, in [3] Andréka, Van Benthem and Németi introduced the so-called *guarded fragment* (GF) of classical logic, which abstracts many of the properties of formulae obtained by structure-preserving translation to clause form for many generalizations of modal logics. The basic properties of the guarded fragment have been studied in [3] and [33]: it has been shown that it has the finite model property, hence it is decidable; Grädel [33] showed that it is decidable in doubly exponential time. Unfortunately not all modal logics can be embedded into the guarded fragment. This is the case for instance if the Kripke-style frames are endowed with a transitive relation, since transitivity cannot be expressed by a guarded formula. Indeed, it has been proved that in the presence of transitivity, decidability is lost [33]. However, Ganzinger, Meyer, and Veanes [19] show that the 2-variable restriction of the guarded fragment is decidable when it is extended by transitive relations; more generally, Swaszt and Tendera showed that the guarded fragment with transitive guards (an extension of GF in which some relations are transitive, transitive relations appear only in guards and equality may appear everywhere) is decidable and complete for deterministic double exponential time [74].

Automated theorem proving by resolution. One of the main advantages of the embedding into first-order logic mentioned above is that it is very suitable to use for automated theorem proving, since proof techniques developed for classical logics can be used for free. Refinements of resolution such as ordered resolution, the use of selection functions, and specially devised calculi to deal with equivalence (or congruence) relations, or with transitive relations proved to be extremely useful in this context. Additionally, strong redundancy notions and criteria for redundancy elimination already developed for automated theorem proving in classical logics can be used.

In what follows we discuss the ideas above in more detail. We start by showing that condition (C) in Section 3 can be used for reducing the problem of checking satisfiability of equalities or inequalities with respect to classes of algebras to checking satisfiability with respect to classes of Kripke-style models.

Theorem 20. *Let $\mathcal{A}$ be a class of (semi)lattices with operators. Assume that there exists a class $\mathcal{R}$ of relational spaces such that condition (C) in Section 3 holds. Then, for any terms t_1, t_2 in the signature O of the algebras in $\mathcal{A}$, with variables in a set Var, the following are equivalent:*

(1) $\mathcal{A} \overset{a}{\models} t_1 \leq t_2$.
(2) For every $X \in \mathcal{R}$ and every $m : \mathsf{Var} \to \mathbb{E}(X)$, $\overline{m}(t_1) \subseteq \overline{m}(t_2)$, where $\overline{m}$ is the unique homomorphic extension of m to formulae.
(3) There exist no $X \in \mathcal{R}$ and no family $\{I_e \mid e$ subterm of t_1 or $t_2\}$, such that:
  (a) $I_e \subseteq \mathbb{E}(X)$ for every subterm e of t_1, t_2;
  (b) $I_{f(t_1,\ldots,t_n)} = f_{\mathbb{E}(X)}(I_{e_1}, \ldots, I_{e_n})$ for every $f \in O$;
  (c) $I_{t_1} \not\subseteq I_{t_2}$.

Proof: (1) $\Rightarrow$ (2) follows from the fact that, by (C), for every $X \in \mathcal{R}$, $\mathbb{E}(X) \in \mathcal{A}$ is an algebra of subsets of X, in which the semilattice order coincides with set inclusion. (2) $\Rightarrow$ (1) Assume that (2) holds. Let $A \in \mathcal{A}$ and $f : \mathsf{Var} \to A$. By condition (C), $\mathbb{D}(A) \in \mathcal{R}$ and there exists an injective homomorphism $\eta_A : A \hookrightarrow \mathbb{E}(\mathbb{D}(A))$. Let $m = \eta_A \circ f : \mathsf{Var} \to \mathbb{E}(\mathbb{D}(A))$. By (2), $\overline{m}(t_1) \subseteq \overline{m}(t_2)$, i.e. $\eta_A(\overline{f}(t_1)) \subseteq \eta_A(\overline{f}(t_2))$, where $\overline{f}$ is the unique homomorphic extension of f to formulae. Then, as η_A is in fact an order embedding, $\overline{f}(t_1) \leq \overline{f}(t_2)$. The equivalence of (2) and (3) is immediate. □

Let $\mathcal{L}$ be a logic which is sound and complete with respect to a class $\mathcal{M}_{\mathcal{L}}$ of matrices of the form (A, D_A) consisting of algebras A (in a class $\mathcal{A}_{\mathcal{L}}$ of (semi)lattices with operators) and sets of distinguished elements of the form $D_A = \{a \in A \mid a \geq c_A\}$, where c is a constant.

We make the following assumptions (A):

(1) We assume that there exists a class $\mathcal{R}$ of relational spaces such that condition (C) in Section 3 holds.
(2) We assume that, based on $\mathcal{R}$, a class $\mathcal{K}_{\mathcal{L}}$ of Kripke-style models can be constructed such that the following conditions hold:
    (i) the class of frames in $\mathcal{K}_{\mathcal{L}}$ is axiomatizable by a set **Dom** of first-order formulae;
    (ii) the properties of the meaning functions of models in $\mathcal{K}_{\mathcal{L}}$ are described by a set **Mng** of first-order formulae.

Theorem 20 justifies the following structure-preserving translation to clause form for formulae in $\mathcal{L}$. Let $\phi \in \mathsf{Fma}(\mathcal{L})$. For every subformula ψ of ϕ a new unary predicate symbol P_ψ is introduced. The fact that the renaming is correct is expressed by a set **Ren** of formulae, which reflects the definition of $\overset{r}{\models}$ (and is, in fact, equivalent to condition (b) in Theorem 20(3)). Proving that ϕ is a theorem of $\mathcal{L}$ can then be reduced to proving that a conjunction of first-order sentences is unsatisfiable. The following corollaries are direct consequences of Theorem 20.

Corollary 5. *Assume that $\mathcal{L}$ is sound and complete with respect to a class $\mathcal{M}_{\mathcal{L}}$ of matrices as discussed above, and that $\mathcal{M}_{\mathcal{L}}$ and $\mathcal{K}_{\mathcal{L}}$ satisfy the conditions in (A). Let $\phi \in \mathsf{Fma}(\mathcal{L})$. Then the following are equivalent:*

(i) $\mathcal{L} \vdash \phi$;
(ii) $\mathcal{A}_{\mathcal{L}} \overset{a}{\models} \phi \geq c$;
(iii) $\mathbf{Dom} \cup \mathbf{Mng} \cup \mathbf{Ren} \cup \{\exists x P_c(x) \wedge \neg P_\phi(x)\}$ *is unsatisfiable.*

If c denotes the greatest element 1 of the algebras in $\mathcal{A}$, we obtain:

Corollary 6. *Assume that $\mathcal{L}$ is sound and complete with respect to the class $\mathcal{A}_{\mathcal{L}}$ of algebras, and that $\mathcal{A}_{\mathcal{L}}$ and $\mathcal{K}_{\mathcal{L}}$ satisfy the conditions in (A). Let $\phi \in \mathsf{Fma}(\mathcal{L})$. Then the following are equivalent:*

(i) $\mathcal{L} \vdash \phi$;
(ii) $\mathcal{A}_{\mathcal{L}} \overset{a}{\models} \phi = 1$;
(iii) $\mathbf{Dom} \cup \mathbf{Mng} \cup \mathbf{Ren} \cup \{\exists x \neg P_\phi(x)\}$ *is unsatisfiable.*

This type of translation has been used for automated theorem proving for various systems of *modal logics* (cf. e.g. [53, 62, 63, 18]).

A more general version of Theorem 20 has been used for deciding *uniform word problems* in classes of distributive lattices with operators [65, 72]. The results in [72] apply to the special problems considered here and yield the following generalization of the results in [66].

Theorem 21. *Let $\mathcal{L}$ be a logic based on a subclass of $\mathsf{D_{01}O_\Sigma}$, sound and complete with respect to an elementary subclass $\mathcal{K}_\mathcal{L}$ of $\mathcal{R}p_\Sigma$ consisting of all preordered Σ-relational structures satisfying a set E of first-order formulae. Let $\phi \in \mathsf{Fma}(\mathcal{L})$. Then $\mathcal{L} \vdash \phi$ iff the following conjunction is unsatisfiable:*

$$\left\{\begin{array}{lll}
\textbf{Neg} & \multicolumn{2}{l}{\exists x P_c(x) \wedge \neg P_\phi(x)} \\
\textbf{Ren} & & \\
 & (1,0) & \forall x P_1(x) \qquad \forall x \neg P_0(x) \\
 & (\vee) & \forall x[P_{\psi_1 \vee \psi_2}(x) \leftrightarrow P_{\psi_1}(x) \vee P_{\psi_2}(x)] \\
 & (\wedge) & \forall x[P_{\psi_1 \wedge \psi_2}(x) \leftrightarrow P_{\psi_1}(x) \wedge P_{\psi_2}(x)] \\
 & (\mathsf{Lh}) & \forall x[P_{h(\psi)}(x) \leftrightarrow P_\psi(h(x))] \\
 & (\mathsf{La}) & \forall x[P_{k(\psi)}(x) \leftrightarrow \neg P_\psi(k(x))] \\
 & (\Sigma_{+1}) & \forall x[P_{f(e_1,\dots,e_n)}(x) \leftrightarrow \exists x_1 \dots x_n(\bigwedge_{i=1}^n P_{e_i}^{\varepsilon_i}(x_i) \wedge R_f(x_1,\dots,x_n,x))] \\
 & (\Sigma_{-1}) & \forall x[P_{g(e_1,\dots,e_n)}(x) \leftrightarrow \\
 & & \quad \leftrightarrow \forall x_1 \dots x_n(R_g(x_1,\dots,x_n,x) \rightarrow (\bigvee_{i=1}^n P_{e_i}^{-\varepsilon_i}(x_i)))] \\
\textbf{Dom} & & \\
 & (\mathsf{R}) & \forall x,\ x \leq x \\
 & (\mathsf{Tr}) & \forall x,y,z\,[(x \leq y \wedge y \leq z) \rightarrow x \leq z] \\
 & (\mathsf{Lh}) & \forall x,y\,[x \leq y \rightarrow h(x) \leq h(y)] \\
 & (\mathsf{La}) & \forall x,y\,[x \leq y \rightarrow k(y) \leq k(x)] \\
 & (\mathsf{Inc}) & \forall x_1,\dots,x_n,x,y\,[x \leq y \wedge R_f(x_1,\dots,x_n,x) \rightarrow R_f(x_1,\dots,x_n,y)] \\
 & (\mathsf{Dec}) & \forall x_1,\dots,x_n,x,y\,[x \leq y \wedge R_g(x_1,\dots,x_n,y) \rightarrow R_g(x_1,\dots,x_n,x)] \\
 & (\mathcal{K}_\mathcal{L}) & E \ \ \textit{(the set of first-order formulae that characterize } \mathcal{K}_\mathcal{L}) \\
\textbf{Mng} & \multicolumn{2}{l}{\forall x,y\,[(x \leq y \wedge P_\psi(x)) \rightarrow P_\psi(y)].}
\end{array}\right.$$

Here the predicates P_ψ are indexed by subformulae ψ of ϕ; f ranges over all elements in $\Sigma_{\varepsilon_1\dots\varepsilon_n \to +1}$, g over all elements in $\Sigma_{\varepsilon_1\dots\varepsilon_n \to -1}$, h over all elements in Lh and k over all elements in La.

For logics based on $\mathsf{HO_\Sigma}$, the class of Heyting algebras with operators in Σ, the following rules for renaming the operators $\Rightarrow$ and $\neg$ have to be added to **Ren**.

$$\begin{array}{ll}
(\Rightarrow) & \forall x[P_{\psi_1 \Rightarrow \psi_2}(x) \leftrightarrow \forall y(x \leq y \wedge P_{\psi_1}(y) \rightarrow P_{\psi_2}(y))] \\
(\neg) & \forall x[P_{\neg\psi}(x) \leftrightarrow \forall y(x \leq y \rightarrow \neg P_\psi(y))]
\end{array}$$

Unsatisfiability of the conjunction above can be checked using (refinements of) resolution. The results in [65] show that, for $\mathcal{K}_\mathcal{L} = \mathsf{D_{01}O_\Sigma}$, the unsatisfiability of set of clauses obtained according to Theorem 21 can be checked in exponential time. (An exponential resolution-based decision procedure is actually obtained for the uniform word problem for distributive lattices with operators, and a doubly exponential decision procedure for the uniform word problem for Heyting algebras.) A similar structure-preserving translation to clause form was used for automated theorem proving for finitely-valued logics based on distributive lattices with operators in [70].

For logics based on residuated distributive lattices, **Ren** contains in addition to the rules for $\vee, \wedge$ and the operators in $\Sigma \cup \{R_\circ\}$ the following rule:

$$(\rightarrow) \quad \forall x[P_{\psi_1 \rightarrow \psi_2}(x) \leftrightarrow \forall y, z(R_\circ(x, y, z) \wedge P_{\psi_1}(y) \rightarrow P_{\psi_2}(z))].$$

Alternatively, we can include in **Ren** the rules for $\vee, \wedge$ and the operators in $\Sigma \cup \{R_\circ, R_\rightarrow\}$, and add to **Dom** the rule

$$\forall x, y, x[R_\rightarrow(x, y, z) \leftrightarrow R_\circ(y, z, x)].$$

A similar idea was used for more general residuation rules in [72] and again yielded an exponential resolution-based decision procedure for a class of bounded distributive lattices with operators satisfying certain generalized residuation rules.

For logics based on non-distributive lattices, like the substructural logics discussed in Section 7.2, the Kripke-style models are semilattices. Accordingly, **Dom** contains in addition the semilattice axioms and **Mng** expresses the fact that meaning functions assign semilattice-filters to formulae, i.e. contains in addition the following formula:

$$\mathbf{Mng}(\wedge) \quad \forall x, y\,[P_\psi(x) \text{ and } P_\psi(y) \rightarrow P_\psi(x \wedge y)].$$

According to the definition of $\overset{r}{\models}$, in this case $\mathbf{Ren}(\vee)$ has to be replaced by:

$$\mathbf{Ren}(\vee)' \quad \forall x[P_{\psi_1 \vee \psi_2}(x) \leftrightarrow \exists y_1, y_2\,[y_1 \wedge y_2 \leq x \text{ and } \textstyle\bigwedge_{i=1,2} (P_{\psi_1}(y_i) \text{ or } P_{\psi_2}(y_i))]]$$

The rules for the operators $\circ$ and $\rightarrow$ correspond to the definition of $\overset{r}{\models}$ in Corollary 4. It still remains to be seen if this embedding into classical first-order logic can be used for obtaining efficient decision procedures for decidable substructural logics.

9 Conclusions

We showed that representation theorems can be used to establish a link between algebraic and Kripke-style models for many non-classical logics. Then, a resolution-based method for automated theorem proving was presented. We pointed out that, in many cases, refinements of resolution yield optimal decision algorithms.

All representation theorems we used were built in a uniform way starting from a general duality theorem, for classes of algebras $\mathcal{V} = ISP(\underline{P})$ in the special case when $\underline{P}$ is a *two element* algebra. This makes it possible to represent algebras with a reduct in $\mathcal{V}$ as algebras of *sets*, and to give Kripke-style models in which statements about the truth of a formula at a possible world are expressed in classical (2-valued) logic. Many classes of algebras with a (distributive) lattice reduct and well-behaved operators are also of the form $ISP(\underline{P})$, where $\underline{P}$ is an algebra with more than two elements. Both representations yield Kripke-like ("fibered") models, with valuations at each possible world *in two-element algebras* or, respectively, *in* $\underline{P}$. Intermediate representations may also exist (cf. e.g. [64] p.311). In future work we would like to understand up to which extent such stronger representations would be of advantage in obtaining decidability proofs or for automated theorem proving.

Other types of representation theorems have also been used for giving Kripke-style models of some substructural logics. For instance, Urquhart's representation theorem was used by Allwein and Dunn in [2] to provide a (3-valued) Kripke-style semantics for linear logic. Similar results are obtained for logics with BCK-implication in [48]. Hartonas' representation theorem for lattices [35] has been used in the study of so-called "normal algebraizable logics", i.e. logics having as algebraic models lattices with "normal operators", similar to the operators we considered here. As examples of such logics he considers structural extensions of the full Lambek calculus.

The Kripke-style semantics we obtained in this paper, as illustrated for the particular case of residuated lattices in Section 7.2, seems to be simpler both than the 3-valued Kripke-style semantics in [2, 48], and than the Kripke-style semantics Hartonas obtains from the representation theorem in [35]. On the other hand, the representation we use does not seem to always lead, in the case of non-distributive lattices, to the so-called "canonical extensions" of lattices with operators, whose construction relies to an idea similar to that in Urquhart's representation theorems. We would like to obtain a better understanding of the relationships between the Kripke-style semantics we provide and the semantics obtained e.g. in [2, 48] or [35].

Acknowledgments. I thank Ewa Orłowska, Matthias Baaz, Harald Ganzinger, Mai Gehrke, Luisa Iturrioz, and Wendy MacCaull for interesting discussions. I also thank the anonymous referee for several suggestions that lead to an improvement of the presentation.

References

1. Anderson, A.R. and Belnap, N.D.: Entailment: The Logic of Relevance and Necessity, volume 1. Princeton University Press, Princeton (1975)

2. Allwein, G. and Dunn, J.M.: Kripke models for linear logic. Journal of Symbolic Logic, 58, (2) (1993) 514–545
3. Andréka, H., van Benthem, J. and Németi, I.: Modal languages and bounded fragments of predicate logic. Journal of Philosophical Logic, 27 (1998) 217–274
4. Blok, W.J. and Ferreirim, I.M.A.: On the structure of hoops. Algebra Universalis, 43 (2000) 233–257
5. Blyth, T.S. and Varlet, J.C.: Ockham Algebras. Oxford Science Publications (1994)
6. Brink, C.: Multisets and the algebra of relevance logic. The Journal of Nonclassical logic, 5 (1988) 75–95
7. Clark, D.M. and Davey, B.A.: Natural dualities for the working algebraist, volume 57 of Cambridge studies in advanced mathematics. Cambridge University Press, 1st edition (1998)
8. Davey, B.: Duality theory on ten dollars a day. In Rosenberg, I.G. and Sabidussi, G. editors, Algebras and Orders (Proceedings of the NATO ASI and Séminaire de mathématiques supérieures on Algebras and Orders, Montréal, Canada, July 29–August 9, 1991, NATO ASI Series, Vol. 389 (1993) 71–111
9. Davey, B.A. and Priestley, H.A.: Introduction to Lattices and Order. Cambridge University Press (1990)
10. Došen, K.: A historical introduction to substructural logics. In Schroeder-Heister P. and Došen, K. editors, Substructural logics, Studies in Logic and Computation 2. Oxford Science Publisher (1993) 63–108
11. Dummet, M.: A propositional calculus with denumerable matrix. The Journal of Symbolic Logic, 24, (2) (1959) 97–106
12. Dunn, J.M.: Partial gaggles applied to logics with restricted structural rules. In Schroeder-Heister, P. and Došen, K. editors, Substructural logics, Studies in Logic and Computation 2. Oxford Science Publisher (1993) 63–108
13. Dunn, J.M.: Positive modal logic. Studia Logica, 55 (1995) 301–317
14. Fitch, F.B.: A system of formal logic without an analogue to the Curry W operator. The Journal of Symbolic Logic, 1 (1936) 92–100
15. Fitting, M.C.: Intuitionistic logic model theory and forcing. Studies in logic and the foundations of mathematics. North-Holland Publishing Company, Amsterdam (1969)
16. Fleischer, I.: Every BCK-algebra is a set of residuables in an integral pomonoid. Journal of Algebra, 119 (1988) 360–365
17. Gabbay, D.: Expressive functional completeness in tense logics. In Mönnich, U. editor, Aspects of Philosophical Logic. Reidel (1981) 91–117
18. Ganzinger, H., Hustadt, U., Meyer, Ch. and Schmidt, R.A.: A resolution-based decision procedure for extensions of K4. In Zakharyaschev, M., Segerberg, K., de Rijke, M. and Wansing, H. editors, Advances in Modal Logic, Volume 2, volume 119 of CSLI Lecture Notes, chapter 9. CSLI, Stanford, USA (2001) 225–246
19. Ganzinger, H., Meyer, C. and Veanes, M. The two-variable guarded fragment with transitive relations. In Proceedings of the Fourteenth Annual IEEE Symposium on Logic in Computer Science. IEEE Computer Society Press (1999) 24–34
20. Gehrke, M.: Personal communication (2001)
21. Gehrke, M. and Priestley, H.A.: Non-canonicity of MV-algebras. Houston Journal of Mathematics. To appear.

22. Gehrke, M. and Harding, J.: Bounded lattice expansions. Journal of Algebra, 238 (2001) 345–371
23. Gehrke, M. and Jónsson, B.: Bounded distributive lattices with operators. Mathematica Japonica, 40, (2) (1994) 207–215
24. Gehrke, M. and Jónsson, B.: Monotone bounded distributive lattice expansions. Mathematica Japonica, 52, (2) (2000) 197–213
25. Gentzen, G.: Untersuchungen über das logische Schließen. Mathematische Zeitschrift (1935)
26. Girard, J.-Y.: Linear logic. Theoretical Computer Science (1987)
27. Grädel, E., Kolaitis, P. and Vardi, M.: On the decision problem for two-variable first-order logic. The Bulletin of Symbolic Logic, 3, (1) (1997) 53–69
28. Goldblatt, R.: Semantic analysis of orthologic. Journal of Philosophical Logic, 3 (1974) 19–35
29. Goldblatt, R.: Orthomodularity is not elementary. The Journal of Symbolic Logic, 49 (1984) 401–404
30. Goldblatt, R.: Varieties of complex algebras. Annals of Pure and Applied Logic, 44, (3) (1989) 153–301
31. Goldblatt, R.: Mathematics of modality, volume 43 of Center for the Study of Language and Information. Univ. of Chicago Press (1993)
32. Goldblatt, R.: Algebraic polymodal logic: A survey. Logic Journal of the IGPL, 8, (4) (2000) 393–450
33. Grädel, E.: On the restraining power of guards. The Journal of Symbolic Logic, 64 (1999) 1719–1742
34. Halmos, P.: Algebraic logic I. Monadic Boolean algebras. Compositio Mathematica, 12 (1955) 217–249
35. Hartonas, C.: Duality for lattice-ordered algebras and for normal algebraizable logics. Studia Logica, 58, (3) (1997) 403–450
36. Hartonas, C. and Dunn, J.M.: Stone duality for lattices. Algebra Universalis, 37, (3) (1997) 391–401
37. Hájek, P.: Metamathematics of Fuzzy Logic. Trends in Logic, Volume 4. Kluwer Academic Publishers, Dordrecht (1998)
38. Iséki, K. and Tanaka, S.: Ideal theory of BCK-algebras. Mathematica Japonica, 21 (1976) 351-366
39. Iturrioz, L.: A topological representation theory for orthomodular lattices. In Colloquia Mathematica Societatis János Bolyai, volume 33. Contributions to Lattice Theory (1980) 503–524
40. Iturrioz, L.: Symmetrical Heyting algebras with operators. Zeitschrift für Mathematische Logik und Grundlagen der Mathematik, 29 (1983) 33–70
41. Iturrioz, L. and Orłowska, E.: A Kripke-style and relational semantics for logics based on Łukasiewicz algebras. Conference in honour of J. Łukasiewicz, Dublin (1996)
42. Jónsson, B. and Tarski, A.: Boolean algebras with operators, Part I. American Journal of Mathematics, 73 (1951) 891–939
43. Jónsson, B. and Tarski, A.: Boolean algebras with operators, Part II. Americal Journal of Mathematics, 74 (1952) 127–162
44. Lambek, J.: The mathematics of sentence structure. American Mathematical Monthly, 65 (1958) 154–170
45. Lambek, J.: On the calculus of syntactic types. In Jacobsen, R. editor, Structure of Language and its Mathematical Aspects, Proceedings of Symposia in Applied Mathematics, XII. American Mathematical Society (1961)

46. Łukasiewicz, J.: Philosophische Bemerkungen zu mehrwertigen Systemen des Aussagenkalküls. Comptes rendus de la Société des Sciences et Lettres de Varsovie, cl.iii, 23 (1930) 51–77. English translation in [47].
47. Łukasiewicz, J.: Selected works (ed. by L. Borkowski). North-Holland (1970)
48. MacCaull, W.: Kripke semantics for logics with BCK implication. Bulletin of the Section of Logic of the University of Lodz, 25, (1) (1996) 41–51
49. MacCaull, W.: A note on a Kripke semantics for residuated logic. Fuzzy sets and systems, 77 (1996) 229–234
50. Martínez, N.G.: A topological duality for some ordered lattice ordered algebraic structures including l-groups. Algebra Universalis, 31 (1994) 516–541
51. Meredith, C.A. and Prior, A.N.: Notes on the axiomatics of the propositional calculus. Notre Dame Journal of Formal Logic, 4 (1963) 171–187
52. Mortimer, M.: On languages with two variables. Zeitschrift für Mathematische Logik und Grundlagen der Mathematik, 21 (1975) 135–140
53. Ohlbach, H.J.: Translation methods for non-classical logics: An overview. Bulletin of the IGPL, 1, (1) (1993) 69–89
54. Ono, H.: Algebraic aspects of logics without structural rules. AMS, Contemporary Mathematics, 131 part 3 (1992) 601–621
55. Ono, H.: Semantics for substructural logics. In Schroeder-Heister, P. and Došen, K. editors, Substructural Logics. Oxford University Press (1993) 259–291
56. Ono, H. and Komori, Y.: Logics without the contraction rule. Journal of Symbolic Logic, 50 (1985) 169–201
57. Orłowska, E.: Personal communication (1999)
58. Pałasiński, M.: An embedding theorem for BCK-algebras. Math. Seminar Notes Kobe Univ., 10 (1982) 749–751
59. Priestley, H.A.: Representation of distributive lattices by means of ordered Stone spaces. Bulletin of the London Mathematical Society, 2 (1970) 186–190
60. Rodríguez, A.J.: Un estudo algebraico de los cálculos proposicionales de Łukasiewicz. PhD thesis, Universidad de Barcelona (1980)
61. Routley, R., Plumwood, R.V., Meyer, K. and Brady, R.T.: Relevant Logics and their Rivals. Ridgeview (1982)
62. Schmidt, R.: Optimized modal translation and resolution. PhD thesis, Max-Planck-Institut für Informatik, Universität des Saarlandes (1997)
63. Schmidt, R.: Decidability by resolution for propositional modal logics. Journal of Automated Reasoning, 22, (4) (1999) 379–396
64. Sofronie-Stokkermans, V.: Fibered Structures and Applications to Automated Theorem Proving in Certain Classes of Finitely-Valued Logics and to Modeling Interacting Systems. PhD thesis, RISC-Linz, J.Kepler University Linz, Austria (1997)
65. Sofronie-Stokkermans, V.: On the universal theory of varieties of distributive lattices with operators: Some decidability and complexity results. In Proceedings of the 16th International Conference on Automated Deduction (CADE-16), volume 1632 of Lecture Notes in Artificial Intelligence, Trento, Italy, Springer (1999) 157–171
66. Sofronie-Stokkermans, V.: Representation theorems and automated theorem proving in non-classical logics. In Proceedings of the 29th IEEE International Symposium on Multiple-Valued Logic, Freiburg im Breisgau, Germany. IEEE Computer Society, IEEE Computer Society Press (1999) 242–247

67. Sofronie-Stokkermans, V.: Duality and canonical extensions of bounded distributive lattices with operators, and applications to the semantics of non-classical logics I. Studia Logica, 64, (1) (2000) 93–132
68. Sofronie-Stokkermans, V.: Duality and canonical extensions of bounded distributive lattices with operators, and applications to the semantics of non-classical logics II. Studia Logica, 64, (2) (2000) 151–172
69. Sofronie-Stokkermans, V.: Priestley duality for SHn-algebras and applications to the study of Kripke-style models for SHn-logics. Multiple-Valued Logic - An International Journal, 5, (4) (2000) 281–305
70. Sofronie-Stokkermans, V.: Automated theorem proving by resolution for finitely-valued logics based on distributive lattices with operators. Multiple-Valued Logic - An International Journal, 6 (2001) 289–344
71. Sofronie-Stokkermans, V.: Representation theorems for distributive lattices with operators. Manuscript (2001)
72. Sofronie-Stokkermans, V.: Resolution-based decision procedures for the universal theory of some classes of distributive lattices with operators. Research Report MPI-I-2001-2-005, Max-Planck-Institut für Informatik, Stuhlsatzenhausweg 85, 66123 Saarbrücken, Germany, September (2001)
73. Stone, M.H.: The theory of representations for Boolean algebras. Transactions of the American Mathematical Society (1936)
74. Szwast, W. and Tendera, L.: On the decision problem for the guarded fragment with transitivity. In Proceedings of the 16th IEEE Annual Symposium on Logic in Computer Science. IEEE Computer Society (2001) 147–156
75. Tarski, A.: Über die Erweiterung der unvollständigen Systeme des Aussagenkalküls. Ergebnisse eines Mathematischen Kolloquiums, 7 (1934/1935) 51–57. English Translation in [76].
76. Tarski, A.: Logic, Semantics, Metamathematics: Papers from 1923 to 1938, (Woodger, J.H. editor). Clarendon Press, Oxford (1956)
77. Urquhart, A.: A topological representation theory for lattices. Algebra Universalis, 8 (1978) 45–58
78. Urquhart, A.: Duality for algebras of relevant logics. Studia Logica, 56, (1,2) (1996) 263–276

Chapter 4
An Algebraic Approach to Entropy and its Generalizations – A Survey

Dan A. Simovici

Univ. of Massachusetts Boston,
Dept. of Computer Science,
Boston, 02125, USA,
dsim@cs.umb.edu

Abstract. We discuss several axiomatizations of entropy and some of its generalizations using an algebraic approach. The entropy is defined for such objects as functions, partitions, and set collections and the axiomatizations use natural operations on such objects. The generalizations of entropy we propose have applications in circuit design, data mining, machine learning, and information retrieval.
Keywords: functions, partitions, collections, entropy, Gini index, square root entropy.

1 Introduction

The notion of entropy was proposed in 1850 by the german physicist Rudolf Clausius as a measure of unused energy when mechanical work is obtained in a closed system from the transformation of termic energy. In the second part of the 19th century another german physicist, Ludwig Eduard Boltzman, the founder of statistical mechanics, re-examined the laws of thermodinamics in probabilistical terms, and interpreted entropy as a measure of the ways in which a macroscopic state of a system can be realized from microscopic states.

In 1948 Claude Shannon [22] introduced the notion of entropy in connection with communication theory following Boltzmann's model of thermodinamical entropy. His work was continued by B. McMillan [18], A. Feinstein [5] and A. I. Khinchin [14], who laid the mathematical foundations of the newly emerging discipline of information theory. The link between the two fundamental points of view (statistical and informational) was established by E. T. Jaynes [11, 12].

The notion of entropy is interesting for engineers and computer scientists who use the concept of entropy for a variety of problems. Engineers (and, in particular, circuit designers) start from the linkage between functional entropy and the power dissipated by circuits that realize functions (cf.[2, 8]) to design circuits that have minimal power consumption. Information measures, especially conditional entropy of a logic function and its variables, have been used for minimization of logic functions (See [16] and [3]). Entropy and some of its generalizations play in important role in machine learning and in data mining, where they are used in designing decision trees [19] or clustering algorithms for categorial data [26].

The interest of mathematicians in entropy is in the axiomatization of this measure and in the possibility of generalizing this concept. Basic contributions in the area of axiomatizations were made by A.I. Khinchin [15], D.K. Faddeev [4], R.S. Ingarden and K. Urbanik [9], and A. Rényi [21].

The axiomatization constructed opens the possibility that some of these measures can be used in new areas of application, and some entirely new characteristics can be used for the same purpose.

The results that we intend to discuss are centered around the idea of introducing the notion of entropy and certain generalizations of this notions using algebraic objects and natural operations involving such objects. This is suggested by the fact that any function $f : A \longrightarrow B$ between the finite sets A and $B = \{b_1, \ldots, b_n\}$ defines naturally a finite random variable:

$$X_f : \begin{pmatrix} b_1 & \cdots & b_n \\ p_1 & \cdots & p_n \end{pmatrix},$$

where $p_i = \frac{|f^{-1}(b_i)|}{|A|}$ for $1 \leq i \leq n$. This leads to the definition of the entropy $\mathcal{H}$ of a function $f : A \longrightarrow B$ between two finite sets as

$$\mathcal{H}(f) = \mathcal{H}(X_f) = \sum_{i=1}^{n} \frac{|f^{-1}(b_i)|}{|A|} \log_2 \frac{|f^{-1}(b_i)|}{|A|}.$$

Conversely, for any random variable

$$X : \begin{pmatrix} b_1 & \cdots & b_n \\ p_1 & \cdots & p_n \end{pmatrix},$$

having rational numbers as probabilities we can define a function f_X between finite sets such that $X_{f_X} = X$; of course, $\mathcal{H}(f_X)$ will equal $\mathcal{H}(X)$. When the probabilities are not rational, we can define, for every $\epsilon > 0$, a function whose entropy differs by less than ϵ from the entropy of X.

The sets $\mathbb{R}, \mathbb{R}_{\geq 0}, \mathbb{Q}, \mathbb{N}$ denote the set of reals, the set of nonnegative reals, the set of rational and the set of natural numbers, respectively. The domain and range of a function f are denoted by $\mathrm{Dom}(f)$ and $\mathrm{Ran}(f)$ respectively.

The k-dimensional simplex is the set $\mathsf{SIMPLEX}_{k-1} = \{(p_1, \ldots, p_k) \in \mathbb{R}^k \mid p_i \geq 0 \text{ and } p_1 + \cdots + p_k = 1\}$. A function $f : \mathbb{R} \longrightarrow \mathbb{R}$ is *concave on a set* $S \subseteq \mathbb{R}$ if $f(\alpha x + (1-\alpha)y) \geq \alpha f(x) + (1-\alpha)f(y)$ for $\alpha \in [0,1]$ and $x, y \in S$. The function f is *sub-additive* (*supra-additive*) on S if $f(x+y) \leq f(x)+f(y)$ ($f(x+y) \geq f(x)+f(y)$) for $x, y \in S$. A concave impurity measure is defined as a concave real-valued function $i : \mathsf{SIMPLEX}_{k-1} \longrightarrow \mathbb{R}$ such that

$i(\alpha\mathbf{p} + (1-\alpha)\mathbf{q}) = \alpha i(\mathbf{p}) + (1-\alpha)i(\mathbf{q})$ implies $\mathbf{p} = \mathbf{q}$ for $\alpha \in [0,1]$ and $\mathbf{p}, \mathbf{q} \in \mathsf{SIMPLEX}_{k-1}$ and, if $\mathbf{p} = (p_1, \ldots, p_k)$, then $i(\mathbf{p}) = 0$ if $p_i = 1$ for some i, $1 \leq i \leq k$.

2 Axiomatization of functional entropy

Classical axiomatizations of entropy (such as Khinchin's [14, 15]) are designed such that they produce the expression of the entropy using rather complicate conditions imposed on certain numerical characteristics of probability distributions. For example, Khinchin system postulates the following conditions for a family of functions

$$H_1(1), H_2(p_1, p_2), \ldots, H_n(p_1, \ldots, p_n), \ldots$$

attached to finite probability distributions with one, two, ..., n, ..., values, respectively:

(KH1) H_n is a non-negative, continuous function on $\mathsf{SIMPLEX}_{n-1}$ for $n \geq 1$.
(KH2) For every $n \geq 1$, H_n is a symmetric function in all its arguments.
(KH3) For $(p_1, \ldots, p_n) \in \mathsf{SIMPLEX}_{n-1}$ we have

$$H_{n+1}(p_1, \ldots, p_n, 0) = H_n(p_1, \ldots, p_n).$$

(KH4) $H_n(p_1, \ldots, p_n) \leq H_n(\frac{1}{n}, \ldots, \frac{1}{n})$.
(KH5) If $p_k = \sum_{\ell=1}^{m} q_{k\ell}$ for $1 \leq k \leq n$, where $q_{k\ell} \geq 0$ for $1 \leq k \leq n$, $1 \leq k \leq n$ and $\sum_{k=1}^{n}\sum_{\ell=1}^{m} q_{k\ell} = 1$, then

$$\begin{aligned} &H_{nm}(q_{11}, q_{12}, \ldots, q_{nm}) \\ &= H_n(p_1, \ldots, p_n) + \sum_{k=1}^{n} p_k H_m\left(\frac{q_{k1}}{p_k}, \ldots, \frac{q_{km}}{p_k}\right). \end{aligned}$$

For this family of functions there is a positive constant λ such that

$$H_n(p_1, \ldots, p_n) = -\lambda \sum_{k=1}^{n} p_k \log p_k.$$

In [24, 10] we proposed an axiomatization for the notion of entropy formulated in terms of functional entropy. As we shall see next, this axiomatization has the advantage of being formulated in terms of functional operations and, actually is expressing properties of circuits that implement finite functions.

Define the partial order $\sqsubseteq$ on $\mathcal{F}$, the class of all functions between finite, nonempty sets, by $f \sqsubseteq g$ if $f : A \longrightarrow B$, $g : A' \longrightarrow B'$, $A \subseteq A'$, $B \subseteq B'$ and $f(a) = g(a)$ for $a \in A$. If A,B,C,D are finite sets, such that $A \cap B = C \cap D = \emptyset$, $f : A \longrightarrow B$, and $g : C \longrightarrow D$, then there exists $\sup\{f, g\}$. We define the function $f \sqcup g = \sup\{f, g\}$ by setting

$$(f \sqcup g)(x) = \begin{cases} f(x) \text{ if } x \in A \\ g(x) \text{ if } x \in B \end{cases}$$

Whenever it is defined, the "$\sqcup$" operation is commutative and associative.

Definition 1. *Let $\mathcal{F}$ be the class of all functions between finite, nonempty sets, and $H : \mathcal{F} \to \mathbb{R}$ be a function assigning a real number to every $f \in \mathcal{F}$. Function H is called entropy of functions between finite sets if and only if it satisfies the following axioms:*

(E1) $H(f\alpha) = H(f)$, *for every function $f : A \to B$ and bijection $\alpha : A' \to A$.*
(E2) $H(gf) \leq H(f)$, *for every $f : A \to B$ and $g : B \to C$.*
(E3) *If A, C are finite sets with $|A| \leq |C|$, $\alpha : A \to B$ and $\beta : C \to D$ are bijections, then $H(\alpha) \leq H(\beta)$.*
(E4) *If $f : A \to B$ and $f : C \to D$ are functions between finite sets and $A \cap C = B \cap D = \emptyset$, then*

$$H(f \sqcup g) = \frac{|A|}{|A \cup C|} H(f) + \frac{|C|}{|A \cup C|} H(g) + H(\chi_{A,C}).$$

(E5) $H(f \times g) = H(f) + H(g)$ *for all functions $f : A \to B$ and $g : C \to D$.*

Axiom **(E5)**, for example, is suggested by the fact that the energy spent by the circuit obtained by the parallel connection between two circuits equals the sum of the energy spent by each of the circuits.

Axioms **(E1)** to **(E5)** imply the following theorem.

Theorem 1. *The entropy of a function $f : A \to B$ between two finite sets, where $B = \{b_1, \ldots, b_m\}$ has the form:*

$$H(f) = -\sum_{i=1}^{m} \frac{|A_i|}{|A|} \log_k \frac{|A_i|}{|A|},$$

where $A_i = \{a \in A | f(a) = b_i\}$, where k is a suitable constant, $k \geq 1$.

The same axiomatic approach can be applied to the notion of conditional entropy. Let $\mathcal{F}$ be the class of all functions between finite, nonempty sets. Denote by $\mathcal{D}$ a class of pairs of functions having identical domains $\mathcal{D} = \{(f, g) \in \mathcal{F} \times \mathcal{F} | \, \mathrm{Dom}(f) = \mathrm{Dom}(g)\}$. Let $H_C : \mathcal{D} \to \mathbb{R}$ be a function assigning a real number to every pair of functions $(f, g) \in \mathcal{D}$. $H_C(f, g)$ will be denoted below as $H(f|g)$.

Definition 2. *Function $H_C(f, g)$ is called conditional entropy of a pair of functions (f, g) if and only if it satisfies the following axioms:*

(CE1) $H(f|c) = H(f)$ *for every constant function c.*
(CE2) $H(f|g_1 \sqcup g_2) = \frac{|A_1|}{|A|} H(f_{A_1}|g_1) + \frac{|A_2|}{|A|} H(f_{A_2}|g_2)$, *for all $f : A \to B$, $g_1 : A_1 \to C_1$, $g_2 : A_2 \to C_2$, $A_1 \cap A_2 = C_1 \cap C_2 = \emptyset$, $A_1 \cup A_2 = A$.*

From these axioms we can derive:

Theorem 2. *Conditional entropy between functions $f : A \to B$ and $g : A \to C$, where $B = \{b_1, \ldots, b_m\}, C = \{c_1, \ldots, c_n\}$, has the following form:*

$$H(f|g) = -\sum_{j=1}^{n}\sum_{i=1}^{m} \frac{|A_j^i|}{|A|} \log \frac{|A_j^i|}{|A_j|},$$

where $A_j^i = \{x \in A | f(x) = b_i \wedge g(x) = c_j\}$, $A_j = \{x \in A | g(x) = c_j\}$.

3 Entropy of partitions

The notion of entropy can be associated to a partition of a finite set in a manner similar to the one used for functions. Indeed, if π is a partition of a finite set A we can consider the natural surjection $f_\pi : A \longrightarrow A/\pi$ that maps an element a onto the equivalence class $[a]$ and define $\mathcal{H}(\pi)$, the entropy of the partition π by $\mathcal{H}(\pi) = \mathcal{H}(f)$. This suggests the idea of formulating an axiomatization of partitions in terms of operations with partitions. Actually, we obtained an axiomatization of a notion that generalizes the classical Shannon entropy.

Let $\mathsf{PART}(A)$ be the set of partitions of the set A; the class of all partitions is denoted by PART. The one-block partition of A is denoted by ω_A. The partition $\{\{a\} \mid a \in A\}$ is denoted by ι_A. If $\pi, \pi' \in \mathsf{PART}(A)$, then $\pi \leq \pi'$ if every block of π is included in a block of π'. Clearly, for every $\pi \in \mathsf{PART}(A)$ we have $\iota_A \leq \pi \leq \omega_A$.

If $A \cap B = \emptyset$, $\pi \in \mathsf{PART}(A)$, $\sigma \in \mathsf{PART}(B)$, where $\pi = \{A_1, \ldots, A_m\}$, $\sigma = \{B_1, \ldots, B_n\}$, then the partition $\pi + \sigma$ is the partition of $A \cup B$ given by

$$\pi + \sigma = \{A_1, \ldots, A_m, B_1, \ldots, B_n\}.$$

Whenever the "+" is defined it is easily seen to be associative. In other words, if A, B, C are pairwise disjoint sets, and $\pi \in \mathsf{PART}(A)$, $\sigma \in \mathsf{PART}(A)$, $\tau \in \mathsf{PART}(A)$, then $\pi + (\sigma + \tau) = (\pi + \sigma) + \tau$.

If $\pi = \{A_1, \ldots, A_m\}$, $\sigma = \{B_1, \ldots, B_n\}$ are partitions of two arbitrary sets, then we denote the partition $\{A_i \times B_j \mid 1 \leq i \leq m, 1 \leq j \leq n\}$ of $A \times B$ by $\pi \times \sigma$.

Definition 3. *The* entropy of partitions *is a non-negative function* $\mathcal{H} : \mathsf{PART} \longrightarrow \mathbb{R}$ *such that:*

(P1) *If* $\pi, \pi' \in \mathsf{PART}(A)$ *and* $\pi \leq \pi'$, *then* $\mathcal{H}(\pi') \leq \mathcal{H}(\pi)$.
(P2) *There is a continuous, non-negative, and monotonic function* $\mu : \mathbb{R} \longrightarrow \mathbb{R}$ *such that* $\mathcal{H}(\iota_A) = \mu(|A|)$.
(P3) *For every disjoint sets* A, B *and partitions* $\pi \in \mathsf{PART}(A)$, *and* $\sigma \in \mathsf{PART}(B)$ *we have:*

$$\mathcal{H}(\pi + \sigma) = \frac{|A|}{|A| + |B|}\mathcal{H}(\pi) + \frac{|B|}{|A| + |B|}\mathcal{H}(\sigma) + \mathcal{H}(\{A, B\}).$$

(P4) *If* $\pi \in \mathsf{PART}(A)$ *and* $\sigma \in \mathsf{PART}(B)$, *then*

$$\mathcal{H}(\pi \times \sigma) = \mathcal{H}(\pi) + \mathcal{H}(\sigma).$$

Choosing A, B in axiom **(P3)** such that $|A| = |B|$ and $\pi = \omega_A$, $\sigma = \omega_B$ we have $\omega_A + \omega_B = \{A, B\}$ which implies $\mathcal{H}(\omega_A) = \mathcal{H}(\omega_B) = 0$. It is clear from **(P1)** that for every $\pi \in \mathsf{PART}(A)$ we have $\mathcal{H}(\pi) \leq \mu(|A|)$. Moreover, the axioms imply that for $\pi = \{A_1, \ldots, A_n\} \in \mathsf{PART}(A)$ we can write:

$$\mathcal{H}(\pi) = \mathcal{H}(\iota_A) - \sum_{i=1}^{n} \frac{|A_j|}{|A|} \mathcal{H}(\iota_{A_j}). \tag{1}$$

Further, note that $\underbrace{\iota_A \times \cdots \times \iota_A}_{n} = \iota_{A^n}$. Therefore, by **(P4)**, we have $\mathcal{H}(\iota_{A^n}) = n\mathcal{H}(\iota_A)$, which means that $\mu(p^n) = n\mu(p)$ for every $p \in \mathbb{N}$. The continuity of μ implies that $\mu(p) = \log_k p$ for some positive k. The equality (1) implies

$$\mathcal{H}(\pi) = \sum_{i=1}^{n} \frac{|A_j|}{|A|} \log_k \frac{|A_j|}{|A|},$$

which is the Shannon entropy.

To extend the above axiomatics to other entropy-like functions we introduce below a system of four axioms satisfied by both Shannon entropy and the Gini index used extensively in statistics and data mining.

Definition 4. *Let* $\beta \in \mathbb{R}_{\geq 0}$, *and let* $\Phi : \mathbb{R}^2_{\geq 0} \longrightarrow \mathbb{R}_{\geq 0}$ *be a continuous function such that* $\Phi(x, y) = \Phi(y, x)$, $\Phi(x, 0) = x$ *for* $x, y \in \mathbb{R}_{x \geq 0}$.

A (Φ, β)*-partition entropy is a function* $\mathcal{H} : \mathsf{PART}(A) \longrightarrow \mathbb{R}_{\geq 0}$ *that satisfies the following conditions:*

(GP1) *If* $\pi, \pi' \in \mathsf{PART}(A)$ *are such that* $\pi \leq \pi'$, *then* $\mathcal{H}(\pi') \leq \mathcal{H}(\pi)$.
(GP2) *If* A, B *are two finite sets such that* $|A| \leq |B|$, *then* $\mathcal{H}(\iota_A) \leq \mathcal{H}(\iota_B)$.
(GP3) *For every disjoint sets* A, B *and partitions* $\pi \in \mathsf{PART}(A)$, *and* $\sigma \in \mathsf{PART}(B)$ *we have:*

$$\begin{aligned} &\mathcal{H}(\pi + \sigma) \\ &= \left(\frac{|A|}{|A| + |B|}\right)^{\beta} \mathcal{H}(\pi) + \left(\frac{|B|}{|A| + |B|}\right)^{\beta} \mathcal{H}(\sigma) \\ &+ \mathcal{H}(\{A, B\}). \end{aligned}$$

(GP4) *If* $\pi \in \mathsf{PART}(A)$ *and* $\sigma \in \mathsf{PART}(B)$, *then*

$$\mathcal{H}(\pi \times \sigma) = \Phi(\mathcal{H}(\pi), \mathcal{H}(\sigma)).$$

It is easy to see from axiom **(GP3)** that $\mathcal{H}(\omega_A) = 0$ for every set A. Further, using Axiom **(GP3)** we can prove the following result:

Theorem 3. *Let A be a set. For every (Φ, β)-entropy and partition $\pi = \{A_1, \dots, A_n\} \in \mathsf{PART}(A)$ we have:*

$$\mathcal{H}(\pi) = \mathcal{H}(\iota_A) - \sum_{j=1}^{n} \left(\frac{|A_j|}{|A|} \right)^{\beta} \mathcal{H}(\iota_{A_j}).$$

If A, B are two sets such that $|A| = |B| > 0$, then, by Axiom **(GP2)**, we have $\mathcal{H}(\iota_A) = \mathcal{H}(\iota_B)$. Therefore, the value of $\mathcal{H}(\iota_A)$ depends only on the cardinality of A, and there exists a function $\mu : \mathbb{N}_1 \longrightarrow \mathbb{R}_{\geq 0}$ such that $\mathcal{H}(\iota_A) = \mu(|A|)$ for every nonempty set A. Axiom **(GP2)** also implies that μ is an increasing function. We will refer to μ as the *core* of the (Φ, β)-system of axioms.

Corollary 1. *Let $\mathcal{H}$ be a (Φ, β)-entropy. For the function μ defined in Axiom* **(P2)** *and every partition $\pi = \{A_1, \dots, A_n\} \in \mathsf{PART}(A)$ we have:*

$$\mathcal{H}(\pi) = \mu(|A|) - \sum_{j=1}^{n} \left(\frac{|A_j|}{|A|} \right)^{\beta} \mu(|A_j|). \tag{2}$$

Axiom **(GP3)** is extended by the next result:

Theorem 4. *Let $A_1, \dots, A_m$ be n nonempty, disjoint sets and let $\pi_i \in \mathsf{PART}(A)$ for $1 \leq i \leq m$. We have*

$$\mathcal{H}(\pi_1 + \cdots + \pi_m) = \sum_{i=1}^{m} \left(\frac{|A_i|}{|A|} \right)^{\beta} \mathcal{H}(\pi_i) + \mathcal{H}(\{A_1, \dots, A_m\}),$$

where $A = A_1 \cup \cdots \cup A_m$.

Using this result and Axiom **(GP4)** we obtain a functional equation satisfied by the core μ:

$$\mu(ab) - \frac{\mu(a)}{b^{\beta-1}} = \mu(b) \tag{3}$$

for every $a, b \in \mathbb{N}$ and $a, b \geq 1$.

An entropy is said to be *non-Shannon* if it defined by a (Φ, β)-system of axioms such that $\beta \neq 1$; otherwise, that is if $\beta = 1$, the entropy will be referred to as a *Shannon* entropy. With this definition we can prove:

Theorem 5. *Let $\mathcal{H}$ be a non-Shannon entropy defined by a (Φ, β)-system of axioms and let μ be the core of this system of axioms.*

There is a constant $k \in \mathbb{R}$ such that $k \cdot (\beta - 1) \geq 0$ and

$$\mu(a) = k \cdot \left(1 - \frac{1}{a^{\beta-1}} \right)$$

for $a \in \mathbb{N}_1$.

Note that for $\beta \neq 1$ we have:

$$k = \begin{cases} \lim_{a \to \infty} \mu(a) \text{ if } \beta > 1 \\ \lim_{a \to 0} \mu(a) \text{ if } \beta < 1. \end{cases} \tag{4}$$

Corollary 2. *If $\mathcal{H}$ is a non-Shannon entropy defined by a (Φ, β)-system of axioms and $\pi \in \mathsf{PART}(A)$, where $\pi = \{A_1, \ldots, A_n\}$, then there exists a constant $k \in \mathbb{R}$ such that*

$$\mathcal{H}(\pi) = k \cdot \left(1 - \sum_{j=1}^{n} \left(\frac{|A_j|}{|A|}\right)^{\beta}\right). \tag{5}$$

The specific form of the function Φ is obtained next:

Theorem 6. *Let $\mathcal{H}$ be the non-Shannon entropy defined by a (Φ, β)-system and let k be as defined by Equality (4), where μ is the core of the (Φ, β)-system of axioms. The function Φ introduced by Axiom* **(P4)** *is given by $\Phi(x, y) = x + y - \frac{1}{k} x \cdot y$ for $x, y \in \mathbb{R}_{\geq 0}$.*

Choosing $k = \frac{1}{\beta - 1}$ in the equality (5) we obtain the Havrda-Charvat entropy (see [17]):

$$\mathcal{H}(\pi) = \frac{1}{\beta - 1} \cdot \left(1 - \sum_{j=1}^{n} \left(\frac{|A_j|}{|A|}\right)^{\beta}\right).$$

The limit case, $\lim_{\beta \to 1} \mathcal{H}(\pi)$ yields the Shannon entropy.

If $\beta = 2$ we obtain the Gini index,

$$\mathcal{H}(\pi) = 1 - \sum_{j=1}^{n} \left(\frac{|A_j|}{|A|}\right)^{2},$$

which is widely used in machine learning and data mining (see [26]).

When $\beta = 1$, by the equality 3 we have:

$$\mu(ab) = \mu(a) + \mu(b)$$

for $a, b \in \mathbb{N}_1$. If $\eta : \mathbb{N}_1 \longrightarrow \mathbb{R}$ is the function defined by $\eta(a) = a\mu(a)$ for $a \in \mathbb{N}_1$, then η is clearly an increasing function and we have

$$\eta(ab) = ab\mu(ab) = b\eta(a) + a\eta(b)$$

for $a, b \in \mathbb{N}_1$. By Theorem A.6 from [24], there exists a constant $c \in \mathbb{R}$ such that $\eta(a) = ca \log_2 a$ for $a \in \mathbb{N}_1$, so $\mu(a) = c \log_2(a)$. Then, equation (2) implies:

$$\mathcal{H}(\pi) = c \cdot \sum_{i=1}^{n} \frac{a_i}{a} \log_2 \frac{a_i}{a},$$

for every partition $\pi = \{A_1, \ldots, A_n\}$ of a set A, where $|A_i| = a_i$ for $1 \leq i \leq n$, and $|A| = a$. This is exactly the expression of Shannon's entropy.

The continuous function Φ is determined, as in the previous case. Indeed, if A, B are two sets such that $|A| = a$ and $|B| = b$, then we must have

$$c \cdot \log_2 ab = \mathcal{H}(\iota_A \times \iota_B) = \Phi(c \cdot \log_2 a, c \cdot \log_2 b)$$

for any $a, b \in \mathbb{N}_1$ and any $c \in \mathbb{R}$. The continuity of Φ implies $\Phi(x, y) = x + y$.

The entropies previously introduced generate corresponding conditional entropies.

Let $\pi \in \mathsf{PART}(A)$ and let $C \subseteq A$. Denote by π_C the "trace" of π on C given by

$$\pi_C = \{B \cap C | B \in \pi \text{ such that } B \cap C \neq \emptyset\}$$

Clearly, $\pi_C \in \mathsf{PART}(C)$; also, if C is a block of π, then $\pi_C = \omega_C$.

Definition 5. *The* conditional entropy *corresponding to the (Φ, β)-entropy $\mathcal{H}$ is the function $\mathcal{C} : \mathsf{PART}^2 \longrightarrow \mathbb{R}_{\geq 0}$ given by:*

$$\mathcal{C}(\pi, \sigma) = \sum_{j=1}^{n} \frac{|C_j|}{|A|} \cdot \mathcal{H}(\pi_{C_j}),$$

where $\pi, \sigma \in \mathsf{PART}(A)$ and $\sigma = \{C_1, \ldots, C_n\}$.

We denote the value of $\mathcal{C}(\pi, \sigma)$ by $\mathcal{H}(\pi|\sigma)$. Note that $\mathcal{H}(\pi|\omega_A) = \mathcal{H}(\pi)$.

The partition $\pi \wedge \sigma$ whose blocks consist of the nonempty intersections of the blocks of π and σ can be written as

$$\pi \wedge \sigma = \pi_{C_1} + \cdots + \pi_{C_n} = \sigma_{B_1} + \cdots + \sigma_{B_m}.$$

Therefore, by Theorem 4, we have:

$$\mathcal{H}(\pi \wedge \sigma) = \sum_{j=1}^{n} \left(\frac{|C_j|}{|A|}\right)^{\beta} \mathcal{H}(\pi_{C_j}) + \mathcal{H}(\sigma).$$

For those entropies with $\beta > 1$ we have

$$\mathcal{H}(\pi \wedge \sigma) \geq \mathcal{H}(\pi|\sigma) + \mathcal{H}(\sigma),$$

while for those having $\beta < 1$, the reverse inequality holds. In the case of Shannon entropy, $\beta = 1$ and

$$\begin{aligned} \mathcal{H}(\pi \wedge \sigma) &= \mathcal{H}(\pi|\sigma) + \mathcal{H}(\sigma) \\ &= \mathcal{H}(\sigma|\pi) + \mathcal{H}(\pi). \end{aligned}$$

If $\mathcal{H}$ is a (Φ, β)-entropy, π, σ are two partitions of a set A such that $\pi = \{B_1, \ldots, B_m\}$ and $\sigma = \{C_1, \ldots, C_n\}$, then the conditional entropy $\mathcal{H}(\pi|\sigma)$ is given by:

$$\begin{aligned}\mathcal{H}(\pi|\sigma) &= \sum_{j=1}^{n} \frac{|C_j|}{|A|}\mu(|C_j|) \\ &= \sum_{j=1}^{n} \frac{|C_j|}{|A|}\mu(|C_j|) \\ &\quad - \sum_{j=1}^{n}\sum_{i=1}^{m} \frac{|B_i \cap C_j|^{\beta}}{|A| \cdot |C_j|^{\beta-1}}\mu(|B_i \cap C_j|).\end{aligned}$$

This equality follows immediately from Corollary 1.

In the case of Shannon entropy, taking $\beta = 1$ and $\mu(n) = \log_2 n$ we obtain the well-known expression of conditional entropy:

$$\mathcal{H}(\pi|\sigma) = -\sum_{i=1}^{m}\sum_{j=1}^{n} \frac{|B_i \cap C_j|}{|A|} \log_2 \frac{|B_i \cap C_j|}{|C_j|}.$$

In the case of the Gini index we have $\beta = 2$ and $\mu(a) = k\left(1 - \frac{1}{a}\right)$ for $a \in \mathbb{N}_1$. Consequently, after some elementary transformations, the conditional Gini index is:

$$\mathcal{H}(\pi|\sigma) = 1 - \sum_{j=1}^{n}\sum_{i=1}^{m} \frac{|B_i \cap C_j|^2}{|A| \cdot |C_j|}.$$

For $\beta = 0.5$, we obtain the "square-root entropy" used by us in clustering of categorial data (see [26]).

4 Entropy of set collections

Another technique for axiomatizing the notion of entropy was introduced in [1]. A *classification* of a collection of objects X is a set of subsets $\{B_1, \ldots, B_n\}$ of X, which in general need not be disjoint. In this note we investigate how one can measure the "information content" of such a classification. Our approach is entirely algebraic, systematically avoiding any reference to notions of probability or decision theory. This yields a more compelling treatment of notions of entropy and information content. We find that the well-known formula for entropy is a consequence of just two natural conditions imposed on a family of measures.

Our treatment of information content is based on the notion of a measure. A *measure* m on a finite set A is a function $m : \mathcal{P}(A) \to \mathbb{R}$ such that for any two disjoint elements $S, T \in \mathcal{P}(A)$, one has that $m(S \cup T) = m(S) + m(T)$.

Definition 6. *A* universal collection *is a collection of nonempty finite sets $\mathcal{U}$ that satisfies the following conditions:*

1. *$\mathcal{U}$ is finitely Cartesian-closed, that is, $X_1, \ldots, X_n \in \mathcal{U}$, imply $X_1 \times \ldots \times X_n \in \mathcal{U}$.*
2. *$\mathcal{U}$ contains sets of arbitrarily large cardinality.*

If $\mathcal{C} = \{B_1, \ldots, B_n\}$ is a collection of subsets of a set X, let $\| \mathcal{C} \| = \sum_{1 \leq i \leq n} |B_i|$. Let $\mathcal{C}, \mathcal{D}$ be two collections of sets, $\mathcal{C} \subseteq \mathcal{P}(X)$ and $\mathcal{D} \subseteq \mathcal{P}(Y)$. Define the collection $\mathcal{C} \times \mathcal{D}$ by

$$\mathcal{C} \times \mathcal{D} = \{C \times D \mid C \in \mathcal{C}, D \in \mathcal{D}\}.$$

It is not difficult to see that $\| \mathcal{C} \times \mathcal{D} \| = \| \mathcal{C} \| \| \mathcal{D} \|$.

We refer to collections of sets that consist of one single set as *singleton collections.* All collections of sets are collections of subsets of a set in $\mathcal{U}$.

For a measure on collections of subsets of a set $m : \mathcal{P}(\mathcal{P}(X)) \to \mathbb{R}$, we define the *normalized measure* as the partial mapping $\mu : \mathcal{P}(\mathcal{P}(X)) \to \mathbb{R}$ given by $\mu(\mathcal{C}) = m(\mathcal{C}) / \| \mathcal{C} \|$ if $\| \mathcal{C} \| \neq 0$.

Let $\alpha : X \to Y$ be a mapping, where $X, Y \in \mathcal{U}$. If $\mathcal{C}$ is a collection, $\mathcal{C} \subseteq \mathcal{P}(X)$, the collection $\alpha(\mathcal{C}) \in \mathcal{P}(\mathcal{P}(Y))$ is

$$\alpha(\mathcal{C}) = \{\alpha(B) \mid B \in \mathcal{C}\}.$$

Definition 7. *Let $\mathcal{U}$ be a universal collection of sets and let $\mathcal{M}$ be a family of measures $\mathcal{M} = \{m_X : \mathcal{P}(\mathcal{P}(X)) \to \mathbb{R} \mid X \in \mathcal{U}\}$. We say that $\mathcal{M}$ is* contracting *if $m_X(\mathcal{C}) \geq m_Y(\alpha(\mathcal{C}))$ for every $X, Y \in \mathcal{U}$, every mapping $\alpha : X \to Y$, and every collection $\mathcal{C} \subseteq \mathcal{P}(X)$.*

Theorem 7. *Let $\mathcal{U}$ be a universal collection of sets and let $\mathcal{M}$ be a contracting family of measures on $\mathcal{U}$. For every two collections $\mathcal{C} = \{B_1, \ldots, B_n\}$, $\mathcal{C}' = \{B'_1, \ldots, B'_n\}$ such that $|B_i| = |B'_i|$ for $1 \leq i \leq n$ we have $m(\mathcal{C}) = m(\mathcal{C}')$.*

Definition 8. *An* entropic family of measures *on $\mathcal{U}$ is a contracting collection of measures $\{m_X : \mathcal{P}(\mathcal{P}(X)) \to \mathbb{R} \mid X \in \mathcal{U}\}$ such that if $\mathcal{C} \in \mathcal{P}(\mathcal{P}(X))$ and $\mathcal{D} \in \mathcal{P}(\mathcal{P}(X))$, then*

$$m_{X \times Y}(\mathcal{C} \times \mathcal{D}) = \| \mathcal{C} \| \, m_Y(\mathcal{D}) + \| \mathcal{D} \| \, m_X(\mathcal{C}).$$

In terms of normalized measures the entropic families are characterized by:

$$\mu_{X \times Y}(\mathcal{C} \times \mathcal{D}) = \mu_X(\mathcal{C}) + \mu_Y(\mathcal{D})$$

for all collections $\mathcal{C}, \mathcal{D}$ such that $\| \mathcal{C} \| > 0$ and $\| \mathcal{D} \| > 0$.

Theorem 8. *Let $\{m_X : \mathcal{P}(\mathcal{P}(X)) \to \mathbb{R} \mid X \in \mathcal{U}\}$ be an entropic family of measures on $\mathcal{U}$. For every collection $\mathcal{C} \in \mathcal{P}(\mathcal{P}(X))$ such that $\mathcal{C} = \{B_i \mid 1 \leq i \leq n\}$ we have*

$$m_X(\mathcal{C}) = k \sum_{1 \leq i \leq n} |B_i| \log_2 |B_i|.$$

for some nonnegative real constant k.

When the constant k above is equal to 1, we write $\mathcal{H}_X$ for the resulting entropic family of measures on $\mathcal{U}$. The measure $\mathcal{H}_X(\mathcal{C})$ will be called the *entropy* or *information content* of the collection $\mathcal{C}$.

The results above generalize easily to the case of multisets We define a *multicollection* of X as a multisubset of $\mathcal{P}(X)$. For a set X, the entropy of a multicollection $\mathcal{C} = \{B_1, \ldots, B_n\}$ of subsets of X is given by

$$\mathcal{H}_X(\mathcal{C}) = \sum_{1 \leq i \leq n} |B_i| \log_2 |B_i|.$$

For a multicollection $\mathcal{C}$ of subsets of X, the possible values that can be taken by $\mathcal{H}_X(\mathcal{C})$ vary over a range that depends, in general, on the cardinalities of X and $\mathcal{C}$ as well as the norm $\| \mathcal{C} \|$. The upper and lower bounds on $\mathcal{H}_X(\mathcal{C})$ are important as a source for *a priori* states from which to compute the increase in information associated with an operation such as looking up a term in an index or retrieving a document from a corpus.

The simplest bounds are obtained when the cardinalities of X and $\mathcal{C}$ are specified.

Proposition 1. *Let X be a set having cardinality l, and let $\mathcal{C}$ be a multicollection of subsets of X such that $|\mathcal{C}| = n$. Then $0 \leq \mathcal{H}_X(\mathcal{C}) \leq nl \log_2 l$.*

The bounds become somewhat more complex when one specifies the norm of $\mathcal{C}$.

Proposition 2. *Let $\mathcal{C}$ be a multicollection of subsets of X such that $|\mathcal{C}| = n$ and $\| \mathcal{C} \| = r$. Then $r \log_2(r/n) \leq \mathcal{H}_X(\mathcal{C})$.*

Proposition 3. *Let X be a set having cardinality l. Let $\mathcal{C}$ be a multicollection of subsets of X such that $\| \mathcal{C} \| = r$. Then $\mathcal{H}_X(\mathcal{C}) \leq r \log_2 l$.*

Note that the upper bound represents the information content of the collection $\{X, \ldots, X\}$ when l divides r. When l does not divide r, the maximum value of $\mathcal{H}_X(\mathcal{C})$ will be strictly less than $r \log_2 l$. However, a more refined analysis can be used to show that the maximum value is approximately equal to $r \log_2 l$. More precisely, let $M(l, r)$ denote the maximum value of $\mathcal{H}_X(\mathcal{C})$. Then

$$\frac{r \log_2 l - M(l, r)}{r \log_2 l} \leq \frac{l}{re \log_e l}.$$

This type of entropy has application in information retrieval; they are discussed in [1].

5 Conclusions and open problems

We examine several types of entropies associated to algebraic objects (functions, partitions, and collections), as well as some generalizations of this concept. The axiomatizations use natural operations between such objects such as function composition, the order on the lattice of partitions of a set, etc. In general, the axiomatizations are simpler that existing axiomatizations of entropy and some axioms have a physical significance.

Several technical problems still remain. For example, in considering the notion of (Φ, β)-entropy of partitions one obtains the Shannon $\mathcal{H}^{sh}$ entropy and the Gini index $\mathcal{H}^{gini}$ as two special cases. However, the behaviour of the conditional entropy is quite different for these two types of entropy. If $\pi, \sigma \in$ PART(A), then we have $\mathcal{H}^{sh}(\pi|\sigma) \leq \mathcal{H}^{sh}(\pi)$, while $\mathcal{H}^{gini}(\pi|\sigma) \geq \mathcal{H}^{gini}(\sigma)$, as the reader can easily verify. It would be interesting to characterize those pairs (Φ, β) for which each of these properties holds.

A common generalization of these axiomatic investigations using the apparatus of category theory is suggested.

References

1. Baclawski, K., Simovici, D.A.: A Characterization of the Information Content of a Classification, Information Processing Letters, vol 57 (1996) 211–214
2. Cheng, K.T. and Agrawal, V.: An Entropy Measure of the Complexity of Multi-Output Boolean Function, Proc. of the 27th Design Automation Conference (1990) 302–305
3. Cheushev, V., Shmerko, V., Simovici, D.A., and Yanushkevich, S.: Functional Entropy and Decision Trees, Proc. of the 28th ISMVL (1998) 257–262
4. Faddeev, D.K.: On the notion of entropy of finite probability distributions, in Usp. Mat. Nauk (in Russian), 11 (1956) 227–231
5. Feinstein, A.: A New Basic Theorem of Information Theory, IRE, 4 (1954) 2–22
6. Guiasu, S.: Mathematical Information Theory (in Romanian), Editura Academiei, Bucuresti (1966)
7. Havrada, J.H., Charvat, F.: Quantification Methods of Classification Processes: Concepts of Structural α Entropy, Kybernetica, 3 (1967) 30-35
8. Hwang, C.H. and Wu, A.C.H.: An Entropy Measure for Power Estimation of Boolean Function, Proc. of Asia and South Pacific Design Automation Conference (ASP-DAC) (1997) 101–106
9. Ingarden, R.S. and Urbanik, K.: Information without Probability, Coll. Math., 1 (1962) 281–304
10. Jaroszewicz, S. and Simovici, D.A.: On Axiomatization of Conditional Entropy of Functions between Finite Sets, Proc. of the 29th ISMVL, Freiburg, Germany (1999) 24–31
11. Jaynes, E.T.: Information Theory and Statistical Mechanics I, Physical Review, 106 (1957) 620–630
12. Jaynes, E.T.: Information Theory and Statistical Mechanics II, Physical Review, 108 (1957) 171–190
13. Kapur, J.N. and Kesavan, H.K.: Entropy Optimization Principles with Applications, Academic Press, San Diego (1992)
14. Khinchin, A.Ia.: On The Fundamental Theorem of Information Theory, in Usp. Mat. Nauk (in Russian), 11 (1956) 17–75
15. Khinchin, A.Ia.: Mathematical Foundations of Information Theory, Dover Publ., New York (1957)

16. Lloris-Ruiz, A., Gomez-Lopera, J.F. and Roman-Roldan, R.: Entropic Minimization of Multiple-Valued Logic Functions, Proc. of the 23rd ISMVL (1993) 24–28
17. Mathai, A.M. and Rathie, P.N.: Basic Concepts in Information Theory and Statistics — Axiomatic Foundations and Applications, Halsted Press, John Wiley & Sons (1975)
18. McMillan, B.: The Basic Theorems of Information Theory, Ann. Math. Statistics, vol. 24, (1953), 196–219
19. Quinlan, J.R.: C4.5: Programs for Machine Learning, Morgan Kaufmann, San Mateo (1993)
20. Reischer, C., Simovici, D.A. and Stojmenovic, I.: Several Remarks on Algebraic Entropy, Proc. of the 24th ISMVL, Boston (1994) 319–322
21. Rényi, A.: On the Dimension and Entropy of Probability Distributions, Acta Math. Acad. Sci. Hung., 10 (1959) 193–215
22. Shannon, C.E.: A Mathematical Theory of Communication, Bell System Technical Journal, 27 (1948) 379–423, 623–659
23. Shmerko, V., Simovici, D.A.: Information Estimations for Logical Functions presented at ACS'97, the 4th Int. Conf. of Applications of Computer Systems, Szczecin, Poland, Nov. (1997)
24. Simovici, D.A. and Reischer, C.: On Functional Entropy, Proc. ISMVL'93 (1993) 100–104
25. Simovici, D.A., Jaroszewicz, S.: On Information-Theoretical Aspects of Relational Databases, in Calude, C. and Paun, Gh. (editors) Finite vs. Infinite, Springer-Verlag (2000) 301–322
26. Simovici, D.A., Cristofor, D., Cristofor, L.: Generalized Entropy and Projection Clustering of Categorial Data, in Principles of Data Mining and Knowledge Discovery, the 4th European Conference, PKDD 2000, Lyon, Lecture Notes in Artificial Intelligence 1910, Springer-Verlag (2000) 619–625
27. Simovici, D.A., Jaroszewicz, S.: An Axiomatization of Partition Entropy, to appear at Transactions on Information Theory, July (2002)

Part II

Proof Theory and Automated Deduction in Multiple-valued Logics

Chapter 5
Classical Gentzen-type Methods in Propositional Many-valued Logics

Arnon Avron

School of Computer Science
Tel-Aviv University
Ramat Aviv 69978, Israel
aa@math.tau.ac.il

Abstract. A classical Gentzen-type system is one which employs two-sided sequents, together with structural and logical rules of a certain characteristic form. A decent Gentzen-type system should allow for direct proofs, which means that it should admit some useful forms of cut elimination and the subformula property. In this survey we explain the main difficulty in developing classical Gentzen-type systems with these properties for many-valued logics. We then illustrate with numerous examples the various possible ways of overcoming this difficulty, and the strong connection between semantic completeness and cut-elimination in each case. Our examples include practically all 3-valued and 4-valued logics, as well as Gödel finite-valued logics and some well-known infinite-valued logics.

1 Introduction

Many-valued logics are, above all, *Logics*. This means that like any other logic, the main issue with which they deal is: what formulas follow from what sets of premises (under certain assumptions). The "many-valued" part of their name refer to the type of semantics which determines the answers they give to this basic question. However, any logic needs efficient *proof* systems in order to actually be applied (as well as for a deeper understanding of its logical properties), and many-valued logics are no exception. The problem of adapting to many-valued logics the usual methods of automated reasoning (like Gentzen-type systems, tableaux, and resolution) has therefore attracted a lot of attention in recent years. The main idea which is used in most of the works on this subject is to use for any particular n some n-valued counterparts of the structures used in the usual proof systems for classical logic (like sequents with n components, tableaux systems with n signs, etc.). This fact is well reflected in two recent survey papers [40] and [19] (see there for extensive list of references). In this paper we concentrate, in contrast, on proofs systems which use the same syntactic structures as the classical ones (with particular emphasis on Gentzen-type systems, on which all other systems are based). This approach has two advantages. First, the implementation of the proofs systems described here can be based on existing systems, because no new data structure is used. Second, two sided sequents (like those used in classical logic) can directly represent the consequence relation of *any* given logic, and this after all is what logics are all about.

A decent Gentzen-type system for a logic should allow for direct proofs, which means that it should admit some useful forms of cut elimination and the subformula property. In the rest of this survey we explain the main difficulty in developing classical Gentzen-type systems with these properties for many-valued logics, and discuss the possible methods of overcoming this difficulty. We then illustrate with numerous examples these methods. Our examples include practically all 3-valued and 4-valued logics, as well as Gödel finite-valued logics and some well-known infinite-valued logics which can be taken to be semi-finite. One of our main goals is to demonstrate the strong connection which exists in all cases between semantic completeness and the admissibility of the cut rule. The various proofs we present (most of which are new) establish therefore both (strong) completeness and (strong) cut-elimination at the same time.

2 Gentzen-type systems — a background

In what follows $\mathcal{L}$ is a propositional language, p, q, r denote atomic formulas, $A, B, C, \psi, \varphi, \phi$ denote arbitrary formulas (of $\mathcal{L}$), and Γ, Δ denote finite sets of formulas of $\mathcal{L}$. A sequent of $\mathcal{L}$ has the form $\Gamma \Rightarrow \Delta$. We use s as a variable for sequents and S as a variable for a finite set of sequents. Following tradition, we write Γ, φ and Γ, Δ for $\Gamma \cup \{\varphi\}$ and $\Gamma \cup \Delta$ (respectively).

Definition 1.

1. [51, 50] *A* (Scott) consequence relation *(*scr *for short) for $\mathcal{L}$ is a binary relation $\vdash$ between (finite) sets of formulas of $\mathcal{L}$ that satisfies the following conditions:*

R	reflexivity*:*	$\{\varphi\} \vdash \{\varphi\}$ *for every* φ.
M	monotonicity*:*	*if* $\Gamma \vdash \Delta$ *and* $\Gamma \subseteq \Gamma'$, $\Delta \subseteq \Delta'$ *then* $\Gamma' \vdash \Delta'$.
C	cut*:*	*if* $\Gamma \vdash \psi, \Delta$ *and* $\Gamma', \psi \vdash \Delta'$ *then* $\Gamma, \Gamma' \vdash \Delta, \Delta'$.

2. *An scr $\vdash$ is called* structural *if $\Gamma \vdash \Delta$ implies $\sigma(\Gamma) \vdash \sigma(\Delta)$ for every uniform substitution σ and for every Γ and Δ. $\vdash$ is* consistent *(or* non-trivial*) if there exist non-empty Γ and Δ s.t. $\Gamma \not\vdash \Delta$.*
3. *A* propositional logic *is a pair $\mathbf{L} = \langle \mathcal{L}, \vdash_{\mathbf{L}} \rangle$, where $\mathcal{L}$ is a propositional language and $\vdash_{\mathbf{L}}$ is a structural consistent scr for $\mathcal{L}$.*

Some notes concerning this Definition are in order:

- The notion of an scr is a natural symmetric generalization of the notion of a *Tarskian* consequence relation (tcr), which is defined similarly, but is a relation between sets of formulas and *formulas* (the monotonicity condition applies in it therefore only to the l.h.s.).
- There are exactly four inconsistent finitary scrs in any given language: the one in which $\Gamma \vdash \Delta$ iff Γ and Δ are non-empty; the one in which $\Gamma \vdash \Delta$ iff Γ is non-empty; the one in which $\Gamma \vdash \Delta$ iff Δ is non-empty; and the one in which $\Gamma \vdash \Delta$ for all Γ and Δ. All of them should be considered trivial, and are excluded from our definition of a *logic*.
- A useful generalization of the concept of an scr is obtained by giving up the monotonicity condition. In what follows we shall call a relation which satisfies conditions **R** and **C** of Definition 1 a *generalized scr*. It is easy to see that any generalized scr $\vdash$ can be extended to an ordinary scr $\vdash^*$ by letting $\Gamma \vdash^* \Delta$ iff there exist $\Gamma' \subseteq \Gamma, \Delta' \subseteq \Delta$ such that $\Gamma' \vdash \Delta'$. There are further, less natural, generalizations of the concept of an scr in which the relation is taken to be between multisets, or even sequences, of formulas rather than sets thereof. We shall have no use here for such generalizations. In fact, we shall see that a many-valued logic always directly defines an scr, and any other entailment relation associated with it is reducible to that scr (at least in all cases I know).

A Gentzen-type calculus ([33]) over $\mathcal{L}$ is an axiomatic system which manipulates higher-level constructs called *sequents*, rather than the formulae themselves. There are several variants of what constitutes a sequent. Here we shall always take it to be a construct of the form $\Gamma \Rightarrow \Delta$, where Γ, Δ are finite *sets* of formulae of $\mathcal{L}$ and $\Rightarrow$ is a new symbol, not occurring in $\mathcal{L}$. In other variants Γ, Δ may be either multisets or sequences of formulae. There is no real difference, if we assume the following *standard structural rules* (permutation, contraction, and expansion, respectively):

$$\frac{\Gamma_1, \varphi, \psi, \Gamma_2 \Rightarrow \Delta}{\Gamma_1, \psi, \varphi, \Gamma_2 \Rightarrow \Delta} \qquad \frac{\Gamma \Rightarrow \Delta_1, \varphi, \psi, \Delta_2}{\Gamma \Rightarrow \Delta_1, \psi, \varphi, \Delta_2}$$

$$\frac{\Gamma_1, \varphi, \varphi, \Gamma_2 \Rightarrow \Delta}{\Gamma_1, \varphi, \Gamma_2 \Rightarrow \Delta} \qquad \frac{\Gamma \Rightarrow \Delta_1, \varphi, \varphi, \Delta_2}{\Gamma \Rightarrow \Delta_1, \varphi, \Delta_2}$$

$$\frac{\Gamma_1, \varphi, \Gamma_2 \Rightarrow \Delta}{\Gamma_1, \varphi, \varphi, \Gamma_2 \Rightarrow \Delta} \qquad \frac{\Gamma \Rightarrow \Delta_1, \varphi, \Delta_2}{\Gamma \Rightarrow \Delta_1, \varphi, \varphi, \Delta_2}$$

Definition 2. *A Gentzen-type system* **G** *is called* standard *if its set of axioms includes the standard axioms:* $\psi \Rightarrow \psi$ *(for all* ψ*), and its set of rules includes (the standard structural rules and) the following* Weakening *and* Cut *rules:*

$$\frac{\Gamma \Rightarrow \Delta}{\Gamma', \Gamma \Rightarrow \Delta, \Delta'} \qquad \frac{\Gamma_1 \Rightarrow \Delta_1, \varphi \qquad \varphi, \Gamma_2 \Rightarrow \Delta_2}{\Gamma_1, \Gamma_2 \Rightarrow \Delta_1, \Delta_2}$$

Definition 3. *A rule of a Gentzen-type system* **G** *is called* pure *([13]), or* multiplicative *([37]), if whenever* $\Gamma \Rightarrow \Delta$ *can be inferred by it from* $\Gamma_i \Rightarrow$

Δ_i ($i = 1, \ldots, n$) then, for arbitrary $\Gamma'_1, \ldots, \Gamma'_n, \Delta'_1, \ldots, \Delta'_n$, the sequent $\Gamma, \Gamma'_1, \ldots, \Gamma'_n \Rightarrow \Delta, \Delta'_1, \ldots, \Delta'_n$ can also be inferred by it from $\Gamma_i, \Gamma'_i \Rightarrow \Delta_i, \Delta'_i$ ($i = 1, \ldots, n$)[1]. ***G** is called pure if all its rules are pure.*

What a Gentzen-type system **G** directly defines is a Tarskian consequence relation between *sequents*: A sequent s follows in **G** from a set S of sequents ($S \vdash_{\mathbf{G}} s$) iff s can be derived from the sequents in S and the axioms of **G** using the rules of **G**. Usually, however, a Gentzen-type system **G** is mainly used as a tool for investigating scrs between the *formulas* of the underlying language. Any non-trivial standard Gentzen-type system **G** naturally defines in fact an scr (which we shall also denote by $\vdash_{\mathbf{G}}$):

Definition 4.

1. *Let **G** be a standard Gentzen-type system. We say that $\Gamma \vdash_{\mathbf{G}} \Delta$ iff $\Gamma' \Rightarrow \Delta'$ is a theorem of **G** for some $\Gamma' \subseteq \Gamma, \Delta' \subseteq \Delta$. $\vdash_{\mathbf{G}}$ is called the* standard scr defined by **G**.
2. *A Gentzen-type system **G** over a language $\mathcal{L}$ is* sound and complete *for a logic $\mathbf{L} = \langle \mathcal{L}, \vdash_{\mathbf{L}} \rangle$ if $\vdash_{\mathbf{L}} = \vdash_{\mathbf{G}}$. **G** is* weakly sound and complete *for **L** if for all ψ, $\vdash_{\mathbf{L}} \psi$ iff $\Rightarrow \psi$ is provable in **G**.*

Proposition 1. *If **G** is standard then the relation $\vdash_{\mathbf{G}}$ is a finitary scr. Moreover: if Γ and Δ are finite then $\Gamma \vdash_{\mathbf{G}} \Delta$ iff $\vdash_{\mathbf{G}} \Gamma \Rightarrow \Delta$.*

We leave the easy proof for the reader. We note only that the first part of this proposition (but not the second) remains true even if the set of rules of **G** does not include the weakening rule.

Proposition 2. *If a Gentzen-type system **G** is both standard and pure then $\varphi_1, \ldots, \varphi_n \vdash_{\mathbf{G}} \psi_1, \ldots, \psi_m$ iff the empty sequent $\Rightarrow$ follows in **G** from*

$$S = \{(\Rightarrow \varphi_1), \ldots, (\Rightarrow \varphi_n), (\psi_1 \Rightarrow), \ldots, (\psi_m \Rightarrow)\}$$

*This remains true even if **G** is not closed under weakening.*

Proof: If some subsequent of $\varphi_1, \ldots, \varphi_n \Rightarrow \psi_1, \ldots, \psi_m$ is provable in **G** then $\Rightarrow$ can be derived from it and the sequents in S using cuts. For the converse, assume that P is a proof in **G** of $\Rightarrow$ from S. Obtain P^* from P by adding (for each i and j) φ_i to the left hand side of all the sequents in P which depend on $\Rightarrow \varphi_i$, and ψ_j to the right hand side of all the sequents which depend on $\psi_j \Rightarrow$. These changes turn the premises from S which are used in P into axioms, while the conclusion $\Rightarrow$ of P is turned into a subsequent of $\varphi_1, \ldots, \varphi_n \Rightarrow \psi_1, \ldots, \psi_m$. Since all the rules of **G** are pure, all inference steps in P remain valid in P^*. Hence P^* is a valid proof in **G** of its conclusion, and so $\varphi_1, \ldots, \varphi_n \vdash_{\mathbf{G}} \psi_1, \ldots, \psi_m$.

[1] In other words: a rule is pure if it is context-free. In practice, this means that there are no side conditions on how it can be applied.

Note: $\vdash_{\mathbf{G}}$ is no doubt the most natural consequence relation associated with a given Gentzen-type system **G**. However, it is not the only possible or useful one. Thus the identification of a formula ψ with the singleton $\Rightarrow \psi$ (implicitly used already in Definition 4) naturally leads to another important *Tarskian* consequence relation frequently associated with **G**, according to which a formula ψ follows from the set $\{\varphi_1, \ldots, \varphi_n\}$ if $\{(\Rightarrow \varphi_1), \ldots, (\Rightarrow \varphi_n)\} \vdash_{\mathbf{G}} \Rightarrow \psi$.

An ideal Gentzen-type system (of which the usual systems for classical logic provide the principal examples) is a pure, standard system in which every logical rule is an introduction rule for one connective. Moreover: it should introduce exactly one occurrence of that connective in its conclusion, no other occurrence of connectives should be mentioned anywhere else in its formulation, and its side formulas should be immediate subformulas of the principal formula. The next definition formulates this idea in exact terms, and provides a method for describing such rules.

Definition 5. ([8])

1. *A* canonical rule *of arity n is an expression of the form $\{\Pi_i \Rightarrow \Sigma_i\}_{1\leq i\leq m}/C$, where $m \geq 0$, C is either $\diamond(p_1, p_2, \ldots, p_n) \Rightarrow$ or $\Rightarrow \diamond(p_1, p_2, \ldots, p_n)$ for some connective $\diamond$ (of arity n), and for $1 \leq i \leq m$, $\Pi_i \Rightarrow \Sigma_i$ is a clause such that $\Pi_i, \Sigma_i \subseteq \{p_1, p_2, \ldots, p_n\}$.*[2]
2. *An application of a canonical rule of the form:*

$$\{\Pi_i \Rightarrow \Sigma_i\}_{1\leq i\leq m} \ / \ \diamond(p_1, \ldots, p_n) \Rightarrow$$

is any inference step of the form:

$$\frac{\{\Gamma_i, \Pi_i^* \Rightarrow \Delta_i, \Sigma_i^*\}_{1\leq i\leq m}}{\Gamma, \diamond(\psi_1, \ldots, \psi_n) \Rightarrow \Delta}$$

where Π_i^* and Σ_i^* are obtained from Π_i and Σ_i (respectively) by substituting ψ_j for p_j (for $1 \leq j \leq n$), Γ_i, Δ_i are any finite sets of formulas, $\Gamma = \bigcup_{i=1}^{m} \Gamma_i$, and $\Delta = \bigcup_{i=1}^{m} \Delta_i$. An application of a canonical rule for introducing a connective on the right hand side is defined similarly.

Notes:

1. Definition 5 allows the case $m = 0$, when canonical rules reduce to canonical *axioms* of the form $\diamond(p_1, p_2, \ldots, p_n) \Rightarrow$ or $\Rightarrow \diamond(p_1, p_2, \ldots, p_n)$. This amounts to allowing canonical propositional constants.

[2] By a clause we mean a sequent which consists of atomic formulas only. When propositional clauses are written in this way, resolution and cut amount to the same thing.

2. We have used in the last definition the multiplicative form of an application of a rule. In what follows we shall usually prefer the *additive* ([37]) version, in which all premises of a rule have the same side formulas. Thus an application of a canonical rule of the form $\{\Pi_i \Rightarrow \Sigma_i\}_{1\leq i\leq m} \ / \ \diamond(p_1,\ldots,p_n) \Rightarrow$ has in the additive version the form:

$$\frac{\{\Gamma, \Pi_i^* \Rightarrow \Delta, \Sigma_i^*\}_{1\leq i\leq m}}{\Gamma, \diamond(\psi_1,\ldots,\psi_n) \Rightarrow \Delta}$$

where Π_i^* and Σ_i^* are as above. The two versions are easily seen to be equivalent in the framework of standard systems (since both weakening and contraction are available in them).

Example 1. The two usual introduction rules for classical conjunction can be formulated as the two canonical rules:

$$\{(p_1, p_2 \Rightarrow)\}/\ p_1 \wedge p_2 \Rightarrow \qquad \{(\Rightarrow p_1), (\Rightarrow p_2)\}\ /\Rightarrow p_1 \wedge p_2$$

Applications of these rules have the form:

$$\frac{\Gamma, \psi, \phi \Rightarrow \Delta}{\Gamma, \psi \wedge \phi \Rightarrow \Delta} \qquad \frac{\Gamma \Rightarrow \Delta, \psi \quad \Gamma' \Rightarrow \Delta', \phi}{\Gamma, \Gamma' \Rightarrow \Delta, \Delta', \psi \wedge \phi}$$

(for the additive versions of these rules see the system GCPL below).

Definition 6. *A Gentzen-type system is called* canonical *if its axioms include the standard axioms, its rules include the standard structural rules, cut and weakening, and all its other rules and axioms are canonical.*

A canonical system is obviously standard and pure. However, in order for $\vdash_{\mathbf{G}}$ to really be an scr, one further condition should be satisfied:

Definition 7. The coherence condition: *Whenever $S_1/\diamond(p_1,p_2,\ldots,p_n) \Rightarrow$ and $S_2/ \Rightarrow \diamond(p_1,p_2,\ldots,p_n)$ are rules of* **G**, *the set of clauses $S_1 \cup S_2$ is classically inconsistent (and so the empty clause can be derived from it using cuts).*

Theorem 1. ([8]) *If* **G** *is a canonical system then $\vdash_{\mathbf{G}}$ is consistent (and so an scr) iff the coherence condition is satisfied. Moreover: any canonical system which satisfies this condition admits cut-elimination.*

Example 2. The two classical rules for conjunction described in Example 1 form a coherent pair of rules, where $S_1 = \{(p_1,p_2 \Rightarrow)\}$, $S_2 = \{(\Rightarrow p_1), (\Rightarrow p_2)\}$. $S_1 \cup S_2$ is here the inconsistent set $\{(p_1,p_2 \Rightarrow), (\Rightarrow p_1), (\Rightarrow p_2)\}$.

3 A classical example

The classical examples of canonical Gentzen-type systems are those for Classical Propositional Logic (CPL) and its fragments. We present now a well-known canonical Gentzen-type systems for CPL. We follow [33] in using the *additive* form of the rules. The reason is that in the classical case (as well as in most of the systems we discuss later) the additive rules are *invertible* (i.e.: their premises can be derived in the system from their conclusion using the other rules of the system, including cut). This property is very useful, and so in describing standard systems we shall use the additive form of a rule whenever it is invertible.

THE SYSTEM GCPL

Axioms: $A \Rightarrow A$

Structural Rules: Cut, Weakening (and the standard rules)

Logical Rules:

$$(\neg\Rightarrow)\quad \frac{\Gamma \Rightarrow \Delta, A}{\neg A, \Gamma \Rightarrow \Delta} \qquad\qquad \frac{A, \Gamma \Rightarrow \Delta}{\Gamma \Rightarrow \Delta, \neg A}\quad (\Rightarrow\neg)$$

$$(\supset\Rightarrow)\quad \frac{\Gamma \Rightarrow \Delta, A \quad B, \Gamma \Rightarrow \Delta}{A \supset B, \Gamma \Rightarrow \Delta} \qquad\qquad \frac{\Gamma, A \Rightarrow \Delta, B}{\Gamma \Rightarrow \Delta, A \supset B}\quad (\Rightarrow\supset)$$

$$(\wedge\Rightarrow)\quad \frac{\Gamma, A, B \Rightarrow \Delta}{\Gamma, A \wedge B \Rightarrow \Delta} \qquad\qquad \frac{\Gamma \Rightarrow \Delta, A \qquad \Gamma \Rightarrow \Delta, B}{\Gamma \Rightarrow \Delta, A \wedge B}\quad (\Rightarrow\wedge)$$

$$(\vee\Rightarrow)\quad \frac{\Gamma, A \Rightarrow \Delta \qquad \Gamma, B \Rightarrow \Delta}{\Gamma, A \vee B \Rightarrow \Delta} \qquad\qquad \frac{\Gamma \Rightarrow \Delta, A, B}{\Gamma \Rightarrow \Delta, A \vee B}\quad (\Rightarrow\vee)$$

The semantic of CPL is of course the classical, two-valued semantics, with the classical interpretation of the connectives. A valuation v in $\{t, f\}$ which respects the classical operations is a model of sentence φ if $v(\varphi) = t$. The corresponding classical scr is defined by: $\Gamma \vdash_{CPL} \Delta$ if every model of all the sentences in Γ is a model of some sentence in Δ. A sequent $\Gamma \Rightarrow \Delta$ is *classically valid* if $\Gamma \vdash_{CPL} \Delta$. The concepts of truth and consequence can be extended to sequents as follows: A valuation v is a model of a sequent $\Gamma \Rightarrow \Delta$ if $v(\varphi) = f$ for some $\varphi \in \Gamma$, or $v(\varphi) = t$ for some $\varphi \in \Delta$. A sequent s

classically follows from a set S of sequents if every model of all the sequents in S is also a model of s. It can easily be seen then that a sequent is classically valid iff it classically follows from $\emptyset$.

It is not difficult to show that $GCPL$ is strongly sound for the classical semantics described above, i.e.: if $S \vdash_{GCPL} s$ than the sequent s classically follows from S (in particular: if s is provable in GCPL then it is classically valid). The completeness of $GCPL$, on the other hand, is one of the two most important theorems concerning this calculus. The other one is Gentzen's celebrated cut-elimination theorem ([33, 57]). We present and simultaneously prove now strengthened forms of both theorems.

Theorem 2. Strong Completeness and Cut-elimination for GCPL: *A sequent s classically follows from $S = \{\Gamma_1 \Rightarrow \Delta_1, \ldots, \Gamma_n \Rightarrow \Delta_n\}$ iff s has a proof in GCPL from S in which* all cuts are done on formulas in $\bigcup_{i=1}^{n} \Gamma_i \cup \bigcup_{i=1}^{n} \Delta_i$.

Proof: We start by introducing some definitions. Let $F = \bigcup_{i=1}^{n} \Gamma_i \cup \bigcup_{i=1}^{n} \Delta_i$. An S-cut is an application of the cut rule in which the cut formula belongs to F. An S-proof of a sequent s is a derivation of s in GCPL in which the sequents of S may be used as extra axioms, and all the cuts are S-cuts. Finally, a sequent $\Gamma^* \Rightarrow \Delta^*$ is called *S-saturated* if:

(1) It has no S-proof.
(2) If $\varphi \in F$ then $\varphi \in \Gamma^* \cup \Delta^*$
(3) If $\varphi \supset \psi \in \Delta^*$ then $\varphi \in \Gamma^*$ and $\psi \in \Delta^*$
If $\varphi \supset \psi \in \Gamma^*$ then either $\varphi \in \Delta^*$ or $\psi \in \Gamma^*$
(4) Similar conditions, corresponding to the other rules of GCPL, obtain for the other connectives.

We shall show now that if s does not have an S-proof then there is a model of S which is not a model of s. This will be done in two stages, reflected in the two lemmas below. The theorem easily follows from these two lemmas.

Lemma 1. If $\Gamma \Rightarrow \Delta$ has no S-proof then it can be extended to an S-saturated sequent $\Gamma^* \Rightarrow \Delta^*$.

Proof of Lemma 1: Let $\Gamma^* \Rightarrow \Delta^*$ be a maximal extension of $\Gamma \Rightarrow \Delta$ which does not have an S-proof, and such that $\Gamma^* \cup \Delta^*$ contains only formulas from F^*, where F^* is the set of subformulas of formulas in $F \cup \Gamma \cup \Delta$ (since F^* is finite, such a maximal extension exists). Obviously, if $\varphi \in F^*$ then $\varphi \notin \Gamma^*$ ($\varphi \notin \Delta^*$) iff $\varphi, \Gamma^* \Rightarrow \Delta^*$ ($\Gamma^* \Rightarrow \Delta^*, \varphi$) has an S-proof. It follows that if $\varphi \in F^*$ but $\varphi \notin \Gamma^* \cup \Delta^*$ then both $\Gamma^* \Rightarrow \Delta^*, \varphi$ and $\varphi, \Gamma^* \Rightarrow \Delta^*$ have S-proofs. φ cannot therefore be an element of F in such a case, since otherwise an S-cut on φ would provide an S-proof of $\Gamma^* \Rightarrow \Delta^*$. Hence $\Gamma^* \Rightarrow \Delta^*$ satisfies the second condition in the definition of an S-saturated sequent. We show now that it satisfies the others as well. So assume, e.g., that $\varphi \supset \psi \in \Delta^*$.

In such a case $\Gamma^* \Rightarrow \Delta^*$ can be derived from $\varphi, \Gamma^* \Rightarrow \Delta^*, \psi$ in a single logical inference step. It follows that $\varphi, \Gamma^* \Rightarrow \Delta^*, \psi$ has no S-proof, and so the maximality property of $\Gamma^* \Rightarrow \Delta^*$ implies that $\varphi \in \Gamma^*$ and $\psi \in \Delta^*$. The other conditions are proved similarly.

Lemma 2. If $\Gamma^* \Rightarrow \Delta^*$ is S-saturated then there is a model of S which is not a model of $\Gamma^* \Rightarrow \Delta^*$.

Proof of Lemma 2: Define

$$v(p) = \begin{cases} t & p \in \Gamma^* \\ f & p \notin \Gamma^* \end{cases}$$

We show by induction on the complexity of φ that $v(\varphi) = t$ for every $\varphi \in \Gamma^*$, and $v(\varphi) = f$ for every $\varphi \in \Delta^*$ (and so v is not a model of $\Gamma^* \Rightarrow \Delta^*$). If p is atomic this follows immediately from the definition of v and the fact that if $p \in \Delta^*$ then $p \notin \Gamma^*$ (otherwise $\Gamma^* \Rightarrow \Delta^*$ would have a trivial S-proof). Assume now that $\varphi = \phi \supset \psi$ (other cases are treated similarly). In such a case if $\varphi \in \Delta^*$ then $\phi \in \Gamma^*$ and $\psi \in \Delta^*$. Hence $v(\phi) = t$ and $v(\psi) = f$ by induction hypothesis, and so $v(\varphi) = f$. If, on the other hand, $\varphi \in \Gamma^*$ then either $\phi \in \Delta^*$ or $\psi \in \Gamma^*$. Hence either $v(\phi) = f$ or $v(\psi) = t$, and in either case $v(\varphi) = t$.

It remains to show that v *is* a model of S. Let $\Gamma_i \Rightarrow \Delta_i$ be an element of S. Then $\Gamma_i \cup \Delta_i \subseteq \Gamma^* \cup \Delta^*$ because $\Gamma^* \Rightarrow \Delta^*$ is S-saturated. It cannot be the case that both $\Gamma_i \subseteq \Gamma^*$ and $\Delta_i \subseteq \Delta^*$, because in such a case $\Gamma^* \Rightarrow \Delta^*$ would have a trivial S-proof (using only weakenings). Hence either $\varphi \in \Delta^*$ for some $\varphi \in \Gamma_i$, or $\varphi \in \Gamma^*$ for some $\varphi \in \Delta_i$. In the first case $v(\varphi) = f$ and so v is a model of $\Gamma_i \Rightarrow \Delta_i$; in the second $v(\varphi) = t$, and again v is a model of $\Gamma_i \Rightarrow \Delta_i$.

Theorem 2 has the following immediate corollaries:

Corollary 1. Completeness of GCPL:

1. *A sequent s classically follows from a set S of sequents iff* $S \vdash_{GCPL} s$.
2. $\Gamma \vdash_{CPL} \Delta$ *iff* $\Gamma \vdash_{GCPL} \Delta$.

Corollary 2. Strong Cut-elimination for GCPL ([38, 14]):

1. $\Gamma \Rightarrow \Delta$ *is derivable in GCPL from* $S = \{\Gamma_1 \Rightarrow \Delta_1, \ldots, \Gamma_n \Rightarrow \Delta_n\}$ *iff it has a proof in GCPL from S in which all cuts are done on formulas in* $\bigcup_{i=1}^{n} \Gamma_i \cup \bigcup_{i=1}^{n} \Delta_i$.
2. $\vdash_{GCPL} \Gamma \Rightarrow \Delta$ *iff it has in GCPL a cut-free proof.*

Note: The proof we have just presented demonstrates the semantic approach to the issue of cut-elimination, and it is paradigmatic: all our proofs below for other systems use similar methods and lines of thought (though the details might be more complicated). It should be noted that Gentzen's original method of proof, in contrast, was completely syntactic: it shows how to constructively eliminate cuts by a double induction on the complexity of the cut-formula and on the sum of the lengths (or of the heights) of the proofs of their premises. This approach has its own advantages, but for our present purpose of demonstrating the deep connection between the admissibility of the cut rule and semantic completeness, the semantic approach is superior.

The (strong) cut-elimination theorem for $GCPL$ has many applications, like: decidability, the interpolation theorem, and the subformula property. It is also the basis for the two main proof search methods for CPL (see [14] for more details):

The Tableaux Method: Try to show that $\Gamma \vdash_{CPL} \Delta$ by searching for a cut-free proof of $\Gamma \Rightarrow \Delta$ in $GCPL$. This is done by applying "backwards" the invertible versions of the rules. In particular: one shows that φ is valid by searching for a cut-free proof of $\Rightarrow \varphi$. This gives either such a proof, or an equivalent set of *clauses* (i.e.: sequents consisting of atomic formulas) which can be translated into a conjunctive normal form for φ.

The Resolution Method: Try to show that $\varphi_1, \ldots, \varphi_n \vdash_{CPL} \psi_1, \ldots, \psi_k$ by showing that the empty sequent $\Rightarrow$ can be derived in $GCPL$ from the set:

$$\{(\Rightarrow \varphi_1), , \ldots, (\Rightarrow \varphi_n), (\psi_1 \Rightarrow), \ldots, (\psi_k \Rightarrow)\}$$

(In particular, show that a formula φ is valid by proving that $(\varphi \Rightarrow) \vdash_{GCPL} \Rightarrow$). For this replace (using tableaux) each $(\Rightarrow \varphi_i)$ and $(\psi_j \Rightarrow)$ by an equivalent set of *clauses*. By the *strong* cut-elimination theorem, if $\Rightarrow$ is derivable from the original set of sequents, then it can be derived from the union of the equivalent sets of clauses using only *cuts*.

4 The problem with many-valued logics

We start by defining in precise terms what we mean by "a finite-valued logic", and, more generally, "a many-valued logic".

Definition 8.

1. *A* matrix $\mathcal{M}$ *for a language* $\mathcal{L}$ *is a triple* $\langle M, D, O\rangle$ *such that:*
   (a) M *is a nonempty set (of "truth-values").*
   (b) D *is a proper, nonempty subset of* M *(the "designated values").*[3]
   (c) O *is a set of operations on* M*, so that for each connective of* $\mathcal{L}$ *there is a corresponding operation on* M*.*

[3] In [58] and elsewhere the only condition concerning D is that it should be a subset of M. We exclude here the two extreme cases ($D = M$ and $D = \emptyset$) because the corresponding scrs (as defined below) are not consistent.

2. *Let $\mathcal{M} =< \mathcal{M}, \mathcal{D}, \mathcal{O} >$ be a matrix for $\mathcal{L}$. A function v from the set of formulas of $\mathcal{L}$ into M is called a valuation in $\mathcal{M}$ if it respects the operations in O. A valuation v is an $\mathcal{M}$-model of a formula φ of $\mathcal{L}$ if $v(\varphi) \in D$. v is an $\mathcal{M}$-model of a set T of formulas if it is an $\mathcal{M}$-model of each element of T.*
3. *Let $\mathcal{M}$ be a matrix for $\mathcal{L}$. $\vdash_{\mathcal{M}}$, the consequence relation induced by $\mathcal{M}$, is defined by: $\Gamma \vdash_{\mathcal{M}} \Delta$ iff every model of Γ is a model of some formula in Δ.*
4. *A sequent $\Gamma \Rightarrow \Delta$ is $\mathcal{M}$-*valid *if $\Gamma \vdash_{\mathcal{M}} \Delta$.*
5. *Let $\mathcal{M}$ be a matrix for $\mathcal{L}$. A valuation v is an $\mathcal{M}$-model of a sequent $\Gamma \Rightarrow \Delta$ if $v(\varphi) \notin D$ for some $\varphi \in \Gamma$, or $v(\varphi) \in D$ for some $\varphi \in \Delta$. A sequent s follows in $\mathcal{M}$ from a set S of sequents if every $\mathcal{M}$-model of all the sequents in S is also an $\mathcal{M}$-model of s.*

It is easy to see that $\vdash_{\mathcal{M}}$ is indeed a consequence relation (this is true in fact for any relation which is based on some notion of a "model" in a way similar to that of $\vdash_{\mathcal{M}}$), that $\langle \mathcal{L}, \vdash_{\mathcal{M}} \rangle$ is a logic (see Definition 1), and that again a sequent is $\mathcal{M}$-valid iff it follows in $\mathcal{M}$ from $\emptyset$. Obviously, the various parts of the last Definition are all straightforward generalizations of the corresponding classical notions.

Definition 9.
Let $n \geq 2$ be a natural number. A logic $\mathbf{L} = \langle \mathcal{L}, \vdash_{\mathbf{L}} \rangle$ is called n-valued if there exists a matrix $\mathcal{M}$ for $\mathcal{L}$ such that M has exactly n elements, and $\vdash_{\mathcal{M}} = \vdash_{\mathbf{L}}$. $\mathbf{L}$ is called weakly *n-valued if there exists an n-valued matrix $\mathcal{M}$ for $\mathcal{L}$ such that for every sentence φ, $\vdash_{\mathcal{M}} \varphi$ iff $\vdash_{\mathbf{L}} \varphi$.*

Notes:

1. A matrix $\mathcal{M}$ is frequently used for defining entailment relations (which are usually not scrs) other than $\vdash_{\mathcal{M}}$. In all cases we know the definitions of these alternative consequence relations can be reduced to the validity of certain formulas in $\mathcal{M}$ (usually connected to what is taken to be the "official" implication connective of the logic), and so to $\vdash_{\mathcal{M}}$. Thus the "consequence relation" usually associated with Łukasiewicz 3-valued logic can be characterized as follows: $\varphi_1, \ldots, \varphi_n \vdash \psi$ iff the formula $\varphi_1 \rightarrow (\varphi_2 \rightarrow (\ldots (\varphi_n \rightarrow \psi) \ldots))$ is valid in Łukasiewicz 3-valued matrix. Similar definitions are used for many other finite-valued "logics". Therefore by finding an appropriate Gentzen-type system which is sound and complete for $\vdash_{\mathcal{M}}$ (or at least weakly complete for it) we almost always solve also the problem of Gentzenizing other logics which are based on $\mathcal{M}$.
2. In the literature the term "many-valued logic" usually has a broader sense than the notion of a "finite-valued logic" which we have just defined. In particular: it includes some logics which do not have a finite characteristic matrix. However, the scope of this notion should be restricted somehow, since if we allow arbitrary characteristic matrices then *every* logic (closed

under substitution) would become "many-valued" by a famous theorem of Łos and Suszko ([44, 58]). Our next definition demarcates the class of "many-valued" logics which are dealt with in this paper:

Definition 10. *A logic* $\mathbf{L} = \langle \mathcal{L}, \vdash_{\mathbf{L}} \rangle$ *is called (weakly)* many-valued *if there exists a matrix* $\mathcal{M}$ *for L such that* $\vdash_{\mathcal{M}} = \vdash_{\mathbf{L}}$ *(*$\vdash_{\mathcal{M}} \varphi$ *iff* $\vdash_{\mathbf{L}} \varphi$ *for every sentence* φ*), and for every finite* Γ *and* Δ *(for every sentence* φ*) there is a finite submatrix* $\mathcal{M}^*$ *of* $\mathcal{M}$ *such that* $\Gamma \vdash_{\mathbf{L}} \Delta$ *(*$\vdash_{\mathbf{L}} \varphi$*) iff* $\Gamma \vdash_{\mathcal{M}^*} \Delta$ *(*$\vdash_{\mathcal{M}^*} \varphi$*).*

The following theorem is the main obstacle to providing decent Gentzen-type calculi for many-valued logics:

Theorem 3. ([8]) *Let* **G** *be a consistent canonical calculus. Then either* **G** *defines a logic which is a fragment of classical logic, or it is not sound and complete for any many-valued logic.*

It follows that a Gentzen-type calculus for a given many-valued logic should use at least one of the following:

- Noncanonical rules and/or axioms which are neither standard nor canonical
- Impure rules (i.e.: rules with side conditions on their applications)
- A nonstandard set of structural rules

In the following sections we shall see examples of all these alternatives.

5 Three-valued logics

We start with the simplest type of non-classical many-valued logics: the three-valued ones. We assume (w.l.o.g.) that those three values are t, f and I, where t is designated and f is not. The 3-valued logics are accordingly divided into two classes: those in which I is designated (i.e. $D = \{t, I\}$), and those in which it is not (i.e. $D = \{t\}$). The logics inside each class differ only with respect to the expressive power of their languages. Hence a logic the language of which includes a functionally complete sets of 3-valued operations contains all other logics in its class. Essentially there are therefore just two 3-valued logics, and all the rest are just fragments of them. Our main strategy in what follows is therefore to select for each of the two cases an appropriate functionally complete set of connectives, and then to find a sound and complete set of rules for that set. Under one minimal requirement from the language (that it includes the most standard three-valued negation), this will allow us to find an adequate set of rules for *every* connective, and so for every 3-valued logic (having this negation).

A crucial guiding line in choosing in each case an appropriate set of connectives is to use, as far as possible, connectives which are obvious counterparts of the common classical connectives. Moreover: we want our systems to closely resemble GCPL (so that their use and implementation would require almost no new efforts). We also want to employ connectives which are actually used in current applications of 3-valued logics. Now the most famous 3-valued logics (including, e.g., Łukasiewicz' Ł$_3$ ([45])) and Kleene's K$_3$ ([43])) employ generalizations of the classical connectives $\neg$, $\vee$ and $\wedge$, in which $\vee$, $\wedge$ are interpreted as the *max* and *min* operations (respectively) according to the order $f \leq I \leq t$, while the interpretation of $\neg$ is given by:

$$\neg t = f \qquad \neg f = t \qquad \neg I = I$$

The above three connectives are not expressive enough for defining (in either of the two cases) an implication connective for which the two classical rules (or just both MP and the deduction theorem) are valid. For this we use the following general construction from [2]:

Definition 11. *Let $\mathcal{M} = \langle M, D, O\rangle$ be a matrix such that $t \in D$. The* natural implication operation *of $\mathcal{M}$ is defined by:*

$$a \supset b = \begin{cases} b & \text{if } a \in D \\ t & \text{if } a \notin D \end{cases}$$

Unlike in the two-valued case, the resulting set of connectives $\{\neg, \vee, \wedge, \supset\}$ is not functionally complete yet (in both cases). In [16] it is proved, however, that by adding the propositional constants f and I (interpreted as the corresponding truth values), one does get a functionally complete set of connectives. In what follows we shall use therefore the set $\{\neg, \vee, \wedge, \supset, f, I\}$ as our basic set of connectives. Note that among the connectives of this set only the propositional constant I is not a counterpart of a standard classical operation and is peculiar to three-valued logics (*any* functionally complete set of 3-valued connectives should contain at least one connective of this sort, of course).

Notes:

1. Unlike the other connectives, the interpretation of $\supset$ depends on the choice of D. Hence we use here in fact two different implications: $\supset_{\{t\}}$ (in the case $D = \{t\}$), and $\supset_{\{t,I\}}$ (in the case $D = \{t, I\}$). The two implications are however definable in terms of each other using negation and the propositional constant f. In fact, $a \supset_{\{t\}} b = \neg(\neg a \supset_{\{t,I\}} f) \vee b$ and similarly $a \supset_{\{t,I\}} b = \neg(\neg a \supset_{\{t\}} f) \vee b$. In what follows we shall usually use just $\supset$ for both, relying on the context for determining which one we have in mind.
2. The connective $\supset_{\{t\}}$ was originally introduced by Słupecki in [54]. It was independently reintroduced in [46, 60, 49] and [12] (see also [20]). The language $\{\neg, \vee, \wedge, \supset_{\{t\}}\}$ is equivalent ([12]) to that used in the logic LPF

of the VDM project ([42]), as well as to the language of Łukasiewicz 3-valued logic $Ł_3$ ([45])).[4]

3. The connective $\supset_{\{t,I\}}$ was first introduced in [22, 21]. It was independently introduced also in [10]. The language $\{\neg, \vee, \wedge, \supset_{\{t,I\}}\}$ is equivalent to that used in the standard 3-valued paraconsistent logic J_3 ([24, 10, 48, 28]. In [12], it is called *Pac*), as well as to that used in the semi-relevant system RM_3 ([4, 5, 26]. See also [10, 12]).[5]

Definition 12.

1. *$\mathcal{M}_3^{\{t\}}$ is the 3-valued matrix for the language $\{\neg, \vee, \wedge, \supset, f, I\}$ in which $D = \{t\}$ and the interpretation of $\supset$ is $\supset_{\{t\}}$.*
2. *$\mathcal{M}_3^{\{t,I\}}$ is the 3-valued matrix for the language $\{\neg, \vee, \wedge, \supset, f, I\}$ in which $D = \{t, I\}$ and the interpretation of $\supset$ is $\supset_{\{t,I\}}$.*

5.1 The use of noncanonical rules and axioms

The most important feature of canonical rules is that they introduce exactly one new occurrence of a connective at a time. Most of the Gentzen-type systems for many-valued logics give up this feature by allowing rules which introduce two occurrences of connectives at the same time. To see how to satisfactorily do it in the present case we check first what rules of GCPL remain valid according to their 3-valued interpretations. It can easily be seen that this is the case with the classical rules for $\vee, \wedge$, and $\supset$. This is true for both $\mathcal{M}_3^{\{t\}}$ and $\mathcal{M}_3^{\{t,I\}}$. The situation with negation is different: in both matrices one of the two rules for $\neg$ is valid, while the other is not. In $\mathcal{M}_3^{\{t\}}$ the rule $(\neg \Rightarrow)$ is valid while $(\Rightarrow \neg)$ is not, and the opposite is true in $\mathcal{M}_3^{\{t,I\}}$. It follows that $\neg$ is the connective which needs noncanonical rules. It seems that the best way to handle negation is to replace, first of all, its two classical rules with standard rules for the *combination* of negation with the classical connectives of the language. This is what is done in the following basic system (from [12]):

THE SYSTEM *GBS*: This is the systems obtained from *GCPL* by deleting the two rules for negation and adding instead the following rules and axioms:

[4] It is in fact the language of all 3-valued operations which are *classically closed* (See [16] for further details and references).

[5] It is the language of all 3-valued operations which are classically closed and *free* ([16]).

$$(\neg\neg \Rightarrow) \quad \frac{A, \Gamma \Rightarrow \Delta}{\neg\neg A, \Gamma \Rightarrow \Delta} \qquad \frac{\Gamma \Rightarrow \Delta, A}{\Gamma \Rightarrow \Delta, \neg\neg A} \quad (\Rightarrow \neg\neg)$$

$$(\neg \supset \Rightarrow) \quad \frac{A, \neg B, \Gamma \Rightarrow \Delta}{\neg(A \supset B), \Gamma \Rightarrow \Delta} \qquad \frac{\Gamma \Rightarrow \Delta, A \qquad \Gamma \Rightarrow, \neg B}{\Gamma \Rightarrow \Delta, \neg(A \supset B)} \quad (\Rightarrow \neg \supset)$$

$$(\neg\vee \Rightarrow) \quad \frac{\Gamma, \neg A, \neg B \Rightarrow \Delta}{\Gamma, \neg(A \vee B) \Rightarrow \Delta} \qquad \frac{\Gamma \Rightarrow \Delta, \neg A \qquad \Gamma \Rightarrow \Delta, \neg B}{\Gamma \Rightarrow \Delta, \neg(A \vee B)} \quad (\Rightarrow \neg\vee)$$

$$(\neg\wedge \Rightarrow) \quad \frac{\Gamma, \neg A \Rightarrow \Delta \qquad \Gamma, \neg B \Rightarrow \Delta}{\Gamma, \neg(A \wedge B) \Rightarrow \Delta} \qquad \frac{\Gamma \Rightarrow \Delta, \neg A, \neg B}{\Gamma \Rightarrow \Delta, \neg(A \wedge B)} \quad (\Rightarrow \neg\wedge)$$

$$(f \Rightarrow) \quad f \Rightarrow \qquad \Rightarrow \neg f \quad (\Rightarrow \neg f)$$

Theorem 4. Strong Soundness of *GBS: If* $S \vdash_{GBS} s$, *then the sequent* s *follows from* S *in any matrix* $\langle M, D, O \rangle$ *(for a language which includes that of GBS) that satisfies the following conditions:*

1. $t \in D$ *and* $f \notin D$
2. *the interpretation of the propositional constant* f *is the truth value* f, *and the interpretation of* $\supset$ *is like in Definition 11.*
3. $a \vee b \in D$ *iff* $a \in D$ *or* $b \in D$
4. $a \wedge b \in D$ *iff* $a \in D$ *and* $b \in D$
5. *The operation* $\neg$ *is an extension of the classical negation which satisfies the De Morgan laws as well as the double negation law, i.e.:*

$$\neg t = f,\ \neg f = t,\ \neg\neg a = a,\ \neg(a \vee b) = \neg a \wedge \neg b,\ \neg(a \wedge b) = \neg a \vee \neg b$$

Proof: We show here the validity of the $(\neg \supset\Rightarrow)$ rule, leaving the other cases for the reader. So assume that v is a model of $A, \neg B, \Gamma \Rightarrow \Delta$. We show that it is also a model of $\neg(A \supset B), \Gamma \Rightarrow \Delta$. This is obvious in case $v(\varphi) \notin D$ for some $\varphi \in \Gamma$ or $v(\varphi) \in D$ for some $\varphi \in \Delta$. Otherwise either $v(A) \notin D$, or $v(A) \in D$ and $v(\neg B) \notin D$. In the first case $v(\neg(A \supset B)) = f$, in the second $v(\neg(A \supset B)) = v(\neg B)$, and so in both $v(\neg(A \supset B)) \notin D$.

It immediately follows from the last theorem that *GBS* is strongly sound for both $\mathcal{M}_3^{\{t\}}$ and $\mathcal{M}_3^{\{t,I\}}$. It is obvious therefore that in order to get corresponding sound and *complete* systems one needs to extend *GBS*.

THE SYSTEM $GM_3^{\{t\}}$: This is *GBS* together with the following axioms:

$$\neg A, A \Rightarrow$$

$$I \Rightarrow \qquad\qquad \neg I \Rightarrow$$

THE SYSTEM $GM_3^{\{t,I\}}$: This is GBS together with the following axioms:

$$\Rightarrow \neg A, A$$

$$\Rightarrow I \qquad\qquad \Rightarrow \neg I$$

Theorem 5. Strong Soundness, Completeness and Cut-elimination for $GM_3^{\{t\}}$ and $GM_3^{\{t,I\}}$: *For $D \in \{\{t\}, \{t, I\}\}$, a sequent $\Gamma \Rightarrow \Delta$ follows in $\mathcal{M}_3^D$ from $S = \{\Gamma_1 \Rightarrow \Delta_1, \ldots, \Gamma_n \Rightarrow \Delta_n\}$ iff $\Gamma \Rightarrow \Delta$ has a proof in GM_3^D from S in which all cuts are done on formulas in $\bigcup_{i=1}^n \Gamma_i \cup \bigcup_{i=1}^n \Delta_i$.*

Proof: We give the proof for $GM_3^{\{t,I\}}$ (the case of $GM_3^{\{t\}}$ is dual).

Since it is easy to check that the extra axioms of $GM_3^{\{t,I\}}$ are valid in $\mathcal{M}_3^{\{t,I\}}$, the strong soundness of $GM_3^{\{t,I\}}$ follows from Theorem 4.

The simultaneous proof of strong completeness and the admissibility of Cut closely follows the proof of Theorem 2. We define the notions of an S-cut, an S-proof, and an S-saturated sequent like there (only in the definition of an S-saturated sequent one should replace, of course, the conditions which corresponds to the classical rules of negation by the conditions induced by the negation rules of GBS. For example: the condition that if $\neg(A \supset B) \in \Gamma^*$ then $A \in \Gamma^*$ and $\neg B \in \Gamma^*$). It remains then to prove the obvious counterparts of lemmas 1 and 2 of that proof. The proof here of Lemma 1 is identical to its proof there. To prove Lemma 2 (i.e.: that if $\Gamma^* \Rightarrow \Delta^*$ is S-saturated, then there is a model of S which is not a model of $\Gamma^* \Rightarrow \Delta^*$) we define:

$$v(p) = \begin{cases} t & \neg p \in \Delta^* \\ I & p \notin \Delta^*, \neg p \notin \Delta^* \\ f & p \in \Delta^* \end{cases}$$

Since $\Rightarrow p, \neg p$ is an axiom of $GM_3^{\{t,I\}}$, and $\Gamma^* \Rightarrow \Delta^*$ is S-saturated, it is impossible that both p and $\neg p$ are elements of Δ^*. Hence v is well defined. We prove now that it has the property that if $\varphi \in \Gamma^*$ then $v(\varphi) \in D$ (which is $\{t, I\}$ in the case we are doing) and if $\varphi \in \Delta^*$ then $v(\varphi) \notin D$. From this the fact that v is model of S which is not a model of $\Gamma^* \Rightarrow \Delta^*$ follows exactly as in the proof of Theorem 2. To show this crucial property of v we use induction on the complexity of φ. The case where φ is a literal (i.e.: an atomic formula or a negation of such a formula) other than $f, \neg f, I$, and $\neg I$ is immediate from the definition of v and the fact that if $\varphi \in \Gamma^*$ then $\varphi \notin \Delta^*$ (since $\Gamma^* \Rightarrow \Delta^*$ is S-saturated). The cases where $\varphi \in \{f, \neg f, I, \neg I\}$ also follow from this fact, using the relevant axioms of $GM_3^{\{t,I\}}$ (for example, if $I \in \Gamma^*$ then trivially $v(\varphi) \in D$, because always $v(\varphi) = I \in D$, while it cannot be the case that $I \in \Delta^*$, because otherwise $\Gamma^* \Rightarrow \Delta^*$ would have had a trivial S-proof from the axiom $\Rightarrow I$). The induction step is very much like in the classical case, and basically follows from the fact that all the logical rules of $GM_3^{\{t,I\}}$ are semantically invertible [6]. We do the case where $\varphi = \neg(A \supset B)$ as an example. Well, if $\neg(A \supset B) \in \Gamma^*$ then also $A \in \Gamma^*$ and $\neg B \in \Gamma^*$. Hence, by induction hypothesis, both $v(A)$ and $v(\neg B)$ are in D. But if $v(A) \in D$ then $v(\neg(A \supset B)) = v(\neg B)$, and so $v(\neg(A \supset B)) \in D$ as well. Similarly, if $\neg(A \supset B) \in \Delta^*$ then either $A \in \Delta^*$ or $\neg B \in \Delta^*$. Hence either $v(A) \notin D$ or $v(\neg B) \notin D$. If $v(A) \notin D$ then $v(\neg(A \supset B)) = f \notin D$. If not, then $v(\neg(A \supset B)) = v(\neg B)$, and so again $v(\neg(A \supset B)) \notin D$.

Corollary 3. Strong Completeness of $GM_3^{\{t\}}$ and $GM_3^{\{t,I\}}$: *Let D be either $\{t\}$ or $\{t, I\}$.*

1. *A sequent s follows in $\mathcal{M}_3^D$ from a set S of sequents iff $S \vdash_{GM_3^D} s$.*
2. *$\Gamma \vdash_{\mathcal{M}_3^D} \Delta$ iff $\Gamma \vdash_{GM_3^D} \Delta$.*

Corollary 4. Strong Cut-elimination for $GM_3^{\{t\}}$ and $GM_3^{\{t,I\}}$: *Let D be either $\{t\}$ or $\{t, I\}$.*

1. *$\Gamma \Rightarrow \Delta$ is derivable in GM_3^D from $S = \{\Gamma_1 \Rightarrow \Delta_1, \ldots, \Gamma_n \Rightarrow \Delta_n\}$ iff it has a proof in GM_3^D from S in which all cuts are done on formulas in $\bigcup_{i=1}^n \Gamma_i \cup \bigcup_{i=1}^n \Delta_i$.*
2. *$\vdash_{GM_3^D} \Gamma \Rightarrow \Delta$ iff it has in GM_3^D a cut-free proof.*

Notes:

1. From the proof of Theorem 5 it is clear that we may restrict the various axioms of $GM_3^{\{t\}}$ and $GM_3^{\{t,I\}}$ to the case in which they contain only literals.
2. The second parts of each of the last two corollaries were proved in [12]. The first parts are new here.

Theorem 5 and its corollaries have the same applications and consequences in the 3-valued case as in the two-valued one. Examples are the interpolation theorem, and an appropriate version of the subformula property (according to which a proof of a sequent contains only subformulas of this sequent, or negations of proper subformulas of it). It also leads to versions of the tableaux and resolution methods which are very similar to those used in the classical case. The differences are as follows:

1. A "clause" in the present context is a sequent which contains only *literals* on both sides. Since all the rules of the systems above are invertible, every sequent can be reduced to an equivalent finite set of clauses (in this sense) by the corresponding tableaux rules.

[6] This is true of course also for $GM_3^{\{t\}}$, and so the proof of the induction step is *identical* in both systems!

2. A clause $\Gamma \Rightarrow \Delta$ is valid not only when $\Gamma \cap \Delta \neq \emptyset$, but also when it contains other axioms of the corresponding system (e.g.: if for some atomic p, $\{p, \neg p\} \subseteq \Gamma$ (if $D = \{t\}$), or $\{p, \neg p\} \subseteq \Delta$ (if $D = \{t, I\}$)).
3. For proving $\varphi_1, \dots, \varphi_n \vdash \psi_1, \dots, \psi_k$ using resolution, the set of clauses obtained from the set $\{(\Rightarrow \varphi_1),, \dots, (\Rightarrow \varphi_n), (\psi_1 \Rightarrow), \dots, (\psi_k \Rightarrow)\}$ should be extended with all nonstandard axioms of the corresponding system which contain atomic formulas occurring in the original sequent (for example: sequents of the form $p, \neg p \Rightarrow$ (in case $D = \{t\}$) or $\Rightarrow p, \neg p$ (in case $D = \{t, I\}$).

Note: In the tableaux method one should basically consider 4 types of signed formulas: $\mathbf{T}\varphi$, $\mathbf{T}\neg\varphi$, $\mathbf{F}\varphi$, and $\mathbf{F}\neg\varphi$. It might be more convenient to use instead four different signs: $\mathbf{T}$, $\mathbf{T_N}$, $\mathbf{F}$ and $\mathbf{F_N}$. The resulting rules become very similar to the classical ones (but $\mathbf{T}\neg\varphi$, e.g., is reduced to $\mathbf{F_N}\varphi$ rather than to $\mathbf{F}\varphi$). There are then also several ways to close a branch, each closely related to the classical case (except those in which I is involved). For example: in the language without the propositional constants a branch is closed iff for some φ, it contains either $\mathbf{T}\varphi$ and $\mathbf{F}\varphi$, or $\mathbf{T_N}\varphi$ and $\mathbf{F_N}\varphi$, or (in case $D = \{t\}$) $\mathbf{T}\varphi$ and $\mathbf{F_N}\varphi$, or (in case $D = \{t, I\}$) $\mathbf{T_N}\varphi$ and $\mathbf{F}\varphi$.

5.2 Systems for Łukasiewicz logic Ł$_3$ and for RM_3

Since the sets of connectives used in the complete systems above are functionally complete, we can use the rules of these systems as a basis for finding an adequate set of rules for *every* connective (and choice of D), and so for every 3-valued logic (provided its language includes $\neg$). As an example, we show how to handle Łukasiewicz 3-valued logic. This logic is in fact equivalent ([16]) to the logic of $\mathcal{M}_3^{\{t\}}$ in the language of $\neg, \vee, \wedge, \supset$. However, instead of $\supset$ another connective is taken as primitive (and as the official "implication" connective of the logic): Łukasiewicz 3-valued implication $\rightarrow_L$ (or just $\rightarrow$, when no confusion may arise). Giving a decent Gentzen-type system for Łukasiewicz 3-valued logic amounts therefore to providing an appropriate set of rules for his implication. Now $\varphi \rightarrow \psi$ is equivalent (in the strong sense of always having the same truth-value) to the formula $(\varphi \supset_{\{t\}} \psi) \wedge (\neg\psi \supset_{\{t\}} \neg\varphi)$. This leads to the following four rules for $\rightarrow$:

$$(\rightarrow\Rightarrow) \quad \frac{\Gamma, \neg\psi \Rightarrow \Delta \quad \Gamma, \varphi \Rightarrow \Delta \quad \Gamma \Rightarrow \psi, \neg\varphi, \Delta}{\Gamma, \psi \rightarrow \varphi \Rightarrow \Delta}$$

$$(\Rightarrow\rightarrow) \quad \frac{\Gamma, \psi \Rightarrow \varphi, \Delta \quad \Gamma, \neg\varphi \Rightarrow \neg\psi, \Delta}{\Gamma \Rightarrow \psi \rightarrow \varphi, \Delta}$$

$$(\neg\rightarrow\Rightarrow) \quad \frac{\Gamma, \psi, \neg\varphi \Rightarrow \Delta}{\Gamma, \neg(\psi \rightarrow \varphi) \Rightarrow \Delta}$$

$$(\Rightarrow\neg\rightarrow) \quad \frac{\Gamma \Rightarrow \psi, \Delta \quad \Gamma \Rightarrow \neg\varphi, \Delta}{\Gamma \Rightarrow \neg(\psi \rightarrow \varphi), \Delta}$$

Let us explain, as an example, how the first (and most complicated) rule in this list is obtained. Well, the sequent $\Gamma, \psi \rightarrow \varphi \Rightarrow \Delta$ is equivalent to the sequent $\Gamma, (\psi \supset \varphi) \wedge (\neg\varphi \supset \neg\psi) \Rightarrow \Delta$. Using the invertibility of the rules of $GM_3^{\{t\}}$ (which can easily be established using cuts or the strong completeness of $GM_3^{\{t\}}$) we find that this sequent is equivalent to the following set of sequents:

$$\{(\varphi, \neg\psi, \Gamma \Rightarrow \Delta), \quad (\varphi, \Gamma \Rightarrow \Delta, \neg\varphi), \quad (\neg\psi, \Gamma \Rightarrow \Delta, \psi), \quad (\Gamma \Rightarrow \Delta, \psi, \neg\varphi)\}$$

Now the second sequent in this list, $\varphi, \Gamma \Rightarrow \Delta, \neg\varphi$, can be replaced by the simpler $\varphi, \Gamma \Rightarrow \Delta$, to which it is equivalent in $GM_3^{\{t\}}$ (the first can be derived from the second using weakening, the second can be derived from the first by using a cut with the axiom $\neg\varphi, \varphi \Rightarrow$). Similarly, the third sequent can be replaced by the equivalent $\neg\psi, \Gamma \Rightarrow \Delta$. After these replacements the first sequent in the list becomes superfluous (since it can be derived from either of the two new sequents using weakening), and can be deleted. We are left with the set of premises used in the rule $(\rightarrow\Rightarrow)$ above.

Note: It should be emphasized again, that the sequents of the Gentzen-type system we have just presented for Łukasiewicz 3-valued logic do *not* reflect the "consequence relation" which is induced by using $\rightarrow$ as the official *implication*. That relation is not even a Tarskian consequence relation, since it is a relation between *multisets* of formulas and formulas, not between sets of formulas and formulas. Indeed, the contraction rule fails for this relation. However, a sentence ψ follows in it from a multiset of sentences $\varphi_1, \ldots, \varphi_n$ iff the singleton sequent $\Rightarrow \varphi_1 \rightarrow (\varphi_2 \rightarrow (\ldots(\varphi_n \rightarrow \psi)\ldots))$ is provable in the system we have just described.

The counterpart of Łukasiewicz 3-valued logic in the case where $D = \{t, I\}$ is the semi-relevant system RM_3 ([4, 5, 26]), which is the strongest logic in the family of the relevant logics. Its language is equivalent in its expressive power to that of $\{\neg, \vee, \wedge, \supset_{\{t,I\}}\}$ (sometimes also f is added), but again instead of $\supset$ another connective is taken as primitive (and as the official "implication" connective of the logic). This time this is Sobociński's implication $\rightarrow_S$ from [55]. Again $\varphi \rightarrow_S \psi$ is strongly equivalent ([10]) to $(\varphi \supset_{\{t,I\}} \psi) \wedge (\neg\psi \supset_{\{t,I\}} \neg\varphi)$, and again this leads to a set of rules for $\rightarrow_S$ which is very similar to that for Łukasiewicz implication. The only difference is that the rule $(\rightarrow\Rightarrow)$ above should be replaced by the following dual (all other rules remain the same):

$$\frac{\Gamma \Rightarrow \Delta, \psi \quad \Gamma \Rightarrow \Delta, \neg\varphi \quad \Gamma, \neg\psi, \varphi \Rightarrow \Delta}{\Gamma, \psi \rightarrow \varphi \Rightarrow \Delta}$$

5.3 A semi-canonical system for Sobociński logic

As we have emphasized above, there can be no canonical system which is sound and complete for a given finite-valued logic (unless it is sound and complete for some two-valued matrix). Nevertheless, in this subsection we present GRM_m, a Gentzen-type system for the important 3-valued logic of Sobociński, which is almost canonical in the sense that all its logical rules are canonical, its axioms are standard, and in addition to these rules and axioms it has only purely structural rules (the only reason we call it "semi-canonical" is that instead of the full weakening rule it has a certain weaker version). One should note, however, that GRM_m is only *weakly* complete (see Definition 4) for the ordinary scr defined by Sobociński's 3-valued matrix, although it is strongly complete for a certain generalized scr (see the third note after Definition 1) which is based on this matrix and is described below.

Sobociński 3-valued matrix was first introduced and weakly axiomatized in [55]. It was later shown ([4]) to be equivalent to RM_m, the purely multiplicative (or "intensional") fragment of Dunn-McCall semi-relevant system RM. The idea behind this logic is that the content of certain sentences may be taken as totally insignificant for certain purposes. When this is the case no real truth-value should be attached to such sentences. This situation is represented by assigning to them the "truth-value" I (meaning "Insignificant", or "Irrelevant", or "don't care"), which should not be taken as a real truth-value. Now the the main principles that guide the semantics of the logic are that a complex is significant iff at least one of its components is significant, and that only significants components should be taken into account in computing truth values and in deductions. Thus a conjunction of sentences should be taken as true iff it has some significant conjunct, and all its *significant* conjuncts are true. It should be taken as false iff at least one of its conjuncts is false (such a conjunct is then significant by definition). This leads to the following interpretation for a binary conjunction, which we denote here by $\otimes$: [7]

$$a \otimes b = \begin{cases} f & \text{if } a = f \text{ or } b = f \\ I & \text{if } a = I \text{ and } b = I \\ t & \text{otherwise} \end{cases}$$

The principles described above imply that a 3-valued n-ary connective $\diamond$ ($n \geq 1$) may be allowed in the language iff it satisfies the following condition:

$$\diamond(a_1, \ldots, a_n) = I \quad \text{iff} \quad a_1 = a_2 = \ldots = a_n = I$$

It can be proved that a connective satisfies this condition iff it is definable in the language of $\{\neg, \otimes\}$ (where $\neg$ is the standard 3-valued negation). We take therefore these connectives as the primitive connectives of the logic (instead of $\otimes$ we could have taken the implication $\to_S$ described at the end of the

[7] This is the notation which was used by Girard in [37] for his multiplicative conjunction, which is characterized by the multiplicative versions the classical canonical rules for conjunction (Example 1 above, and the system below). Nowadays this is the common notation for multiplicative conjunction in all substructural logics, including relevant logics.

previous subsection, since it is easy to see that $\otimes$ and $\rightarrow_S$ are definable from each other using $\neg$ exactly like the conjunction and implication of CPL).

The principles described above determine also the logic's concept of validity, which takes into account only valuations which are *relevant* to the formulas or sequents under consideration:

Definition 13.

1. *A valuation v is* relevant *to a formula φ iff it assigns a real truth value (either t or f) to at least one atomic formula which occurs in φ (this happens iff $v(\varphi) \in \{t, f\}$). v is* relevant *to a sequent $\Gamma \Rightarrow \Delta$ iff it is relevant to at least one formula in $\Gamma \cup \Delta$ (iff $v(p) \in \{t, f\}$ for some atomic formula which occurs in $\Gamma \Rightarrow \Delta$).*
2. *A sequent $\Gamma \Rightarrow \Delta$ is called RM_m-*valid *iff for every v which is relevant to it we have that $v(\varphi) = f$ for some $\varphi \in \Gamma$ or $v(\varphi) = t$ for some $\varphi \in \Delta$.*

It can easily be seen that a sequent $\Gamma \Rightarrow \Delta$ is RM_m-valid if *every* valuation v *agrees* with it, when we say that v agrees with $\Gamma \Rightarrow \Delta$ if either $v(\varphi) = f$ for some $\varphi \in \Gamma$, or $v(\varphi) = t$ for some $\varphi \in \Delta$, or $v(\varphi) = I$ for all $\varphi \in \Gamma \cup \Delta$ [8]. It follows that a sequent of the form $\Rightarrow \varphi$ is RM_m-valid iff $v(\varphi) \in \{t, I\}$ for every valuation v, i.e.: iff φ is valid in $\mathcal{M}_3^{\{t,I\}}$. Hence a Gentzen-type system which proves exactly the RM_m-valid sequents would be weakly complete for the multiplicative (i.e.: the $\{\neg, \rightarrow_S\}$-) fragment of RM_3. We present now such a Gentzen-type system.

THE SYSTEM GRM_m

Axioms: $A \Rightarrow A$

Structural Rules: Cut, and the following *Mingle* [9] rule:

$$\frac{\Gamma_1 \Rightarrow \Delta_1 \qquad \Gamma_2 \Rightarrow \Delta_2}{\Gamma_1, \Gamma_2 \Rightarrow \Delta_1, \Delta_2}$$

Logical Rules:

$$(\neg \Rightarrow)\ \frac{\Gamma \Rightarrow \Delta, A}{\neg A, \Gamma \Rightarrow \Delta} \qquad\qquad \frac{A, \Gamma \Rightarrow \Delta}{\Gamma \Rightarrow \Delta, \neg A}\ (\Rightarrow \neg)$$

$$(\otimes \Rightarrow)\ \frac{\Gamma, A, B \Rightarrow \Delta}{\Gamma, A \otimes B \Rightarrow \Delta} \quad \frac{\Gamma_1 \Rightarrow \Delta_1, A \qquad \Gamma_2 \Rightarrow \Delta_2, B}{\Gamma_1, \Gamma_2 \Rightarrow \Delta_1, \Delta_2, A \otimes B}\ (\Rightarrow \otimes)$$

[8] It is easy to see that by defining $\Gamma \vdash_{Sob} \Delta$ iff this condition obtains for all v we get a generalized scr $\vdash_{Sob}$.

[9] Also called "mix" in the literature on Linear Logic ([37]).

Notes:

1. Recall that Γ, Δ etc. are here finite *sets* of formulas, and that we should in principle have written $\{A\}$ rather than just A.
2. As in GCPL, it is easy to see that it suffices to take as axioms only sequents of the form $p \Rightarrow p$ where p is atomic. In the rest of this section we assume that this is the case.

Theorem 6. Soundness of GRM_m: *If* $\vdash_{GRM_m} s$ *then* s *is* RM_m*-valid.*

Proof: From the definition of validity it follows that any axiom of GRM_m is valid, because given a valuation v, the three possibilities ($v(p) = f$, $v(p) = t$, and $v(p) = I$) exactly match the three cases in which v agrees with $p \Rightarrow p$. We next prove for every rule of GRM_m that if a valuation v agrees with its premises then it agrees also with its conclusion. We do here the two more difficult cases, leaving the other cases for the reader.

- *The case of Cut*: Suppose that v agrees with both $\Gamma_1 \Rightarrow \Delta_1, \varphi$ and $\varphi, \Gamma_2 \Rightarrow \Delta_2$. We show that it also agrees with $\Gamma_1, \Gamma_2 \Rightarrow \Delta_1, \Delta_2$. This is obvious if $v(A) = f$ for some $A \in \Gamma_1, \Gamma_2$ or $v(A) = t$ for some $A \in \Delta_1, \Delta_2$. If this is not the case then $v(\varphi)$ cannot be t (since otherwise v would not agree with $\varphi, \Gamma_2 \Rightarrow \Delta_2$), and it is not f either (since otherwise v would not agree with $\Gamma_1 \Rightarrow \Delta_1, \varphi$). It follows that $v(\varphi) = I$, and the only possibility that remains for v to agree with the two premises is that $v(A) = I$ for all $A \in \Gamma_1, \Gamma_2, \Delta_1, \Delta_2$. But in this case it again agrees with $\Gamma_1, \Gamma_2 \Rightarrow \Delta_1, \Delta_2$.
- *The case of* $(\Rightarrow \otimes)$: Suppose that v agrees with $\Gamma_1 \Rightarrow \Delta_1, \varphi$ as well as with $\Gamma_2 \Rightarrow \Delta_2, \psi$. We show that it also agrees with $\Gamma_1, \Gamma_2 \Rightarrow \Delta_1, \Delta_2, \varphi \otimes \psi$. This is obvious if $v(A) = f$ for some $A \in \Gamma_1, \Gamma_2$ or $v(A) = t$ for some $A \in \Delta_1, \Delta_2$. If this is not the case then there are four possibilities: $v(\varphi) = t$ and $v(\psi) = t$, $v(\varphi) = t$ and $v(A) = I$ for all $A \in \Gamma_2 \cup \Delta_2 \cup \{\psi\}$, $v(\psi) = t$ and $v(A) = I$ for all $A \in \Gamma_1 \cup \Delta_1 \cup \{\varphi\}$, and $v(A) = I$ for all $A \in \Gamma_1 \cup \Gamma_2 \cup \Delta_1 \cup \Delta_2 \cup \{\varphi, \psi\}$. In the first three cases $v(\varphi \otimes \psi) = t$. In the forth $v(A) = I$ for every formula A in $\Gamma_1 \cup \Gamma_2 \cup \Delta_1 \cup \Delta_2 \cup \{\varphi \otimes \psi\}$.

Our next goal is to prove the completeness of GRM_m and the cut-elimination theorem for it. As usual in this paper, this will be done simultaneously.

Notations: For the rest of this section "$\vdash$" and "provable" mean "provable in GRM_m without a cut". $A(X)$ denotes the sets of atomic formulas which occurs in X (X may be a formula or a sequent).

The following Lemma shows that an important special case of weakening is admissible in GRM_m:

Lemma 1: If $\vdash \Gamma \Rightarrow \Delta$ and $A(\varphi) \subseteq A(\Gamma \Rightarrow \Delta)$ then $\vdash \varphi, \Gamma \Rightarrow \Delta$ and $\vdash \Gamma \Rightarrow \Delta, \varphi$.

Proof of Lemma 1: By induction on the complexity of φ. The base case (where φ is atomic) is done by induction on the length of the proof of $\Gamma \Rightarrow \Delta$. The base case of this inner induction uses the special form we use for the axioms of $GRMI_m$, while both the induction step of the inner induction and the induction step of the main one are easy consequence of the fact that all rules of GRM_m are pure (multiplicative).

Definition 14. *Let $\Gamma \Rightarrow \Delta$ be a sequent such that $\not\vdash \Gamma \Rightarrow \Delta$. $\Gamma \Rightarrow \Delta$ is called saturated if it also satisfies the following conditions:*

(i) If $\neg\varphi \in \Gamma$ then $\varphi \in \Delta$
(ii) If $\neg\varphi \in \Delta$ then $\varphi \in \Gamma$
(iii) If $\varphi \otimes \psi \in \Gamma$ then $\varphi \in \Gamma$ and $\psi \in \Gamma$
(iv) If $\varphi \otimes \psi \in \Delta$ and $\not\vdash \Gamma \Rightarrow \Delta, \varphi$ then $\varphi \in \Delta$
(v) If $\varphi \otimes \psi \in \Delta$ and $\not\vdash \Gamma \Rightarrow \Delta, \psi$ then $\psi \in \Delta$.

Lemma 2: If $\not\vdash \Gamma \Rightarrow \Delta$ then there exists a saturated sequent $\Gamma^* \Rightarrow \Delta^*$ such that $\Gamma \subseteq \Gamma^*$, $\Delta \subseteq \Delta^*$, $\not\vdash \Gamma^* \Rightarrow \Delta^*$ and $A(\Gamma^* \Rightarrow \Delta^*) = A(\Gamma \Rightarrow \Delta)$.

Proof of Lemma 2: If $\not\vdash \Gamma \Rightarrow \Delta$ and $\Gamma \Rightarrow \Delta$ is not saturated then it is possible to properly extend $\Gamma \Rightarrow \Delta$ by some of its subformulas without making the new sequent provable (this is obvious and standard if one of the conditions (i)–(iii) is violated by $\Gamma \Rightarrow \Delta$, and trivial in the special cases (iv)–(v)). Since $\Gamma \Rightarrow \Delta$ has only finitely many subformulas, this process must stop with a saturated sequent.

Lemma 3: If $\Gamma \Rightarrow \Delta$ is saturated and $\not\vdash \Gamma \Rightarrow \Delta$ then $\Gamma \Rightarrow \Delta$ has a countermodel (i.e.: a valuation v which does not agree with it).

Proof of Lemma 3: Assume that $\Gamma \Rightarrow \Delta$ has these properties. Define:

$$I(\Gamma \Rightarrow \Delta) = \{p \in A(\Gamma \Rightarrow \Delta) \mid p \in \Gamma \cap \Delta\}$$

$$v(p) = \begin{cases} I & p \in I(\Gamma \Rightarrow \Delta) \\ t & p \in \Gamma, p \notin \Delta \\ f & p \notin \Gamma \end{cases}$$

We show that this v is a countermodel of $\Gamma \Rightarrow \Delta$. For this we first show by induction on the complexity of φ that if $\varphi \in \Gamma$ then $v(\varphi) \neq f$, and if $\varphi \in \Delta$ then $v(\varphi) \neq t$. This is obvious in case φ is atomic. In case $\varphi = \neg\psi$ the claim follows easily from the induction hypothesis and conditions (i)–(ii) from Definition 14. If $\varphi = \psi_1 \otimes \psi_2$ and $\varphi \in \Gamma$ then the claim follows from the induction hypothesis concerning ψ_1 and ψ_2 and condition (iii) of Definition 14. Finally assume that $\varphi = \psi_1 \otimes \psi_2$ and $\varphi \in \Delta$. Had both $\Gamma \Rightarrow \Delta, \psi_1$ and $\Gamma \Rightarrow \Delta, \psi_2$ been provable, so $\Gamma \Rightarrow \Delta$ would have been (since $\varphi \in \Delta$). Hence one of those sequents is unprovable. Assume, e.g., that $\not\vdash \Gamma \Rightarrow \Delta, \psi_1$. Then

$\psi_1 \in \Delta$ by condition (iv) of Definition 14. Hence $v(\psi_1) \neq t$ by induction hypothesis. If $v(\psi_1) = f$ then $v(\varphi) = f \neq t$. Assume, therefore, that $v(\psi_1) = I$. Hence $v(p) = I$ for every $p \in A(\psi_1)$, and so $A(\psi_1) \subseteq I(\Gamma \Rightarrow \Delta)$. Now $\vdash A(\psi_1) \Rightarrow A(\psi_1)$ because of the Mingle rule, and so $\vdash A(\psi_1) \Rightarrow A(\psi_1), \psi_1$ by Lemma 1. It is not possible therefore that $\vdash \Gamma \Rightarrow \Delta, \psi_2$, since otherwise we would have that $\vdash \Gamma, A(\psi_1) \Rightarrow \Delta, A(\psi_1), \psi_1 \otimes \psi_2$, and so that $\vdash \Gamma \Rightarrow \Delta$ (since $\psi_1 \otimes \psi_2 = \varphi \in \Delta$, $A(\psi_1) \subseteq \Gamma$, and $A(\psi_1) \subseteq \Delta$). It follows by condition (v) of Definition 1 that $\psi_2 \in \Delta$, and so $v(\psi_2) \neq t$ by induction hypothesis. Now if $v(\psi_2) = f$ then $v(\varphi) = f \neq t$, while if $v(\psi_2) = I$ then $v(\varphi) = I \otimes I = I \neq t$.

To show that v is a countermodel of $\Gamma \Rightarrow \Delta$ it remains now to eliminate the possibility that $v(\varphi) = I$ for all $\varphi \in \Gamma \cup \Delta$. Well, had this been the case we would have that $v(p) = I$ for all $p \in A(\Gamma \Rightarrow \Delta)$, and so that $A(\Gamma \Rightarrow \Delta) = I(\Gamma \Rightarrow \Delta)$. Since $I(\Gamma \Rightarrow \Delta) \subseteq \Gamma$, $I(\Gamma \Rightarrow \Delta) \subseteq \Delta$, and $\vdash I(\Gamma \Rightarrow \Delta) \Rightarrow I(\Gamma \Rightarrow \Delta)$ (using mingle), Lemma 1 would have implied then that $\vdash \Gamma \Rightarrow \Delta$. A contradiction.

Theorem 7. Completeness and Cut-elimination for GRM_m: *A sequent $\Gamma \Rightarrow \Delta$ in the language of GRM_m is RM_m-valid iff it has a proof in GRM_m without cuts.*

Proof: Assume that $\nvdash \Gamma \Rightarrow \Delta$. By lemma 2 there exists an unprovable saturated sequent $\Gamma^* \Rightarrow \Delta^*$ such that $A(\Gamma^* \Rightarrow \Delta^*) = A(\Gamma \Rightarrow \Delta)$, $\Gamma \subseteq \Gamma^*$, and $\Delta \subseteq \Delta^*$. These last 3 properties are easily seen to entail that every model of $\Gamma \Rightarrow \Delta$ is also a model of $\Gamma^* \Rightarrow \Delta^*$.[10] Hence the countermodel of $\Gamma^* \Rightarrow \Delta^*$ given by lemma 3 is also a countermodel of $\Gamma \Rightarrow \Delta$.

Corollary 5. *(**Completeness of** GRM_m)*

1. *A sequent in the language of GRM_m is RM_m-valid iff it is provable in GRM_m*
2. *A sentence φ in the language of GRM_m is valid in $\mathcal{M}_3^{\{t,I\}}$ iff $\Rightarrow \varphi$ is provable in GRM_m*

Corollary 6. *The system GRM_m admits cut-elimination.*

Note: The last two corollaries were proved (using two unrelated proofs) already in [11] (but were known to relevance logicians much before).

6 Four-valued logics

There are basically 3 four-valued logics, differing according to the number of designated truth-values in their matrices (1, 2, or 3). The most important of them is by far the one in which there are exactly two. We devote most of this section to this useful logic.

[10] Note that the first one is crucial here, since ordinary monotonicity fails.

6.1 Belnap's four-valued logic and its extensions

The methods used above for three-valued logics can be extended with very slight changes to four-valued logics in which there are exactly two designated elements. Let the truth values of these logics be $t, f, \top$, and $\bot$, where t and f are the classical values. According to Belnap's suggestion in [18, 17], $\top$ should represent the truth-value of formulas about which there is inconsistent data (such a formula is "both true and false"), while $\bot$ is the truth-value of formulas on which no data at all is available ("neither true nor false"). This intuition is the basis of what is known as Belnap four-valued logic and of various extensions suggested in the literature [11]. Obviously, according to these interpretations $\top$ is the four-valued counterpart of the 3-valued designated I, while $\bot$ is the counterpart of the 3-valued non-designated I. The corresponding four-valued matrix may therefore be taken as a combination of the two 3-valued matrices. We make our choice of connectives and their interpretations accordingly. In particular: the partial order $\leq_t$ we use for defining conjunction and disjunction is simply the union of the partial orders which are used for this purpose in $\mathcal{M}_3^{\{t\}}$ and $\mathcal{M}_3^{\{t,I\}}$ (where I is replaced, respectively, by $\bot$ and $\top$). $\leq_t$ is defined therefore by: $f \leq_t \top, \bot \leq_t t$.

Definition 15. *Let* $L_4 = \{\neg, \vee, \wedge, \supset, f, \bot, \top\}$.

The matrix $\mathcal{M}_4 = \langle M_4, D_4, O_4 \rangle$ *for* L_4 *is defined as follows:*

- $M_4 = \{t, f, \top, \bot\}$
- $D_4 = \{t, \top\}$
- *The operations in* O_4 *are defined by:*
  1. $\neg t = f, \quad \neg f = t, \quad \neg\top = \top, \quad \neg\bot = \bot$
  2. $\supset$ *is defined like in Definition 11*
  3. $a \vee b = sup_{\leq_t}(a, b), \quad a \wedge b = inf_{\leq_t}(a, b)$
  4. *The propositional constants* $f, \bot$ *and* $\top$ *are interpreted by the corresponding truth values.*

We present now a Gentzen-type system which is strongly sound and complete for $\mathcal{M}_4$, and for which the (strong) cut-elimination theorem obtains. Since L_4 is a functionally complete set of connectives for 4-valued logics ([16]), this again allows us to find a complete, cut-free Gentzen-type system for any 4-valued logic in which $D = \{t, \top\}$.

THE SYSTEM GM_4: This is the systems obtained from GBS by adding to it the following axioms:

$$\bot \Rightarrow \qquad\qquad \neg\bot \Rightarrow$$

$$\Rightarrow \top \qquad\qquad \Rightarrow \neg\top$$

[11] Belnap's structure is nowadays known also as the basic (distributive) *bilattice*, and its logic — as the basic logic of (distributive) bilattices (see [35, 36, 30, 29, 31, 32, 1–3]).

Theorem 8. Strong Soundness, Completeness and Cut-elimination for GM_4: *A sequent s follows in $\mathcal{M}_4$ from $S = \{\Gamma_1 \Rightarrow \Delta_1, \ldots, \Gamma_n \Rightarrow \Delta_n\}$ iff s has a proof in GM_4 from S in which all cuts are done on formulas in $\bigcup_{i=1}^n \Gamma_i \cup \bigcup_{i=1}^n \Delta_i$.*

Proof: The new axioms concerning $\top$ and $\bot$ are obviously valid in $\mathcal{M}_4$. Hence the strong soundness of GM_4 easily follows from Theorem 4.

The simultaneous proof of the strong completeness and of the strong Cut-elimination Theorem closely follows the proofs of Theorems 2 and 5. The main difference is that this time to prove Lemma 2 (that if $\Gamma^* \Rightarrow \Delta^*$ is S-saturated, then there is a model of S which is not a model of $\Gamma^* \Rightarrow \Delta^*$) we define:

$$v(p) = \begin{cases} t & p \in \Gamma^*, \neg p \notin \Gamma^* \quad or \quad p \notin \Delta^*, \neg p \in \Delta^* \\ \bot & p \in \Delta^*, \neg p \in \Delta^* \\ \top & p \in \Gamma^*, \neg p \in \Gamma^* \\ f & otherwise \end{cases}$$

As usual, we next prove that v is well-defined, and that if $\varphi \in \Gamma^*$ then $v(\varphi) \in D_4$, while if $\varphi \in \Delta^*$ then $v(\varphi) \notin D_4$. Details are left for the reader.

Corollary 7. Strong Completeness of GM_4:

1. *A sequent s follows in $\mathcal{M}_4$ from a set S of sequents iff $S \vdash_{GM_4} s$.*
2. *$\Gamma \vdash_{\mathcal{M}_4} \Delta$ iff $\Gamma \vdash_{GM_4} \Delta$.*

Corollary 8. (Strong Cut-elimination for GM_4:

1. *s is derivable in GM_4 from $S = \{\Gamma_1 \Rightarrow \Delta_1, \ldots, \Gamma_n \Rightarrow \Delta_n\}$ iff it has a proof there from S in which all cuts are on formulas in $\bigcup_{i=1}^n \Gamma_i \cup \bigcup_{i=1}^n \Delta_i$.*
2. *$\vdash_{GM_4} \Gamma \Rightarrow \Delta$ iff it has in GM_4 a cut-free proof.*

Notes:

1. Again from the proof of Theorem 8 it is clear that one may restrict the axioms of GM_4 to the case in which they contain only literals.
2. The second parts of each of the last two corollaries were claimed in [12] and proved in [1]. The first parts are new here.
3. The results of this section have the same applications here as in the 3-valued case. In fact, both of the resulting tableaux and resolution methods are simpler here than in the 3-valued case, because the set of axioms is simpler in the present system than in its 3-valued counterparts.

6.2 Other four-valued logics

Except for the class just dealt with, there are two other possible classes of 4-valued logics: those with $D = \{t\}$, and those with $D = \{t, \top, \bot\}$. These two classes are dual to each other: it is easy to see that a sequent $\Gamma \Rightarrow \Delta$ is valid according to one iff $\neg\Delta \Rightarrow \neg\Gamma$ is valid according to the other (where $\neg$ is the connective defined in the previous subsection, and $\neg\{A_1, \ldots, A_n\} = \{\neg A_1, \ldots, \neg A_n\}$). In principle it suffices therefore to find an appropriate system for one of these cases. This can indeed be done by methods which are similar to those used above. We demonstrate this claim for the case $D = \{t\}$ in subsection 7.2 below, using the four-valued Gödel's logic as a basis. We note, however, that at present the known systems for the two other classes are more complicated and less elegant than those in the case $D = \{t, \top\}$.

7 Infinite-valued logics and related logics

Up to now, we have only considered three- or four-valued logics. In this section we want to show that our methods may be applicable to n-valued logics with n having bigger values, and even to infinite-valued logics.

7.1 Gödel-Dummett logics

In [39] Gödel introduced a sequence $\{\mathcal{G}_n\}$ ($n \geq 2$) of n-valued matrices. He used these matrices to show some important properties of intuitionistic logic. An infinite-valued matrix $\mathcal{G}_\omega$ in which all the $\mathcal{G}_n$s can be embedded was later introduced by Dummett in [25]. The logic of $\mathcal{G}_\omega$ was axiomatized in the same paper, and has been known since then as Gödel-Dummett's *LC*. It is probably the most important intermediate logic, which turns up in several places, such as the provability logic of Heyting's Arithmetics ([59]), and relevance logic ([23]). Recently it has again attracted a lot of attention because of its recognition as one of the three most basic fuzzy logics ([41]).

Definition 16. *Let* $L_{LC} = \{\rightarrow, \vee, \wedge, f\}$.

1. *The matrix* $\mathcal{G}_\omega = \langle G_\omega, D_{LC}, O_{LC}\rangle$ *for* L_{LC} *is defined as follows:*
   - $G_\omega = N \cup \{t, f\}$ [12]
   - $D_{LC} = \{t\}$
   - *The operations in* O_{LC} *are defined by:*
     (a) $a \rightarrow b = \begin{cases} t & a \leq b \\ b & a \not\leq b \end{cases}$
     Here $\leq$ *is the usual order on* N *extended by a greatest element* t *and a smallest element* f.
     (b) $a \vee b = max_{\leq}(a, b)$ *and* $a \wedge b = min_{\leq}(a, b)$.
     (c) *The propositional constant* f *is interpreted by the corresponding truth value.*
2. $LC = \langle L_{LC}, \vdash_{\mathcal{G}_\omega}\rangle$
3. *for* $n \geq 2$ *the matrix* $\mathcal{G}_n$ *is defined like* $\mathcal{G}_\omega$, *but its the set of truth values is just* $\{1, ..., n-2\} \cup \{t, f\}$.

[12] Instead of adding f one may identify it with the number 0.

Notes:

1. The matrices we have just defined are not those given by Gödel and Dummett, but their duals. We note also that for the application as a fuzzy logic it is more useful ([41]) to use instead of $N \cup \{t, f\}$ the real interval [0,1], with 1 playing the role of t (and 0 that of f). This makes a difference only when we consider inferences from infinite theories.
2. It is not difficult to see that LC is indeed a many-valued logic according to Definition 10, since $\Gamma \vdash_{\mathcal{G}_\omega} \Delta$ iff $\Gamma \vdash_{\mathcal{G}_k} \Delta$, where k is the number of atomic formulas occurring in $\Gamma \cup \Delta$.

A cut-free Gentzen-type formulation for LC was first given by Sonobe in [56]. His approach was improved in [6] and [27]. All those systems have, however, the serious drawback of using the following rule, which introduces an arbitrary number of implications, and has an arbitrary number of premises, all of which contain formulas of essential importance for the inference:

$$\frac{\{\Gamma, \varphi_i \Rightarrow \psi_i, \Delta^i\}}{\Gamma \Rightarrow \Delta} R \rightarrow$$

Here Δ contains exactly $m > 0$ implicational formulas $\varphi_i \rightarrow \psi_i, i = 1, \ldots, m$ (Δ may also contain some other kinds of formulas). Δ^i denotes the multiset consisting of exactly the $m - 1$ implicational formulas of Δ other than $\varphi_i \rightarrow \psi_i$.

An alternative sound and complete system GLC_{RS} for LC has been given in [7]. Like the former systems, GLC_{RS} admits cut-elimination, but it does not have a rule with arbitrary number of premises. The idea behind GLC_{RS} is again to use noncanonical invertible rules which decompose formulas in a sequent into simpler ones, until a set of "clauses" equivalent to the original sequent is reached. However, unlike in the 3-valued and 4-valued logics described above, in GLC_{RS} a "clause" is defined as a sequent consisting only of atomic formulas or implications between atomic formulas. Accordingly, instead of having for each connective rules for its combinations with negation, we have rules for all its possible combinations with $\rightarrow$. The four needed rules for $\rightarrow$ itself are:

$$(\Rightarrow\rightarrow(\rightarrow)) \quad \frac{\Gamma \Rightarrow \Delta, \psi_1 \rightarrow \psi_2, \varphi \rightarrow \psi_2}{\Gamma \Rightarrow \Delta, \varphi \rightarrow (\psi_1 \rightarrow \psi_2)}$$

$$(\rightarrow(\rightarrow)\Rightarrow) \quad \frac{\Gamma, \psi_1 \rightarrow \psi_2 \Rightarrow \Delta \qquad \Gamma, \varphi \rightarrow \psi_2 \Rightarrow \Delta}{\Gamma, \varphi \rightarrow (\psi_1 \rightarrow \psi_2) \Rightarrow \Delta}$$

$$(\Rightarrow(\rightarrow)\rightarrow) \quad \frac{\Gamma \Rightarrow \Delta, \varphi_2 \rightarrow \psi \qquad \Gamma, \varphi_1 \rightarrow \varphi_2 \Rightarrow \Delta, \psi}{\Gamma \Rightarrow \Delta, (\varphi_1 \rightarrow \varphi_2) \rightarrow \psi}$$

$$((\rightarrow)\rightarrow\Rightarrow) \quad \frac{\Gamma, \varphi_2 \rightarrow \psi \Rightarrow \Delta, \varphi_1 \rightarrow \varphi_2 \qquad \Gamma, \psi \Rightarrow \Delta}{\Gamma, (\varphi_1 \rightarrow \varphi_2) \rightarrow \psi \Rightarrow \Delta}$$

The number of rules needed for each binary connective other than $\rightarrow$ is six. The rules for $\vee$, for example, are:

$$(\Rightarrow \vee)\ \frac{\Gamma \Rightarrow \Delta, \varphi, \psi}{\Gamma \Rightarrow \Delta, \varphi \vee \psi}$$

$$(\vee \Rightarrow)\ \frac{\Gamma, \varphi \Rightarrow \Delta \qquad \Gamma, \psi \Rightarrow \Delta}{\Gamma, \varphi \vee \psi \Rightarrow \Delta}$$

$$(\Rightarrow \vee \rightarrow)\ \frac{\Gamma \Rightarrow \Delta, \varphi_1 \rightarrow \psi \qquad \Gamma \Rightarrow \Delta, \varphi_2 \rightarrow \psi}{\Gamma \Rightarrow \Delta, (\varphi_1 \vee \varphi_2) \rightarrow \psi}$$

$$(\vee \rightarrow \Rightarrow)\ \frac{\Gamma, \varphi_1 \rightarrow \psi, \varphi_2 \rightarrow \psi \Rightarrow \Delta}{\Gamma, (\varphi_1 \vee \varphi_2) \rightarrow \psi \Rightarrow \Delta}$$

$$(\Rightarrow \rightarrow \vee)\ \frac{\Gamma \Rightarrow \Delta, \varphi \rightarrow \psi_1, \varphi \rightarrow \psi_2}{\Gamma \Rightarrow \Delta, \varphi \rightarrow (\psi_1 \vee \psi_2)}$$

$$(\rightarrow \vee \Rightarrow)\ \frac{\Gamma, \varphi \rightarrow \psi_1 \Rightarrow \Delta \qquad \Gamma, \varphi \rightarrow \psi_2 \Rightarrow \Delta}{\Gamma, \varphi \rightarrow (\psi_1 \vee \psi_2) \Rightarrow \Delta}$$

Like in the previous cases, these rules are not sufficient for completeness. We need to have also means for determining which "clauses" (or "basic sequents") are valid. The straightforward way of doing so is again to add appropriate non-standard axioms. Unfortunately, the needed axioms are here much more complicated than those employed in the 3-valued and 4-valued cases described above:

Definition 17. *Let $\Gamma \Rightarrow \Delta$ be a clause of GLC_{RS}.*

- *We say that $(p \leq q) \in (\Gamma \Rightarrow \Delta)$ iff $(p \rightarrow q) \in \Gamma$.*
- *We say that $(t \leq q) \in (\Gamma \Rightarrow \Delta)$ iff $q \in \Gamma$.*
- *We say that $(p < q) \in (\Gamma \Rightarrow \Delta)$ iff $(q \rightarrow p) \in \Delta$.*
- *We say that $(q < t) \in (\Gamma \Rightarrow \Delta)$ iff $q \in \Delta$.*
- *Let p, q be either atomic formulas or t* [13]*. We say that $(p \triangleleft q) \in (\Gamma \Rightarrow \Delta)$ iff either $(p \leq q) \in (\Gamma \Rightarrow \Delta)$ or $(p < q) \in (\Gamma \Rightarrow \Delta)$.*
- *A sequence $q_1, \ldots, q_l$ (where q_i is either atomic or t) is called a* strictly increasing *sequence for $\Gamma \Rightarrow \Delta$ if $(q_j \triangleleft q_{j+1}) \in (\Gamma \Rightarrow \Delta)$ for $1 \leq j \leq l-1$, and either $(q_i < q_{i+1}) \in (\Gamma \Rightarrow \Delta)$ for some $1 \leq i \leq l-1$, or $q_1 = t$, $q_l = f$.*

The axioms of the system GLC_{RS} [14] *are those clauses for which there exists a strictly increasing sequence $q_1 \ldots q_l$ such that either $q_1 = q_l$, or $q_1 = t$, or $q_l = f$.*

As we noted above, the system GLC_{RS} is sound and complete for LC, and the cut-elimination theorem obtains for it. The proofs given to this in [7] can easily be extended to proofs of the corresponding *strong* versions of these facts by applying the method used in this paper. We omit the details.

To get a system GLC^k_{RS} with similar properties for the *finite* k-valued Gödel-Dummett logic, one needs to add as axioms all basic sequents s having

[13] Note that in this paper t is *not* an official symbol of the language of LC. Here, however, it is used in the metalanguage.

[14] In [7] this version with new axioms is called GLC^*_{RS}.

a strictly increasing sequence $q_1 \dots q_l$ such that for at least k different q_i's, $(q_i < q_{i+1}) \in s$.

Instead of enriching the set of axioms, an alternative approach presented in [7] is to employ special analytic simplification rules. Those needed for the infinite-valued LC are:

Transitivity: $$\frac{\Gamma, p \to q, q \to r, p \to r \Rightarrow \Delta}{\Gamma, p \to q, q \to r \Rightarrow \Delta}$$

Left maximality: $$\frac{\Gamma, p \to q, p, q \Rightarrow \Delta}{\Gamma, p \to q, p \Rightarrow \Delta}$$

Right maximality: $$\frac{\Gamma \Rightarrow \Delta, q \to p, p}{\Gamma \Rightarrow \Delta, q \to p}$$

Linearity: $$\frac{\Gamma, p \to q \Rightarrow \Delta, q \to p}{\Gamma \Rightarrow \Delta, q \to p}$$

Minimality of f: $$\frac{\Gamma, p \to f, f \to p \Rightarrow \Delta}{\Gamma, p \to f \Rightarrow \Delta}$$

Note: All the *logical* rules of the systems described in this subsection have what might be called the *semi-subformula property*: written in Polish notation, every formula in their premises either appears in their conclusion or is obtained from some formula there by deleting some of its symbols. This is not very different from the usual subformula property. The simplification rules, in contrast, do not have this property. Nevertheless, each of them is still *analytic* in the sense that its possible premises (for any potential conclusion) are determined by its conclusion, they are finite in number, and can be found effectively. These rules are close in nature to *analytic cuts* (see section 7.3), but they are simpler.

7.2 The four-valued logic with a single designated element

The matrix $\mathcal{G}_4$ is a four-valued matrix in which $D = \{t\}$. The corresponding system GLC^4_{RS} can therefore be used as a basis for an appropriate Gentzen-type system for the universal four-valued logic in which there is exactly one designated truth-value. The main problem in doing so is that the set of connectives of this matrix is not functionally complete. It can however be shown that by adding to L_{LC} the connective $\neg$ as well as the propositional constants $\top$ and $\bot$, and by giving these new connectives the same interpretations they have in $\mathcal{M}_4$, one does obtain a language with a functionally complete set of connectives. Let us call the corresponding matrix $\mathcal{M}_4^{\{t\}}$ [15]. To get a sound and complete system for $\mathcal{M}_4^{\{t\}}$, all we need to do is to add to GLC^4_{RS} appropriate rules and axioms for the new connectives. Now GLC^4_{RS} is based on the idea of using invertible rules for reducing any sequent to a set of "clauses"

[15] Note that the interpretations of $\vee$ and $\wedge$ in $\mathcal{M}_4^{\{t\}}$ is *not* identical to interpretations they have in $\mathcal{M}_4$!

in the sense of GLC_{RS}. The connective $\neg$ should therefore be treated here exactly as $\vee$ and $\wedge$ (and so its role here is completely different from the one it has in GM_4). In other words: we should find for it a full set of decomposition rules (i.e.: invertible sound rules which make it possible to get rid of it in backward reasoning). It is not difficult to show that the following six rules constitute such a set:

$$(\neg\Rightarrow)\quad \frac{\Gamma, A\rightarrow f\Rightarrow\Delta}{\Gamma,\neg A\Rightarrow\Delta}$$

$$(\Rightarrow\neg)\quad \frac{\Gamma\Rightarrow\Delta, A\rightarrow f}{\Gamma\Rightarrow\Delta,\neg A}$$

$$(\neg\rightarrow\Rightarrow)\quad \frac{\Gamma, A\Rightarrow\Delta\quad \Gamma, B\Rightarrow\Delta\quad \Gamma, A\rightarrow B\Rightarrow\Delta, A\rightarrow f}{\Gamma,(\neg A)\rightarrow B\Rightarrow\Delta}$$

$$(\Rightarrow\neg\rightarrow)\quad \frac{\Gamma, A\rightarrow f\Rightarrow\Delta, B\quad \Gamma\Rightarrow\Delta, A, A\rightarrow B}{\Gamma\Rightarrow\Delta,(\neg A)\rightarrow B}$$

$$(\rightarrow\neg\Rightarrow)\quad \frac{\Gamma, A\rightarrow f\Rightarrow\Delta\quad \Gamma, B\rightarrow f\Rightarrow\Delta\quad \Gamma, A\rightarrow B\Rightarrow\Delta, B}{\Gamma, A\rightarrow\neg B\Rightarrow\Delta}$$

$$(\Rightarrow\rightarrow\neg)\quad \frac{\Gamma, B\Rightarrow\Delta, A\rightarrow f\quad \Gamma\Rightarrow\Delta, A\rightarrow B, B\rightarrow f}{\Gamma\Rightarrow\Delta, A\rightarrow\neg B}$$

Note: These rules for $\neg$ are sound and invertible also in $\mathcal{G}_\omega$, provided the interpretation of $\neg$ is given by:

$$\neg a = \begin{cases} f & a = t \\ t & a = f \\ a & \text{otherwise} \end{cases}$$

It follows that by adding these rules to GLC_{RS} and GLC^k_{RS} we get systems which are strongly sound and complete for the corresponding extended matrices, and in which the strong cut elimination theorem obtains.

Returning to $\mathcal{M}_4^{\{t\}}$, we should take into account also the fact that its language includes the constants $\top$ and $\bot$. Their presence is not relevant to the decomposition rules, but *is* relevant to the question what "clauses" (i.e.: basic sequents) should be taken as axioms (because now the set of atomic formulas includes these two new constants). It is not difficult to see that what we should do is to add to the axioms of GLC^4_{RS} all basic sequents s for which there exists a strictly increasing sequence $q_1 \dots q_l$ satisfying one of the following conditions:

1. $q_1 = \top$ and there are at least two different i's such that $(q_i < q_{i+1}) \in s$
2. $q_l = \bot$ and there are at least two different i's such that $(q_i < q_{i+1}) \in s$

3. $q_1 = \perp$ and there are at least three different i's such that $(q_i < q_{i+1}) \in s$
4. $q_l = \top$ and there are at least three different i's such that $(q_i < q_{i+1}) \in s$

By a straightforward adaption of the proofs given in [7] one can show now that the resulting system $GM_4^{\{t\}}$ is strongly sound and complete for $\mathcal{M}_4^{\{t\}}$, and that the strong cut elimination theorem obtains for it.

7.3 The modal $S5$

In this subsection we present one important example of the use of an impure rule: the standard Gentzen-type system for the modal logic $S5$. Like most other modal logics, $S5$ has a well-known (particularly simple) possible-worlds semantics: A frame for $S5$ is a pair $\langle W, v \rangle$ where W is a nonempty set (of "possible-worlds") and v is a function which assigns to each sentence in the language of $S5$ a subset of W, so that $v(A \vee B) = v(A) \cup v(B)$, $v(A \wedge B) = v(A) \cap v(B)$, $v(\neg A) = W - v(A)$, and $v(\Box A) = W$ if $v(A) = W$, $\emptyset$ otherwise. A sentence A is *true* in a world $w \in W$ if $w \in v(A)$, and *valid* in W if it true in all worlds of W (i.e.: if $v(A) = W$). It is obvious that a frame may be viewed as a pair of a matrix and a valuation in it. In fact, a slight generalization of this semantical characterization of $S5$ had already been given in [52] (much before the Kripke-style semantics of this system was discovered). This many-valued semantics of $S5$ is given in the next definition.

Definition 18. *Let* $L_\Box = \{\neg, \vee, \wedge, \Box\}$.

A matrix $\mathcal{A} = \langle A, D_A, O_A \rangle$ *for* $L_\Box$ *is called an* $S5$-matrix *if:*

- $\{0,1\} \subseteq A$ $(0 \neq 1)$
- $D_A = \{1\}$
- $\langle A, \vee, \wedge, \neg, 1, 0 \rangle$ *is a Boolean Algebra*
- $\Box a = \begin{cases} 1 & a = 1 \\ 0 & a \neq 1 \end{cases}$

It has been proved in [52] that (the Hilbert-type formulation of) $S5$ is sound and complete for the class of $S5$-matrices. Moreover: if a sentence φ in the language of $S5$ is not a theorem of $S5$, then it is refutable in some *finite* structure of this type (with at most 2^{2^n} elements, where n is the number of atomic formulas occurring in φ). All these finite matrices are embeddable in one denumerable matrix, which is characteristic for $S5$. Hence $S5$ is many-valued logic according to Definition 10[16].

We present now $GS5$, the standard Gentzen-type system for $S5$ (originally introduced in [47]). This system is obtained from $GCPL$ by adding to it the following two simple rules:

[16] As far as I know, this is not true for other famous Modal Logics.

$$(\Box \Rightarrow) \ \frac{\varphi, \Gamma \Rightarrow \Delta}{\Box\varphi, \Gamma \Rightarrow \Delta} \qquad\qquad \frac{\Gamma \Rightarrow \Delta, \varphi}{\Gamma \Rightarrow \Delta, \Box\varphi} \ (\Rightarrow \Box)$$

There is, however, a side condition on the application of $(\Rightarrow \Box)$, which makes this rule impure: all the side formulas (i.e., the formulas of Γ and Δ) should begin with $\Box$ (an equivalent version demands them only to be *essentially modal*, in the sense that each occurrence of an atomic formula should be within the scope of a $\Box$).

Definition 19.

1. *Let s be the sequent $A_1, \ldots, A_n \Rightarrow B_1, \ldots, B_k$. φ_s, the standard interpretation of s, is the sentence $\neg A_1 \vee \ldots \vee \neg A_n \vee B_1 \vee \ldots \vee B_k$.*
2. *A* model *of a sequent s in the language $L_\Box$ is a pair $\langle \mathcal{A}, v\rangle$ where $\mathcal{A}$ is an $S5$-matrix, and v is a valuation in $\mathcal{A}$ such that $v(\varphi_s) = 1$ (in particular: $\langle \mathcal{A}, v\rangle$ is a model of $\Rightarrow \varphi$ iff it is a model of φ).*
3. *A sequent s follows in $S5$ from a set S of sequents if every model of S is also a model of s.*

Theorem 9. Strong Soundness of $GS5$: *If a sequent s has a proof in $GS5$ from a set S of sequents then it follows in $S5$ from S.*

We leave the easy proof to the reader. As usual, we want next to turn to a simultaneous proof for $GS5$ of strong completeness and strong cut elimination. There is a problem, though: the cut elimination theorem is *not* valid for $GS5$. However, we have a satisfactory substitute: the possibility to eliminate *non-analytic* cuts:

Definition 20. *A cut in a proof of a sequent s from a set S of sequents is called analytic if the cut formula is a subformula of some formula in $S \cup \{s\}$.*

Theorem 10. Strong Completeness and Analytic Cut-elimination for $GS5$: *A sequent s follows in $S5$ from $S = \{\Gamma_1 \Rightarrow \Delta_1, \ldots, \Gamma_n \Rightarrow \Delta_n\}$ iff it has a proof in $GS5$ from S in which every cut is done either on a formula in $\bigcup_{i=1}^{n} \Gamma_i \cup \bigcup_{i=1}^{n} \Delta_i$, or on a subformula of the form $\Box A$ of some formula in $S \cup \{s\}$.*

Proof: Again the proof follows to a great extent that of Theorem 2. Given S we define first an S-proof as a proof which is permitted by the formulation of the theorem. We then define when a a sequent $\Gamma^* \Rightarrow \Delta^*$ is S-saturated like in the proof of Theorem 2, but with two more conditions:

(A) If $\Box\varphi \in \Gamma^*$ then $\varphi \in \Gamma^*$.
(B) If $\Box\varphi$ is a subformula of a formula in $\Gamma^* \cup \Delta^*$ then $\Box\varphi \in \Gamma^* \cup \Delta^*$.

It is easy to prove that if a sequent s has no S-proof then it can be extended to an S-saturated sequent consisting only of subformulas of formulas in $S \cup \{s\}$. It remains therefore to show that if $s = \Gamma^* \Rightarrow \Delta^*$ is S-saturated then there is a model of S which is not a model of s. For this let W be the set of all S-saturated sequents $\Gamma \Rightarrow \Delta$ which consist only of subformulas of formulas in s, and which satisfy the following condition for every formula A:

$$(*) \quad \Box A \in \Gamma \text{ iff } \Box A \in \Gamma^*$$

(by condition (B) this implies that also $\Box A \in \Delta$ iff $\Box A \in \Delta^*$). W is obviously finite, and $P(W)$ is a Boolean Algebra with the usual operations of union, intersection etc. This Boolean Algebra can be extended to an S-matrix in a unique way. Let $\mathcal{A}$ be this S-matrix. Define a valuation v in it by:

$$v(p) = \{\Gamma \Rightarrow \Delta \in W \mid p \in \Gamma\}$$

We show next by induction on complexity of formulas that for every $\Gamma \Rightarrow \Delta \in W$ and every sentence φ, if $\varphi \in \Gamma$ then $\Gamma \Rightarrow \Delta \in v(\varphi)$, while if $\varphi \in \Delta$ then $\Gamma \Rightarrow \Delta \notin v(\varphi)$. Most of the induction steps are straightforward. We do here the case $\varphi = \Box\psi$:

- Assume $\Box\psi \in \Gamma$. Then by the definition of W, $\Box\psi$ belongs to the antecedents of all the sequents in W. Hence by condition (A) so does ψ itself. It follows that $v(\psi) = W$, and so also $v(\Box\psi) = W$. In particular: $\Gamma \Rightarrow \Delta \in v(\Box\psi)$.
- Assume $\Box\psi \in \Delta$. Let $\Box(\Gamma^*)$ and $\Box(\Delta^*)$ be the sets of boxed formulas in Γ^* and Δ^* (respectively). The sequent $\Box(\Gamma^*) \Rightarrow \Box(\Delta^*), \psi$ does not have an S-proof (since otherwise so would have $\Gamma^* \Rightarrow \Delta^*$, using an application of $(\Rightarrow \Box)$ and weakenings), and so it can be extended to an S-saturated sequent $\Gamma^\# \Rightarrow \Delta^\#$, consisting only of subformulas of formulas in $\Gamma^* \Rightarrow \Delta^*$. Since $\Gamma^* \Rightarrow \Delta^*$ satisfies condition (B), $\Gamma^\# \Rightarrow \Delta^\#$ cannot have new boxed formulas in addition to those in $\Box(\Gamma^*) \Rightarrow \Box(\Delta^*), \psi$, and so $\Gamma^\# \Rightarrow \Delta^\#$ is in W. By induction hypothesis we have therefore that $\Gamma^\# \Rightarrow \Delta^\# \notin v(\psi)$, and so $v(\Box\psi) = \emptyset$. In particular: $\Gamma \Rightarrow \Delta \notin v(\Box\psi)$.

It can easily be seen that $\langle \mathcal{A}, v\rangle$ is a model of $A_1, \ldots, A_n \Rightarrow B_1, \ldots, B_k$ iff for every $s \in W$, either $s \in v(B_i)$ for some i, or $s \notin v(A_j)$ for some j. Hence what we have just shown imply that $\langle \mathcal{A}, v\rangle$ is a model of S which is not a model of $\Gamma^* \Rightarrow \Delta^*$ (the proof that this entails the validity in $\langle \mathcal{A}, v\rangle$ of the sequents in S is similar to the way this is done in the proof of Theorem 2).

Corollary 9. Strong Completeness of $GS5$:

1. *A sequent s follows in $S5$ from a set S of sequents iff $S \vdash_{GS5} s$.*
2. *A formula φ is valid in $S5$ iff $\vdash_{GS5} \Rightarrow \varphi$.*

Corollary 10. (Strong Analytic Cut-elimination for $GS5$:

1. *A sequent s follows in $GS5$ from $S = \{\Gamma_1 \Rightarrow \Delta_1, \ldots, \Gamma_n \Rightarrow \Delta_n\}$ iff it has a proof in $GS5$ from S in which every cut is done either on a formula in $\bigcup_{i=1}^n \Gamma_i \cup \bigcup_{i=1}^n \Delta_i$, or on a subformula of the form $\Box A$ of some formula in $S \cup \{s\}$.*

2. *$\vdash_{GS5}$ s iff there exists a proof in GS5 of s in which all the cuts are analytic and are performed on formulas of the form $\Box\psi$.*

This corollary suffices for the subformula property, and for developing a good tableaux system for $S5$.

Note: $\vdash_{GS5}$, the standard consequence relation defined by $GS5$, is different from that induced by the many-valued semantics described above (for example: $\varphi \vdash_{S5} \Box\varphi$ according to that semantics, but the sequent $\varphi \Rightarrow \Box\varphi$ is not provable in $GS5$). What we do have is that $\Gamma \vdash_{GS5} \Delta$ iff $\varphi_{\Gamma\Rightarrow\Delta}$ is valid according to this semantics.[17]

7.4 The infinite-valued purely relevant logic

Substructural logics ([53]) are logics defined by Gentzen-type systems with an irregular set of structural rules (the "regular" set being the one consisting of the standard structural rules, weakening, and cut). The official consequence relation associated with a logic of this kind is therefore not a Scott consequence relation in the sense of Definition 1. Its set of valid formulas may still correspond, however, to some many-valued logic. In subsection 5.3 we have already seen an example in which this happens with a 3-valued logic. We now shortly review another example, this time with an infinite-valued logic.

The system $GRMI_m$ is obtained from $GCPL$ by simply deleting the weakening rule[18] (or from GRM_m of subsection 5.3 by deleting the mingle rule. Note that again we assume that the two sides of a sequent consists of *sets*. In other words: although we do not have weakening, the expansion rule is still available). It is easy to prove, using Gentzen's original purely syntactic method, that cut-elimination still obtains for this system. Moreover: a strong form of the interpolation theorem holds for it: $\Rightarrow \varphi \to \psi$ is provable only if there exists an interpolant ϕ *containing only atomic formulas common to φ and ψ* [19]. This puts the corresponding logic, RMI_m, in the family of Relevance logics (in which framework it was originally introduced and investigated). In [9] it was proved that it is also weakly many-valued according

[17] $\vdash_{GS5}$ corresponds to the "truth" consequence relation of modal logic, while the Tarskian consequence relation between singleton sequents which is induced by $GS5$ corresponds to its "validity" consequence relation (see [13] for these notions). The fact that these CRs are different is due to the impurity of the rules of $GS5$.

[18] In the literature on relevance logic (including [9], from which the results below are taken) the symbols $\sim$, $\to$, $\circ$, $+$ are used instead of $\neg$, $\supset$, $\wedge$, $\vee$ (respectively). We use therefore these symbols below, except that instead of $\circ$ we use $\otimes$, which is more common today (recall that we did the same in section 5.3).

[19] Unlike in classical logic, such atomic formulas necessarily exist in case $\Rightarrow \varphi \to \psi$ is provable in $GRMI_m$. This is called the "variable sharing property" in the literature on relevance logics ([4, 5, 26]), and it is the most characteristic feature of these logics.

to Definition 10. The corresponding matrix, called A_ω in [9], is the following:

Truth Values: $\{t, f, I_1, I_2, I_3 \ldots\}$.

Designated Values: All elements except f.

Operations: $\sim t = f$, $\sim f = t$, $\sim I_j = I_j$ $(1 \leq j < \infty)$.

$$a \to b = \begin{cases} t & a = f \text{ or } b = t \\ I_j & a = b = I_j \\ f & \textit{otherwise} . \end{cases}$$

$$a + b = \sim a \to b \qquad a \otimes b = \sim (a \to \sim b)$$

Like G_ω, A_ω can be viewed as the "limit" of its finite substructures. Indeed, let A_n be the substructure of A_ω consisting of $\{t, f, I_1, \ldots, I_n\}$. Then we have:

Theorem 11. *A formula φ is valid in A_ω (or in all of the finite structures A_n) iff $\vdash_{GRMI_m} \Rightarrow \varphi$. Moreover, if φ contains n atomic formulas and $\not\vdash_{GRMI_m} \Rightarrow \varphi$ then there is a valuation which refutes φ already in A_n.*

It follows that the logic of the matrix A_ω is weakly many-valued according to Definition 10, and that $GRMI_m$ is sound and weakly complete relative to it.

Notes:

1. Again we have here only *weak* completeness. Indeed, $GRMI_m$ is not strongly complete for A_ω (thus $A \otimes B \vdash_{A_\omega} A$, but the corresponding sequent is not provable in $GRMI_m$).
2. In [15] it is proved that the logic of the matrix A_ω is in fact many-valued according to Definition 10 (not only weakly so).
3. The proofs in [9] of the completeness of the system and of the cut elimination theorem for it are completely separated. It is possible however to find a simultaneous proof of both in this case as well.

References

1. Arieli, O. and Avron, A.: Logical bilattices and inconsistent data, in Proc. 9th IEEE Annual Symp. on Logic in Computer Science (LICS'94), IEEE Press (1994) 468–476
2. Arieli, O. and Avron, A.: Reasoning with logical bilattices, J. of Logic, Language and Information, vol. 5, no. 1 (1996) 25–63
3. Arieli, O. and Avron, A.: The value of four values, Artificial Intelligence, vol. 102, no. 1 (1998) 97–141

4. Anderson, A.R. and Belnap, N.D.: Entailment, vol. I. Princeton University Press (1975)
5. Anderson, A.R. and Belnap, N.D.: Entailment, vol. II. Princeton University Press (1992)
6. Avellone, A., Ferrari, M. and Miglioli, P.: Duplication-free tableaux calculi together with cut-free and contraction-free sequent calculi for the interpolable propositional intermediate logics, Logic Journal of IGPL, vol. 7 (1999) 447–480
7. Avron, A. and Konikowska, B.: Decomposition proof systems for Gödel-Dummett logics, Studia Logica, vol. 69 (2001) 197–219
8. Avron, A. and Lev, I.: Canonical propositional gentzen-type systems, in *Proceedings of the 1st International Joint Conference on Automated Reasoning (IJCAR 2001)* (Goré, R., Leitsch, A. and Nipkow, T. eds.), Springer Verlag, LNAI 2083 (2001) 529–544
9. Avron, A.: Relevant entailment - semantics and formal systems, Journal of Symbolic Logic, vol. 49 (1984) 334–342
10. Avron, A.: On an implication connective of RM, Notre Dame Journal of Formal Logic, vol. 27 (1986) 201–209
11. Avron, A.: A constructive analysis of RM, Journal of Symbolic Logic, vol. 52 (1987) 939–951
12. Avron, A.: Natural 3-valued logics: Characterization and proof theory, J. of Symbolic Logic, vol. 56, no. 1 (1991) 276–294
13. Avron, A.: Simple consequence relations, Information and Computation, vol. 92, no. 1 (1991) 105–139
14. Avron, A.: Gentzen-type systems, resolution and tableaux, Journal of Automated Reasoning, vol. 10 (1993) 265–281
15. Avron, A.: Multiplicative Conjunction as an Extensional Conjunction, Logic Journal of the IGPL, vol. 5 (1997) 181–208
16. Avron, A.: On the expressive power of three-valued and four-valued languages, Journal of Logic and Computation, vol. 9 (1999) 977–994
17. Belnap, N.D.: How computers should think, in Contemporary Aspects of Philosophy (Ryle, G. ed.), Oriel Press, Stocksfield, England (1977) 30–56
18. Belnap, N.D.: A useful four-valued logic, in Modern Uses of Multiple-Valued Logic (Epstein, G. and Dunn, J.M. eds.), Reidel, Dordrecht (1977) 7–37
19. Baaz, M., Fermüller, C.G. and, Salzer, G.: Automated deduction for many-valued logics, in Handbook of Automated Reasoning (Robinson, A. and Voronkov, A. eds.), Elsevier Science Publishers (2000)
20. Busch, D.: Sequent formalizations of three-valued logic, in Partiality, Modality, and Nonmonotonicity, Studies in Logic, Language and Information, CSLI Publications (1996) 45–75
21. da Costa, N.C.A.: On the theory of inconsistent formal systems, Notre Dame Journal of Formal Logic, vol. 15 (1974) 497–510
22. D'Ottaviano, I.L.M. and da Costa, N.C.A.: Sur un problemè de Jaśkowski, C.R. Acad. Sc. Paris, vol. 270, Sèrie A (1970) 1349–1353
23. Dunn, J.M. and Meyer, R.K.: Algebraic completeness results for dummett's lc and its extensions, Z. math. Logik und Grundlagen der Mathematik, vol. 17 (1971) 225–230
24. D'Ottaviano, I.L.M.: The completeness and compactness of a three-valued first-order logic, Revista Colombiana de Matematicas, vol. XIX, no. 1-2 (1985) 31–42
25. Dummett, M.: A propositional calculus with a denumerable matrix, Journal of Symbolic Logic, vol. 24 (1959) 96–107

26. Dunn, J.M.: Relevance logic and entailment, in [34], vol. III, ch. 3 (1986) 117–224
27. Dyckhoff, R.: A deterministic terminating sequent calculus for Gödel-Dummett logic, Logic Journal of IGPL, vol. 7 (1999) 319–326
28. Epstein, R.L.: The semantic foundation of logic, vol. I: propositional logics, ch. IX. Kluwer Academic Publisher (1990)
29. Fitting, M.: Bilattices in logic programming, in 20th Int. Symp. on Multiple-Valued Logic (Epstein, G. ed.), IEEE Press (1990) 238–246
30. Fitting, M.: Kleene's logic, generalized, Journal of Logic and Computation, vol. 1 (1990) 797–810
31. Fitting, M.: Bilattices and the semantics of logic programming, Journal of Logic Programming, vol. 11, no. 2 (1991) 91–116
32. Fitting, M.: Kleene's three-valued logics and their children, Fundamenta Informaticae, vol. 20 (1994) 113–131
33. Gentzen, G.: Investigations into logical deduction, in The Collected Works of Gerhard Gentzen (Szabo, M.E. ed.), North Holland, Amsterdam (1969) 68–131
34. Gabbay, D.M. and Guenthner, F.: Handbook of Philosophical Logic. D. Reidel Publishing company (1986)
35. Ginsberg, M.L.: Multiple-valued logics, in Readings in Non-Monotonic Reasoning (Ginsberg, M.L. ed.), Los-Altos, CA (1987) 251–258
36. Ginsberg, M.L.: Multivalued logics: A uniform approach to reasoning in AI, Computer Intelligence, vol. 4 (1988) 256–316
37. Girard, J.-Y.: Linear logic, Theoretical Computer Science, vol. 50 (1987) 1–102
38. Girard, J.-Y.: Proof Theory and Logical Complexity. Bibliopolis (1987)
39. Gödel, K.: On the intuitionistic propositional calculus, in Collected Work of K. Gödel (Feferman, S. and all, eds.), vol. I, Oxford University Press (1986) 222–225
40. Hähnle, R.: Tableaux for multiple-valued logics, in Handbook of Tableau Methods (D'Agostino, M., Gabbay, D.M., Hähnle, R. and Posegga, J. eds.), Kluwer Publishing Company (1999) 529–580
41. Hajek, P.: Metamathematics of Fuzzy Logic. Kluwer Academic Publishers (1998)
42. Jones, C.B.: Systematic Software Development Using VDM. Prentice-Hall International, U.K. (1986)
43. Kleene, S.C.: Introduction to metamathematics. Van Nostrad (1950)
44. Łoś, J. and Suszko, R.: Remarks on sentential logics, Indagationes Mathematicae, vol. 20 (1958) 177–183
45. Łukasiewicz, J.: On 3-valued logic, in Polish Logic (McCall, S. ed.), Oxford University Press (1967)
46. Monteiro, A.: Construction des algebres de Łukasiewicz trivalentes dans les algebres de Boole monadiques, i Mat. Jap., vol. 12 (1967) 1–23
47. Ohnishi, M. and Matsumoto, K.: Gentzen methods in modal calculi, Osaka Mathematical Journal, vol. 9 (1957) 113–130
48. Rozoner, L.I.: On interpretation of inconsistent theories, Information Sciences, vol. 47 (1989) 243–266
49. Schmitt, P.H.: Computational aspects of three-valued logic, in Proceedings of the 8th Conference on Automated Deduction, Springer Verlag, LNCS 230 (1986) 190–198

50. Scott, D.S.: Completeness and axiomatization in many-valued logics, in Proc. of the Tarski symposium, vol. XXV of Proc. of Symposia in Pure Mathematics, (Rhode Island), American Mathematical Society (1974) 411–435
51. Scott, D.S.: Rules and derived rules, in Logical Theory and Semantical Analysis (Stenlund, S. ed.), Reidel, Dordrecht (1974) 147–161
52. Scroggs, S.J.: Extensions of the Lewis system s5, Journal of Symbolic Logic, vol. 16 (1951) 112–120
53. Schroeder-Heister, P. and Došen, K. eds.: Substructural Logics. Oxford University Press (1993)
54. Słupecki, J.: Der volle dreiwertige aussagenkalkül, Com. rend. Soc. Sci. Lett. de Varsovie, vol. 29 (1936) 9–11
55. Sobociński, B.: Axiomatization of partial system of three-valued calculus of propositions The Journal of Computing Systems, vol. 11, no. 1 (1952) 23–55
56. Sonobe, O.: A Gentzen-type formulation of some intermediate propositional logics, J. of Tsuda College, vol. 7 (1975) 7–14
57. Troelstra, A.S. and Schwichtenberg, H.: Basic Proof Theory, 2nd Edition. Cambridge University Press (2000)
58. Urquhart, A.: Many-valued logic, in [34], vol. III, ch. 2 (1986) 71–116
59. Visser, A.: On the completeness principle: A study of provability in Heyting's arithmetic, Annals of Mathematical Logic, vol. 22 (1982) 263–295
60. Wójcicki, R.: Lectures on Propositional Calculi. Warsaw: Ossolineum (1984)

Chapter 6
Sequent of Relations Calculi: A Framework for Analytic Deduction in Many-valued Logics

Matthias Baaz, Agata Ciabattoni*, Christian G. Fermüller

Technische Universität Wien
A-1040 Vienna, Austria
[baaz,agata,chrisf]@logic.at

Abstract. We present a general framework that allows to construct systematically analytic calculi for a large family of (propositional) many-valued logics — called projective logics — characterized by a special format of their semantics. All finite-valued logics as well as infinite-valued Gödel logic are projective. As a case-study, sequent of relations calculi for Gödel logics are derived. A comparison with some other analytic calculi is provided.

1 Introduction

A central task of logic in computer science is to provide analytic — "Gentzen-style" — calculi for a wide range of non-classical logics. Such calculi serve as a basis for automated deduction and allow the extraction of more information from proofs (compared to traditional Hilbert-style systems). A large number of Gentzen-style systems for many-valued logics have been introduced since the 1950s (see, e.g., the handbook article [9] for an overview). In particular, it is now well understood how to construct efficient analytic proof systems in a uniform manner for the whole family of *finite-valued* logics, using many-placed (or labeled) sequents or, equivalently, signed tableaux. These systems not only allow to handle connectives defined by arbitrary finite matrices, but have also been extended to so-called distribution quantifiers, a rather general concept of quantification in many-valued logic. However, in contrast to the finite-valued case, there is much less literature on analytic Gentzen-style systems for *infinite-valued* logics. In particular, many important infinite-valued logics, such as two of the three fundamental formal representations of fuzzy logic — namely Łukasiewicz [21] and Product logic [20] — still seem to escape satisfactory characterizations through elegant and useful analytic proof systems. An important exception is the third main formalization of fuzzy logic, namely Gödel logic G_∞. Analytic systems for this logic — also known as (Gödel-)Dummett logic, since Dummett [14] presented the first

* Research supported by EC Marie Curie fellowship HPMF–CT–1999–00301.

axiomatization matching Gödel's matrix characterization — can be found, e.g., in [25, 1–3, 16, 4, 5]. The interest in G_∞ is well motivated by the fact that it naturally turns up in a number of different contexts. Already in the 1930s Gödel [18] used it in investigations of intuitionistic logic; later, Dunn and Meyer [15] pointed out its relevance for relevance logic; Visser [29] employed it in investigations of the provability logic of Heyting arithmetic; and eventually it was recognized as one of the most important formal representations of fuzzy logic. In particular, the connectives of G_∞ correspond to the min-max t-norm and its residuum (see [19]).

The purpose of this contribution is not just to introduce yet another analytic proof system for G_∞, but to present a *general* framework that allows to construct analytic calculi for a large family of (propositional) many-valued logics, including G_∞ as a prime example. Our calculi — called sequents of relations — are systematically derived from a specific presentation of the semantics of the connectives involved. This format of presentation inspires us to call the corresponding class of logics "*projective*". Our main example of a projective logic is G_∞. But also G_∞^Δ, i.e., G_∞ enriched with the Δ projection modality introduced in [22] (see also [6]), as well as all finite-valued logics can be considered projective, as we will explain. In particular, this allows us to transfer the obvious semantic connection between G_∞ and the finite-valued Gödel logics to corresponding proof systems.

The main concepts and results of this paper summarize the two conference contributions [8] and [7].

The paper is organized as follows: in Section 2 we define the class of projective logics and we show that the family of (finite- and infinite-valued) Gödel logics is a particular case of them (see Section 2.1). Section 2.2 relates finite- and infinite-valued projective logics. Sequents of relations calculi are introduced in Section 3. There we also shown how to translate a given specification of a projective logic into such a calculus in a systematic, even mechanizable, way. Soundness and completeness of the obtained calculi is proved in Section 3.1. Admissibility of cut rules as well as of other structural rules in sequent of relations calculi is discussed in Section 3.2. The special case of Gödel logics is presented in detail in Section 4. Sequent of relations calculi for finite and infinite-valued Gödel logic, as well as G_∞^Δ, are derived. Relationships between sequents of relations and some other analytic calculi for Gödel logics are discussed in Section 4.1.

2 Projective logics

The syntax of the propositional logics considered here is very general. The *(object) language* for a logic is based on an infinite supply of *propositional variables*, a finite set of *connectives* (with fixed arity), and a finite number of *truth constants*. (Truth constants will also be considered as 0-ary connectives.) As usual, the *formulas* of such a language are build up from its variables, constants and connectives.

The logics under investigation are characterized by a special format of the definition of their semantics. Again, we take a very general approach. To specify a semantics we refer to some (classical, first order) theory **T** — called *semantic theory* — whose intended range of discourse is a set of truth values V. **T** can, e.g., be the theory of linear orders or lattices or any other class of relational structures. It can also be the (first order) specification of a single structure. The only requirements we put on **T** are as follows:

(F) **T** is based on a function free language with finite signature. I.e., the atomic formulas are of form $R(t_1, \ldots, t_k)$, where the t_i are variables or constants.

(D) The set of Π_1-formulas that are valid in **T** is decidable.

We use the notation "$\mathcal{M} \models A[\sigma]$" to denote that the formula A is satisfied in a model $\mathcal{M}$ (of **T**) under the assignment σ of elements of the domain of $\mathcal{M}$ to the free variables of A. The domain of $\mathcal{M}$ is called *set of truth values.* By "$\mathbf{T} \models A$" we mean that A is valid in **T**, i.e. $A[\sigma]$ is satisfied in *all* models of **T** for all assignments σ.

Constants of **T** denote truth values and are identified with truth constants of the object language.

Quantifier and negation free formulas of **T**, i.e., formulas built up from atomic formulas using conjunction and disjunction only, will
play a special rôle. Let us call such formulas *simple.*

We call an n-ary connective $\Box$ *projective* (with respect to an interpretation based on a semantic theory **T**) if the corresponding *truth function* $\tilde{\Box}$ can be written in the following form:

$$\tilde{\Box}(x_1, \ldots, x_n) = \begin{cases} t_1 & \text{if } A_1 \\ \vdots & \vdots \\ t_m & \text{if } A_m \end{cases}$$

where each t_i is either a truth constant or in $\{x_1, \ldots, x_n\}$. Moreover, the conditions A_i are simple formulas of the underlying semantic theory **T** whose free variables are among $\{x_1, \ldots, x_n\}$. Since $\tilde{\Box}$ is a total function these conditions have to satisfy the following properties:

Totality: $\mathbf{T} \models \forall x_1 \cdots \forall x_n \bigvee_{1 \le i \le m} A_i$

Functionality: for all $i \neq j$; $i, j \in \{1, \ldots, m\}$: $\mathbf{T} \models \forall x_1 \cdots \forall x_n \neg(A_i \wedge A_j)$

To specify a *logic* we also need a *notion of designated truth values* or, shorter, *designating predicate.* Any simple formula $Des(x)$ of **T** with exactly one free variable x may be chosen for this purpose.[1]

[1] If one prefers to exclude the empty logic and the inconsistent "top logics" — consisting of *all* formulas of a particular signature — from the realm of logics then one should insist on $\mathbf{T} \models \exists x\, Des(x)$ and $\mathbf{T} \models \exists x \neg\, Des(x)$, respectively.

In general, for a propositional many-valued logic, an *interpretation* $\mathcal{I}$ is a mapping from the set of propositional variables $\mathcal{PV}$ into some set V of *truth values*. Given *truth functions* $\widetilde{\Box}$ for all connectives $\Box$ of the language, an interpretation $\mathcal{I}$ extends to an *evaluation function* $val_{\mathcal{I}}$, that maps all formulas into truth values, as follows:

$$val_{\mathcal{I}}(F) = \mathcal{I}(F) \quad \text{if } F \in \mathcal{PV}$$

$$val_{\mathcal{I}}(\Box(F_1, \ldots, F_n)) = \widetilde{\Box}(val_{\mathcal{I}}(F_1), \ldots, val_{\mathcal{I}}(F_n))$$

Note that this definition is independent of the particular semantic framework described above.

In the context of a semantic theory $\mathbf{T}$, logics arise at two *different levels*:

(1) The set of truth values is understood as the domain of a *model of* $\mathbf{T}$. The semantics depends on the corresponding interpretation of the conditions A_i of the (projective) truth functions. In this sense, any model $\mathcal{M}$ of $\mathbf{T}$ determines a logic $\mathcal{L}_{\mathcal{M}}$: the *projective logic* of $\mathcal{M}$ (with respect to given projective truth functions for all connectives of the language). We call a formula F *valid* in $\mathcal{L}_{\mathcal{M}}$ if for all interpretations $\mathcal{I}$ the designating predicate is satisfied in $\mathcal{M}$ whenever the value of F is assigned to the only free variable of *Des*. $\mathcal{L}_{\mathcal{M}}$ is identified with the set of valid formulas, i.e.:

$$\mathcal{L}_{\mathcal{M}} = \{F \mid \mathcal{M} \models Des[val_{\mathcal{I}}(F)/x] \text{ for all } \mathcal{I}\}.$$

(2) There is yet another useful interpretation of this semantic machinery under which the semantic *theory itself* determines a logic. Namely, instead of evaluating the conditions A_i in a particular model of $\mathbf{T}$, we may check whether the relevant instances A_i' of A_i and Des' of Des are satisfied in *all models* of $\mathbf{T}$. This way we do not have to fix the set of truth values in speaking of the *projective logic* $\mathcal{L}_{\mathbf{T}}$ associated with $\mathbf{T}$ and some projective connectives (possibly including truth constants). This allows us to speak, e.g., of *the* projective logic of, say, partial orders (with respect to a fixed set of projective connectives). Formally,

$$\mathcal{L}_{\mathbf{T}} = \{F \mid \mathcal{M} \models Des[val_{\mathcal{I}}(F)/x] \text{ for all } \mathcal{I} \text{ and } \mathcal{M} \text{ of } \mathbf{T}\}$$

Example 1. To see that every finite-valued propositional logic is projective we only have to consider *monadic* semantic theories.

Let the language of $\mathbf{T}$ contain a monadic predicate symbol C_i for each truth value c_i. In addition, we take the truth values as constants of the language. For any n-valued logic, $\mathbf{T}$ can be axiomatized by:

$$\begin{array}{l} \forall x : \bigvee_{1 \leq i \leq n} C_i(x) \\ \text{for all } i \neq j, 1 \leq i, j \leq n : \ \forall x : \neg(C_i(x) \wedge C_j(x)) \\ \text{for all } i, 1 \leq i \leq n : \ C_i(c_i) \end{array}$$

Any particular n-valued logic is determined by truth tables for all connectives and a choice of designated truth values, as usual. Every entry

$$\widetilde{\Box}(c_{i_1}, \ldots, c_{i_n}) = c_j$$

in the truth table for the n-ary connective $\Box$ translates into the part

$$\widetilde{\Box}(x_1, \ldots, x_n) = c_j \quad \text{if } C_{i_1}(x_1) \wedge \ldots \wedge C_{i_n}(x_n)$$

of the definition of the truth function as above. If $\{c_{d_1}, \ldots, c_{d_m}\}$ is the set of designated truth values, obviously $C_{d_1}(x) \vee \ldots \vee C_{d_m}(x)$ serves as designating predicate. It is easy do see that here $\mathcal{L}_{\mathbf{T}} = \mathcal{L}_{\mathcal{M}}$, for every model $\mathcal{M}$ of $\mathbf{T}$.

Alternatively, we can base the semantics on a theory containing only the equality predicate and all truth constants: just replace $C_i(t)$ by $c_i = t$. This approach obviously results in the same logics as above, but excludes interpretations based on "non-intended" sets of truth values with cardinality $> n$ for n-valued logics.

As we shall see below, even in the case of finite-valued logics, it may be advantageous to choose a more expressive semantic theory to define the truth functions for its connectives.

2.1 Gödel logics

Our main example of projective logics is the family of Gödel logics. To formulate their semantics, we assume the set of truth values to be linearly ordered and equipped with a minimal element 0 and a maximal element 1 (distinct from 0). A standard axiomatization of the corresponding semantic theory is given by:

$\forall x : \neg(x < x)$	(Irrefl$_<$)	$\forall x : x = 0 \vee 0 < x$	(Min$_<$)
$\forall x \forall y \forall z : (x < y \wedge y < z) \supset x < z$	(Trans$_<$)	$\forall x : x = 1 \vee x < 1$	(Max$_<$)
$\forall x \forall y : x = y \vee x < y \vee y < x$	(Linear$_<$)	$0 < 1$	(Distinct)

Although one could derive a "sequent of relations calculus" (see Section 3) directly from this theory we prefer an alternative formulation of it. We do not want to have to consider "=" as a basic relation, but rather base the semantics theory, denoted by $\mathbf{T}_{\leq,<}$, on the relation symbols "$\leq$" and "<" by adding the following to the axioms Irrefl$_<$, Trans$_<$, and Distinct:

$\forall x : x \leq x$	(Refl$_\leq$)	$\forall x : 0 \leq x$	(Min$_\leq$)
$\forall x \forall y \forall z : (x \leq y \wedge y \leq z) \supset x \leq z$	(Trans$_\leq$)	$\forall x : x \leq 1$	(Max$_\leq$)
$\forall x \forall y : x \leq y \vee y \leq x$	(Linear$_\leq$)	$\forall x \forall y : x < y \vee y \leq x$	(Conn.)
$\forall x \forall y : x < y \supset \neg(y \leq x)$	(Strict)		

Remark 1. Efficient decision procedures for $\mathbf{T}_{\leq,<}$ can be found, e.g., in [11, 12]. These results, in particular, imply that $\mathbf{T}_{\leq,<}$ fulfills condition **(D)** on semantic theories (i.e., decidability for Π_1-formulas).

We can now state the truth functions for disjunction ($\vee$ – maximum), conjunction ($\wedge$ – minimum), implication ($\supset$), and negation ($\neg$) in such a way that it gets clear that these connectives are projective with respect to $\mathbf{T}_{\leq,<}$.

$$\tilde{\vee}(x,y) = \begin{cases} x & \text{if } y \leq x, \\ y & \text{if } x < y \end{cases} \qquad \tilde{\wedge}(x,y) = \begin{cases} x & \text{if } x \leq y, \\ y & \text{if } y < x \end{cases}$$

$$\tilde{\supset}(x,y) = \begin{cases} 1 & \text{if } x \leq y, \\ y & \text{if } y < x \end{cases} \qquad \tilde{\neg} = \begin{cases} 1 & \text{if } x \leq 0, \\ 0 & \text{if } 0 < x \end{cases}$$

Negation can be treated as derived connective by defining $\neg A := A \supset 0$.

"1" is intended to be the only designated truth value. Therefore we take "$1 \leq x$" as designating predicate.

Infinite-valued Gödel logic G_∞ is the logic of models of $\mathbf{T}_{\leq,<}$, that contain infinite "<-chains". This condition on models can be enforced by adding to $\mathbf{T}_{\leq,<}$ the density axiom:

$$\forall x \forall y \exists z : x < y \supset (x < z \wedge z < y) \qquad \text{(Dense)}$$

However, G_∞ is not only the logic[2] of the theory axiomatized by $\mathbf{T}_{\leq,<}$ extended with (Dense), or the logic $\mathcal{L}_\mathcal{M}$ for any dense infinite model $\mathcal{M}$ of $\mathbf{T}_{\leq,<}$, but also the logic $\mathcal{L}_{\mathbf{T}_{\leq,<}}$ itself.

If we restrict attention to finite models of $\mathbf{T}_{\leq,<}$ we obtain

the family of finite-valued Gödel logics G_n. Let, e.g., $\mathcal{M}$ be the (up to isomorphism) unique model of $\mathbf{T}$ with 4 elements, in which also $\exists x \exists y : 0 < x \wedge x < y \wedge y < 1$ is satisfied, then $\mathcal{L}_\mathcal{M}$ is the 4-valued Gödel logic G_4. Instead of focusing directly on particular models $\mathcal{M}$ one may equivalently augment $\mathbf{T}_{\leq,<}$ to become the unique (first order) theory of $\mathcal{M}$. An even simpler way to obtain finite-valued Gödel logics consists in adding the following axiom to $\mathbf{T}_{\leq,<}$:

$$\exists x_1 \cdots \exists x_n \forall y : y \equiv x_1 \vee \ldots \vee y \equiv x_n \qquad (\text{Finite}_n)$$

where $x \equiv y$ abbreviates $(x \leq y \wedge y \leq x)$. This results in $\mathcal{L}_{\mathbf{T}_{\leq,<}} = G_n$.

Remark 2. One might intend to include conditions $\neg(x_i \equiv x_j)$, for $i \neq j$, in Finite_n; however, since *G_m contains G_n* if $m < n$, these conditions are redundant.

2.2 A relation between finite- and infinite-valued logics

It is well known that G_∞ is the intersection of all finite-valued Gödel logics. The concept of projective connectives allows us to grasp the connection between logics corresponding to finite and arbitrary models of a semantic theory, respectively, at a more general level.

[2] There is only one infinite-valued *propositional* Gödel logic. On the first order level different topologies on the set of truth values induce different logics.

Proposition 1. *Let* $\mathbf{T}$ *be a universal theory (i.e., axiomatized by* Π_1*-formulas only) and* F *be any formula of a projective logic over* $\mathbf{T}$*. If a formula* F *is valid in* $\mathcal{L}_{\mathcal{M}}$ *for all finite models* $\mathcal{M}$ *of* $\mathbf{T}$ *then* F *is valid in* $\mathcal{L}_{\mathcal{M}}$ *for* all *models* $\mathcal{M}$ *of* $\mathbf{T}$*. More concisely:*

$$\Big(\bigcap_{\mathcal{M} \text{ finite}} \mathcal{L}_{\mathcal{M}} \Big) = \mathcal{L}_{\mathbf{T}}.$$

Proof. Let $\mathcal{M}$ be an arbitrary model of $\mathbf{T}$ such that $\mathcal{M} \not\models Des[val_{\mathcal{I}}(F)/x]$ forsome interpretation $\mathcal{I}$. Since the connectives of F are projective its evaluation only depends on the elements of $\mathcal{M}$ assigned by $\mathcal{I}$ to the propositional variables of F and the constants of $\mathbf{T}$. That is: we can filtrate $\mathcal{M}$ into a model $\mathcal{M}'$ with domain $\{\mathcal{I}(p) \mid p \text{ occurs in } F\} \cup \{c \mid c \text{ is the value of some constant of } \mathbf{T}\}$. (Remember that the signature of any semantic theory is finite.) Therefore, if F is valid in all finite models it must be valid in arbitrary models, i.e. in $\mathcal{L}_{\mathbf{T}}$. ∎

Observe that the proof provides a bound for the size of models (i.e. number of truth values) that we have to consider if we want to check whether a formula is valid in $\mathcal{L}_{\mathbf{T}}$. In the case of G_∞ we obtain: $F \in \mathrm{G}_{|F|+2}$ implies $F \in \mathrm{G}_\infty$ where $|F|$ is the number of (distinct) propositional variables occurring in F.

3 Sequent calculi of relations

There are quite different ways to interpret Gentzen's classical sequent calculus **LK** [17]. These lead to different types of generalizations of Gentzen's calculus. One — very useful — interpretation of a sequent

$$F_1, \ldots, F_n \longrightarrow G_1, \ldots, G_m$$

is to understand it as expressing the assertion that either one of the F_i ($1 \leq i \leq n$) is false or one of the G_j ($1 \leq j \leq m$) is true. In this view a classical sequent can be identified with a sequence

$$False(F_1), \ldots, False(F_n), True(G_1), \ldots, True(G_m)$$

of (monadic) atomic formulas referring to the usual semantic theory. It is well known how this leads to the formulation of sequent calculi for all finite-valued logics (see, e.g., [23, 26, 10]).

However, one may prefer to think of the sequent arrow in

$$F \longrightarrow G$$

as associated with the *binary* semantic predicate "F implies G". In the context of a many-valued logic with an ordered set of truth value this can, e.g., be understood as

$$val_{\mathcal{I}}(F) \leq val_{\mathcal{I}}(G)$$

for all interpretations $\mathcal{I}$.

The concept of "hypersequents" (as investigated extensively by A. Avron in, e.g., [2, 3], see Section 4.1) extends the range of logics for which analytic Gentzen style systems can be given. Hypersequents are sequences of sequents understood as disjunctively connected (at the external level). If external contraction and external weakening are present and a "splitting rule" (which is an instance of Avron's communication rule, introduced in [2] to define a hypersequent calculus for G_∞ based on Gentzen's sequent calculus **LJ** for intuitionistic logic) is admissible, then the hypersequent

$$\ldots \mid F_1, \ldots, F_n \longrightarrow G \mid \ldots$$

is equivalent to the hypersequent

$$\ldots \mid F_1 \longrightarrow G \mid \ldots \mid F_n \longrightarrow G \mid \ldots$$

This hypersequent can again be viewed as a sequence of (binary) atomic formulas referring to a semantic theory. (For this one needs truth constants that correspond to an empty left or right hand sight of the sequents.)

The connection to the semantic framework described in Section 2 is manifested in the following definition:

Let $R_1, \ldots, R_n$ be the predicate symbols of a semantic theory $\mathbf{T}$, then a *sequent of relations* is a finite multiset[3] written in form

$$R_{i_1}(F_1^1, \ldots, F_{r_1}^1) \mid \ldots \mid R_{i_k}(F_1^k, \ldots, F_{r_k}^k)$$

where for all $1 \leq j \leq k$: $i_j \in \{1, \ldots, n\}$, r_ℓ is the arity of R_{i_ℓ} and all F_j^i are formulas of a logic. (Strictly speaking, the relational symbols R_j just *correspond* to symbols of the language of $\mathbf{T}$, since the terms of $\mathbf{T}$ are not formulas but variables and constants for truth values.)

We are now going to define the *sequent calculus of relations* $\mathbf{R}\mathcal{L}_\mathbf{T}$ for a projective logic $\mathcal{L}_\mathbf{T}$ defined with respect to a semantic theory $\mathbf{T}$.

Axiom sequents

Let $\mathbf{T} \models \forall \bar{x} \bigvee_{1 \leq j \leq n} B_j$ where the B_j are atomic formulas and $\bar{x}$ are the free variables in $\bigvee_{1 \leq j \leq n} B_j$. Let θ be any substitution of formulas for the variables $\bar{x}$. Then

$$B_1\theta \mid \ldots \mid B_n\theta$$

is an axiom of $\mathbf{R}\mathcal{L}_\mathbf{T}$.

Remark 3. Since $\mathbf{T}$ decides all Π_1-formulas the set of axioms is recursive.

Remark 4. Instead of taking all valid disjunctions of atomic formulas to define axioms one may just consider *minimal* valid disjunctions. I.e., one reduces the set of axioms modulo the (provability) equivalence relation induced by the structural rules described below.

[3] If one prefers sequences over multisets as basic objects of inference then a permutation rule has to be added to the calculus. This approach was used [8].

Structural rules

As already mentioned above, the structural rules for relational sequents should capture the intended interpretation of "|" as disjunction. Therefore we have the following rules in $\mathbf{R}\mathcal{L}_\mathbf{T}$.

$$\frac{\mathcal{H}}{A \mid \mathcal{H}} \text{ weakening} \qquad \frac{A \mid A \mid \mathcal{H}}{A \mid \mathcal{H}} \text{ contraction}$$

where A, B are arbitrary atomic relations on formulas and $\mathcal{H}$ is an arbitrary (possible empty) *side sequent.*

We call these rules *external* since they manipulate whole components (i.e., relations) of sequent but do not change formulas within components.

Logical rules

Let $\Box$ be an n-ary projective connective with the following truth function:

$$\widetilde{\Box}(x_1, \ldots, x_n) = \begin{cases} t_1 & \text{if } A_1(x_1, \ldots, x_n) \\ \vdots & \vdots \\ t_m & \text{if } A_m(x_1, \ldots, x_n) \end{cases}$$

For each predicate symbol R of arity r and each position p, where $1 \leq p \leq r$, we obtain a rule $(\Box : R : p)$ for introducing $\Box$ at position p into an R-component of a relational sequent. For this one considers the formula

$$\alpha_{\Box:R:p} = \bigvee_{1 \leq \ell \leq m} A_\ell(x_1, \ldots, x_n) \wedge R(z_1, \ldots, z_r)\{t_\ell / z_p\}$$

Take any conjunction of disjunctions of atomic formulas $\bigwedge_{1 \leq j \leq s} \bigvee_{1 \leq k \leq u_j} B_{j,k}$ that is equivalent in $\mathbf{T}$ to $\alpha_{\Box:R:p}$. Then we have the rule

$$\frac{B_{1,1}\theta \mid \ldots \mid B_{1,u_1}\theta \mid \mathcal{H} \quad \ldots \quad B_{s,1}\theta \mid \ldots \mid B_{s,u_s}\theta \mid \mathcal{H}}{R(z_1, \ldots, z_r)\{\Box(x_1, \ldots, x_n)/z_p\}\theta \mid \mathcal{H}} \ (\Box : R : p)$$

where θ substitutes formulas for the variables $\{x_1, \ldots, x_n\} \cup \{z_1, \ldots, z_r\} - \{z_p\}$, and $\mathcal{H}$ is the side sequent of the rule.

Remark 5. We make use of the fact that the conditions A_i are simple, i.e. negation and quantifier free.

Remark 6. In general there are *many* conjunctive normal forms equivalent to $\alpha_{\Box:R:p}$. To obtain compact rules it is often essential to apply simplifications justified by $\mathbf{T}$-valid formulas.

Remark 7. The $\alpha_{\Box:R:p}$ are Π_1-formulas of $\mathbf{T}$. Since $\mathbf{T}$ decides all Π_1-formulas, the transformation of the specification of a truth function into a logical rule for sequents of relations can — in principle — be automatized.

Example 2. Continuing Example 1 we arrive at a sequent calculus of (monadic) relations for each finite-valued logic if we follow the above definitions. In fact, because of the presence of the standard structural rules, these calculi are just notational variants of the many-placed sequent calculi or signed calculi as described, e.g., in [23, 10]. (The special case of classical logic — **LK** [17] — was already sketched at the beginning of the section.) We can even get rid of the truth constants in the formulation of the calculi. The reason for this is that, obviously, any atomic formula of **T** that contains a constant can only be of form $C_i(c_j)$, and thus is either simply true or false. For the axioms and rules this means that formulas $C_i(c_j)$ where $i \neq j$ are deleted from the sequents, and sequents containing $C_i(c_j)$ where $i = j$ are discarded, altogether.

3.1 Correctness, completeness, decidability

A sequent (of relations) $\mathcal{S}$ is called *provable* in $\mathbf{R}\mathcal{L}_\mathbf{T}$ if there is an upward tree of sequents rooted in $\mathcal{S}$, such that every leaf (topmost sequent) is an axiom and every other sequent is obtained from the ones standing immediately above it by application of one of the rules of $\mathbf{R}\mathcal{L}_\mathbf{T}$.

For any sequent

$$\mathcal{S} = R_1(F_{1,1}, ..., F_{1,r_1}) \mid \ldots \mid R_n(F_{n,1}, ..., F_{n,r_n})$$

let

$$\beta_\mathcal{S} = \bigvee_{1 \leq i \leq n} R_i(t_{i,1}, ..., t_{i,r_i})$$

be the **T**-*formula corresponding to* $\mathcal{S}$, where $t_{i,j}$ is identical to $F_{i,j}$ if $F_{i,j}$ is a truth constant[4] and is a new variable $x_{i,j}$ otherwise ($x_{i,j} = x_{k,\ell}$ iff $F_{i,j} = F_{k,\ell}$).

Since the designating predicate Des is a simple formula it is equivalent to a formula Des' of form $\bigwedge_{1 \leq i \leq p} \bigvee_{1 \leq j \leq q_i} A_{i,j}$ where the $A_{i,j}$ are atomic formulas with at most one free variable x. By

$$\mathcal{D}_1\{x/F\}, \ldots, \mathcal{D}_p\{x/F\}$$

we denote the sequence of sequents that correspond to the conjuncts of Des' if x is replaced by the formula F.

For the following statements let **T** be any semantic theory and $\mathcal{L}_\mathbf{T}$ be the logic determined by **T**, an object language, projective truth functions for this language and a designating predicate Des. $\mathbf{R}\mathcal{L}_\mathbf{T}$ is the corresponding sequent calculus of relations as defined above.

Theorem 1 (Correctness). *If all sequents* $\mathcal{D}_1\{x/F\}, \ldots, \mathcal{D}_p\{x/F\}$ *are provable in* $\mathbf{R}\mathcal{L}_\mathbf{T}$ *then* F *is valid in* $\mathcal{L}_\mathbf{T}$.

[4] Remember that we identify the constants of **T** with truth constants of the object language.

Proof. We show by induction on the length of proofs that for all models $\mathcal{M}$ of $\mathbf{T}$ and all interpretations $\mathcal{I}$: $\mathcal{M} \models \beta_{\mathcal{S}}[\sigma_{\mathcal{I}}]$ if $\mathcal{S}$ is provable, where $\sigma_{\mathcal{I}}$ assigns $val_{\mathcal{I}}(F_{i,j})$ to the corresponding variable $x_{i,j}$. From this the theorem follows by the definition of $\mathcal{D}_1\{x/F\}, \ldots, \mathcal{D}_p\{x/F\}$ and the fact that F is valid if for all $\mathcal{M}$ and $\mathcal{I}$: $\mathcal{M} \models Des[val_{\mathcal{I}}(F)/x]$.

For axioms the claim immediately follows from their definition.

For applications of structural rules with premiss $\mathcal{S}$ and conclusion $\mathcal{S}'$ we have $\beta_{\mathcal{S}}$ implies $\beta_{\mathcal{S}'}$ by the fact that the $\mathbf{T}$-formulas corresponding to sequents are classical disjunctions.

For the application of a logical rule $(\Box : R : p)$ it suffices to observe that, by definition, for any σ: $\mathcal{M} \models \alpha_{\Box:R:p}[\sigma]$ implies that $\mathcal{M} \models R(z_1, \ldots, z_r)[\sigma']$, where σ' is as σ except for assigning $\widetilde{\Box}(\sigma(x_1), \ldots, \sigma(x_n))$ to the only variable z_p that does not already occur in $\alpha_{\Box:R:p}$. ∎

Theorem 2 (Completeness). *If F is valid in $\mathcal{L}_{\mathbf{T}}$ then all sequents $\mathcal{D}_1\{x/F\}$, ..., $\mathcal{D}_p\{x/F\}$ are provable.*

Proof. (Sketch) We employ Schütte's method of reduction trees [24]. That is, we construct a reduction tree RT for every sequent $\mathcal{S}$ such that either a proof of $\mathcal{S}$ or a model in which $\beta_{\mathcal{S}}$ is not valid can be extracted from RT.

The construction of the upward tree of sequents RT for the sequent $\mathcal{S}$ is in stages as follows:

Stage 0: Write $\mathcal{S}$ at the root of RT.

Stage k: If the topmost sequent $\mathcal{S}'$ of a branch contains only propositional variables (as arguments of its relations) then stop the reduction for this branch. Otherwise $\mathcal{S}'$ contains a relation $R(F_1, \ldots, F_r)$ where $F_p \equiv \Box(G_1, \ldots, G_n)$ for some $1 \leq p \leq r$. If the indicated occurrence of $\Box(G_1, \ldots, G_n)$ is not the result of a reduction at this stage and has not yet been reduced on this branch then replace $\mathcal{S}'$ by

$$\frac{B_{1,1}\theta \mid \ldots \mid B_{1,t_1}\theta \mid \mathcal{S}' \quad \ldots \quad B_{s,1}\theta \mid \ldots \mid B_{s,t_s}\theta \mid \mathcal{S}'}{\mathcal{S}'}$$

where the $B_{i,j}$ are as in the definition of rule $(\Box : R : p)$ and θ is given by $R(z_1, \ldots, z_r)\{\Box(x_1, \ldots, x_n)/z_p\}\theta = R(F_1, \ldots, F_r)$.

Since every occurrence of a formula is only reduced once in a branch, the construction of RT stops after finitely many steps.

We say that a sequent $\mathcal{S}'$ *contains* an axiom if it can be derived from an axiom using structural rules only. If each leaf of RT contains an axiom of $\mathbf{R}\mathcal{L}_{\mathbf{T}}$ then a proof of $\mathcal{S}$ is easily constructed from RT by inserting weakenings and contractions.

Otherwise there is a leaf sequent $\mathcal{R}$ that does not contain an axiom. Let $\beta_{\mathcal{R}}$ be the $\mathbf{T}$-formula corresponding to $\mathcal{R}$ (by replacing the propositional variables f_i occurring in $\mathcal{R}$ by variables x_i of the language of $\mathbf{T}$).

By definition of the set of axioms, there is a model $\mathcal{M}$ of $\mathbf{T}$ and an assignment σ such that $\mathcal{M} \not\models \beta_{\mathcal{R}}[\sigma]$. The assignment σ of truth values to the x_i induces an interpretation $\mathcal{I}$ of the corresponding propositional variables f_i. By going down the branch from $\mathcal{R}$ to the root $\mathcal{S}$ one can augment $\mathcal{I}$ to an interpretation of all propositional variables occurring in $\mathcal{S}$ such that $\mathcal{M} \not\models \beta_{\mathcal{S}}[val_{\mathcal{I}}(H_1)/x_1, \ldots, val_{\mathcal{I}}(H_k)/x_k]$, where the H_i are the formulas in $\mathcal{S}$. (For this, of course, one has to use the corresponding truth functions as interpreted in $\mathcal{M}$.)

F is valid in $\mathcal{L}_{\mathbf{T}}$ iff for all $\mathcal{M}$ and $\mathcal{I}$, $\mathcal{M} \models \forall x\, Des[val_{\mathcal{I}}(F)/x]$. Since $\mathbf{T} \models \forall x\, Des \Leftrightarrow \bigwedge_{1 \leq i \leq p} \mathcal{D}_i$ it follows that all leaves of a reduction tree RT_i for a sequent $\mathcal{D}_i\{x/F\}$ contain axioms. Therefore all sequents $\mathcal{D}_1\{x/F\}, \ldots, \mathcal{D}_p\{x/F\}$ are provable in $\mathbf{R}\mathcal{L}_{\mathbf{T}}$ if F is valid. ∎

Since the construction of the reduction trees is effective we obtain:

Corollary 1. *All projective logics $\mathcal{L}_{\mathbf{T}}$ are decidable.*

Remark 8. The construction of the reduction tree can be seen as the search for a proof in tableau format. Here, the atomic elements of the tableau are not just formulas but (atomic) relations between formulas. The reduction of compound formulas corresponds to the introduction rules of the sequent calculus. The tableau closure rules correspond to the axioms. The close relationship between reduction trees (i.e., tableaux) and sequent proofs relies on the fact that, like in classical logic, we can view sequents as sets (i.e., modulo contraction) and can move all weakenings up to the axioms.

3.2 Extended structural rules

So far we only considered (analytic) rules for introducing connectives and traditional forms of structural rules. Let us now investigate
which types of cut rules or more general forms of structural rules
(that possibly allow to exchange formulas from different relations in sequents) are admissible in our calculi. Although we know — by completeness — that such rules are not needed for proof search, one should keep in mind that vast speedups (at least with respect to proof length) can be gained by applying such rules. This is already well known for the "simplest" case of a projective logic, namely classical logic.

We call a rule *extended structural rule* if it is of the form:

$$\frac{\mathcal{H} \mid \Gamma_1\theta \qquad \ldots \qquad \mathcal{H} \mid \Gamma_n\theta}{\mathcal{H} \mid \Gamma\theta}$$

where $\Gamma_1, \ldots, \Gamma_n, \Gamma$ are sequences of atomic formulas of $\mathbf{T}$ (separated by "|"), θ is a substitution of variables by formulas and $\mathcal{H}$ a side sequent.

Remark 9. Because of the presence of weakening and contraction, there is no loss of generality in considering only identical side sequents in the premisses. Indeed this "additive" version of rules (see [28]) is more suitable in the context of tableau style proof search.

An extended structural rule is *admissible* in $\mathbf{R}\mathcal{L}_\mathbf{T}$ if

$$\mathbf{T} \models \forall \bar{x}(\widehat{\Gamma_1} \wedge \ldots \wedge \widehat{\Gamma_n}) \supset \widehat{\Gamma}$$

where $\bar{x}$ is the vector of all variables occurring in $\Gamma_1, \ldots, \Gamma_n, \Gamma$ and $\widehat{\Delta}$ is the disjunction of the atomic formulas Δ consists of. ($\widehat{\Delta} \equiv True$ for empty Δ.)

It follows from this definition that, indeed, all sequents provable in $\mathbf{R}\mathcal{L}_\mathbf{T}$ augmented by admissible extended structural rules are already provable in $\mathbf{R}\mathcal{L}_\mathbf{T}$ without these rules.

It is important to notice that admissibility is a decidable property of rules, because we required all Π_1-sentences to be decidable in $\mathbf{T}$.

Let $vars(\Delta)$ denote the set of variables occurring in the sequent Δ. We call an extended structural rule *cut rule* if $vars(\Gamma) \subsetneq \bigcup_{1 \leq i \leq n} vars(\Gamma_i)$. (That is at least one formula is "cut out" from the premisses.) If $vars(\Gamma) \supseteq \bigcup_{1 \leq i \leq n} vars(\Gamma_i)$ we speak of an *analytic structural rule*.

Remark 10. In general, *many* different extended structural rules are admissible. They constitute an open list of possible (i.e., admissible) but (for completeness) not necessary extensions of the analytic calculi defined in Section 3. Some examples are as follows:

Example 3. If $\mathbf{T}$ contains a transitive relation "$\prec$" — e.g., "$<$" and "$\leq$" in the semantic theory $\mathbf{T}_{\leq,<}$ of Gödel logics — then

$$\frac{F \prec G \mid \mathcal{H} \qquad G \prec H \mid \mathcal{H}}{F \prec H \mid \mathcal{H}} \text{ (tr-cut)}$$

is an admissible cut rule, called *transitivity-cut*.

If the partial ordering $\prec$ has a minimal element 0 and a maximal element 1 distinct from 0 — again Gödel logics are concrete examples — then

$$\frac{F \prec 0 \mid \mathcal{H} \qquad 1 \prec F \mid \mathcal{H}}{\mathcal{H}}$$

is another admissible cut rule.

If $\prec$ is irreflexive — as "$<$" for Gödel logics — also the unary cut rules

$$\frac{1 \prec F \mid \mathcal{H}}{\mathcal{H}} \quad \text{and} \quad \frac{F \prec 0 \mid \mathcal{H}}{\mathcal{H}}$$

are admissible.

Example 4. Let "$\leq$" and 0, 1 be as in the semantic theory $\mathbf{T}_{\leq,<}$ for Gödel logics. Then

$$\frac{1 \leq F \mid \mathcal{H}}{G \leq F \mid \mathcal{H}}\ (w:\leq:l)\quad , \quad \frac{G \leq 0 \mid \mathcal{H}}{G \leq F \mid \mathcal{H}}\ (w:\leq:r) \quad \text{and} \quad \frac{1 \leq 0 \mid \mathcal{H}}{\mathcal{H}}$$

are examples of admissible analytic structural rules. The first two correspond to (internal) weakening in standard sequent calculi.

An important analytic structural rule, admissible for Gödel logics, is:

$$\frac{F \leq G \mid \mathcal{H} \qquad H \leq I \mid \mathcal{H}}{H \leq G \mid F \leq I \mid \mathcal{H}}$$

It corresponds to an instance of Avron's communication rule as we shall see in Section 4.1

4 A sequent of relations calculus for Gödel logics

As an illustration of the proof theoretic framework of the last section we present sequent of relations calculi for the sequent of relations calculus $\mathbf{RG}_\infty$ for infinite-valued Gödel logic G_∞. Calculi for G_n and for G_∞^Δ are obtained from $\mathbf{RG}_\infty$ by adding suitable axioms or rules, as explained in Remarks 11 and 12, respectively.

The rules of $\mathbf{RG}_\infty$ are derived using the semantic theory $\mathbf{T}_{\leq,<}$ (see Section 2.1) as described in Section 3.

Remember, that the (external) **structural rules** are the same for all sequent of relations calculi. For $\mathbf{RG}_\infty$ they can be written as follows:

$$\frac{\mathcal{H}}{A \lhd B \mid \mathcal{H}}\ \text{weakening} \qquad \frac{A \lhd B \mid A \lhd B \mid \mathcal{H}}{A \lhd B \mid \mathcal{H}}\ \text{contraction}$$

where $\lhd \in \{<, \leq\}$.

The **axioms of** $\mathbf{RG}_\infty$ are all sequents that contain a sequent

$$A_1 \lhd_1 A_2 \mid A_2 \lhd_2 A_3 \mid \ldots \mid A_k \lhd_k A_1$$

for $k \geq 1$, where $\lhd_i \in \{<, \leq\}$ for all $1 \leq i \leq k$, but $\lhd_i \equiv \leq$ for at least one i. In addition, all sequents that are obtained from the above ones by deleting relations of form

$$A < 0, \qquad 1 < A, \qquad \text{or} \qquad 1 \leq 0$$

are axioms.

In [7] the following explicit description of the above axioms has been introduced:

(a) $A_1 \lhd_n A_n \mid \ldots \mid A_3 \lhd_2 A_2 \mid A_2 \leq A_1$, where $\lhd_i \in \{<, \leq\}$ and the case $n = 1$ is defined as $A_1 \leq A_1$,

(b) $A_n \leq A_{n-1} \mid A_{n-1} < A_{n-2} \mid \ldots \mid A_1 < 1$, where the case $n = 1$ is defined as $A_1 \leq 1$,
(c) $0 < A_n \mid \ldots \mid A_3 < A_2 \mid A_2 \leq A_1$, where the case $n = 1$ is defined as $0 \leq A_1$,
(d) $0 < A_1 \mid A_1 < A_2 \mid \ldots \mid A_n < 1$, where the case $n = 0$ is defined as $0 < 1$.

The above description, corresponds to the original set of axioms up to (external) weakening.

Theorem 3. *Axioms of type (a) – (d) are, up to (external) weakening, all and only the atomic sequents valid in* $\mathbf{T}_{\leq,<}$.

It is easy to check that all of the above axioms correspond to valid statements in $\mathbf{T}_{\leq,<}$. However, to guarantee completeness we also have to show the converse: namely, that all valid atomic sequents are obtained from these axioms using external weakening only. For this purpose it is better to consider the dual form of the axioms. I.e., we make use of the fact that $val_{\mathcal{I}}(A) < val_{\mathcal{I}}(B)$ iff $\neg[val_{\mathcal{I}}(B) \leq val_{\mathcal{I}}(A)]$, and thus may consider conjunctions of components instead of disjunctions.

Definition 1. *A set of components is called* dual to axioms *if it does not contain any subset of one of the following forms:*

(a) (anti-cycle) $\{A_1 < A_2, A_2 \lhd_2 A_3, \ldots, A_n \lhd_n A_1\}$, *where* $\lhd_i \in \{<, \leq\}$ *and the case* $n = 1$ *is defined as* $\{A_1 < A_1\}$,
(b) (anti-1-chain) $\{1 \leq A_1, \ldots, A_{n-2} \leq A_{n-1}, A_{n-1} < A_n\}$, *where the case* $n = 1$ *is defined as* $\{1 < A_1\}$,
(c) (anti-0-chain) $\{A_1 < A_2, A_2 \leq A_3, \ldots, A_n \leq 0\}$, *where the case* $n = 1$ *is defined as* $\{A_1 < 0\}$,
(d) (anti-0-1-chain) $\{1 \leq A_1, A_2 \leq A_3, \ldots, A_n \leq 0\}$, *where the case* $n = 0$ *is defined as* $\{1 \leq 0\}$.

It suffices to prove the following:

Theorem 4. *Let* Γ *be a finite set of components* $A \lhd B$, $\lhd \in \{<, \leq\}$, *where* A *and* B *are either propositional variables or truth constants. If* Γ *is dual to axioms then* Γ *is satisfiable; i.e., there exists an interpretation that satisfies all components of* Γ.

To prove Theorem 4 we extend any Γ that is dual to axioms to a "maximal" set Γ^* that is still dual to axioms. Let us write $B \in [A] \iff \{A \leq B, B \leq A\} \subseteq \Gamma^*$. It will follow from Proposition 2 and Lemma 1, below, that this is an equivalence relation and that the set of equivalence classes $\overline{\Gamma^*} = \{[A] : A \text{ occurs in } \Gamma^*\}$ is totally ordered with respect to $[A] < [B] \iff A < B \in \Gamma^*$. The minimal element of the ordering is $[0]$ and its maximal element is $[1]$ (if 0 and 1 occur in Γ). The ordering thus allows to match equivalence classes with truth values in a way that induces an interpretation satisfying Γ^* and therefore also Γ.

We first add $A \leq B$ to Γ whenever $A < B \in \Gamma$. This is justified by the following simple observation:

Proposition 2. *If Γ is dual to axioms then $\Gamma \cup \{A \leq B : A < B \in \Gamma\}$ is dual to axioms, too.*

The existence of $\overline{\Gamma^*}$ follows from the following:

Lemma 1. *If Γ is dual to axioms then either $\Gamma \cup \{A < B\}$ or $\Gamma \cup \{B \leq A\}$ is dual to axioms, too.*

Proof. The proof proceeds by case distinctions:

(1) $\Gamma \cup \{A < B\}$ contains an anti-cycle. Then either already Γ contains an anti-cycle or $\{B \leq U_1, \ldots, U_n \leq A\} \subseteq \Gamma$. From this it follows that $\Gamma \cup \{B \leq A\}$ is dual to axioms iff Γ is dual to axioms.

(2) $\Gamma \cup \{B \leq A\}$ contains an anti-cycle. Then either already Γ contains an anti-cycle or $\{A \leq U_1, \ldots, U_k < U_{k+1}, \ldots U_n \leq B\} \subseteq \Gamma$. From this it follows that $\Gamma \cup \{A < B\}$ is dual to axioms iff Γ is dual to axioms.

(3) Neither $\Gamma \cup \{A < B\}$ nor $\Gamma \cup \{B \leq A\}$ contains an anti-cycle.

(3.1) $\Gamma \cup \{A < B\}$ contains an anti-1-chain W.l.o.g., the anti-1-chain is not already contained in Γ. Therefore **(a)**: $\{1 \leq V_1, \ldots, V_{n-1} \leq A\} \subseteq \Gamma$.

(3.1.1) $\Gamma \cup \{B \leq A\}$ contains an anti-1-chain that is not already contained in Γ. Therefore $\{1 \leq U_1, \ldots, U_{k-1} \leq B\} \subseteq \Gamma$ and $\{A \leq U_{k+1}, \ldots, U_{k+m-1} < U_{k+m}\} \subseteq \Gamma$. The latter subset can be combined with **(a)** to an anti-1-chain in Γ.

(3.1.2) $\Gamma \cup \{B \leq A\}$ contains an anti-0-chain that is not already contained in Γ. Therefore $\{U_1 < U_2, \ldots, U_{k-1} \leq B\} \subseteq \Gamma$ and $\{A \leq U_{k+1}, \ldots, U_{k+m} \leq 0\} \subseteq \Gamma$. The latter subset can be combined with **(a)** to an anti-0-1-chain in Γ.

(3.1.3) $\Gamma \cup \{B \leq A\}$ contains an anti-0-1-chain that is not already contained in Γ. Therefore $\{1 \leq U_1, \ldots, U_{k-1} \leq B\} \subseteq \Gamma$ and $\{A \leq U_{k+1}, \ldots, U_{k+m} \leq 0\} \subseteq \Gamma$. The latter subset can be combined with **(a)** to an anti-0-1-chain in Γ.

(3.2) $\Gamma \cup \{B \leq A\}$ contains an anti-1-chain that is not already contained in Γ. Therefore
(b1): $\{1 \leq V_1, \ldots, V_{k-1} \leq B\} \subseteq \Gamma$ and **(b2)**: $\{A \leq V_{k+1}, \ldots, V_{k+m-1} < V_{k+m}\} \subseteq \Gamma$.

(3.2.1) $\Gamma \cup \{A < B\}$ contains an anti-1-chain. This case was already settled in (3.1.1).

(3.2.2) $\Gamma \cup \{A < B\}$ contains an anti-0-chain that is not already contained in Γ. Then $\{B \leq U_2, \ldots, U_n \leq 0\} \subseteq \Gamma$. This subset can be combined with **(b1)** to an anti-0-1-chain in Γ.

(3.3) Neither $\Gamma \cup \{A < B\}$ nor $\Gamma \cup \{A \leq B\}$ contain an anti-1-chain.

(3.3.1) $\Gamma \cup \{A < B\}$ contains an anti-0-chain that is not already contained in Γ. Then **(c)** $\{B \leq V_2, \ldots, V_n \leq 0\} \subseteq \Gamma$.

(3.3.1.1) $\Gamma \cup \{B \leq A\}$ contains an anti-0-chain that is not already contained in Γ. Therefore $\{U_1 < U_2, \ldots, U_{k-1} \leq B\} \subseteq \Gamma$ and $\{A \leq U_{k+1}, \ldots, U_{k+m} \leq 0\} \subseteq \Gamma$. The first subset can be combined with **(c)** to an anti-0-chain in Γ.

(3.3.1.2) $\Gamma \cup \{B \leq A\}$ contains an anti-0-1-chain that is not already contained in Γ. Therefore $\{1 < U_2, \ldots, U_{k-1} \leq B\} \subseteq \Gamma$ and $\{A \leq U_{k+1}, \ldots, U_{k+m} \leq 0\} \subseteq \Gamma$. The first subset can be combined with **(c)** to an anti-0-1-chain in Γ.

Finally observe that if $\Gamma \cup \{A < B\}$ contains an anti-0-1-chain then this anti-0-1-chain is already contained in Γ. It is easy to check that this settles all remaining cases. ∎

The **logical rules of $\mathbf{RG}_\infty$** — i.e., the rules for introducing connectives at any place of a sequent — are easily derived as described in Section 3 above by manipulating the rule-defining formulas $\alpha_{\Box:R:p}$ in $\mathbf{T}_{\leq,<}$. Here $R \in \{<, \leq\}$; $p \in \{l, r\}$ for the left and right argument position of the binary relations, respectively, and $\Box \in \{\supset, \wedge, \vee\}$.

$$\begin{aligned}
\alpha_{(\supset:<:r)} &\equiv (x \leq y \wedge z < 1) \vee (y < x \wedge z < y) \\
&\Longleftrightarrow (x \leq y \vee y < x) \wedge (x \leq y \vee z < y) \wedge \\
&\qquad (z < 1 \vee y < x) \wedge (z < 1 \vee z < y) \\
&\Longleftrightarrow (x \leq y \vee z < y) \wedge (z < 1)
\end{aligned}$$

The rule for introducing implication at the right argument place of "<" in a relational sequent can therefore be stated as:

$$\frac{A \leq B \mid C < B \mid \mathcal{H} \qquad C < 1 \mid \mathcal{H}}{C < (A \supset B) \mid \mathcal{H}} \; (\supset:<: r)$$

Similarly we have in $\mathbf{T}_{\leq,<}$

$$\begin{aligned}
\alpha_{(\supset:<:l)} &\equiv (x \leq y \wedge 1 < z) \vee (y < x \wedge y < z) \\
&\Longleftrightarrow y < x \wedge y < z
\end{aligned}$$

Thus a rule for introducing implication at the left argument place of "<" is:

$$\frac{B < A \mid \mathcal{H} \qquad B < C \mid \mathcal{H}}{(A \supset B) < C \mid \mathcal{H}} \; (\supset:<: l)$$

For implication at the right hand side of the $\leq$-relations we obtain:

$$\begin{aligned}
\alpha_{(\supset:\leq:r)} &\equiv (x \leq y \wedge z \leq 1) \vee (y < x \wedge z \leq y) \\
&\Longleftrightarrow (x \leq y \vee y < x) \wedge (x \leq y \vee z \leq y) \\
&\Longleftrightarrow x \leq y \vee z \leq y
\end{aligned}$$

This induces the rule

$$\frac{A \leq B \mid C \leq B \mid \mathcal{H}}{C \leq (A \supset B) \mid \mathcal{H}} \; (\supset:\leq: r)$$

A compact rule for introducing implication at the left hand side of the $\leq$-relations is obtained by the following derivation in $\mathbf{T}_{\leq,<}$:

$$\begin{aligned}\alpha_{(\supset:\leq:l)} &\equiv (x \leq y \wedge 1 \leq z) \vee (y < x \wedge y \leq z)\\ &\Longleftrightarrow (x \leq y \vee y < x) \wedge (x \leq y \vee y \leq z) \wedge\\ &\quad\ (1 \leq z \vee y < x) \wedge (1 \leq z \vee y \leq z)\\ &\Longleftrightarrow (1 \leq z \vee y < x) \wedge (y \leq z)\end{aligned}$$

This induces the rule

$$\frac{1 \leq C \mid B < A \mid \mathcal{H} \qquad B \leq C \mid \mathcal{H}}{(A \supset B) \leq C \mid \mathcal{H}} \ (\supset:\leq:l)$$

Observe that this is the only $\leq$-rule exhibiting "<" in the premisses. (This is of importance for the connection to Avron's hypersequent calculus $\mathbf{G}LC$ for G_∞; see Section 4.1 below.)

Computing the rules for disjunction and conjunction is easy. They take the same form for both relations. We therefore let $\lhd$ stand for either $<$ or $\leq$ (uniformly in each rule):

$$\frac{C \lhd A \mid \mathcal{H} \qquad C \lhd B \mid \mathcal{H}}{C \lhd (A \wedge B) \mid \mathcal{H}} \ (\wedge : \lhd : r) \qquad \frac{A \lhd C \mid B \lhd C \mid \mathcal{H}}{(A \wedge B) \lhd C \mid \mathcal{H}} \ (\wedge : \lhd : l)$$

$$\frac{C \lhd A \mid C \lhd B \mid \mathcal{H}}{C \lhd (A \vee B) \mid \mathcal{H}} \ (\vee : \lhd : r) \qquad \frac{A \lhd C \mid \mathcal{H} \qquad B \lhd C \mid \mathcal{H}}{(A \vee B) \lhd C \mid \mathcal{H}} \ (\vee : \lhd : l)$$

Remark 11. To obtain a calculus for n-valued Gödel logic G_n one has to add to $\mathbf{RG}_\infty$ all axioms of form

$$A_1 \lhd_1 A_2 \mid A_2 \lhd_2 A_3 \mid \ldots \mid A_l \lhd_l A_{l+1}$$

where $\lhd_i \equiv \leq$ for at least n different $i \in \{1, \ldots, l\}$, where $l > n$. Below, we refer to this calculus as $\mathbf{RG}_n$.

Remark 12. In [6] G_∞ is extended with the "projection modality" Δ. In the resulting logic G_∞^Δ, one can make "fuzzy" statements "crisp", since a formula ΔP is mapped to the distinguished truth value 1 if the value of P equals 1, and to 0 otherwise. Clearly, Δ is projective with respect to the semantic theory $\mathbf{T}_{\leq,<}$ of Section 2.1. Indeed, its truth function can be defined as:

$$\widetilde{\Delta}(x) = \begin{cases} 1 & \text{if } 1 \leq x \\ 0 & \text{if } x < 1 \end{cases}$$

To obtain a calculus for G_∞^Δ one has to add to $\mathbf{RG}_\infty$ suitable rules for introducing Δ. These rules, defined using the algorithm described in Section 3, are as follows:

$$\frac{A < 1 \mid 1 \leq B}{\Delta A \leq B} \ (\Delta :\leq: l) \qquad \frac{B \leq 0 \mid 1 \leq A}{B \leq \Delta A} \ (\Delta :\leq: r)$$

$$\frac{0 < B \quad A < 1}{\Delta A < B}\ (\Delta :<: l) \qquad \frac{B < 1 \quad 1 \leq A}{B < \Delta A}\ (\Delta :<: r)$$

Remark 13. As pointed out in Section 3.2, different forms of cuts are admissible in $\mathbf{RG}_\infty$. In [7] the following version of the cut rule has been considered

$$\frac{\mathcal{H} \mid A \leq B \qquad \mathcal{H} \mid B < A}{\mathcal{H}}\ (\text{cut}_{</\geq})$$

and a constructive proof for the elimination of $(\text{cut}_{</\geq})$ from given derivations has been presented.

Focusing on $(\text{cut}_{</\geq})$ is motivated by the fact that it allows to simulate other forms of cut straightforwardly. E.g., the *transitivity-cut* (see Section 3.2)

$$\frac{A \leq B \mid \mathcal{H} \qquad B \leq C \mid \mathcal{H}}{A \leq C \mid \mathcal{H}}\ (\text{tr-cut}_{\leq})$$

can be derived from a 3-component-cycle by applying $(\text{cut}_{</\geq})$ twice in the following way:

$$\frac{\dfrac{C < B \mid B < A \mid A \leq C \qquad B \leq C \mid \mathcal{H}}{B < A \mid A \leq C \mid \mathcal{H}} \qquad A \leq B \mid \mathcal{H}}{A \leq C \mid \mathcal{H}}$$

Similar admissible rules involving $<$ instead of $\leq$ can be treated analogously.

As we shall see in Section 4.1 below, $(\text{cut}_{</\geq})$ also allows to derive a version of Avron's communication rule.

A different type of admissible rule, related to cut, is the so-called Takeuti-Titani rule. This rule, which expresses the density of the set of truth values, has been used in [27] to axiomatize first-order G_∞:

$$\frac{F \leq p \mid p \leq G \mid \mathcal{H}}{F \leq G \mid \mathcal{H}}\ (\text{tt})$$

where p is a propositional variable not occurring in the lower sequent. It is interesting to observe that (tt) cannot be derived in $\mathbf{RG}_\infty$ since — in contrast to $(\text{cut}_{</\geq})$ and the other rules of $\mathbf{RG}_\infty$ — (tt) is not strongly sound, e.g., in finite-valued Gödel logics.

4.1 Relationships with hypersequent calculi and decomposition proof systems for Gödel logics

We first relate our approach to *hypersequent calculi* for Gödel logics.

As already mentioned in Section 3, a hypersequent is a finite sequence of sequents (see [3] for an overview). In [2] Avron introduced a hypersequent calculus for G_∞. This calculus — called **G***LC* — is defined by embedding Gentzen's **LJ**-sequents into hypersequents. Suitable structural rules are added to manipulate the additional layer of structure to the basic objects of inferences. More precisely, these structural rules are external permutation, external weakening, external contraction and the communication rule:

$$\frac{\Gamma, \Gamma' \Rightarrow U \mid \mathcal{H} \quad \Gamma_1, \Gamma_1' \Rightarrow V \mid \mathcal{H}}{\Gamma, \Gamma_1 \Rightarrow U \mid \Gamma', \Gamma_1' \Rightarrow V \mid \mathcal{H}} \ (com)$$

This rule increases the expressive power of **G***LC* compared to **LJ** but is problematic from a computational point of view.

G*LC* can be viewed as a calculus of relations. For this purpose we use an equivalent formulation, where the hypersequents are fully split (see Section 3). This version of Avron's calculus, consequently, consists of:

- axioms $A \leq A$
- (external) structural rules
- internal weakening rules $(w :\leq: l)$ and $(w :\leq: r)$ (see Example 4)
- all $\leq$-rules of $\mathbf{RG}_\infty$ with the exception of $(\supset:\leq: l)$, which is replaced by

$$\frac{D \leq A \mid \mathcal{H} \qquad B \leq C \mid \mathcal{H}}{(A \supset B) \leq C \mid D \leq C \mid \mathcal{H}} \ (\supset:\leq: l)^*$$

- the communication rule:

$$\frac{A_1 \leq U \mid \ldots \mid A_n \leq U \mid \mathcal{H} \qquad B_1 \leq V \mid \ldots \mid B_m \leq V \mid \mathcal{H}}{A_1 \leq V \mid \ldots \mid A_n \leq V \mid B_1 \leq U \mid \ldots \mid B_m \leq U \mid \mathcal{H}} \ (\text{rcom})$$

It is not hard to see that $\mathbf{RG}_\infty$ and (the above version of) **G***LC* are equivalent. Indeed, $(\supset:\leq: l)^*$ is derivable from $(\supset:\leq: l)$ in $\mathbf{RG}_\infty$ using transitivity-cuts which can be eliminated from proofs (see Remark 13 and [7]):

$$\frac{\dfrac{\dfrac{\dfrac{D \leq A \mid \mathcal{H} \qquad A \leq B \mid B < A}{D \leq B \mid B < A \mid \mathcal{H}} \ (\text{tr-cut}) \qquad B \leq C \mid \mathcal{H}}{D \leq C \mid B < A \mid \mathcal{H}} \ (\text{tr-cut})}{1 \leq C \mid B < A \mid D \leq C \mid \mathcal{H}} \ (\text{weak.}) \qquad B \leq C \mid \mathcal{H}}{A \supset B \leq C \mid D \leq C \mid \mathcal{H}} \ (\supset:\leq: l)$$

Moreover $(rcom)$ is derivable in $\mathbf{RG}_\infty$ using 4-component-cycles. E.g., in $(rcom)$, let $n = m = 1$, one has:

$$\frac{B_1 \leq V \mid \mathcal{H} \qquad \dfrac{A_1 \leq U \mid \mathcal{H} \qquad U < A_1 \mid A_1 \leq V \mid V < B_1 \mid B_1 \leq U}{A_1 \leq V \mid V < B_1 \mid B_1 \leq U \mid \mathcal{H}}}{A_1 \leq V \mid B_1 \leq U \mid \mathcal{H}}$$

Consequently, $\mathbf{RG}_\infty$ simulates Avron's **G***LC*.

In the other direction, it easy to show that pure $\leq$-sequents derivable in $\mathbf{RG}_\infty$ are also derivable in (the above version of) **G***LC*: Let $(A < B)^d \equiv 1 \leq ((B \supset A) \supset B)$ and $(A \leq B)^d \equiv 1 \leq A \supset B$. Then all ^{d}-translations of derivable $\mathbf{RG}_\infty$-sequents are derivable in **G***LC*, since the translations of $\mathbf{RG}_\infty$-rules are derivable in **G***LC*.

Concerning efficient proof search, the most important feature of $\mathbf{RG}_\infty$ is the fact that all the rules are invertible. In particular, we do not have to use the communication rule (or any similar rule destroying the "locality" of tableau style proof search). On the other hand, even if we enrich $\mathbf{G}LC$ by the (more general) axioms of $\mathbf{RG}_\infty$ the valid sequent

$$A \supset B \leq C \mid A \leq B \mid C \leq B$$

is *not* cut-free provable *without* using the communication rule. In other words: the communication rule cannot be avoided if "$<$" is eliminated from the signature. This is connected with the fact that this rule cannot be permuted over the rule for introducing $\supset$ on the left while can be permuted over all the remaining rules of $\mathbf{G}LC$.

Remark 14. A hypersequent calculus $\mathbf{G}LC_n$ for n-valued Gödel logic, with $n \geq 2$, has been introduced in [13]. This calculus is obtained by replacing, in $\mathbf{G}LC$, the communication rule with the following one:

$$\frac{\Gamma_1, \Gamma_2 \Rightarrow U_1 \mid \mathcal{H} \quad \Gamma_2, \Gamma_3 \Rightarrow U_2 \mid \mathcal{H} \quad \ldots \quad \Gamma_{n-1}, \Gamma_n \Rightarrow U_{n-1} \mid \mathcal{H}}{\Gamma_1 \Rightarrow U_1 \mid \ldots \mid \Gamma_{n-1} \Rightarrow U_{n-1} \mid \Gamma_n \Rightarrow \ \mid \mathcal{H}} \ (G_n)$$

In analogy to the case above, it is not hard to see that the sequent of relations calculus $\mathbf{RG}_n$ (see Remark 11) simulates $\mathbf{G}LC_n$ (and vice versa). As an example consider the following (sequent of relations) version of the (G_3) rule

$$\frac{A_1 \leq U_1 \mid A_2 \leq U_1 \mid \mathcal{H} \quad A_2 \leq U_2 \mid A_3 \leq U_2 \mid \mathcal{H}}{A_1 \leq U_1 \mid A_2 \leq U_2 \mid A_3 \leq 0 \mid \mathcal{H}} \ (G_3)$$

which can be derived in $\mathbf{RG}_3$ as follows:

$$\dfrac{A_2 \leq U_2 \mid A_3 \leq U_2 \qquad \dfrac{A_1 \leq U_1 \mid A_2 \leq U_1 \quad A_1 \leq U_1 \mid U_1 < A_2 \mid A_2 \leq U_2 \mid U_2 < A_3 \mid A_3 \leq 0}{A_1 \leq U_1 \mid A_2 \leq U_2 \mid U_2 < A_3 \mid A_3 \leq 0}}{A_1 \leq U_1 \mid A_2 \leq U_2 \mid A_3 \leq 0}$$

$\mathbf{RG}_\infty$ is also closely related with the *decomposition proof system* GLC^*_{RS} introduced in [5]. This calculus has been defined following the general Rasiowa-Sirkorski methodology for constructing analytic proof systems for semantically defined logics. GLC^*_{RS} and $\mathbf{RG}_\infty$ are interderivable using simple notational transformations. Indeed, in one direction, a sequent $A_1, \ldots, A_n \Rightarrow B_1, \ldots, B_m$ in GLC^*_{RS} can be translated into a sequent of relations $A_1 \leq B_1 \mid \ldots \mid A_1 \leq B_m \mid A_n \leq B_1 \mid \ldots \mid A_n \leq B_m$. One can easily check that the (translated versions of the) axioms and rules of GLC^*_{RS} are derivable in $\mathbf{RG}_\infty$. As an example we consider the following version of the rules $(\Rightarrow\supset(\supset))$ and $(\supset(\supset)\Rightarrow)$ of GLC^*_{RS}, respectively:

$$\frac{X \leq B \supset C \mid X \leq A \supset C \mid \mathcal{H}}{X \leq A \supset (B \supset C) \mid \mathcal{H}} \qquad \frac{B \supset C \leq X \mid \mathcal{H} \quad A \supset C \leq X \mid \mathcal{H}}{A \supset (B \supset C) \leq X \mid \mathcal{H}}$$

The above rules can be derived in $\mathbf{RG}_\infty$ as follows (we omit side sequents that are not involved in the derivations):

$$\dfrac{X \leq B \supset C \mid X \leq A \supset C \qquad \dfrac{\dfrac{\dfrac{\dfrac{\dfrac{C < A \mid A \leq C}{1 \leq C \mid C < A \mid A \leq C}\,(\text{weak.}) \qquad \dfrac{C \leq C}{C \leq C \mid A \leq C}\,(\text{weak.})}{A \supset C \leq C \mid A \leq C}\,(\supset:\leq:\text{l})}{A \supset C \leq C \mid B \leq C \mid A \leq C \mid B \leq C}\,(\text{weak.})}{A \supset C \leq C \mid B \leq C \mid A \leq B \supset C}\,(\supset:\leq:\text{r})}{A \supset C \leq B \supset C \mid A \leq B \supset C}\,(\supset:\leq:\text{r})}{\dfrac{\dfrac{X \leq B \supset C \mid A \leq B \supset C \mid X \leq B \supset C}{X \leq B \supset C \mid A \leq B \supset C}\,(\text{contr.})}{X \leq A \supset (B \supset C)}\,(\supset:\leq:\text{r})}\,(\text{tr-cut})$$

and

$$\dfrac{\dfrac{\dfrac{A \supset C \leq X \qquad \dfrac{\dfrac{A \leq C \mid C < A}{1 \leq C \mid A \leq C \mid C < A}\,(\text{weak.})}{1 \leq A \supset C \mid C < A}\,(\supset:<:\text{r})}{1 \leq X \mid C < A}\,(\text{tr-cut}) \qquad \dfrac{B \supset C \leq X \qquad \dfrac{\dfrac{B \leq C \mid C < B}{1 \leq C \mid B \leq C \mid C < B}\,(\text{weak.})}{1 \leq B \supset C \mid C < B}\,(\supset:<:\text{r})}{1 \leq X \mid C < B}\,(\text{tr-cut})}{1 \leq X \mid B \supset C < A}\,(\supset:<:\text{l}) \qquad B \supset C \leq X}{A \supset (B \supset C) \leq X}\,(\supset:\leq:\text{l})$$

In the other direction, a sequent G in $\mathbf{RG}_\infty$ can be translated into a sequent $\Gamma \Rightarrow \Delta$ in GLC^*_{RS} as follows (see Section 4 of [5]):

1. If $1 \leq B$ is a component of G, then $B \in \Delta$
2. If $B \leq 1$ is a component of G, then $B \supset B \in \Delta$
3. If $1 < B$ is a component of G, then $B \supset B \in \Gamma$
4. If $B < 1$ is a component of G, then $B \in \Gamma$
5. If $A \leq B$ is a component of G $(A, B \neq 1)$, then $A \supset B \in \Delta$
6. If $A < B$ is a component of G $(A, B \neq 1)$, then $B \supset A \in \Gamma$

One can easily check that this translation makes the axioms and rules of $\mathbf{RG}_\infty$ identical to those of GLC^*_{RS}.

We emphasize that $\mathbf{RG}_\infty$ is just a special instance of a more general framework for theorem proving in many-valued logics. Our approach sheds light on the connections between G_∞ and other projective logics. In particular, it allows to understand the various formulations of "signed calculi" for finite-valued logics (see, e.g., [9]) as instances of the same methodological principle that is used to derive $\mathbf{RG}_\infty$.

References

1. Avellone, A., Ferrari, M. and Miglioli, P.: Duplication-free tableau calculi together with cut-free and contraction free sequent calculi for the interpolable propositional intermediate logics. Logic J. of the IGPL, 7, (4) (1999) 447–480
2. Avron, A.: Hypersequents, logical consequence and intermediate logics for concurrency. Annals of Mathematics and Artificial Intelligence, 4 (1991) 225–248
3. Avron, A.: The method of hypersequents in the proof theory of propositional nonclassical logics. In Logic: from Foundations to Applications, European Logic Colloquium. Oxford Science Publications. Clarendon Press. Oxford (1996) 1–32
4. Avron, A.: A tableau system for Gödel-Dummett logic based on a hypersequential calculus. In Automated Reasoning with Tableaux and Related Methods (Tableaux'2000), volume 1847 of Lectures Notes in Artificial Intelligence (2000) 98–112
5. Avron, A. and Konikowska, B.: Decomposition proof systems for Gödel logics. Studia Logica, 69 (2001) 197–219
6. Baaz, M.: Infinite-valued Gödel logics with 0-1-projections and relativizations. In Gödel 96. Kurt Gödel's Legacy, volume 6 of LNL (1996) 23–33
7. Baaz, M., Ciabattoni, A. and Fermüller, C.: Cut-elimination in a sequents-of-relations calculus for Gödel logic. In International Symposium on Multiple Valued Logic (ISMVL'2001). IEEE (2001) 181–186
8. Baaz, M. and Fermüller, C.: Analytic calculi for projective logics. In Automated Reasoning with Tableaux and Related Methods (Tableaux'99), volume 1617 of Lectures Notes in Artificial Intelligence (1999) 36–51
9. Baaz, M., Fermüller, C. and Salzer, G.: Automated deduction for many-valued logic. In Handbook of Automated Reasoning. Elsevier (2001)
10. Baaz, M., Fermüller, C. and Zach, R.: Elimination of cuts in first-order finite-valued logics. J. Inform. Process. Cybernet. **EIK**, 29, (6) (1994) 333–355
11. Bachmair, L. and Ganzinger, H.: Ordered chaining for total orderings. In CADE'94, volume 814 of Lecture Notes in Computer Science (1994) 435–450
12. Bachmair, L. and Ganzinger, H.: Ordered chaining calculi for first-order theories of transitive relations. J. of the ACM, 45, (6) (1998) 1007–1049
13. Ciabattoni, A. and Ferrari, M.: Hypersequent calculi for some intermediate logics with bounded Kripke models. J. of Logic and Computation, 2, (11) (2001) 283–294
14. Dummett, M.: A propositional logic with denumerable matrix. J. of Symbolic Logic, 24 (1959) 96–107
15. Dunn, J.M. and Meyer, R.K.: Algebraic completeness results for Dummett's LC and its extensions. Z. Math. Logik Grundlagen Math, 17 (1971) 225–230
16. Dyckhoff, R.: A deterministic terminating sequent calculus for Gödel Dummett logic. Logic Journal of the IGPL, 7 (1999) 319–326
17. Gentzen, G.: Untersuchungen über das logische Schliessen I, II. Mathematische Zeitschrift, 39 (1934) and (1935) 176–210 and 405–431
18. Gödel, K.: Zum intuitionistischen Aussagenkalkül. Anz. Akad. Wiss. Wien, 69 (1932) 65–66
19. Hájek, P.: Metamathematics of Fuzzy Logic. Kluwer (1998)
20. Hájek, P., Godo, L. and Esteva, F.: A complete many-valued logic with product-conjunction. Archive for Mathematical Logic, 35 (1996) 191–208

21. Łukasiewicz, J.: Philosophische Bemerkungen zu mehrwertigen Systemen der Aussagenlogik. Comptes Rendus de la Societé des Science et de Lettres de Varsovie (1930) 51–77
22. Moisil, G.: Essais sur les logiques non chrysipiennes. Editions de l'Academie de la Republique Socialiste de Roumanie, Bucarest (1972)
23. Rousseau, G.: Sequents in many valued logic I. Fund. Math., 60 (1967) 23–33
24. Schütte, K.: Proof Theory. Springer, Berlin and New York (1977)
25. Sonobe, O.: A Gentzen-type formulation of some intermediate propositional logics. J. of Tsuda College, 7 (1975) 7–14
26. Takahashi, M.: Many-valued logics of extended Gentzen style I. Science Reports of the Tokyo Kyoiku Daigaku, 9 (1967) 95–116
27. Takeuti, G. and Titani, T.: Intuitionistic fuzzy logic and intuitionistic fuzzy set theory. J. of Symbolic Logic, 49 (1984) 851–866
28. Troelstra, A.S. and Schwichtenberg, H.: Basic Proof Theory. Cambridge University Press (1996)
29. Visser, A.: On the completeness principle: a study of provability in Heyting's Arithmetic. Annals of Math. Logic, 22 (1982) 263–295

Chapter 7
Polarity-based Stochastic Local Search Algorithms for Non-clausal Satisfiability

Zbigniew Stachniak

York University, Toronto M3J 1P3, Canada,
zbigniew@cs.yorku.ca

Abstract. This paper discusses the use of polarity to guide a stochastic local search for a satisfying assignment for a free-form formula of a finitely-valued propositional logic.

1 Introduction

In 1992, Selman, Levesque, and Mitchell published a paper entitled *A New Method for Solving Hard Satisfiability Problems*, [14]. In their paper, an incomplete algorithm GSAT for the satisfiability problem for the clausal fragment of classical propositional logic (SAT) is described. Since then, stochastic local search algorithms for SAT became the subject of vigorous research in the AI and Theoretical Computer Science communities (cf. [12, 3]). This outburst of research stems not only from the theoretical importance of SAT, but also from the surprising effectiveness of the local search algorithms for SAT in finding solutions to 'realistic' instances of hard problems.

The satisfiability problem is NP-complete for a number of MV-logics known from the literature (e.g., for the n-valued and infinitely-valued sentential calculi of Łukasiewicz, cf. [9]; see also the results for signed multiple-valued logics in [5]). While the steady progress in the performance of the state-of-the-art SAT solvers for the clausal fragment of classical logic is impressive, there are disciplines within computer science that can benefit from a similar advancement in the design of satisfiability algorithms for multiple-valued logics (MV-logics). One might also expect that stochastic local search satisfiability algorithms can prove as rewarding in the multiple-valued setting as in the case of classical logic. The works such as [1, 2] on SAT problem for regular logics have already provided ample evidence in support of this claim.

In this paper, I propose two local search satisfiability algorithms for a class of finitely-valued MV-logics. The algorithms are non-clausal; they exploit polarity–the semantic property of truth-functions reflected in the syntax of formulas that define them–in their search for a satisfying assignment. The first algorithm proposed in this paper–polGSAT–is a non-clausal counterpart of GSAT. The other algorithm–polSAT–guides the search for a satisfying assignment for an input formula α by exploring the syntactic structure of α, aided by the relationship between polarity values and truth-values of subformulas of α. This second algorithm demonstrated good performance during the experimental tests.

In his 1982 paper, *Completely Non-Clausal Theorem Proving,* [10], Neil Murray constructed a refutational proof system for classical logic whose main inference rule–the resolution rule–was defined in terms of polarity. Since then, polarity has been exploited in a number of ways in areas ranging from proof theory and knowledge representation to automated reasoning and logic programming (cf. [16, 17]). This paper is an attempt to expand the scope of polarity-based research to include the non-clausal satisfiability problem (NC-SAT) and the development of competitive NC-SAT algorithms for MV-logics.

The remainder of this paper is structured as follows. I begin with a brief overview of the relationship between polarity and satisfiability. I then discuss two polarity-based local search algorithms for NC-SAT: polGSAT and polSAT. The performance of these algorithms is tested against the non-clausal local search satisfiability algorithms described in [13]. Throughout the survey, I shall use classical propositional logic and the 3-valued propositional logic of Łukasiewicz to illustrate the notions and empirical results presented in the paper.

2 Polarity and satisfiability

The classical notion of polarity of a variable p_i in a formula $\alpha(p_0, \ldots, p_m)$ (cf. [6]) captures syntactically the monotonic behavior of the Boolean function $f_\alpha(p_0, \ldots, p_m)$, defined by α, over its i-th argument: p_i is positive (resp. negative) in α, if f_α is monotonically non-decreasing (resp. non-increasing) over its i-th argument (assuming that the truth-value *true* (1) is greater than *false* (0)).

This classical view of polarity can be adapted to the many-valued setting effortlessly. For, let $\mathcal{P} = \langle L, \vdash \rangle$ be a finitely-valued extensional propositional logic, where L is its set of all well-formed formulas and $\vdash$ is its inference operation. Furthermore, let $TV = \{0, \ldots, n\}$ be the set of truth-values of $\mathcal{P}$ and let $\leq_{TV}$ be a binary relation on TV such that $\langle TV, \leq_{TV} \rangle$ is a lattice. As usually, some of the truth-values of TV are selected as *designated.*

Definition 1 (Referential Polarity). *Let $\alpha(p_0, \ldots, p_m)$ be a formula of $\mathcal{P}$ and let $f_\alpha(p_0, \ldots, p_m)$ be the interpretation of α in TV, i.e. a function such that for every valuation h of L into TV, $h(\alpha) = f_\alpha(h(p_0), \ldots, h(p_m))$. A variable p_i has positive polarity, '+' (resp. negative polarity, '−') in α, if f_α is monotonically non-decreasing (resp. non-increasing) over its i-th argument with respect to $\leq_{TV}$.*

The lattice ordering $\leq_{TV}$ imposes the 'degree of truth' structure on the space of truth-values; if such an interpretation of truth-values is either inappropriate for a given logic, or void of meaning, then one might define $\leq_{TV}$ to hold between undesignated and designated truth-values only, i.e., one might require that $x \leq_{TV} y$ if and only if $x = y$ or y is a designated and x is an undesignated truth-value.

The definition of referential polarity of variables can be extended to occurrences of subformulas: if β is an occurrence of some subformula of α, then the polarity of β in α can be determined by regarding β as a new propositional variable that does not occur in α.

The referential polarity is a useful concept in the development of logical tools that explicitly refer to, or manipulate, truth-values. Other applications, such as proof-theoretic studies of MV-logics, may resort to different notions of polarity (such as the inferential polarity defined in [16]).

If one is to take advantage of the information conveyed by the polarity values of variables for tasks such as, for instance, speeding up the deductive process in an automated reasoning system or the search for a satisfying truth-value assignment, then it is requisite that such values are computed efficiently. Since an efficient and exact labeling of variables with referential polarity values is, in general, not feasible, efficient but incomplete polarity assignment algorithms are sought and used. Such algorithms might label only some variables in accordance with the referential polarity and assign 'no polarity' to the remaining variables. This means that the positive or negative polarities of some of the variables might be ignored. One of such algorithms is presented in the following example.

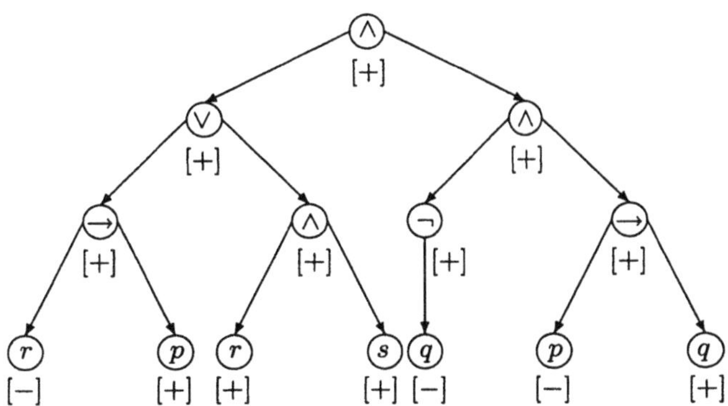

Fig. 1. Tree representation of $((r \to p) \vee (r \wedge s)) \wedge (\neg q \wedge (p \to q))$ of classical logic labeled with polarity values of subformulas (indicated by the labels '[+]' and '[−]').

Example 1. (Cf. [17]) One of the efficient but incomplete algorithms for assigning polarity values to variables in formulas of classical logic $\mathcal{P}_2$ works as follows. Let p be a variable in a formula α. First, the algorithm assigns polarity values to every occurrence of p in α. It assigns '+' (resp. '−') to an occurrence of p, if this occurrence is in the scope of an even (resp. odd) number of negation connectives. (For the purpose of the polarity assignment, any implication $\beta \rightarrow \gamma$ is regarded as the disjunction $\neg\beta \vee \gamma$.) If this occurrence is in the scope of an equivalence, then it is labeled as both '+' and '−'. Then, the algorithm assigns the polarity value to p: it assigns '+' (resp. '−') to p, if every occurrence of p is '+' (resp. '−') in α. This variable is said to be of no polarity, if it is neither '+' nor '−' in α. Hence, p is '+', q is '−', and r is of no polarity in $r \rightarrow (q \rightarrow (r \vee p))$. Figure 1 shows the tree representation of $((r \rightarrow p) \vee (r \wedge s)) \wedge (\neg q \wedge (p \rightarrow q))$. In this formula, s is the only positive variable while the remaining variables are of no polarity.

This incomplete polarity assignment algorithm can also be used for every n-valued Łukasiewicz logic $\mathcal{P}_n$, $n > 2$ (under the assumption that $\leq_{TV}$ is the usual $\leq$). Other polarity assignment algorithms for finitely-valued logics (such as verifier or verifier-operator polarity assignment algorithms) are defined and discussed in [17]. □

Let $\mathcal{P}$ be an arbitrary finitely-valued logic that conforms to the description given at the beginning of this section. Let us assume that the truth-values 0 (the bottom of the lattice of truth-values of $\mathcal{P}$) and n (the top of the lattice of truth-values of $\mathcal{P}$) are definable by formulas F and T, respectively. If there are any positive or negative variables in a formula α, then the search for a satisfying truth-value assignment for α can be simplified by assigning the truth-value n to positive and 0 to negative variables. Indeed, the following proposition (whose proof is a simple consequence of the definition of polarity) justifies the correctness of such an assignment:

Proposition 1. *Let $\alpha(p)$ be a formula of $\mathcal{P}$ that contains a variable p and let h be a truth-value assignment such that $h(\alpha) = n$. Then:*

(i) *if p is positive in α, then $h[p/n](\alpha) = n$;*
(ii) *if p is negative in α, then $h[p/0](\alpha) = n$.*

In Proposition 1 and in the remainder of this paper, $h[p/v]$ denotes the truth-value assignment obtained from h by changing the truth-value of the variable p from $h(p)$ to v. By Proposition 1, during the search for a satisfying assignment, the positive variables can be simply replaced by the formula T while negative by the formula F. In some cases, a series of such replacements may even eliminate the search all together. Indeed, let us look again at the formula α of classical logic depicted in Figure 1. While this formula can only be simplified to the formula of Figure 2 (by substituting T for s and, then, replacing $r \wedge T$ with r), the formula $\alpha^* = ((r \rightarrow s) \vee (r \wedge s)) \wedge (\neg q \wedge (p \rightarrow q))$, obtained from the formula of Figure 1 by replacing the first occurrence of p by s, can be reduced all the way to T. To this end, substitute T for positive s and F for negative p in α^* and, after the simplification, obtain the formula $\neg q$. Now, q is negative in $\neg q$ and its replacement by F reduces the entire formula to T. Hence, every truth-value assignment h such that $h(p) = 0, h(s) = 1$, and $h(q) = 0$ satisfies α^* in classical logic.

In classical logic, the simplification of a formula that results from the replacement of positive and negative variables with logical constants T and F, respectively, is analogous to the pure-literal rule of the clausal Davis-Putnam procedure. This simplification process reduces the search for a satisfying assignment for α to the same task but for a formula with a fewer variables than α.

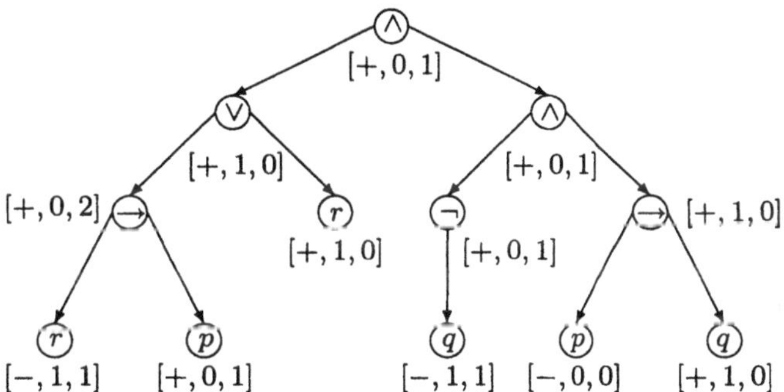

Fig. 2. Tree representation of $((r \to p) \vee r) \wedge (\neg q \wedge (p \to q))$ labeled with polarity values, truth-values, and cumulative clash values of subformulas.

3 Polarity and NC-SAT

Stochastic local-search algorithms for non-clausal propositional satisfiability (NC-SAT) for finitely-valued propositional logics can be derived from the generic form of the clausal local search satisfiability algorithm for classical logic given in Figure 3 (cf. [14]).

Given a set of clauses C as input, xSAT starts by generating a random assignment h of truth-values to the variables in clauses of C. Then, it locally modifies h by selecting a variable p (function *select_variable*(C, h)) and changing ('flipping') its truth-value from $h(p)$ to $1 - h(p)$. Such selections and flips are repeated until either h satisfies all the clauses in C or the allocated time to modify h into a satisfying assignment has elapsed ($MaxFlips$). The process is repeated (if needed) up to the specified $MaxTries$ times. It is the selection heuristic *select_variable* that determines the computational characteristics of a particular SAT solver (cf. [12]). In GSAT, the selection of the variable *select_variable*(C, h) is based on the objective to minimize the number of unsatisfied clauses in C (cf. [14]).

```
procedure xSAT(C)
    for i := 1 to MaxTries do
        h := random_assignment(C)
        for j := 1 to MaxFlips do
            if h(C) = {1} then return h
            else p := select_variable(C, h)
                h(p) := 1 − h(p)
        end for
    end for
    return "satisfying assignment for C not found"
```

Fig. 3. Generic stochastic local search algorithm for SAT.

In classical logic, there is a straightforward way to extend $xSAT$ from clausal to completely non-clausal algorithm in cases when the implementation of $select\_variable(C, h)$ resorts to the operations of random selection of unsatisfiable clauses and literals in such clauses, and of counting the number of clauses false under a given assignment (as in GSAT [14], WalkSAT [15], or Novelty [7]). Indeed, let $CNF(\alpha)$ denote the set of clauses obtained from α using the equivalence preserving CNF translation. Given a formula α and an assignment h, one can compute efficiently the number of clauses in $CNF(\alpha)$ false under h, without the explicit conversion of α into $CNF(\alpha)$ (cf. [13]). The same goes for the random selection of a literal in a random clause of $CNF(\alpha)$ false under h. And this is where the attractiveness of this approach ends: such NC-SAT solvers are still immersed in clausal reality and implicitly manipulate literals and clauses of $CNF(\alpha)$ in the way GSAT, WalkSAT, or Novelty solvers would do. The deeper structure of an input formula α is not exploited.

In this paper I propose a different line of research that exploits polarity for the purpose of defining the variable selection heuristic $select\_variable(\alpha, h)$, without resorting, of course, to the explicit or implicit translation of α into clauses. To explain informally the key idea behind the new variable selection process, let us suppose that a formula α of classical logic $\mathcal{P}_2$ is false under an assignment h. If some occurrence β of a subformula of α has positive polarity and $h(\beta) = 1$, then the falsehood of α must be caused by occurrences of subformulas other than β and, hence, the search for a variable to be flipped should sidestep β. The same happens when β is negative and $h(\beta) = 0$. It is precisely that type of information gathered by confronting the polarity values of subformulas with their truth-values which I shall use to guide the search for a satisfying assignment.

Definition 2. *Let α be a formula of a finitely-valued logic $\mathcal{P}$, let h be a truth-value assignment, and let β be an occurrence of a subformula in α. The* *polarity of β is said to clash with the truth-value $h(\beta)$, if either the polarity of β is '+' and $h(\beta) \neq n$, or the polarity of β is '−' and $h(\beta) \neq 0$.*

Definition 3. *Let α, β, and h be as in Definition 2. The cumulative clash value for β, denoted by $clash(\beta, h)$, is defined recursively as follows:*

(i) if β is a variable occurrence whose polarity clashes with $h(\beta)$, then $clash(\beta, h) = 1$, else $clash(\beta, h) = 0$;
(ii) if β is a logical constant, then $clash(\beta, h) = 0$;
(iii) if $\beta = f(\gamma_1, \ldots, \gamma_k)$, where f is a k-ary connective, and if the polarity of β clashes with $h(\beta)$, then $clash(\beta, h) = \sum_{i=1}^{k} clash(\gamma_i, h)$, else $clash(\beta, h) = 0$;

Intuitively, $clash(\beta, h)$ estimates the number of variable occurrences in β whose polarity values clash with their truth-values. The larger the value, the more difficult it could be to eliminate the clash between the polarity of β and $h(\beta)$ in a single flip. Let us look again at the formula depicted in Figure 2. Every node β in the tree (which represents a subformula occurrence) is labeled with three values: $[polarity, h(\beta), clash(\beta, h)]$. For instance, the leftmost '→'-node on level 2 (assuming that the root of the tree is on level 0) is labeled with $[+, 0, 2]$, since the polarity of this occurrence is '+', the truth-value is 0 (the consequence of the choice of truth-values for the variables r and p), and since '+' clashes with the truth-value. The cumulative clash value of this node (2) indicates that there are two occurrences of variables whose flip may change the truth-value of this '→'-node and, hence, eliminate the clash. On the other hand, the '∨'-node on level 1 is labeled with $[+, 1, 0]$; the third element of the label indicates that there is no need to be concerned with variables in this region of the formula since there is no clash for this node. And, finally, the label of the root is $[+, 0, 1]$, since: every formula is positive in itself, and the entire formula is false under h. The cumulative clash value 1 indicates that there is only one variable that should be considered for a flip. The discussion concerning possible ways to find such a variable is the subject of the following sections.

3.1 polGSAT defined

The non-clausal variant of the GSAT algorithm (cf. [14]) for a finitely-valued logic $\mathcal{P}$ can be obtained effortlessly by replacing the variable selection function *select_variable*(α, h) of GSAT with the function *select_assignment*(α, h) that returns a variable p and a truth-value $v \neq h(p)$ based on the objective to minimize the value of $clash(\alpha, h[p/v])$:

polGSAT$(\mathcal{P})$: *select_assignment*(α, h) returns a variable p and a truth-value v with the smallest value $clash(\alpha, h[p/v])$; in the case of tie, p is selected randomly.

When p and v have been selected, the assignment h is replaced by $h[p/v]$. For classical logic $\mathcal{P}_2$, polGSAT assumes the form:

> **polGSAT($\mathcal{P}_2$)**: *select_assignment*(α, h) returns the variable p with the smallest value $clash(\alpha, h[p/1-h(p)])$; in the case of tie, p is selected randomly.

To test the effectiveness of polGSAT one might try to compare the performance of polGSAT($\mathcal{P}_2$) with GSAT on random formulas of classical logic. There is one difficulty though: the equivalence preserving CNF translation of a random formula α may result in a formula in conjunctive normal form of length exponentially larger than α. On the other hand, a structure preserving CNF translation of α introduces new variables and, hence, increases the size of the search space. And that would force the comparison of the algorithms on quite different formulas. For these reasons, I compared polGSAT($\mathcal{P}_2$) with another non-clausal local search algorithm for classical logic, NC-GSAT, proposed in [13]. NC-GSAT adopts the same objective as GSAT: to minimize the number of unsatisfied clauses in $CNF(\alpha)$. As mentioned at the beginning of Section 3, the number of such clauses can be calculated efficiently and without converting α into $CNF(\alpha)$. In other words, NC-GSAT is equivalent to GSAT without the explicit conversion of an input formula into clauses.

> **NC-GSAT**: *select_variable*(α, h) returns the variable p that results in the smallest number of clauses in $CNF(\alpha)$ which are false under $h[p/1 - h(p)]$; in the case of tie, p is selected randomly.

While NC-GSAT is certainly not the state-of-the-art NC-SAT solver, its choice for the comparison with polGSAT($\mathcal{P}_2$) is based on shared characteristics: both algorithms are non-clausal, greedy, stochastic local search procedures that reduce to GSAT when inputs are restricted to formulas in conjunctive normal form.

polGSAT($\mathcal{P}_2$) was compared with NC-GSAT on random formulas of lengths $\leq 2,300$. Such formulas were generated by constructing their tree representations in top down and breadth-first manner. Every node in a partially constructed tree was randomly labeled with a variable (one of $p_1, \ldots, p_{300}$) or a connective name and only nodes labeled with the connectives were further expanded. If k is the number of logical connectives in the language, then a node was labeled with a specific connective or with a variable with probability $1/(k+1)$. When the node count reached $2,000$, the only labels that could be assigned to nodes were variables.

In most cases, polSAT($\mathcal{P}_2$) significantly outperformed NC-GSAT. Since the difference in performance of the algorithms can be attributed to the fact that the implementation of polGSAT that was used for testing included the formula simplifier that replaces positive and negative variables in an input formula with appropriate logical constants and then simplifies the resulting formula (as described in Section 2), further tests were done on random formulas that contained neither positive nor negative variables. (Such formulas are obtained from randomly generated formulas by either negating some randomly selected occurrences of polarized variables or, if a polarized variable has just one occurrence in a formula, by replacing the variable by one of the randomly selected variables of the formula.) On this restricted class of random formulas, the performance of both algorithms (measured in the number of flips) was remarkably similar, i.e., either both algorithms failed to find a satisfying assignment or performed insignificantly different number of flips to find such an assignment. The results remained similar when the testing was restricted to 'hard' instances of NC-SAT, i.e., to random formulas α (generated as above) such that the ratio of clauses to variables for $CNF(\alpha)$ was near the satisfiability threshold of the phase transition (cf. [8]).

As we shall see shortly, the use of $clash(\alpha, h)$ parameter in aiding the search for a satisfying assignment can be more rewarding in other local search satisfiability algorithms. (See also the conclusion.)

3.2 polSAT defined

A different approach to the implementation of the $select\_variable(\alpha, h)$ heuristic is to use the clash parameter to guide the search from the root of the tree representing the input formula α to one of its leaves (i.e., one of the variable occurrences). The polSAT algorithm for NC-SAT, described below, explores the structure of the formula tree by selecting regions of α (i.e., subformulas of α) with the smallest possible, but > 0, cumulative clash values. There is some empirical evidence (discussed at the end of this section) that flipping a variable from such a region is advantageous. Formally:

> **polSAT($\mathcal{P}$):** *select_variable*(α, h) constructs a branch of the tree representing α starting from the root and always branches into a child node with the smallest non-zero cumulative clash value, or randomly selects a child node when all the children have the same cumulative clash values; *select_variable*(α, h) returns the leaf of the constructed branch. To escape local minima, *select_variable*(α, h) constructs a branch of the tree from the root by randomly branching into a child node with non-zero clash value.

To complete the definition of polSAT($\mathcal{P}$) one has to specify the way the truth-value of a selected variable is to be 'flipped': if *select_variable*(α, h) returns an occurrence of a variable p of polarity *pol*, then:

> **case** *pol* **of**
> '+': **if** $h(p) < n$, **then return** $h(p) + 1$
> **else return** random truth-value $\neq h(p)$;
> '−': **if** $h(p) > 0$ **then return** $h(p) - 1$
> **else return** random truth-value $\neq h(p)$;
> **end case**

To empirically test the performance of polSAT I first compared its classical logic instance, polSAT($\mathcal{P}_2$), with a variant of the non-clausal local search satisfiability algorithm of Sebastiani (cf. [13], p. 313) which roughly resembles the clausal WalkSAT algorithm of Selman, Kautz, and Cohen (cf. [15]). For reasons that have already been discussed in Section 3.1, the direct comparison between polSAT($\mathcal{P}_2$) and WalkSAT would be rather difficult.

NC-RSAT: *select_variable*(α, h) constructs a branch of the tree representing α starting from the root and randomly branching into a child node with non-zero score value; *select_variable*(α, h) returns the leaf of the constructed branch. To escape local minima, *select_variable*(α, h) repeats the variable selection procedure.

In NC-RSAT, the score of an occurrence β of a subformula is the number of clauses in CNF(β) false under h. Intuitively, *select_variable*(α, h) of NC-RSAT randomly selects a literal in a random clause of CNF(α) false under h (without, of course, converting α into CNF(α)). polSAT($\mathcal{P}_2$) and NC-RSAT were executed on random formulas which contained neither positive nor negative variables (for the same reason as in the case of polGSAT and NC-GSAT). The results convincingly favor the polSAT($\mathcal{P}_2$) algorithm. For 98% of the test instances, polSAT($\mathcal{P}_2$) performed at least as well as NC-RSAT; for 7% of the test formulas, NC-RSAT failed to find a satisfying assignment while polSAT($\mathcal{P}_2$) was successful at generating such an assignment. [1] Similar results were obtained when NC-RSAT was modified to branch into nodes with the lowest (or, alternatively, the highest) score.

Further tests were done on graph coloring problem for the class $G_{n,0.5}$ of random graphs (of n vertices and edges generated with the probability 0.5 for every pair of distinct vertices, cf. [4]. Already for the class $G_{20,0.5}$ NC-RSAT was no match for polSAT($\mathcal{P}_2$), while polSAT($\mathcal{P}_2$) was still exhibiting good performance for graphs in $G_{80,0.5}$. The SAT encoding used for the graph coloring problems is a non-clausal variant of the encoding described in [14]. For instance, the constraint that no two adjacent vertices can have the same color is encoded as the conjunction of formulas of the form

$$\bigwedge_{j \leq k} \{p_{ij} \rightarrow \neg(p_{i_1 j} \vee \ldots \vee p_{i_m j})\}$$

where: p_{vc} encodes 'the color of v-th vertex is c', and $i_1, \ldots, i_m$ are the indices of all the vertices adjacent to the i-th vertex, and k is the number of colors.

Finally, to test the effectiveness of the selection heuristic *select_variable* (α, h), the performance of polSAT($\mathcal{P}$) was tested against one of its variants

[1] polSAT($\mathcal{P}_2$) and NC-RSAT were compared (in terms of the number of flips performed) on 100,000 random non-tautological formulas of lengths $\leq 8,500$ and containing $\leq 1,050$ variables. The formulas were generated in the way described in Section 3.1. Cutoff for both algorithms: $MaxTries \times MaxFlips = 20,000$ flips.

that, similarly to NC-RSAT, selects a variable by constructing a branch of the tree representing an input formula by branching into a random child node with non-zero cumulative clash value; it flips the truth-value of the selected variable in the way described for polSAT($\mathcal{P}$). The tests were done on random formulas of length $\leq 1,200$ and containing ≤ 150 variables, using sample size of 100,000. polSAT outperformed its variant by the factor between 4 and 4.5 for the 3-valued Łukasiewicz logic $\mathcal{P}_3$ (the performance was measured in terms of the number of flips performed to find a satisfying assignment), and by the factor between 5 and 5.5 for classical logic. The random formulas of $\mathcal{P}_3$ were generated using the method described in Section 3.1.

4 Conclusions

In this paper I discussed one of the possible ways in which information conveyed by the structure of a free-form formula can be used for the design of non-clausal local search algorithms for a class of finitely-valued propositional logics. Although the empirical results concerning the polarity-based polSAT algorithm are promising, there are many ways in which it could be significantly improved. One is to incorporate more of the stochastic magic into the polSAT algorithm (as, for instance, in Novelty solver, [7]). Another promising new technique is based on the observation that a single call to *select_variable*(α, h) may allow to flip the truth-value of not just the selected variable but of other variables as well. Indeed, if the call to *select_variable*(α, h) returns a variable p and a truth-value v, then the substitution of the logical constant that defines v for p in α (if such a constant exists) may render other variables positive or negative in α. Hence, as discussed in Section 2, the truth-values of such positive variables can be set to n while negative to 0.

Acknowledgments

I would like to acknowledge the support from the Natural Sciences and Engineering Research Council of Canada. I would also like to thank an anonymous reviewer for helpful and insightful comments.

References

1. Béjar, R.: Systematic and Local Search Algorithms for Regular-SAT. PhD thesis, Universitat Autònoma de Barcelona (2000)
2. Béjar, R. and Manyà, F.: Solving combinatorial problems with regular local search algorithms. Proc. 6th Int. Conf. on Logic for Programming and Automated Reasoning, LPAR-99, Springer LNAI 1705 (1999) 33–43

3. Du, D., Gu, J. and Pardalos, P. editors: Satisfiability Problem: Theory and Applications. DIMACS Series in Discrete Mathematics and Theoretical Computer Science, vol. 35, American Mathematical Society (1997)
4. Johnson, D., Aragon, C. and McGeoch, L.: Optimization by simulated annealing: an experimental evaluation; part ii, graph coloring and number partioning. Operations Research, 39 (1991) 378–406
5. Hähnle, R.: Complexity of Many-Valued Logics. Proc. of the 31st IEEE Int. Symp. on Multiple-Valued Logic (2001) 137–146
6. Manna, Z. and Waldinger, R.: The Deductive Foundations of Computer Programming. Eddison-Wesley (1993)
7. McAllester, D., Selman, B. and Kautz, H.: Evidence for invariants in local search. Proc. AAAI-97 (1997) 321–326
8. Mitchell, D., Selman, B. and Levesque, H.: Hard and Easy Distribution of SAT Problems. Proc. AAAI-92 (1992) 459–465
9. Mundici, D.: Satisfiability in Many-Valued Sentential Logics is NP-Complete. Theoretical Computer Science, 52 (1987) 145–153
10. Murray, N.: Completely Non-Clausal Theorem Proving. Artificial Intelligence, 18 (1982) 67–85
11. Plaisted, D. and Greenbaum, S.: A structure-preserving clause form translation. J. of Symbolic Computation, 2 (1986) 293–304
12. Schuurmans, D. and Southey, F.: Local search characteristics of incomplete SAT procedures. Proc. AAAI-2000 (2000) 297–302
13. Sebastiani, R.: Applying GSAT to non-clausal formulas. J. of Artificial Intelligence Research, 1 (1994) 309–314
14. Selman, B., Levesque, H. and Mitchell, D.: A new method for solving hard satisfiability problems. Proc. AAAI-92 (1992) 440–446
15. Selman, B., Kautz, H. and Cohen, B.: Noise strategies for improving local search. Proc. AAAI-94 (1994) 337–343
16. Stachniak, Z.: Exploiting Polarity in Multiple-Valued Inference Systems. Proc. of the 31st IEEE Int. Symp. on Multiple-Valued Logic (2001) 149–156
17. Stachniak, Z.: Resolution Proof Systems: An Algebraic Theory. Kluwer (1996)

Chapter 8
Model Checking for Multi-valued Computation Tree Logics

Beata Konikowska[1], Wojciech Penczek*[12]

[1] Institute of Computer Science, PAS
01-237 Warsaw, ul. Ordona 21, Poland
{beatak,penczek}@ipipan.waw.pl
[2] Akademia Podlaska
Institute of Informatics, Siedlce, Poland

Abstract. A multi-valued version of CTL* (mv-CTL*), where both the propositions and the accessibility relation are multi-valued taking values in a finite quasi-Boolean algebra, is defined. A translation from mv-CTL* model checking to CTL* model checking is investigated. First, the case where the elements of quasi-Boolean algebras are totally ordered is considered. Secondly, it is shown how to design a translation algorithm for the two most commonly applied quasi-Boolean algebras. This construction suggests the way one can deal with more complex quasi-Boolean algebras if necessary.

1 Introduction

Model checking is one of the most popular methods used in automated verification of concurrent systems like hardware circuits, communication protocols, and distributed programs [13]. It consists in verifying that a finite state program P satisfies a property φ (denoted $P \models \varphi$). When P is represented by its model M_P and the property φ is given by a temporal logic formula, one checks whether $M_P \models \varphi$. The process of generating M_P for P and checking that $M_P \models \varphi$ is automated. The complexity of model checking methods strongly depends on the translation from P to M_P and on the type of the temporal logic that the formula φ belongs to.

In this paper we consider model checking algorithms for Computation Tree Logic* (CTL*) [18] and two of its sublogics: CTL [12] and LTL [23]. We are not concerned with the problem of getting a model M_P from a program P and just assume that such a model is provided. The simplest model checking algorithm that can be applied for CTL consists in the standard labelling of the model's states with subformulas of a given CTL formula [12]. For LTL and CTL*, a more involved algorithm based on automata theory [7] has to

* Partly supported by the State Committee for Scientific Research under the grant No. 7T11C 00620.

be used. A temporal formula is translated to an automaton A that accepts all its models, and then the product of A with the model (automaton) M_P is checked for non-emptiness.

Two-valued models and logics, like sublogics of CTL*, are sufficient for proving properties of standard concurrent systems. However, there is a number of problems for which we need to reason under uncertainty or to use models containing inconsistency. Uncertainty can occur either when complete information is not known or cannot be obtained, or when this information has been abstracted away [14, 15]. On the other hand, models can be inconsistent because they combine conflicting points of view or contain components developed by different designers [17].

Multi-valued modal logics provide a solution to both reasoning under uncertainty and reasoning under inconsistency. Several model checkers based on 3- and 4-valued logics have been already defined [3, 4, 14, 22]. Moreover, there seems to be some interest in model checkers dealing with temporal logics using more than 4 logical values [11]. We discuss the existing model checking algorithms in detail in Section 2 of our paper.

There are two different approaches to extending modal logics to multi-valued versions. Either solely the interpretation of propositional variables is extended to a multi-valued one, or in addition the accessibility relation becomes a multi-valued one [20, 21]. Moreover, there can also be various approaches to selecting the domain of logical values to be used. Here some kind of a (partially) ordered set is usually considered, with the most common choices being either a finite totally ordered set or, more generally, a finite quasi-Boolean lattice. In order to define a model checking algorithm for a selected multi-valued modal logic mv-L (which extends a two-valued modal logic L), one can either design a new model checking algorithm for mv-L by extending an existing algorithm for L or define a translation from model checking for mv-L to model checking for L.

The latter approach is usually more efficient. Indeed: with a simple translation, it allows us to use the benefits of well-developed and verified, standard model checking algorithms for L without having to rediscover things from the start. Clearly, the latter is usually connected with a greater work outlay, plus some practically inevitable faults and/or inefficiencies of the resulting new algorithm. Moreover, such an approach provides greater flexibility, since it is not restricted to any specific model checking method for the logic L. Therefore, it allows us to use any existing (present or future) model checking algorithm for L to provide model checking for mv-L.

This is the approach we follow in this paper. We start with explaining our approach for a simple multi-valued modal logic mv-K. Then, in Section 4 we define mv-CTL*, where both the propositions and the accessibility relation are multi-valued and take values in a finite quasi-Boolean algebra. In Section 5, we define a translation from mv-CTL* model checking to CTL* model checking. To this aim, we first consider the case where the elements of quasi-Boolean algebras are totally ordered. Secondly, we show how to design a translation algorithm for the two most commonly applied quasi-Boolean algebras. This construction suggests the way one can deal with more complex quasi-Boolean algebras if necessary. In Section 6, we discuss a model checking method for CTL* that exploits alternating tree automata. Other model checking methods and reduction techniques are discussed in Section 7.

2 Related work

A translation from a 3-valued model checking problem for CTL* and the modal μ-calculus to a standard model checking problem has been defined in [3, 4]. The authors have considered a restricted version of the logics with the 2-valued interpretation of the accessibility-relation. Our translation can be viewed as a generalization of these results for CTL*.

Another approach has been taken by Chechik et al. A new model checking algorithm for a multi-valued CTL has been defined exploiting mv-BDD's [8] for unrestricted interpretations and MTBDD's [9] for distributive quasi-Boolean algebras. Moreover, a model checker for mv-LTL under restricted interpretations (a 2-valued accessibility relation and totally ordered sets for the propositions) has been implemented, based on a translation to (mv)-Büchi automata [10].

3 Standard modal logic *K*

To start with, we explain our approach for a simple multi-valued modal logic mv-K, which is an extension of the standard modal logic K.

3.1 Two valued modal logic - *K*

Let $PV = \{p_1, p_2, \ldots\}$ be a set of propositional variables and let $\mathcal{L} = (\mathrm{L}, \leq)$, where $\mathrm{L} = \{\mathbf{F}, \mathbf{T}\}$ is the standard two-valued Boolean lattice with the operations $\sim, \sqcap, \sqcup$.

Definition 1. *A* **model** *is a four-tuple* $M = (S, R, V, \mathcal{L})$, *where*

- S *is a set, called the set of* states (worlds),
- $R : S \times S \longrightarrow L$ *is a total function, called the* accessibility function,
- $V : S \times PV \longrightarrow L$ *is a function, called the* valuation function, *and*
- $\mathcal{L}$ *is the 2-valued Boolean lattice.*

We call a model *finite state* if the set S of its states is finite.

Our modal language contains only a single modality K and the standard logical connectives: $\neg, \wedge$, and $\vee$. The semantics of formulas in a model M is a function $[\cdot]^M$: $\mathbf{Form} \times S \longrightarrow \mathcal{L}$ (where **Form** is the set of all formulas), with the semantics of φ at a state s of the model M being denoted by $[\varphi]_s^M$. Usually, M is omitted when it is understood.

$$\begin{array}{l} \bullet\ [p]_s = V(s,p), \\ \bullet\ [\neg\varphi]_s = \sim [\varphi]_s, \\ \bullet\ [\varphi \wedge \psi]_s = [\varphi]_s \cap [\psi]_s, \\ \bullet\ [\varphi \vee \psi]_s = [\varphi]_s \cup [\psi]_s, \\ \bullet\ [K\varphi]_s = \bigcap_{t \in S}(\sim R(s,t) \cup [\varphi]_t) \end{array} \tag{1}$$

Given a finite state model M, a state ι of S, and a formula φ, the model checking problem is usually defined as checking whether $[\varphi]_\iota^M = \mathbf{T}$.

In this simple case, the semantics of K defines the model checking algorithm, which consists in labelling the states of M with the subformulas of φ that hold over these states. The algorithm starts with the most deeply nested subformulas of φ and then inductively (on the complexity of φ) labels the states of the model. As a result, $[\varphi]_\iota^M = \mathbf{T}$ if the beginning state ι has been labelled with φ.

Before we define the multi-valued version of K, we transform K to its positive form, which is much more useful for our further considerations. The idea is to extend the language by an operator dual to K, denote it $\overline{K}$, and to restrict the language so that negation can only be applied to propositional variables. It is easy to check that this does not change the expressive power of the language. The semantics of the positive K is given by:

$$\begin{array}{l} \bullet\ \text{Eqs. (1), with the negation clause applied to } p \in PV \text{ only,} \\ \bullet\ [\overline{K}\varphi]_s = \bigcup_{t \in S}(R(s,t) \cap [\varphi]_t). \end{array} \tag{2}$$

In what follows we consider the positive version of K rather than K itself.

The next section deals with quasi-Boolean lattices and finite totally ordered sets, which are used as logical domains of interpretations for the formulas of our multi-valued logics.

3.2 Quasi-boolean lattices

Definition 2. *A* lattice *is a partial order $\mathcal{L} = (L, \leq)$ satisfying the following conditions:*

- *for any two elements $x, y \in L$ there exists their greatest lower bound $(x \cap y)$ and their least upper bound $(x \cup y)$.*[3]

Definition 3. *A lattice $\mathcal{L} = (L, \leq)$ is quasi-Boolean if there exists a unary operator $\sim$ having the following properties for $x, y \in L$:*

[3] Obviously, from antisymmetry of a partial order it follows that both the greatest lower bound and least upper bound of x, y are uniquely determined elements of L.

- $\sim (x \cap y) = \sim x \cup \sim y$,
- $\sim (x \cup y) = \sim x \cap \sim y$,
- $x \leq y$ *iff* $\sim y \leq \sim x$,
- $\sim\sim x = x$.

$\mathcal{L}$ is distributive if the following conditions hold for $x, y, z \in L$:

- $z \cup (x \cap y) = (z \cup x) \cap (z \cup y)$,
- $z \cap (x \cup y) = (z \cap x) \cup (z \cap y)$.

Definition 4. *A lattice $\mathcal{L} = (L, \leq)$ is a finite total order (FTO) if the following conditions hold:*

- *L is a finite set,*
- *for each two elements $x, y \in L$, either $x \leq y$ or $y \leq x$.*

Example 1. In Figure 1 we show FTO's denoted $\mathbf{2}, \mathbf{3}, \ldots, \mathbf{K+1}$ representing uncertainty, and two well-known finite lattices, the use of which is motivated by clear practical intuitions: the lattice $\mathbf{2} \times \mathbf{2}$ representing disagreement, and the 6-valued lattice $\mathbf{2} + \mathbf{4}$ representing both uncertainty and disagreement.

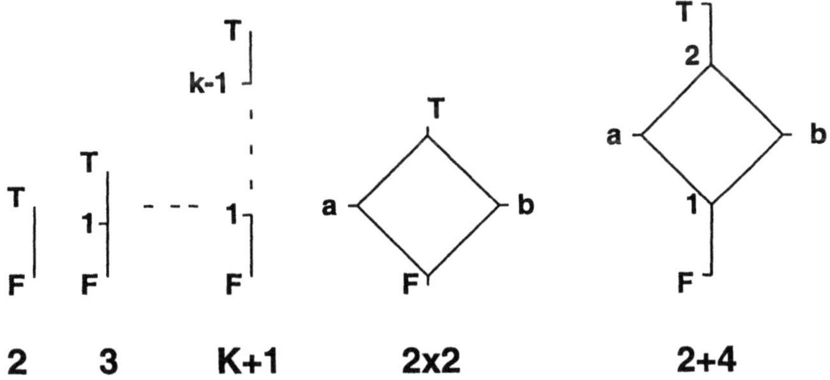

Fig. 1. Standard multi-valued lattices

3.3 Multi-valued positive K

Let $\mathcal{L} = (\mathrm{L}, \mathcal{D}, \leq)$ be a finite, quasi-boolean lattice with the operations $\sim, \cap, \cup$, where $\mathcal{D} \subset \mathrm{L}$ is a set of *designated values.* closed w.r.t. $\cap$ and $\cup$.

The semantics of the multi-valued positive K is similar to the semantics of the positive K given by (2) except for the fact that now the values of propositional variables are taken from the quasi-Boolean lattice $\mathcal{L}$.

Example 2. In Figure 2 we show an example of a multi-valued model representing our requirements for a simple coffee machine (a modified example from [8]). The model uses multiple truth values to distinguish between transitions and propositions that **must**, **should**, **should not**, or **must not** be true. Two types of unkown values are used: **do not know** for values controlled by the system, but not yet decided, and **do not care** for values controlled by the environment. We assume that the set of designated values is $\mathcal{D} = \{\mathrm{T}\}$. Labelling a state s with $p = x$ means that $V(s,p) = x$, where p is a proposition and x is a truth value. Labelling a transition with x means that its truth value is x. In the figure we omit all the transitions labelled **F**, i.e., "prohibited". The coffee machine starts in the state **OFF**, where it is irrelevant whether a cup is provided or not, which is modeled by $cup = \mathbf{DC}$. Since it has not been decided yet whether the machine should be able to go directly to IDLE, the corresponding transition is labelled **DK**. On the other hand, all the transitions labelled **T** represent the required changes of states. Note that the transitions from the states **Delivers ...** to **OFF** are labelled N. Since they are dependent on the environment they cannot be prohibited, i.e., labelled **F**. The value **DK** of the proposition power in the state **IDLE** represents the fact that it has not yet been decided whether the power should be ON or OFF on idle.

There are several properties that can be specified and then checked in the model:

1. The machine must eventually be able to deliver coffee,
2. Coffee is always delivered in a cup,
3. After coffee has been delivered, coffee cannot be immediately delivered again.

Given a finite state model M, a state ι of S, and a formula φ, the model checking problem is now defined as checking whether $[\varphi]_{\iota}^{M} \in \mathcal{D}$. Notice that also in this case, the semantics of mv-K gives rise to a model checking algorithm. We directly compute the value $[\varphi]_{\iota}^{M}$ by labelling the states of the model with pairs consisting of subformulas of φ and the values they take at these states. Then, we check whether the value assigned to φ at ι is in $\mathcal{D}$.

However, for some different logic, say L, more expressive than K, the situation can get more complicated due to the following reasons:

- labelling of the states of a model for mv-L cannot be defined as an easy extension of the corresponding algorithm for L,
- a model checking algorithm for L is not based on labelling of the states of a model.

The former applies to CTL, whereas the latter to CTL* and LTL. Therefore, the following sections deal with the above mentioned temporal logics and their multi-valued interpretations.

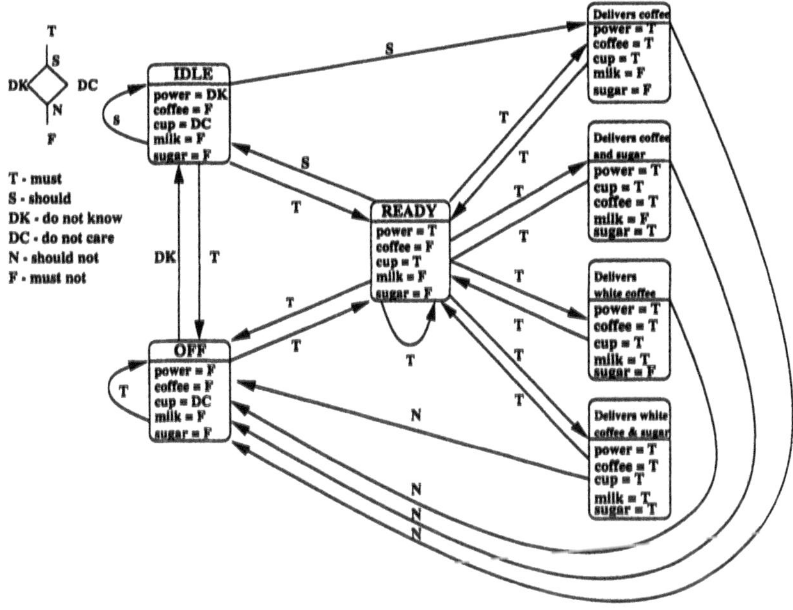

Fig. 2. The lattice **2** + **4** and a coffee machine model over this lattice

4 Mv-temporal logics: mv-CTL*, mv-CTL, mv-LTL

Syntax of mv-CTL*

Let PV be a finite set of propositions containing the symbol *true*, and let $\overline{PV} = \{\neg p \mid p \in PV\}$. First, we give the syntax of mv-CTL*, and then restrict it to standard sublanguages. As explained before, we define the positive version of the language. The set of state formulas and the set of path formulas of mv-CTL* are defined inductively:

S1. every element of either PV or $\overline{PV}$ is a state formula,
S2. if φ and ψ are state formulas, then so are $\varphi \vee \psi$ and $\varphi \wedge \psi$,
S3. if φ is a path formula, then $\mathrm{A}\varphi$ and $\mathrm{E}\varphi$ are state formulas,
P1. any state formula φ is also a path formula,
P2. if φ, ψ are path formulas, then so are $\varphi \wedge \psi$ and $\varphi \vee \psi$,
P3. if φ, ψ are path formulas, then so are $X\varphi$, $\mathrm{U}(\varphi, \psi)$, and $\overline{\mathrm{U}}(\varphi, \psi)$.

The modal operator A has the intuitive meaning "for all paths", E "there is a path", U denotes the standard Until, whereas $\overline{\mathrm{U}}$ denotes Release (dual to Until). The language of mv-CTL* consists of the set of all state formulas.

The following abbreviations will be used:

- $false \stackrel{def}{=} \neg true$,

- $\mathrm{G}\varphi \stackrel{def}{=} \overline{\mathrm{U}}(\varphi, false)$,
- $\mathrm{F}\varphi \stackrel{def}{=} \mathrm{U}(true, \varphi)$.

Sublogics of mv-CTL*

mv-CTL: The sublogic of mv-CTL* in which the state modalities A, E and the path modalities X, U and $\overline{\mathrm{U}}$ may only appear paired in the combinations AX, AU, $\mathrm{A}\overline{\mathrm{U}}$ and EX, EU, $\mathrm{E}\overline{\mathrm{U}}$.

mv-LTL: Restriction to formulas of the form $\mathrm{A}\varphi$, where φ contains neither A nor E. We usually write φ instead of $\mathrm{A}\varphi$ if misunderstanding is unlikely.

Semantics of mv-CTL*

Definition 5. *A* **model** *is a triple* $M = (S, R, V, \mathcal{L})$, *where*

- S *is a set, called the set of* states (worlds),
- $R : S \times S \longrightarrow L$ *is a function, called the* accessibility function,
- $V : S \times (PV \cup \overline{PV}) \longrightarrow L$ *is a function, called the* valuation function,
- $\mathcal{L} = (L, \mathcal{D}, \leq)$ *is a finite quasi Boolean-lattice, with a set of designated elements* $\mathcal{D}$, *the least element* $\mathbf{F}$ *and the greatest element* $\mathbf{T}$.

We assume that R *satisfies the condition* $(\forall s \in S)(\exists s' \in S)\ R(s, s') > \mathbf{F}$.

The latter requirement for R is aimed at keeping the semantics of mv-CTL* as close as possible to that of CTL*. Define

$$\Pi(s) = \{s_0 s_1 \ldots \mid s_0 = s \wedge (\forall i \geq 0)\ R(s_i, s_{i+1}) > \mathbf{F}\},\ \Pi = \bigcup_{s \in S} \Pi(s)$$

to be the set of all the paths starting at s in M, and the set of all the paths in M, respectively.

For $\pi = s_0 s_1 \ldots \in \Pi(s)$, let $\pi_i = s_i s_{i+1} \ldots$ denote the i-th suffix of π and let $\pi(i) = s_i$ be the i-th state of π.

Then, $[\varphi]_s^M$ and $[\varphi]_\pi^M$ is defined inductively as follows (as before, we omit M when it is understood):

S1. $[true]_s = \mathbf{T}$, $[q]_s = V(s, q)$, $[\neg q]_s = V(s, \neg q)$ for $q \in PV$,

S2. $[\varphi \wedge \psi]_s = [\varphi]_s \cap [\psi]_s$, $[\varphi \vee \psi]_s = [\varphi]_s \cup [\psi]_s$,

S3. $[\mathrm{A}\varphi]_s = \bigcap_{\pi \in \Pi(s)} [\varphi]_\pi$, $[\mathrm{E}\varphi]_s = \bigcup_{\pi \in \Pi(s)} [\varphi]_\pi$,

P1. $[\varphi]_\pi = [\varphi]_{\pi(0)}$ for any state formula φ,

P2. $[\varphi \wedge \psi]_\pi = [\varphi]_\pi \cap [\psi]_\pi$, $[\varphi \vee \psi]_\pi = [\varphi]_\pi \cup [\psi]_\pi$,

P3. $[X\varphi]_\pi = R(\pi(0), \pi(1)) \cap [\varphi]_{\pi_1}$,
$[\mathrm{U}(\varphi, \psi)]_\pi =$
$\bigcup_{i>0}([\psi]_{\pi_i} \cap \bigcap_{0<j<i}(R(\pi(j-1), \pi(j)) \cap [\varphi]_{\pi_j}) \cap [\varphi]_{\pi_0}) \cup [\psi]_{\pi(0)}$,
$[\overline{\mathrm{U}}(\varphi, \psi)]_\pi =$
$\bigcap_{i\geq 0}((R(\pi(i), \pi(i+1)) \cap [\psi]_{\pi_i}) \cup (\bigcup_{0<j<i}(R(\pi(j-1), \pi(j)) \cap [\varphi]_{\pi_j}) \cup [\varphi]_{\pi_0})$,

Given a finite-state model M, a state ι of S, and a formula φ, the model checking problem is defined again as that of checking whether $[\varphi]_\iota^M \in \mathcal{D}$.

Example 3. The properties of the coffee machine model of Example 2 can be specified by the following formulas:

1. EF*coffee*,
2. AG($\neg$*coffee* $\vee$ *cup*),
3. AG($\neg$*coffee* $\vee$ AX($\neg$*coffee*)).

5 Model checking for mv-CTL*

In this section we show that the model checking problem for mv-CTL* can be translated to the model checking problem for CTL*. This will be shown first for mv-CTL* interpreted over FTO's, and then for some typical finite pseudo-Boolean lattices.

5.1 Translation for FTO's

We assume that $\mathcal{L}$ is an FTO such that $\mathrm{L} = \{\mathbf{F}, 1, \ldots, k-1, \mathbf{T}\}$. Let $M = (S, R, V, \mathcal{L})$ be an L-valued model, and φ be an mv-CTL* formula. For each $1 \leq i \leq \mathbf{T}$, define the model $M_i = (S, R^i, V^i, \mathcal{L})$ by replacing in M all the L-values $j < i$ (returned by R and V) by $\mathbf{F}$ and all the values $j \geq i$ by $\mathbf{T}$. Formally, M_i is defined as follows:

- $Z^i(x, y) = \mathbf{F}$ iff $Z(x, y) < i$,
- $Z^i(x, y) = \mathbf{T}$ iff $Z(x, y) \geq i$,
  where $Z \in \{R, V\}$.

Thus, each model M_i is two-valued over $\{\mathbf{F}, \mathbf{T}\}$.

In order to apply standard induction based proofs, we need the following lemma for dealing with the cases of $[\mathrm{U}(\varphi, \psi)]_\pi$, $[\overline{\mathrm{U}}(\varphi, \psi)]_\pi$, $[\mathrm{A}\varphi]_s$, and $[\mathrm{E}\varphi]_s$.

Lemma 1. *Let $x_i \in L$ for all $i \geq 0$. Then, the following conditions hold:*

- $\bigcup_{i\geq 0} x_i = max_{i\geq 0}\{x_i\}$,
- $\bigcap_{i\geq 0} x_i = min_{i\geq 0}\{x_i\}$.

Our translation algorithm is based on the following theorem.

Theorem 1. *The following implications hold for each $i \leq \mathbf{T}$:*

- *If* $[\varphi]_\iota^M = i$, *then* $[\varphi]_\iota^{M_i} = \mathbf{T}$,
- *If* $[\varphi]_\iota^{M_i} = \mathbf{T}$, *then* $[\varphi]_\iota^M \geq i$.

Proof. By structural induction on φ using Lemma 1. The proof is very similar to the proof of Theorem 2, which we present in detail below.

The value of $[\varphi]_\iota^M$ is computed by the following algorithm, where the output 0 is interpreted as $\mathbf{F}$ and the output k as $\mathbf{T}$:

```
for i := k downto 1
  if [φ]_ι^{M_i} = T,
   then return i;
return 0;
```

Notice that the order of performing model checking by going from M_k to M_1 is crucial for the above algorithm. Correctness of the algorithm follows directly from Theorem 1.

5.2 Multi-valued logics over quasi-Boolean lattices

Up to now, we have only considered many-valued logics with the set of logical values L coming from a lattice $\mathcal{L}$ which is an FTO. However, some literature on the subject (see [9]) also discusses the case when $\mathcal{L}$ is an arbitrary lattice. This might be justified in some practical applications (see Example 2). However, any decision to use lattice-based semantics should be taken with great caution, since such a semantics has some pitfalls which might never be noticed by some authors having no experience with many-valued logics.

One such pitfall is that in lattice logics disjunction of two non-designated values can be designated, which might lead to paradoxical evaluation of formulae, inconsistent with either our intuition or practical considerations of model-checking for concurrent system.

To better illustrate this problem, consider a lattice being a set-theoretical union of two chains: $\mathbf{F}, a_1, \ldots, a_n, \mathbf{T}$ and $\mathbf{F}, b_1, \ldots, b_m, \mathbf{T}$, where $a_i \neq b_j$ for all i, j. Further, suppose that only $\mathbf{T}$ is a designated value. As in this lattice a_i, b_j are incomparable for any i, j, then $a_i \cup b_j = \mathbf{T}$. In consequence, if on a path π such that $R(\pi(i), \pi(i+1)) = \mathbf{T}$ for every i, we have $[\varphi]_{\pi_i} = a_1, [\varphi]_{\pi_j} = b_1$ for some $i \neq j$, and $[\varphi]_{\pi_k} = \mathbf{F}$ for all $k \neq i, j$, then

$$[\mathrm{F}\varphi]_\pi = \bigcup_{k \geq 0} [\varphi]_{\pi_k} = [\varphi]_{\pi_i} \cup [\varphi]_{\pi_j} = a_1 \cup b_1 = \mathbf{T}$$

Clearly, this result is quite paradoxical: if on each suffix of π the formula can only take either the value $\mathbf{F}$ or the least undesigned value greater than $\mathbf{F}$ on either chain, how can we explain the fact that $F\varphi$ takes the maximal, designated value $\mathbf{T}$ on that path?

Obviously, we do not want to encounter this kind of situations in any practical cases. Hence, when choosing a many-valued, lattice-based logic for specifying or describing any concurrent system, we must take great care to ensure that that such paradoxes cannot occur. The minimal requirement is that they cannot occur with respect to any formulae describing significant properties of the system that we might be interested in. This is, in fact, a basic test for the practical value of any lattice-based many-valued logics for a concurrent system we might possibly adopt.

It can be shown that the example of a simple model of a coffee machine shown in Figure 2 and a lattice based logic for its descriptions satisfy the above requirements.

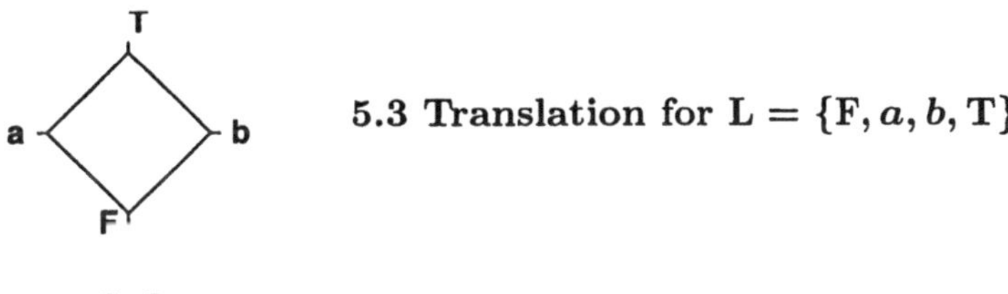

5.3 Translation for $\mathbf{L} = \{\mathbf{F}, a, b, \mathbf{T}\}$

Let $M = (S, R, V, \mathcal{L})$ be a model over the 4-valued pseudo-Boolean lattice $\mathcal{L}$, and let φ be an mv-CTL* formula. Let M' denote the 2-valued model obtained from M by replacing the value a by $\mathbf{T}$, and the value b by $\mathbf{F}$. Similarly, let M'' denote the 2-valued model obtained from M by replacing the value a by $\mathbf{F}$ and the value b by $\mathbf{T}$. Formally, M' and M'' are defined as follows:

$M' = (S, R', V', \mathcal{L})$, where
- $Z'(x, y) = \mathbf{T}$ iff $Z(x, y) \geq a$,
- $Z'(x, y) = \mathbf{F}$ iff $Z(x, y) \leq b$,
with $Z \in \{V, R\}$,

$M'' = (S, R'', V'', \mathcal{L})$, where
- $Z''(x, y) = \mathbf{T}$ iff $Z(x, y) \geq b$,
- $Z''(x, y) = \mathbf{F}$ iff $Z(x, y) \leq a$,
with $Z \in \{V, R\}$.

Analogously to the case of FTO interpretations, in order to apply standard induction based proofs we need the following lemma for dealing with the cases of $[\mathrm{U}(\varphi, \psi)]_\pi$, $[\overline{\mathrm{U}}(\varphi, \psi)]_\pi$, $[\mathrm{A}\varphi]_s$, and $[\mathrm{E}\varphi]_s$.

Lemma 2. *Let $x_i \in L$ for all $i \geq 0$. The following conditions hold:*

- $\bigcup_{i \geq 0} x_i = \mathbf{T}$ *iff*
  - $(\exists i \geq 0)\ x_i = \mathbf{T}$ *or*
  - $(\exists j \geq 0)\ x_j = a$ *and* $(\exists k \geq 0)\ x_k = b$,
- $\bigcup_{i \geq 0} x_i = a$ *iff* $(\exists j \geq 0)\ x_j = a$ *and* $(\forall k \neq j)\ x_k \leq a$,

- $\bigcup_{i\geq 0} x_i = b$ *iff* $(\exists j \geq 0)\ x_j = b$ *and* $(\forall k \neq j)\ x_k \leq b$,
- $\bigcup_{i\geq 0} x_i = \mathbf{F}$ *iff* $(\forall i \geq 0)\ x_i = \mathbf{F}$,
- $\bigcap_{i\geq 0} x_i = \mathbf{T}$ *iff* $(\forall i \geq 0)\ x_i = \mathbf{T}$,
- $\bigcap_{i\geq 0} x_i = a$ *iff* $(\exists j \geq 0)\ x_j = a$ *and* $(\forall k \neq j)\ x_k \geq a$,
- $\bigcap_{i\geq 0} x_i = b$ *iff* $(\exists j \geq 0)\ x_j = b$ *and* $(\forall k \neq j)\ x_k \geq b$,
- $\bigcap_{i\geq 0} x_i = \mathbf{F}$ *iff*
  - $(\exists j \geq 0)\ x_j = \mathbf{F}$ *or*
  - $(\exists j \geq 0)\ x_j = a$ *and* $(\exists k \geq 0)\ x_k = b$.

Next, we formulate the main theorem of this section. In what follows, $[\varphi]_x$ denotes $[\varphi]_x^M$, $[\varphi]'_x$ denotes $[\varphi]_x^{M'}$, whereas $[\varphi]''_x$ denotes $[\varphi]_x^{M''}$.

Theorem 2. *The following conditions hold for every x, which belongs to either S or Π:*

1. $[\varphi]_x = \mathbf{T}$ *iff* $[\varphi]'_x = \mathbf{T}$ *and* $[\varphi]''_x = \mathbf{T}$,
2. $[\varphi]_x = a$ *iff* $[\varphi]'_x = \mathbf{T}$ *and* $[\varphi]''_x = \mathbf{F}$,
3. $[\varphi]_x = b$ *iff* $[\varphi]'_x = \mathbf{F}$ *and* $[\varphi]''_x = \mathbf{T}$,
4. $[\varphi]_x = \mathbf{F}$ *iff* $[\varphi]'_x = \mathbf{F}$ *and* $[\varphi]''_x = \mathbf{F}$.

Proof. See the Appendix.

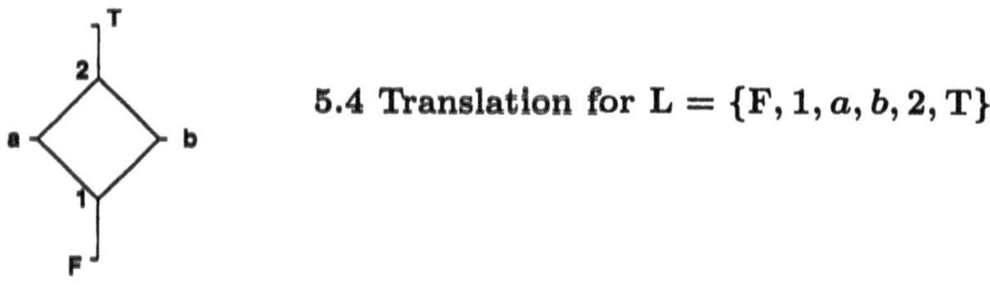

5.4 Translation for $\mathbf{L} = \{\mathbf{F}, 1, a, b, 2, \mathbf{T}\}$

Let $M = (S, R, V, \mathcal{L})$ be a 6-valued model and let φ be an mv-CTL* formula. We show that the model checking problem over M can be translated to the model checking problem over a 3-valued model M' and a 4-valued model M''. This, by Theorems 1 and 2, yields a translation to standard two-valued models.

Let M' denote the 3-valued model obtained from M by replacing the values $1, a, b, 2$ of R and V by 1. Let M'' denote the 4-valued model obtained from M by replacing the value 1 of R and V by $\mathbf{F}$ and the value 2 of R and V by $\mathbf{T}$. Formally, M' and M'' are defined as follows:

$M' = (S, R', V', \mathcal{L})$, where
- $Z'(x,y) = Z(x,y)$ for $Z(x,y) \in \{\mathbf{F}, \mathbf{T}\}$,
- $Z'(x,y) = 1$ iff $1 \leq Z(x,y) \leq 2$,

with $Z \in \{V, R\}$.

$M'' = (S, R'', V'', \mathcal{L})$, where

- $Z''(x,y) = Z(x,y)$ for $Z(x,y) \in \{a,b\}$,
- $Z''(x,y) = \mathbf{F}$ iff $Z(x,y) \leq 1$,
- $Z''(x,y) = \mathbf{T}$ iff $Z(x,y) \geq 2$,

with $Z \in \{V, R\}$.

Theorem 3. *The following conditions hold:*

- $[\varphi]_\iota = \mathbf{T}$ *iff* $[\varphi]'_\iota = \mathbf{T}$,
- $[\varphi]_\iota = \mathbf{F}$ *iff* $[\varphi]'_\iota = \mathbf{F}$.
- *If* $[\varphi]_\iota \in \{1, a, b, 2\}$, *then*
  - $[\varphi]_\iota = a$ *iff* $[\varphi]''_\iota = a$,
  - $[\varphi]_\iota = b$ *iff* $[\varphi]''_\iota = b$,
  - $[\varphi]_\iota = 1$ *iff* $[\varphi]''_\iota = \mathbf{F}$,
  - $[\varphi]_\iota = 2$ *iff* $[\varphi]''_\iota = \mathbf{T}$.

Proof. By structural induction on φ using Lemmas 1 and 2. [4]

Notice that M' is a 3-valued model which preserves the values $\mathbf{T}$ and $\mathbf{F}$ of φ over M, but bundles together all the other values of φ over M, putting them down as the value 1. On the other hand if φ does evaluate to 1 over M', then M'' is used to identify the "precise" value of φ. It is easy to see that the 4-valued model M'' preserves the values $1, a, b, 2$ of φ over M.

Complexity of mv-CTL* model checking depends on the structure of a lattice $\mathcal{L}$ and the complexity of the model checking problem for CTL* (see the next section). It is easy to see that if deciding whether $[\varphi]_\iota^M = \mathbf{T}$ for a two valued model M, requires time $O(Y)$, then the value of $[\varphi]_\iota^M$ can be found in time:

- $O((k-1) \times Y)$ for M over the lattice $\mathbf{K}$,
- $O(2 \times Y)$ for M over the lattice $\mathbf{2} \times \mathbf{2}$,
- $O(4 \times Y)$ for M over the lattice $\mathbf{2} + \mathbf{4}$.

6 Model checking for CTL*

So far we have shown that the model checking problem for mv-CTL* can be translated to the model checking problem for CTL*, but we have not discussed how the latter problem is solved.

A traditional CTL model checking algorithm [12] decides whether $[\varphi]_\iota^M = \mathbf{T}$ in time $O((|S| + |R|) \times |\varphi|)$ by labelling the states of the model M with subformulas of the formula φ.

[4] When considering the proof of the backward implication for the last four clauses, one should keep in mind that they are conditional upon the assumption that $[\varphi]_\iota \in \{1, a, b, 2\}$. In consequence, all other values of φ, i.e., $\mathbf{F}$ and $\mathbf{T}$, in the original 6-valued model are excluded.

A standard model checking algorithm for LTL and CTL* is based on the automata theory. Since the formulas of these logics have translations to non-deterministic Büchi alternating word or tree automata, the model checking problem is reduced to checking the non-emptiness of the product of the automaton representing the model and the automaton for the formula [7]. Such an algorithm decides whether $[\varphi]_{\iota}^{M} = \mathbf{T}$ in time $O((|S| + |R|) \times 2^{|\varphi|})$.

7 Conclusions and future work

We have shown that a model checking algorithm for a selected multi-valued version of CTL* can be obtained by making several queries to a standard model checking algorithm for CTL*. Such an approach does not restrict us to using any of the specific model checking methods for CTL*.

In our future work we are going to consider the possibility of applying one of the most recently defined model checking methods, called bounded model checking (BMC) [2, 1], for multi-valued temporal logics. Notice that BMC is available for LTL [2] and for the universal fragment of CTL (ACTL) [26]. Therefore, it can be applied for mv-LTL and mv-ACTL only. The idea consists in translating a model checking problem to the problem of checking satisfiability of a propositional formula. Thus, a temporal formula and the transition relation of a model are translated to propositional formulas, the conjunction of which is then tested for satisfiability. For a multi-valued temporal logic, the translation of the formula would not change. However, the transition relation of the model would be translated several times, for any two-valued model obtained via the translation from the many-valued model checking to 2-valued model checking.

The possibility of selecting any of the available model checking methods is not the only advantage of our approach. Another superiority is demonstrated by a possibility of applying different model reduction techniques directly. This is very important since practical applicability of model checking is strongly restricted by the state explosion problem. Thus, each of the two-valued models obtained via the translation can be reduced using the following methods: partial order reductions [24, 27, 25], symmetry reductions [19], abstraction techniques [16], and BDD-based symbolic storage methods [5, 6]. Moreover, there is no need to make any changes to the above methods in order to use them in our approach.

8 Appendix

Proof of Theorem 2.

(=>) By induction on the length of the formula. The hypothesis holds for propositional variables. Now, assume that it is true for length k, and that φ is of length $k + 1$. Then:

1. (*a*) $\varphi = \varphi_1 \vee \varphi_2$ is a state formula. Then there are two possible cases:
   - If $[\varphi_i]_s = \mathbf{T}$ for some $i \in \{1,2\}$, we have $[\varphi_i]'_s = [\varphi_i]''_s = \mathbf{T}$ by structural induction and Lemma 2.
   - If $[\varphi_1]_s = a$ and $[\varphi_2]_s = b$ (or symmetric), we have $[\varphi_1]'_s = \mathbf{T}$ and $[\varphi_2]''_s = \mathbf{T}$ by 2., 3., structural induction and Lemma 2.

   (*b*) $\varphi = \varphi_1 \wedge \varphi_2$ is a state formula. Then $[\varphi_i]_s = \mathbf{T}$ for each $i = 1,2$. Hence $[\varphi_i]'_s = \mathbf{T}$ and $[\varphi_i]''_s = \mathbf{T}$ for $i = 1,2$ by 1., and structural induction. This implies that $[\varphi]'_s = \mathbf{T}$ and $[\varphi]''_s = \mathbf{T}$.

   (*c*) $\varphi = \mathrm{A}\psi$. Then $[\mathrm{A}\psi]_s = \mathbf{T}$ iff $[\psi]_\pi = \mathbf{T}$ for each $\pi \in \Pi(s)$. Hence $[\psi]'_\pi = \mathbf{T}$ and $[\psi]''_\pi = \mathbf{T}$ for each $\pi \in \Pi(s)$ by 1., and structural induction. This implies that $[\varphi]'_s = \mathbf{T}$ and $[\varphi]''_s = \mathbf{T}$.

   (*d*) $\varphi = \mathrm{E}\psi$. Then there are two possible cases:
   - If $[\psi]_\pi = \mathbf{T}$ for all $\pi \in \Pi(s)$, we have $[\psi]'_\pi = \mathbf{T}$ and $[\psi]''_\pi = \mathbf{T}$ by structural induction and Lemma 2. Hence $[\varphi]'_s = \mathbf{T}$ and $[\varphi]''_s = \mathbf{T}$.
   - If $[\psi]_{\pi'} = a$ and $[\psi]_{\pi''} = b$ for some $\pi', \pi'' \in \Pi(s)$, we have $[\psi]'_{\pi'} = \mathbf{T}$, $[\psi]''_{\pi'} = \mathbf{F}$, and $[\psi]'_{\pi''} = \mathbf{F}$, $[\psi]''_{\pi''} = \mathbf{T}$ by 2.,3., structural induction and Lemma 2. Thus, $[\varphi]'_s = \mathbf{T}$ and $[\varphi]''_s = \mathbf{T}$.

   (*e*) $\varphi = \varphi_1 \vee \varphi_2$ is a path formula. This case is similar to (*a*).

   (*f*) $\varphi = \varphi_1 \wedge \varphi_2$ is a path formula. This case is similar to (*b*).

   (*g*) $\varphi = X\psi$. Then, $[X\psi]_\pi = \mathbf{T}$ iff $R(\pi(0), \pi(1)) = \mathbf{T}$ and $[\psi]_{\pi_1} = \mathbf{T}$. So, $R'(\pi(0), \pi(1)) = \mathbf{T}$ and $[\psi]'_{\pi_1} = \mathbf{T}$, $R''(\pi(0), \pi(1)) = \mathbf{T}$ and $[\psi]''_{\pi_1} = \mathbf{T}$ by 1. and structural induction. This implies that $[\varphi]'_\pi = \mathbf{T}$ and $[\varphi]''_\pi = \mathbf{T}$.

   (*h*) $\varphi = \mathrm{U}(\varphi_1, \varphi_2)$. Similar to case (*d*), follows from Lemma 2 and structural induction.

   (*i*) $\varphi = \overline{\mathrm{U}}(\varphi_1, \varphi_2)$. Similar to case (*c*), follows from Lemma 2 and structural induction.

2. (*a*) $\varphi = \varphi_1 \vee \varphi_2$ is a state formula. Then there are two possible cases:
   - If $[\varphi_i]_s = a$ for each $i \in \{1,2\}$, we have $[\varphi_1]'_s = [\varphi_2]'_s = \mathbf{T}$ and $[\varphi_1]''_s = [\varphi_2]''_s = \mathbf{F}$ by 2., and structural induction. Whence $[\varphi]'_s = \mathbf{T}$ and $[\varphi]''_s = \mathbf{F}$.
   - If $[\varphi_1]_s = a$ and $[\varphi_2]_s = \mathbf{F}$ (or conversely), then $[\varphi_1]'_s = \mathbf{T}$ and $[\varphi_1]''_s = \mathbf{F}$, $[\varphi_2]'_s = [\varphi_2]''_s = \mathbf{F}$ by 2., 4., and structural induction. Thus, $[\varphi]'_s = \mathbf{T}$ and $[\varphi]''_s = \mathbf{F}$.

   (*b*) $\varphi = \varphi_1 \wedge \varphi_2$ is a state formula. Then there are two possible cases:
   - If $[\varphi_i]_s = a$ for each $i \in \{1,2\}$, the proof is as above.
   - If $[\varphi_1]_s = a$ and $[\varphi_2]_s = \mathbf{T}$, then $[\varphi_1]'_s = \mathbf{T}$ and $[\varphi_1]''_s = \mathbf{F}$, $[\varphi_2]'_s = [\varphi_2]''_s = \mathbf{T}$ by 2., 1., and structural induction. Thus $[\varphi]'_s = \mathbf{T}$ and $[\varphi]''_s = \mathbf{F}$.

   (*c*) $\varphi = \mathrm{A}\psi$. Similar to case (*b*), follows from Lemma 2 and structural induction.

   (*d*) $\varphi = \mathrm{E}\psi$. Similar to case (*a*), follows from Lemma 2 and structural induction.

   (*e*) $\varphi = \varphi_1 \vee \varphi_2$ is a path formula. This case is similar to (*a*).

   (*f*) $\varphi = \varphi_1 \wedge \varphi_2$ is a path formula. This case is similar to (*b*).

   (*g*) $\varphi = X\psi$. Then there are two possible cases

- If $R(\pi(0), \pi(1)) = a$ and $[\psi]_{\pi_1} \geq a$, we have $R'(\pi(0), \pi(1)) = \mathbf{T}$ and $[\psi]'_{\pi_1} = \mathbf{T}$, $R''(\pi(0), \pi(1)) = \mathbf{F}$ and $[\psi]''_{\pi_1} \geq \mathbf{F}$ by structural induction. This implies that $[\varphi]'_{\pi} = \mathbf{T}$ and $[\varphi]''_{\pi} = \mathbf{F}$.
- If $R(\pi(0), \pi(1)) \geq a$ and $[\psi]_{\pi_1} = a$, we have $R'(\pi(0), \pi(1)) = \mathbf{T}$ and $[\psi]'_{\pi_1} = \mathbf{T}$, $R''(\pi(0), \pi(1)) \geq \mathbf{F}$ and $[\psi]''_{\pi_1} = \mathbf{F}$ by structural induction. This implies that $[\varphi]'_{\pi} = \mathbf{T}$ and $[\varphi]''_{\pi} = \mathbf{F}$.

(*h*) $\varphi = \mathrm{U}(\varphi_1, \varphi_2)$. Similar to case (*d*), follows from Lemma 2 and structural induction.

(*i*) $\varphi = \overline{\mathrm{U}}(\varphi_1, \varphi_2)$. Similar to case (*c*), follows from Lemma 2 and structural induction.

3. By symmetry to 2.,
4. (*a*) $\varphi = \varphi_1 \vee \varphi_2$ is a state formula. Then $[\varphi_1]_s = [\varphi_2]_s = \mathbf{F}$. So $[\varphi_i]'_s = [\varphi_i]''_s = \mathbf{F}$ for each $i = 1, 2$ by structural induction. Hence $[\varphi]'_s = [\varphi]''_s = \mathbf{F}$.

(*b*) $\varphi = \varphi_1 \wedge \varphi_2$ is a state formula. Then there are two possible cases:
- If $[\varphi_i]_s = \mathbf{F}$ for some $i \in \{1, 2\}$, we have $[\varphi_i]'_s = [\varphi_i]''_s = \mathbf{F}$ by structural induction. Hence $[\varphi]'_s = [\varphi]'_s = \mathbf{F}$.
- If $[\varphi_1]_s = a$ and $[\varphi_2]_s = b$ (or conversely), we have $[\varphi_1]'_s = \mathbf{T}$, $[\varphi_1]''_s = \mathbf{F}$, and $[\varphi_2]'_s = \mathbf{F}$, $[\varphi_2]''_s = \mathbf{T}$ by 2., 3., and structural induction. Thus $[\varphi]'_s = \mathbf{F}$ and $[\varphi]''_s = \mathbf{F}$.

(*c*) $\varphi = \mathrm{A}\psi$. Then there are two possible cases:
- If $[\psi]_{\pi} = \mathbf{F}$ for some $\pi \in \Pi(s)$, we have $[\psi]'_{\pi} = \mathbf{F}$ and $[\psi]''_{\pi} = \mathbf{F}$ by structural induction. Hence $[\varphi]'_s = \mathbf{F}$ and $[\varphi]''_s = \mathbf{F}$.
- If $[\psi]_{\pi'} = a$ and $[\psi]_{\pi''} = b$ for some $\pi', \pi'' \in \Pi(s)$, we have $[\psi]'_{\pi'} = \mathbf{T}$, $[\psi]''_{\pi'} = \mathbf{F}$, and
$[\psi]'_{\pi''} = \mathbf{F}$, $[\psi]''_{\pi''} = \mathbf{T}$ by 2., 3., and structural induction. Thus $[\varphi]'_s = \mathbf{F}$ and $[\varphi]''_s = \mathbf{F}$.

(*d*) $\varphi = \mathrm{E}\psi$. Then $[\mathrm{E}\psi]_s = \mathbf{F}$ iff $[\psi]_{\pi} = \mathbf{F}$ for each $\pi \in \Pi(s)$. Hence $[\psi]'_{\pi} = \mathbf{F}$ and $[\psi]''_{\pi} = \mathbf{F}$ for each $\pi \in \Pi(s)$ by 1. and structural induction. This implies that $[\varphi]'_s = \mathbf{F}$ and $[\varphi]''_s = \mathbf{F}$.

(*e*) $\varphi = \varphi_1 \vee \varphi_2$ is a path formula. This case is similar to (*a*).

(*f*) $\varphi = \varphi_1 \wedge \varphi_2$ is a path formula. This case is similar to (*b*).

(*g*) $\varphi = X\psi$. Then, $[X\psi]_{\pi} = \mathbf{F}$ iff $R(\pi(0), \pi(1)) \wedge [\psi]_{\pi_1} = \mathbf{F}$. There are three possible cases:
- If $R(\pi(0), \pi(1)) = \mathbf{F}$ and $[\psi]_{\pi_1} = \mathbf{F}$, we have $R'(\pi(0), \pi(1)) = \mathbf{F}$ and $[\psi]'_{\pi_1} = \mathbf{F}$, $R''(\pi(0), \pi(1)) = \mathbf{F}$ and $[\psi]''_{\pi_1} = \mathbf{F}$ by 1., and structural induction. This implies that $[\varphi]'_{\pi} = \mathbf{F}$ and $[\varphi]''_{\pi} = \mathbf{F}$.
- If $R(\pi(0), \pi(1)) = a$ and $[\psi]_{\pi_1} = b$, we have $R'(\pi(0), \pi(1)) = \mathbf{T}$ and $[\psi]'_{\pi_1} = \mathbf{F}$, $R''(\pi(0), \pi(1)) = \mathbf{F}$ and $[\psi]''_{\pi_1} = \mathbf{T}$ by 1., and structural induction. This implies that $[\varphi]'_{\pi} = \mathbf{F}$ and $[\varphi]''_{\pi} = \mathbf{F}$.
- If $R(\pi(0), \pi(1)) = b$ and $[\psi]_{\pi_1} = a$, we have $R'(\pi(0), \pi(1)) = \mathbf{F}$ and $[\psi]'_{\pi_1} = \mathbf{T}$, $R''(\pi(0), \pi(1)) = \mathbf{T}$ and $[\psi]''_{\pi_1} = \mathbf{F}$ by 1., and structural induction. This implies that $[\varphi]'_{\pi} = \mathbf{F}$ and $[\varphi]''_{\pi} = \mathbf{F}$.

(*h*) $\varphi = \mathrm{U}(\varphi_1, \varphi_2)$. Similar to case (*d*), follows from Lemma 2 and structural induction.

(*i*) $\varphi = \overline{\mathrm{U}}(\varphi_1, \varphi_2)$. Similar to case (*c*), follows from Lemma 2 and structural induction.

(<=) Follows easily from (=>) since the latter proof exhausts all possible combinations of the values of $[\varphi]_x, [\varphi]'_x$, and $[\varphi]''_x$.

References

1. Biere, A., Cimatti, A., Clarke, E., Fujita, M. and Zhu, Y.: Symbolic model checking using SAT procedures instead of BDDs, Proc. of ACM/IEEE Design Automation Conference (DAC'99) (1999) 317–320
2. Biere, A., Cimatti, A., Clarke, E. and Zhu, Y.: Symbolic model checking without BDDs, Proc. of TACAS'99, LNCS, vol. 1579, Springer-Verlag (1999) 193–207
3. Bruns, G. and Godefroid, P.: Model checking partial state spaces, Proc. of CAV'99, LNCS, vol. 1633, Springer-Verlag (1999) 274–287
4. Bruns, G. and Godefroid, P.: Generalized model checking: reasoning about partial state spaces, Proc. of CONCUR'00, LNCS, vol. 1877, Springer-Verlag, (2000) 168–182
5. Bryant, R.: Graph-based algorithms for boolean function manipulation, IEEE Transaction on Computers 35, (8) (1986) 677–691
6. Bryant, R.: Binary Decision Diagrams and beyond: Enabling technologies for formal verification, Proc. of Int. Conf. on Computer-Aided Design (ICCAD'95) (1995) 236–243
7. Bernholtz, O., Vardi, M.Y. and Wolper, P.: An automata-theoretic approach to branching-time model checking, Proc. of CAV'94, LNCS, vol. 818, Springer-Verlag (1994) 142–155
8. Chechik, M., Devereux, B. and Easterbrook, S.: Implementing a multi-valued symbolic model-checker, Proc. of TACAS'01, LNCS, vol. 2031, Springer-Verlag (2001) 404–419
9. Chechik, M., Devereux, B., Easterbrook, S., Lai, A.Y.C. and Petrovykh, V.: Efficient multiple-valued model-checking using lattice representations, Proc. of CONCUR'01, LNCS, vol. 2154, Springer-Verlag (2001) 441–455
10. Chechik, M., Devereux, B. and Gurfinkel, A.: Model checking infinite state-spaces systems with fine-grained abstractions using spin, Proc. of SPIN'01, LNCS, vol. 2057, Springer-Verlag (2001) 16–36
11. Chechik, M., Easterbrook, S. and Petrovykh, V.: Model checking over multi-valued temporal logics, Proc. of FME'01, LNCS, vol. 2001, Springer-Verlag (2001) 72–98
12. Clarke, E.M., Emerson, E.A. and Sistla, A.P.: Automatic verification of finite state concurrent systems using temporal logic specifications: A practical approach, ACM Transactions on Programming Languages and Systems 8, (2) (1986) 244–263
13. Clarke, E.M., Grumberg, O. and Peled, D.: Model Checking, MIT Press (1999)
14. Chechik, M.: On interpreting results of model-checking with abstraction, Tech. Report 417, University of Toronto (2000)
15. Dams, D., Gerth, R., Dohmen, G., Herrmann, R., Kelb, P. and Pargmann, H.: Model checking using adaptive state and data abstraction, Proc. of CAV'94, LNCS, vol. 818, Springer-Verlag (1994) 455–467
16. Dams, D., Grumberg, O. and Gerth, R.: Abstract interpretation of reactive systems: Abstractions preserving ACTL*, ECTL* and CTL*, Proceedings of the IFIP Working Conference on Programming Concepts, Methods and Calculi (PROCOMET'94), Elsevier Science Publishers (1994)

17. Easterbrook, S. and Chechik, M.: A framework for multi-valued reasoning over inconsistent viewpoints, Proc. of ICSE'01 (2001) 411–420
18. Emerson, E.A.: Handbook of Theoretical Computer Science, vol. B: Formal Methods and Semantics, ch. Temporal and Modal Logic, Elsevier (1990) 995–1067
19. Emerson, E.A. and Sistla, A.P.: Symmetry and model checking, Formal Methods in System Design 9 (1995) 105–131
20. Fitting, M.: Many-valued modal logics, Fundamenta Informaticae 15, (3-4) (1991) 335–350
21. Fitting, M.: Many-valued modal logics II, Fundamenta Informaticae 17 (1992) 55–73
22. Hazelhurst, S.: Compositional model checking of partially ordered state spaces, Ph.D. thesis, University of British Columbia (1996)
23. Lichtenstein, O. and Pnueli, A.: Checking that finite-state concurrent programs satisfy their linear specification, Proc. of the 12th ACM Symposium on Principles of Programming Languages (POPL'84), ACM Press (1984) 97–107
24. Peled, D.: Partial order reduction: Linear and branching temporal logics and process algebras, Proc. of Partial Order Methods in Verification (POMIV'96), ACM/AMS DIMACS Series, vol. 29, Amer. Math. Soc. (1996) 79–88
25. Penczek, W., Szreter, M., Gerth, R. and Kuiper, R.: Improving partial order reductions for universal branching time properties, Fundamenta Informaticae 43 (2000) 245–267
26. Penczek, W. and Woźna, B.: Towards bounded model checking for Timed Automata, Proc. of the Int. Workshop on Concurrency Specification and Programming (CS&P'01) (2001) 195–209
27. Valmari, A.: Stubborn sets for reduced state space generation, Proc. of the Int. Conf. on Applications and Theory of Petri Nets (ICATPN'89), LNCS, vol. 483, Springer-Verlag (1989) 491–515

Chapter 9
Complexity of Many-valued Logics*

Reiner Hähnle

Chalmers University of Technology, Dept. of Computing Science
S-41296 Gothenburg, Sweden, reiner@cs.chalmers.se

Abstract. As is the case for other logics, a number of complexity-related questions can be posed in the context of many-valued logic. Some of these, such as the complexity of the sets of satisfiable and valid formulas in various logics, are completely standard; others only make sense in a many-valued context. In this overview I concentrate on two kinds of complexity problems related to many-valued logic: first, I discuss the complexity of the membership problem in various languages, such as the satisfiable, respectively, the valid formulas in some well-known logics. Second, I discuss the size of representations of many-valued connectives and quantifiers, because this has a direct impact on the complexity of many kinds of deduction systems. I include results on both propositional and on first-order logic.

1 Introduction

As it is the case for classical logic and other non-standard logics, a number of complexity-related questions can be asked in the context of many-valued logic. Some questions, such as the complexity of the sets of satisfiable and valid formulas in various logics, are completely standard; others, such as the maximal size of representations of many-valued connectives, only make sense in a many-valued context.

Such investigations have several rewards that motivate them: First, comparison of the complexity of many-valued problems with corresponding classical problems gives insight into whether and when many-valuedness adds descriptive expressivity to a language. The results can be surprising: for example, deciding satisfiability of the well-known infinite-valued propositional logics has the same complexity as classical satisfiability, this changes drastically on the first-order level. Second, upper complexity bounds are important to evaluate implemented algorithms in applications. The latter should realize the best known upper bounds. In many cases, proofs of upper bounds are constructive and suggest algorithms themselves. Third, when many-valued logic is used as a modeling language in applications, it is crucial to know

* This article is an extended and revised version of an invited tutorial at ISMVL'01 [37]. It contains as well material from [36] in updated and revised form.

the complexity of deciding satisfiability. For example, the results in Section 4 indicate that small changes in the language syntax and semantics can mean the difference between tractability and untractability. Complexity results tell then, which language constructs are expensive to realize in tools and should be avoided in modeling.

In this overview I concentrate mainly on two kinds of complexity problems related to many-valued logics: in Section 6 I discuss the complexity of the membership problem in various languages, such as the satisfiable, respectively, the valid formulas in some well-known logics. Two basic proof techniques are presented in some detail: a reduction of many-valued logic to mixed integer programming and a reduction to classical logic. I do not mention results on complexity of algorithms, in particular, of constructing decision diagrams, because they are a quickly moving target. Instead, in Section 7, I discuss the size of representations of many-valued connectives and quantifiers, because this has a direct impact on the complexity of many kinds of deduction systems. I consider results on both propositional and first-order logic throughout the article. The following section defines a suitable framework for the discussion of a wide class of many-valued logics. This is followed by three brief sections that introduce into Signed Logics, Mixed Integer Programming, and Polynomial Expressions, respectively.

This article is intended to be mainly self-contained, but the diversity of the material imposes some limits on this. Therefore, I added references for background reading, where appropriate. General overviews of many-valued logic, including its many applications, are [28, 36].

2 A general framework for many-valued logics

In many-valued logic we work with standard propositional and first-order languages, but since we have different and/or additional connectives and quantifiers it is convenient to regard these not as fixed, but to parameterize many-valued logic languages with sets of connectives and possibly quantifiers. Apart from this, the following definitions are fairly standard.

2.1 Propositional logic

As usual, a countably infinite set of propositional variables $\Sigma = \{p, q, r, \ldots\}$ is a **propositional signature**. Let Θ be a finite set of operator names, called **connectives**; the **arity** of $\theta \in \Theta$ is a non-negative integer given by a function α. Connectives with arity 0 are called **logical constants**. Given a **propositional language** $\mathbf{L}^0 = \langle \Theta, \alpha \rangle$ and a signature Σ, the set of $\mathbf{L}^0_\Sigma$-**formulas** is defined inductively as usual:

1. Members of Σ and logical constants are $\mathbf{L}^0_\Sigma$-formulas.
2. If $\theta \in \Theta$, $\alpha(\theta) = r > 0$, and $\varphi_1, \ldots, \varphi_r$ are $\mathbf{L}^0_\Sigma$-formulas, then $\theta(\varphi_1, \ldots, \varphi_r)$ is a $\mathbf{L}^0_\Sigma$-formula as well.

The concrete signature usually is not relevant, and then we omit it from the index.

Example 1. Throughout this article, $\overline{0}$ is a logical constant, all connectives of the form $\neg_x$ are unary, while $\rightarrow_x$, $\vee$, $\wedge$, $\oplus_x$, $\odot_x$ are binary ones. The language $\mathbf{L}^0_c$ of **classical propositional logic** contains the connectives $\overline{0}, \wedge, \rightarrow$. The language $\mathbf{L}^0_Ł$ of **Łukasiewicz logic** contains the connectives $\overline{0}, \odot_Ł, \rightarrow_Ł$. The language $\mathbf{L}^0_G$ of **Gödel logic** contains the connectives $\overline{0}, \wedge, \rightarrow_G$. The language $\mathbf{L}^0_\Pi$ of **product logic** contains the connectives $\overline{0}, \odot_\Pi, \rightarrow_\Pi$.

We say **negation** to the connectives $\neg_x$, **implication** to $\rightarrow_x$, **conjunction** to $\wedge$, **disjunction** to $\vee$, **product** to $\odot_x$, and **sum** to $\oplus_x$. This can be qualified with the name of a logical language, for example, if one writes "Łukasiewicz implication", then the connective $\rightarrow_Ł$ is meant, etc.

A **propositional matrix** $\mathbf{A}^0 = \langle N, (A_\theta)_{\theta\in\Theta}\rangle$ for a propositional language $\mathbf{L}^0 = \langle \Theta, \alpha\rangle$ consists of a non-empty set of **truth values** N and a collection $(A_\theta)_{\theta\in\Theta}$ of operations on N such that $A_\theta : N^{\alpha(\theta)} \rightarrow N$ for each $\theta \in \Theta$. $|N|$ denotes the cardinality of N.

A pair $\mathcal{L}^0 = \langle \mathbf{L}^0, \mathbf{A}^0\rangle$, where $\mathbf{L}^0 = \langle \Theta, \alpha\rangle$ is a propositional language and $\mathbf{A}^0 = \langle N, (A_\theta)_{\theta\in\Theta}\rangle$ is a matrix for $\mathbf{L}^0$, is called (N-valued) **propositional logic**.

Example 2. For each $n \geq 2$ let $\mathbf{n} = \{0, \frac{1}{n-1}, \ldots, \frac{n-2}{n-1}, 1\}$ be the truth value set of cardinality n consisting of equidistant rational numbers. With $[0,1]$ we denote the closed real unit interval. In the following, let N be either $[0,1]$ or $\mathbf{n}$ for some n. In all logics considered below, $A_{\overline{0}} = 0$, $A_\wedge = \min$, and $A_\vee = \max$. Moreover, implication is always defined as $A_{\rightarrow_x}(i,j) = \sup\{k \mid A_{\odot_x}(i,k) \leq j\}$ with the help of product. This process is called **residuation**. Note that residuated implication is always 1, whenever $i \leq j$ and $\odot_x \leq \min$.

In **classical propositional logic** $\mathcal{L}^0_c$, $n = 2$ (hence, $N = \{0,1\}$). If residuation is based on $\wedge$, the usual definitions result. In **Łukasiewicz logic** [45] $\mathcal{L}^0_Ł$, $A_{\odot_Ł}(i,j) = \max\{0, i+j-1\}$. In **Gödel logic** [26] $\mathcal{L}^0_G$, $A_\wedge = A_{\odot_G} = \min$. In **product logic** [39] $\mathcal{L}^0_\Pi$, $A_{\odot_\Pi} = \cdot$ (multiplication). Product logic is only defined for $N = [0,1]$, because the sets $\mathbf{n}$ are not closed under $A_{\odot_\Pi}$ for $n > 2$.

In fuzzy theory one frequently considers $\overline{0}$ and $\odot_x$ as primitive operators (and implication obtained by residuation), in algebraic treatments $\neg$, and $\rightarrow$ or $\oplus$ are more usual. The base is not really relevant, though: in each of the logics mentioned above, for example, negation and conjunction connectives are definable as follows:

$$\neg_x\varphi \quad \text{abbreviates} \quad \varphi \rightarrow_x \overline{0} \tag{1}$$

$$\varphi \wedge_x \psi \quad \text{abbreviates} \quad \varphi \odot_x (\varphi \rightarrow_x \psi) \tag{2}$$

An N-valued matrix $\mathbf{A}^0$ is called **functionally complete**, if every m-ary function $f : N^m \rightarrow N$ can be defined using operators from $\mathbf{A}^0$ alone. The matrix of classical logic is functionally complete. None of the other matrices defined in Example 2 is functionally complete.

A **propositional interpretation** $\mathbf{I}$ determines the truth value of each variable of a given signature Σ. Hence, it is simply a mapping $\mathbf{I} : \Sigma \rightarrow N$. For each propositional many-valued logic $\mathcal{L}^0$ its matrix $\mathbf{A}^0$ uniquely determines the extension of a given Σ-interpretation $\mathbf{I}$ to a **propositional valuation** function on $\mathbf{L}^0_\Sigma$ (denoted with the same symbol) by means of the usual homomorphy property:

$$\mathbf{I}(\theta(\varphi_1, \ldots, \varphi_r)) = A_\theta(\mathbf{I}(\varphi_1), \ldots, \mathbf{I}(\varphi_r)) \ . \tag{3}$$

Let D be any subset of N and Ψ be a set of $\mathbf{L}^0_\Sigma$-formulas. We say that Ψ is **D-satisfiable**, if there is a Σ-interpretation $\mathbf{I}$ such that $\mathbf{I}(\varphi) \in D$ for all $\varphi \in \Psi$. Such an $\mathbf{I}$ is called **D-model** of Ψ, in symbolic notation $\mathbf{I} \models_D \Psi$. The set Ψ is **D-valid**, if all Σ-interpretations are D-models of Ψ, in symbols $\models_D \Psi$. We usually write $\models_D \varphi$ instead of $\models_D \{\varphi\}$. Finally, a formula φ is a **D-consequence** of Ψ if every D-model of Ψ is a D-model of φ, in symbols $\Psi \models_D \varphi$.

The well-known duality in classical logic between satisfiability and validity extends as follows to the many-valued case: φ is D-valid iff it is not $N \backslash D$-satisfiable.

$D = \emptyset$ is not excluded for technical reasons; obviously, no formula is $\emptyset$-satisfiable. The notions of D-validity and -satisfiability first appeared in [43]. In many cases, the set D is considered to be fixed in a given many-valued logic. It is then called the set of **designated truth values**, and one writes "satisfiable" instead of "D-satisfiable", "$\models$" instead of "$\models_D$", etc.

Example 3. For the logics of Example 2 the usual choice for the set of designated truth values is $D = \{1\}$. The following are valid formulas in each of the logics defined in Example 2:

$$(\varphi \rightarrow_x \psi) \rightarrow_x ((\psi \rightarrow_x \chi) \rightarrow_x (\varphi \rightarrow_x \chi))$$
$$(\varphi \odot_x \psi) \rightarrow_x \varphi.$$

2.2 First-order logic

A **first-order signature** Σ is a triple $\langle \mathbf{P}_\Sigma, \mathbf{F}_\Sigma, \alpha_\Sigma \rangle$, where $\mathbf{P}_\Sigma$ is a non-empty family of **predicate symbols**, $\mathbf{F}_\Sigma$ a possibly empty family of **function symbols** disjoint from $\mathbf{P}_\Sigma$, and α_Σ assigns an arity (non-negative integer) to each member of $\mathbf{P}_\Sigma \cup \mathbf{F}_\Sigma$.

Let Term_Σ be the set of Σ-**terms**, defined inductively as usual over a countably infinite set of **object (or individual) variables** $\mathrm{Var} = \{x_0, x_1, \ldots\}$:

1. Members of Var and $c \in \mathbf{F}_\Sigma$ with $\alpha_\Sigma(c) = 0$ are Σ-terms.
2. If $f \in \mathbf{F}_\Sigma$, $\alpha_\Sigma(f) = r > 0$, and $t_1, \ldots, t_r$ are Σ-terms, then $f(t_1, \ldots, t_r)$ is a Σ-term.

$\mathrm{At}_\Sigma = \{p(t_1, \ldots, t_r) \mid p \in \mathbf{P}_\Sigma,\ \alpha_\Sigma(p) = r,\ t_i \in \mathrm{Term}_\Sigma\}$ is the set of **atoms**, Term^0_Σ is the set of variable-free terms.

A **first-order language** is a triple $\mathbf{L} = \langle \Theta, \Lambda, \alpha \rangle$, where $\langle \Theta, \alpha \rangle$ is a propositional language and Λ is a finite set of **first-order quantifier** symbols.

Example 4. All of the propositional languages $\mathbf{L}^0_x$ defined in Example 1 give rise to first-order languages named $\mathbf{L}_x$ by adding the quantifier set $\Lambda = \{\forall, \exists\}$.

The set of $\mathbf{L}_\Sigma$**-formulas** of a first-order language $\mathbf{L}$ over a first-order signature Σ is inductively defined:

1. The atoms over Σ are $\mathbf{L}_\Sigma$-formulas.
2. If $\theta \in \Theta$, $\alpha(\theta) = r > 0$, and $\varphi_1, \ldots, \varphi_r$ are $\mathbf{L}_\Sigma$-formulas, then $\theta(\varphi_1, \ldots, \varphi_r)$ is a $\mathbf{L}_\Sigma$-formula as well.
3. If $\lambda \in \Lambda$, $\varphi \in \mathbf{L}_\Sigma$, and $x \in \mathrm{Var}$, then $(\lambda x)\varphi$ is an $\mathbf{L}_\Sigma$-formula. The subformula φ is the **scope** of $(\lambda x)\varphi$, and the occurrence of x in (λx) is called a **defining occurrence** of x in $(\lambda x)\varphi$.

A variable $x \in \mathrm{Var}$ occurs **bound** in a formula φ, if φ contains a subformula of the form $(\lambda x)\psi$ such that x occurs in ψ not as a defining occurrence. A variable $x \in \mathrm{Var}$ occurs **free** in φ, if there is an occurrence of x in φ that is not a defining occurrence and not in the scope of a defining occurrence of x. A variable can occur both free and bound in the same formula. An occurrence can be neither free nor bound (as, for example, x in $(\forall x)p$).

If $\mathbf{L} = \langle \Theta, \Lambda, \alpha \rangle$ is a first-order language, then a **first-order matrix** is a triple $\mathbf{A} = \langle N, (A_\theta)_{\theta \in \Theta}, (Q_\lambda)_{\lambda \in \Lambda} \rangle$, where $\langle N, (A_\theta)_{\theta \in \Theta} \rangle$ is a propositional matrix for $\langle \Theta, \alpha \rangle$ and $Q_\lambda : \mathcal{P}^+(N) \to N$ for each $\lambda \in \Lambda$ with $\mathcal{P}^+(N)$ being the non-empty subsets of N. Q_λ is called the **distribution function** of the quantifier λ.

A pair $\mathcal{L} = \langle \mathbf{L}, \mathbf{A} \rangle$, where $\mathbf{L} = \langle \Theta, \Lambda, \alpha \rangle$ is a first-order language and $\mathbf{A} = \langle N, (A_\theta)_{\theta \in \Theta}, (Q_\lambda)_{\lambda \in \Lambda} \rangle$ is a first-order matrix for $\mathbf{L}$, is called (N-valued) **first-order logic**.

Example 5. From each first-order language $\mathbf{L}_x$ of Example 4 one obtains a first-order logic named $\mathcal{L}_x$ over $N = [0, 1]$ or $N = \mathbf{n}$ by taking the distribution functions $Q_\forall = \inf$ and $Q_\exists = \sup$. Here, it would not do to have merely rational numbers as truth values, because the rationals are not closed under inf and sup. The reader is invited to check that $\mathcal{L}_c$ is classical first-order logic.

Given a first-order signature Σ, a **first-order structure** $\mathbf{M} = \langle \mathbf{D}, \mathbf{I} \rangle$ fixes a non-empty set $\mathbf{D}$, the **domain** of discourse, and the meaning of function and predicate symbols via a **first-order interpretation** $\mathbf{I}$ that maps each function symbol $f \in \mathbf{F}_\Sigma$ of arity $r = \alpha_\Sigma(f)$ to a function $\mathbf{I}(f) : \mathbf{D}^r \to \mathbf{D}$, and each predicate symbol $p \in \mathbf{P}_\Sigma$ of arity $s = \alpha_\Sigma(p)$ to a function $\mathbf{I}(p) : \mathbf{D}^s \to N$. Observe that, if $\mathbf{F}_\Sigma = \emptyset$ and $\mathbf{P}_\Sigma$ contains only 0-ary predicate symbols, then $\mathbf{I}$ reduces to a propositional interpretation.

Like in the propositional case, for each first-order logic $\mathcal{L}$ its matrix $\mathbf{A}$ uniquely determines for any first-order structure $\mathbf{M}$ a valuation function on arbitrary terms and $\mathbf{L}_\Sigma$-formulas. In addition, one must specify the meaning of object variables that might occur within formulas. This is done as usual with a **variable assignment** $\beta : \mathrm{Var} \rightarrow \mathbf{D}$.

For given $\mathbf{M}$ and β, the **first-order valuation** function $v_{\mathbf{M},\beta}$ maps terms into $\mathbf{D}$ and $\mathbf{L}_\Sigma$-formulas into N. For $t \in \mathrm{Term}_\Sigma$, one writes $t^{\mathbf{M},\beta}$ instead of $v_{\mathbf{M},\beta}(t)$. The definition is by structural induction:

$$\begin{aligned} x^{\mathbf{M},\beta} &= \beta(x),\ x \in \mathrm{Var} \\ f(t_1,\ldots,t_r)^{\mathbf{M},\beta} &= \mathbf{I}(f)(t_1^{\mathbf{M},\beta},\ldots,t_r^{\mathbf{M},\beta}),\ f \in \mathbf{F}_\Sigma,\ \alpha_\Sigma(f) = r \\ v_{\mathbf{M},\beta}(p(t_1,\ldots,t_r)) &= \mathbf{I}(p)(t_1^{\mathbf{M},\beta},\ldots,t_r^{\mathbf{M},\beta}),\ p \in \mathbf{P}_\Sigma,\ \alpha_\Sigma(p) = r \\ v_{\mathbf{M},\beta}(\theta(\varphi_1,\ldots,\varphi_r)) &= A_\theta(v_{\mathbf{M},\beta}(\varphi_1),\ldots,v_{\mathbf{M},\beta}(\varphi_r)),\ \theta \in \Theta,\ \alpha(\theta) = r \\ v_{\mathbf{M},\beta}((\lambda x)\varphi) &= Q_\lambda(\{v_{\mathbf{M},\beta_x^d}(\varphi) \mid d \in \mathbf{D}\}),\ \lambda \in \Lambda \end{aligned}$$

The expression $\{v_{\mathbf{M},\beta_x^d}(\varphi) \mid d \in \mathbf{D}\}$ in the previous line is the **distribution of φ at x**.

Satisfaction is defined in analogy to the propositional case, with the exception that the presence of assignments gives rise to one more semantic concept, just as in classical logic.

Let Σ be a first-order signature, $\mathcal{L} = \langle \mathbf{L}, \mathbf{A} \rangle$ an N-valued first-order logic, D a subset of N, and Ψ be a set of $\mathbf{L}_\Sigma$-formulas. We say that Ψ is D-**satisfiable** if there is a first-order structure $\mathbf{M}$ over Σ and a variable assignment β such that $v_{\mathbf{M},\beta}(\varphi) \in D$ for all $\varphi \in \Psi$, for short write $\mathbf{M}, \beta \models_D \Psi$. When $\mathbf{M}, \beta \models_D \Psi$ for all variable assignments β, Ψ is said to be D-**true** in $\mathbf{M}$, for short $\mathbf{M} \models_D \Psi$, and $\mathbf{M}$ is called D-**model** of Ψ. The formula set Ψ is D-**valid** if it is true in all first-order Σ-structures, in symbols $\models_D \Psi$. Logical consequence is defined in analogy to the propositional case, that is, $\Psi \models_D \varphi$ iff every D-model of Ψ is a D-model of φ. The conventions about dropping D when it is obvious and identifying $\{\varphi\}$ with φ are as before.

Example 6. For the logics of Example 5 the usual choice for the designated truth values is $D = \{1\}$. The following formulas are
valid in each of these logics:

$(\forall x)(\varphi \rightarrow_x \psi) \rightarrow_x (\varphi \rightarrow_x (\forall x)\psi)$, whenever x not free in φ,
$(\forall x)(\varphi \rightarrow_x \psi) \rightarrow_x ((\exists x)\varphi \rightarrow_x \psi)$, whenever x not free in ψ.

The semantics of quantifiers in many-valued logic is not straightforward. In logics having disjunction- and conjunction-like connectives $\vee$ and $\wedge$, these can be used to define existential and universal quantifiers $\exists$ and $\forall$, where $v_{\mathbf{M},\beta} \models (\exists x)\varphi$ holds iff $\bigvee_{d \in \mathbf{D}}(v_{\mathbf{M},\beta_x^d} \models \varphi)$ and similar for $\forall$.[1] More generally,

[1] In abuse of notation, the symbol $\vee$ here is used to denote meta-level disjunction.

whenever N is a complete lattice with operations $\sqcap$ and $\sqcup$, universal and existential quantifiers Π and Σ can be defined as above, and they are characterized by $Q_\Pi(S) = \bigsqcap_{i \in S} i$, respectively, by $Q_\Sigma(S) = \bigsqcup_{i \in S} i$, see [70, 5, 34]. Here, we take a more general stance by admitting arbitrary distributions of truth values.

The idea of considering distributions of values is encountered in classical two-valued logic as well: generalized two-valued quantifiers ξ are obtained from $Q_\xi : 2^{\mathbf{D}} \to \{\text{true}, \text{false}\}$ based on the distribution $\{d \in \mathbf{D} \mid v_{\mathbf{M},\beta_x^d}(\varphi) = \text{true}\}$. Our notion of a many-valued quantifier based on distribution functions is due to Rosser & Turquette [59, Chapter IV], but it appears implicitly already in [52].

2.3 McNaughton's theorem

Denote with $\mathbf{I}_{p_1,\ldots,p_r}^{i_1,\ldots,i_r}$ a propositional interpretation that fixes $\mathbf{I}(p_j) = i_j$ for $1 \leq j \leq r$. Any propositional formula φ over r variables, say $\{p_1, \ldots, p_r\}$, determines for any given logic in a natural way a function $f_\varphi : N^r \to N$ via $f_\varphi(i_1, \ldots, i_r) = \mathbf{I}_{p_1,\ldots,p_r}^{i_1,\ldots,i_r}(\varphi)$ for all $i_1, \ldots, i_r \in N$. Each $\mathbf{L}_{Ł}^0$-formula φ, in particular, determines for Łukasiewicz logic a function $f_\varphi : [0,1]^r \to [0,1]$. It is easy to prove (or see [20, 3.1.8]) that for every $\varphi \in \mathbf{L}_{Ł}^0$ over r variables, f_φ has the following properties:

1. f_φ is continuous;
2. there is a finite set $\mathcal{P}$ of linear polynomials with integral coefficients in r variables over $[0,1]$ such that for each $\boldsymbol{i} \in [0,1]^r$, there is a polynomial $P \in \mathcal{P}$ with $f_\varphi(\boldsymbol{i}) = P(\boldsymbol{i})$.

McNaughton [50] showed that, conversely, for every function $f : [0,1]^r \to [0,1]$ satisfying the above properties, there is a formula φ of Łukasiewicz logic such that $f = f_\varphi$. McNaughton originally gave an indirect argument, but as shown in [54], the formula φ can be effectively constructed from f.

3 Mixed integer programming

It is well-known (see, for example, [41, 42]) that propositional classical formulas in conjunctive normal form correspond to certain 0-1 integer programs. More precisely, given a set Γ of classical disjunctive clauses over the signature Σ one transforms each clause of the form $C = p_1 \vee \cdots \vee p_k \vee \neg p_{k+1} \vee \cdots \vee \neg p_{k+m}$ into the linear inequation

$$\textstyle\sum_{i=1}^{k} p_i - \sum_{j=k+1}^{k+m} p_j \geq 1 - m \qquad (4)$$

Here, the variables of a propositional signature Σ are interpreted as function variables with domain $\{0,1\}$. It is easy to see that the resulting set of inequations is solvable iff Γ is satisfiable: recall that $A_\vee(i,j) = \max\{i,j\}$ and $A_\neg(i) = 1 - i$, so clause C is satisfiable iff $f_C \geq 1$ iff (4) holds.

McNaughton's theorem suggests that this embedding of logic into integer programming can be generalized to cover Łukasiewicz logic (after all, a McNaughton function is defined by linear polynomials) and, indeed, it turns out to be possible [30, 33]. To aid the presentation of this result, let us recall some facts and definitions about Mixed Integer Programming (MIP). With the expression **linear inequation** I mean in the following always a term of the form $a_1p_1 + \cdots + a_mp_m \geq c$, where $a_1p_1 + \cdots + a_mp_m$ is a linear polynomial over variables $\{p_1, \ldots, p_m\}$ and integral coefficients $\{a_1, \ldots, a_m, c\}$. The domain of the variables can be any truth value set N as defined in Example 2. The expression $a_1p_1 + \cdots + a_mp_m$ is called a **linear term**.

Let $\mathbf{J}$ be a finite set of linear inequations and K a linear term. Let Σ be the set of variables occurring in $\mathbf{J}$ and K. Let the domain N of each variable be finite. Then $\langle \mathbf{J}, K\rangle$ is a **bounded integer program** (IP) with **cost function** K.[2] If the domain of the variables in Σ is one of $\{0,1\}$ and $[0,1]$, then one has a **bounded 0-1 mixed integer program** (MIP). When all variables in Σ run over infinite N, the result is a **bounded linear program** (LP).

A variable assignment $\sigma : \Sigma \to [0,1]$ that respects the domain of each variable and, moreover, satisfies all inequations in $\mathbf{J}\sigma$ (the MIP that results from replacing each $x \in \Sigma$ in $\mathbf{J}$ with $\sigma(x)$) is called a **feasible solution** of $\langle \mathbf{J}, K\rangle$. A variable assignment σ such that the value of $K\sigma$ is minimal among all feasible solutions is called an **optimal solution.** $\langle \mathbf{J}, K\rangle$ is **feasible** iff there are feasible solutions.

Only the feasibility part of (M)IPs/LPs is required in the following, cost functions are not considered.

Let $M \subseteq [0,1]^k$. M is a **MIP-representable** set if there is an MIP $\mathbf{J}$ over variables $\Sigma' = \{x_1, \ldots, x_k\}$ with domain $[0,1]$ and variables Σ'' with domain $\{0,1\}$ such that

$$M = \{(\rho(x_1), \ldots, \rho(x_k)) \mid \rho \text{ feasible solution of } \mathbf{J}\sigma \text{ for some } \sigma : \Sigma'' \to \{0,1\}\}$$

A many-valued logic is **MIP-representable** iff for all its connectives $\theta \in \Theta$ the function A_θ (in relational form) is an MIP-representable subset of $[0,1]^{\alpha(\theta)+1}$. The variable in a relational MIP-representation of a function that holds the function value is called **output variable**, the variables that hold the function arguments are called **argument variables**.

All finite-valued logics are MIP-representable, simply because A_θ is a finite set of rational vectors in $[0,1]^{\alpha(\theta)+1}$. It is more interesting that Łukasiewicz logic is MIP-representable. To see this, consider the following MIP:

$$\begin{array}{llcr}
(i) & \mathrm{in}^1 + \mathrm{in}^2 + z - \mathrm{out} & \geq & 1 \\
(ii) & -\mathrm{in}^1 - \mathrm{in}^2 + z + \mathrm{out} & \geq & -1 \\
(iii) & -\mathrm{in}^1 - \mathrm{in}^2 - z & \geq & -2 \\
(iv) & \mathrm{in}^1 + \mathrm{in}^2 + z & \geq & 1 \\
(v) & \qquad\qquad - z - \mathrm{out} & \geq & -1
\end{array}$$

[2] The adjective *integer* is justified, because the elements of $N = \mathbf{n}$ can without loss of generality be assumed to have the form $\{0, 1, \ldots, n-1\}$.

This is an MIP-representation of $\odot_{\mathrm{L}}$, where out $= A_{\odot_{\mathrm{L}}}(\mathrm{in}^1, \mathrm{in}^2) = \max\{0, \mathrm{in}^1 + \mathrm{in}^2 - 1\}$, so in^1 and in^2 are argument variables, out is output variable and z is an additional variable with domain $\{0,1\}$: to justify this claim, first set $z = 0$, then lines (i) and (ii) say that out $= \mathrm{in}^1 + \mathrm{in}^2 - 1$, lines (iii) and (v) are trivially true, and (iv) says that $\mathrm{in}^1 + \mathrm{in}^2 \geq 1$, thus specifying the area, where the non-zero value is taken on. The case $z = 1$ is similar. An MIP-representation of $\neg_{\mathrm{L}}$ is straightforward:

$$\begin{array}{rcr} -\mathrm{in} - \mathrm{out} & \geq & -1 \\ \mathrm{in} + \mathrm{out} & \geq & 1 \end{array}$$

All other connectives are definable with $\odot_{\mathrm{L}}$ and $\neg_{\mathrm{L}}$.

An important property of MIP-representable functions is that their composition is again MIP-representable. This makes it possible to construct an MIP-representation of f_φ for a given formula φ in $\mathcal{O}(\varphi)$ steps: assume we have already MIP-representations of f_{φ_1} and f_{φ_2}. Then construct an MIP-representation of $f_{\theta(\varphi_1,\varphi_2)}$ with constant overhead (here θ is binary, but the generalisation is straightforward):

1. Let MIP_1 be the MIP-representation of f_{φ_1} with fresh variables out_1, in_1^j.
2. Let MIP_2 be the MIP-representation of f_{φ_2} with fresh variables out_2, in_2^k.
3. Let MIP_θ be the MIP-representation of A_θ with fresh variables out_θ, in_θ^l $(l = 1, 2)$.
4. Add equations $\mathrm{in}_\theta^1 = \mathrm{out}_1$, $\mathrm{in}_\theta^2 = \mathrm{out}_2$ to connect MIP_1, MIP_2, MIP_θ.

Now McNaughton's Theorem can be strengthened to provide a direct link between MIP-representable logics and MIP:

Theorem 1 ([33]).

1. *If $\varphi(\mathbf{p})$ is a formula of an MIP-representable logic then there is an MIP $\mathbf{J}_\varphi$ with argument variables $\mathbf{p}$ and output variable y whose feasible solutions restricted to $(\mathbf{p}, y)$ are the function $f_\varphi(\mathbf{p})$. Moreover, the size of $\mathbf{J}_\varphi$ is linear in the size of φ.*
2. *Let $\mathbf{J}$ be an MIP over variables Σ. Then there is a Σ-formula $\varphi_{\mathbf{J}} \in \mathbf{L}_L^0$ which is satisfiable iff $\mathbf{J}$ is feasible.*

Part 1 follows immediately from the construction outlined above. It does not require integrity of coefficients of the associated functions f_φ in any crucial way, hence it works for logics corresponding to generalized McNaughton functions with possibly non-integral coefficients. To determine the algebraic and logical counterpart of this class of functions is ongoing research. Part 2 is proven in [33].

Non-continuous (but linear) connectives, such as Gödel implication and negation, can easily be handled by MIPs with strict inequalities. The connectives of product logic, on the other hand, lead outside MIP and into non-linear programming.

4 Signed logic

Signed logics are characterized by having capability for semantic reflexion: elements of their semantic domain, that is, truth values can be directly expressed by syntactical entities: signs. Semantic reflexion does, in principle, not depend on the presence of signs. For example, it is easy to express in classical logic that a formula evaluates to a certain truth value. Consider the formula $(\overline{1} \leftrightarrow (p \vee \neg p)) \wedge (\varphi \leftrightarrow \overline{1})$ that first defines the logical constant $\overline{1}$, and then says that φ always evaluates to truth value 1. In signed logics there is a separate syntactic category for truth values (called *signs*). Instead of $\varphi \leftrightarrow \overline{1}$ one writes $1{:}\varphi$, etc.

In classical logic, the difference between signed and unsigned formulas is only notational. In non-classical logics, however, truth values may be undefinable or only at great cost. Therefore, signed logics are used in various deductive approaches to reasoning in non-classical logics, in particular, in many-valued logics [60, 29, 5]. In general, one takes *truth value sets* as signs, because this can speed up proof search exponentially [29].

A **signed formula** of an N-valued logic $\mathcal{L}$ over language $\mathbf{L}$ is an expression of the form $S{:}\varphi$, where $S \subseteq N$, and $\varphi \in \mathbf{L}$. Satisfiability, validity, and consequence of signed formulas are defined with the help of D-satisfiability, etc., by identifying $\models_S \varphi$ with $\models S{:}\varphi$. If φ is atomic, $S{:}\varphi$ is called a **signed atom**. If S is a singleton, one has a **monosigned formula**.

A signed formula $S{:}\varphi$ is an implicit disjunction over the statements $\mathbf{I}(\varphi) = i$ for $i \in S$: $\mathbf{I} \models S{:}\varphi$ iff $\mathbf{I}(\varphi) = i$ for some $i \in S$. This additional level in the language leads to more compact syntactic characterizations of connectives and quantifiers.

Not every subset of N is necessarily admitted as a sign. Several systems of signs are considered. The most usual ones are these:

$$\mathcal{S}_{\text{mono}} = \{\{i\} \mid i \in N\} \tag{5}$$
$$\mathcal{S}_{\text{full}} = \mathcal{P}^{+}(N) - \{N\} \tag{6}$$

Further, let N be equipped with a partial order $\leq$ and denote with $\uparrow i = \{j \in N \mid j \geq i\}$ the **upset** of $i \in N$ and with $\downarrow i = \{j \in N \mid j \leq i\}$ the **downset** of $i \in N$. Then the following system of signs is defined:

$$\begin{aligned} \mathcal{S}_{\text{regular}} = {} & \{\uparrow i \mid i \in N,\ \uparrow i \neq N\} \cup \\ & \{\overline{\uparrow i} \mid i \in N,\ \uparrow i \neq N\} \end{aligned} \tag{7}$$

This is the set of non-trivial order filters of N and their complements that are generated by single elements. It is defined for any partially ordered set of truth values. For totally ordered N, these are exactly the prime ideals/filters and their complements and were called **regular sign** in [29]. The name is kept for the present, more general, definition.

A **signed conjunctive normal form (CNF) formula** has the form

$$(\forall x_1)\cdots(\forall x_r)\bigwedge_{k=1}^{M}\bigvee_{l=1}^{J_k} S_{kl}{:}p_{kl} \ ,$$

in which the $S_{kl}{:}p_{kl}$ are signed atoms and $\{x_1,\ldots,x_t\}$ are the free variables in the scope. For any $1 \leq k \leq M$, the expression

$$(\forall y_1)\cdots(\forall y_m)(S_{k1}{:}p_{k1} \vee \cdots \vee S_{kJ_k}{:}p_{kJ_k}) \tag{8}$$

where $\{y_1,\ldots,y_m\}$ are the free variables in the scope, is a **signed clause**. Like in classical logic, the quantifier prefix is often not written explicitly and a signed CNF formula is identified with the set of its clauses.

Any signed formula in any finite-valued first-order logic has an at most polynomially larger signed CNF, which is satisfiability equivalent [31].

Signed CNF formulas are generic or "logic-free" in the sense that their syntax and semantics are fixed and independent of any logic. Signed CNF formulas do not contain *any* many-valued connective and are simply a generic and flexible language for denoting many-valued interpretations.

Regular CNF formulas are signed CNF formulas, where all occurring signs are regular as in (7). Regular CNF formulas are a natural formalism whose syntactical expressivity is between classical CNF and bounded IP. Many combinatorial optimization problems can be formulated easily as Regular CNF instances, and there is experimental evidence that a regular logic problem formulation can be more efficiently solvable than a classical one [13].

If N is equipped with an ordering, there is a natural notion of a Horn formula. Recall that in classical logic a CNF formula is a **Horn formula** iff each clause contains at most one positive literal. Signed atoms are, in general, neither positive nor negative in any sense, but a natural notion of **polarity** is present in regular signs (7): literals of the form $\uparrow i{:}p$ are **positive** while literals of the form $\overline{\uparrow i}{:}p$ are **negative**. With this convention, a **regular Horn formula** is defined exactly as in the classical case.

5 Polynomial expressions

Decision diagrams (DD) are a family of data structures originally developed for efficient representation and manipulation of Boolean formulas, but now successfully used for many purposes in computer science, in particular in circuit design tools [16]. To understand this section, some background in decision diagrams is helpful. A good introduction is [21].

Many-valued decision diagrams can be computed with the help of a generalized Shannon expansion:

$$\varphi \text{ is replaced with } \begin{cases} \texttt{switch } p \\ \texttt{case } 0: & \varphi\{p/0\}; \\ \texttt{case } \frac{1}{n-1}: & \varphi\{p/\frac{1}{n-1}\}; \\ \dots & \dots \\ \texttt{case } 1: & \varphi\{p/1\} \end{cases} \tag{9}$$

The notation $\varphi\{p/i\}$ means to replace every occurrence of the variable p in φ by i. Recursive application of this rule yields an expression in which variables occur merely at the `switch` position. This is called a **many-valued decision diagram** (MDD).

Equation (9) is based on an $(n+1)$-ary `switch` connective in n-valued logic with the following semantics:

$$A_{\texttt{switch}}(i, j_0, \dots, j_{n-1}) = \begin{cases} j_0 & \text{if } i = 0 \\ j_1 & \text{if } i = \frac{1}{n-1} \\ \dots & \dots \\ j_{n-1} & \text{if } i = 1 \end{cases}$$

If $n = 2$, the usual **`if-then-else`** connective and the Shannon expansion, on which binary decision diagrams are based, is obtained.

Now write (9) as $(\{0\}{:}p \wedge \varphi\{p/0\}) \vee \dots \vee (\{1\}{:}p \wedge \varphi\{p/1\})$. If, in addition, we write a signed atom of the form $\{i\}{:}p$ as p^i, "$\wedge$" as "$\cdot$", and "$\vee$" as "+", then we obtain for $n = 2$, for example,

$$\begin{aligned} p \to q &\equiv p^1 \cdot (1 \to q) + p^0 \cdot (0 \to q) \\ &\equiv p^1 \cdot (q^1 \cdot (1 \to 1) + q^0 \cdot (1 \to 0)) + p^0 \cdot (q^1 \cdot (0 \to 1) + q^0 \cdot (0 \to 0)) \\ &\equiv p^1 \cdot q^1 \cdot 1 + p^1 \cdot q^0 \cdot 0 + p^0 \cdot q^1 \cdot 1 + p^0 \cdot q^0 \cdot 1 \end{aligned}$$

Assume $\{p_1, \dots, p_k\}$ are the variables occurring in a formula. In switching theory [63], a polynomial expression as above is known as **sum-of-products** expression or **SOP**, the products $p_1^{i_1} \cdots p_k^{i_k}$ are called **minterm**, and the coefficients from N **discriminant**. Many readers will have realized that an SOP in classical logic is also a disjunctive normal form (DNF), because minterms with discriminant 0 can be deleted and discriminants 1 are absorbed.

Replacing disjunction "+" with exclusive or "$\oplus$" yields an **exclusive-or-of-products** expression or **ESOP**, which is of great practical value [63]. In the two-valued case, the Shannon expansion yields SOP expressions. In this case, $p \wedge q \equiv 0$ iff $p \oplus q \equiv p \vee q$, so one obtains ESOP expressions in the same way.

There are many kinds of polynomial representations. Most of them, for example, general SOPs and ESOPs are not canonical (that is, there is more than one (E)SOP that represents the same function). On the other hand, certain restrictions of (E)SOPs are canonical, for example, **positive polarity Reed-Muller** (PPRM) expressions [63]. Canonical expressions are an elegant way to check satisfiability: for an unsatisfiable classical formula its canonical polynomial representation must be 0. Polynomial expressions can be immediately realized with two-level networks.

In the many-valued case one considers arbitrary signed atoms $S{:}p$ as constituents of minterms. Written p^S, they are known in logic design under the name **set literal** or **universal literal** [61, 23].

6 Satisfiability and validity

Given an N-valued (propositional or first-order) logic $\mathcal{L}$ and a truth value set $D \subseteq N$, let us denote with D-SAT$_\mathcal{L}$ the D-satisfiable formulas, with D-VAL$_\mathcal{L}$ the D-valid formulas of $\mathcal{L}$. Further, let SAT$_\mathcal{L}$=$\{1\}$-SAT$_\mathcal{L}$ and VAL$_\mathcal{L}$= $\{1\}$-VAL$_\mathcal{L}$.

Theorem 2. *SAT$_\mathcal{L}$ is NP-complete and VAL$_\mathcal{L}$ is co-NP-complete for each logic $\mathcal{L} \in \{\mathcal{L}^0_c, \mathcal{L}^0_G, \mathcal{L}^0_Ł, \mathcal{L}^0_\Pi\}$ and any choice of N.*

Proof. (Sketch) For $\mathcal{L}^0_c$ this is, of course, Cook's Theorem.

For hardness of the problems associated to a logic $\mathcal{L}^0 \in \{\mathcal{L}^0_G, \mathcal{L}^0_Ł, \mathcal{L}^0_\Pi\}$, first observe that by virtue of equations (1), (2) one may assume that $\mathbf{L}^0 \supseteq \mathbf{L}^0_c$. Now it is sufficient to define for each $\varphi \in \mathbf{L}^0_c$ an $\mathbf{L}^0$-formula φ^* that restricts $\mathcal{L}^0$-interpretations of φ to values in $\{0,1\}$. Then $\varphi \in \mathrm{SAT}_{\mathcal{L}^0_c}$ iff $\varphi \wedge \varphi^* \in \mathrm{SAT}_{\mathcal{L}^0}$. In the case of $\mathcal{L}^0_Ł$, for example, one can use $\varphi^* = \bigwedge_{p \text{ occurs in } \varphi}(p \vee \neg p)$, in the other logics similar formulas work.

For finite-valued logics, NP-, respectively, co-NP-easyness is straightforward, so assume $N = [0,1]$ from now.

$\mathrm{SAT}_{\mathcal{L}^0_c}$=$\mathrm{SAT}_{\mathcal{L}^0_\Pi}$=$\mathrm{SAT}_{\mathcal{L}^0_G}$ is shown by giving straightforward polynomial-size embeddings of the latter two into the first, see [38].

$\mathrm{VAL}_{\mathcal{L}^0_G}$ [8]: assume there is an interpretation $\mathbf{I}$ of $\varphi \in \mathbf{L}^0_G$ over variables $p_1, \ldots, p_m$ such that $\mathbf{I}(\varphi) < 1$. There is a trivial order-isomorphism o mapping $\mathbf{I}(p_1), \ldots, \mathbf{I}(p_m)$ into $\mathbf{m+2} = \{0, \frac{1}{m+1}, \ldots, \frac{m}{m+1}, 1\}$. All Gödel operations f have the property that $f(i_1, \ldots, i_{\alpha(f)}) \in \{i_1, \ldots, i_{\alpha(f)}\} \cup \{0,1\}$, hence, $o(\mathbf{I})(\varphi) < 1$; now it suffices to guess such an interpretation over $\mathbf{m+2}$ and check that $o(\mathbf{I})(\varphi) < 1$, which can obviously be done in polynomial time.

$\mathrm{VAL}_{\mathcal{L}^0_\Pi}$: there is a polynomial embedding of $\mathcal{L}^0_\Pi$ into $\mathcal{L}^0_Ł$, see [8].

$\mathrm{VAL}_{\mathcal{L}^0_Ł}$, $\mathrm{SAT}_{\mathcal{L}^0_Ł}$: an immediate consequence of Theorem 1, part 1. □

Co-NP-completeness of $\mathrm{VAL}_{\mathcal{L}^0_Ł}$ was first shown by Mundici [53]. The polynomial embedding of MIP-representable logics described in Section 3 also yields NP-easiness of Gödel logic and of the extension of Łukasiewicz logic that is characterized by piecewise linear functions with rational coefficients (discussed at the end of Section 3). In [38] also the complexity of the sets $N\backslash\{0\}$-SAT$_\mathcal{L}$ and $N\backslash\{0\}$-VAL$_\mathcal{L}$ is considered.

All of the logics mentioned in the previous theorem are examples of so-called triangular logics or **t-norm** logics as introduced in fuzzy theory. In these logics, the product $\odot_t$ is an associative, commutative, monotonic, and continuous function on $[0,1]$ with $A_{\odot_t}(0,i) = 0$ and $A_{\odot_t}(1,i) = i$. Implication is defined by residuation. Let $\mathrm{VAL}_{\mathcal{L}_t^0}$ be the formulas that are valid in every continuous t-norm logic. Every continuous t-norm can be characterized as a, possibly infinite, ordinal sum of the products of Łukasiewicz, Gödel, and product logic (and this partly explains the prominent position of these logics). Nevertheless, it turns out that $\mathrm{VAL}_{\mathcal{L}_t^0}$ is co-NP-complete and $\mathrm{SAT}_{\mathcal{L}_t^0}$=$\mathrm{SAT}_{\mathcal{L}_c^0}$ is NP-complete [7].

The mixed integer programming technique used in the proof of Theorem 1 was also used in proving NP-completeness of the satisfiability problem of fuzzy probability logic over Łukasiewicz logic [40].

While the decision problems of propositional infinite-valued Łukasiewicz, Gödel and product logic have the same complexity as two-valued logic, the situation changes drastically in the first-order case:

Theorem 3. *$VAL_{\mathcal{L}_G}$ is Σ_1-complete, $VAL_{\mathcal{L}_L}$ is Π_2-complete, and $VAL_{\mathcal{L}_\Pi}$ is Π_2-hard in the case $N = [0,1]$.*

The proofs are rather technical and must be omitted (they can be found in [38]). Π_2-completeness of $\mathcal{L}_{\text{Ł}}$ was first shown in [65, 58]. An embedding of $\mathcal{L}_{\text{Ł}}$ into $\mathcal{L}_\Pi$ [38] can be used to show Π_2-hardness of the latter; it is an open question, whether $\mathcal{L}_\Pi$ is Π_2-complete. Σ_1-easiness of $\mathcal{L}_{\mathrm{G}}$ follows from the existence of a complete first-order axiomatization [38]. Surprisingly, $\mathrm{VAL}_{\mathcal{L}_{\mathrm{G}}}$ is *not* recursively enumerable anymore, for example, over the truth value set $N_* = \{\frac{1}{n+1} \mid n \in \mathbb{N}\} \cup \{0\}$ [9]. So the order type of the truth value set can drastically change the complexity of infinite-valued first-order logic.

The complexity of deciding logical consequence depends on the availability and form of deduction theorems for a given logic. In the following theorem φ^m stands for $\varphi \odot_x \cdots \odot_x \varphi$ (m copies of φ).

Theorem 4. *Let $\mathcal{L}^0$ be any of the logics of Example 2. If Ψ is any set of $\mathbf{L}^0$-formulas and $\varphi, \rho \in \mathbf{L}^0$, then $\Psi \cup \{\rho\} \models \varphi$ iff $\Psi \models \rho^m \rightarrow_x \varphi$ for some $m \in \mathbb{N}$.*

For classical logic, this is a well-known result (with $m = 1$). For Łukasiewicz logic the theorem was proven in [57] and improved in [19, 3] by giving a concrete upper bound for the number m, depending on Ψ, φ, ρ (simply exponential in the number of variable occurrences in the formulas). For Gödel logic, again, $m = 1$ is sufficient, that is, the classical deduction theorem holds for Gödel logic. More general results are in [38].

More fine-grained investigations were made into the complexity of satisfiability problems associated with signed CNF formulas, defined in equation (8). Let CNF-SAT be the set of satisfiable propositional signed CNF formulas, let 2-CNF-SAT be the restriction of CNF-SAT, where signed clauses contain exactly two signed atoms, and let HORN-SAT be the satisfiable regular Horn formulas. If the truth value set N, together with a partial order, is fixed, this is denoted with CNF-SAT$_N$, 2-CNF-SAT$_N$, and HORN-SAT$_N$.

Like in classical logic, CNF-SAT is NP-complete, but some of its subclasses are polynomially solvable. NP-hardness of CNF-SAT is trivial, because classical SAT is the same as CNF-SAT$_{\{0,1\}}$ up to notation; NP-easiness of the problem CNF-SAT for all finite N is straightforward, see above. Further results are summarized in Table 1 (NPC means NP-complete, all truth value sets N are finite) and are discussed in [12, 14].

Table 1. Complexity of signed SAT problems.

	CNF	2-CNF	HORN
classical	NPC	linear [25]	linear [22]
monosigned	NPC	linear [48]	—
regular, N totally ordered	NPC	$\|\Gamma\| \log \|\Gamma\|$ [14]	$\|\Gamma\| \log \|\Gamma\|$ [32, 14]
regular, N distributive lattice, signs of form $\uparrow i$ *and* $\overline{\uparrow i}$	NPC	NPC [11]	$\|\Gamma\|\|N\|^2$ [67]
regular, N lattice, signs of form $\uparrow i$ *and* $\overline{\uparrow i}$	NPC	NPC	polynomial [10]
regular, N lattice, signs of form $\uparrow i$ *and* $\downarrow i$	NPC	polynomial [11]	—
regular (arbitrary)	NPC	NPC	—
signed (arbitrary)	NPC	NPC [46]	—

2-CNF-SAT$_N$ for any $|N| \geq 3$ and, hence, 2-CNF-SAT is NP-hard [46, 48] (in contrast to classical 2-CNF-SAT that can be solved in linear time) by a reduction of the graph 3-colorability problem. A direct reduction of classical CNF-SAT to 2-CNF-SAT is in [10].

Even *regular* 2-CNF-SAT is NP-complete for general partial orders, which can be shown by reducing unrestricted signed 2-CNF-SAT to regular 2-CNF-SAT [11]. Under certain restrictions, however, membership in regular 2-CNF-SAT can be checked in polynomial time: for formulas Γ over totally ordered truth value sets [14] give a $|\Gamma| \log |\Gamma|$ procedure via a reduction to classical 2-CNF-SAT (see also below). A result for the more general case when N is a lattice and all occurring signs are of the form $\uparrow i$ or $\downarrow i$ is in [11]. A linear-time procedure for solving monosigned 2-CNF-SAT is described in [48].

If N is totally ordered, the problem of deciding whether a regular Horn formula Γ is satisfiable can be solved in time linear in $|\Gamma|$, if $|N|$ is fixed, and in $|\Gamma| \log |\Gamma|$ time, otherwise [32, 14]. An algorithm for a particular subclass of regular Horn formulas appeared first in [24].

Many of these results can be proven via a reduction to classical logic: as an example, reduce regular CNF-SAT to classical SAT [14]. We define a mapping δ from instances Γ of regular CNF-SAT over a signature Σ to satisfiability equivalent instances Γ^δ of SAT.

The signature of Γ^δ is $\Sigma^\delta = \{\uparrow i : p \mid i \in N, p \in \Sigma\}$, and Γ^δ is formed by (i) the clauses of Γ over signature Σ^δ (that is, the clauses of Γ, but interpreted classically) and (ii) an additional set of clauses Γ' ensuring that two inconsistent regular literals of the form $\uparrow i : p$ and $\overline{\uparrow j} : p$ with $j \leq i$ are also *classically* inconsistent. If these clauses are not added, then only literals of the form $\uparrow j : p$ and $\overline{\uparrow j} : p$ are *classically* inconsistent.

Let $N_\Gamma \subseteq N$ be the set of truth values that occur as a value in signs of literals in Γ, and let $N_\Gamma(p) \subseteq N_\Gamma$ be the set of truth values that occur as a value in signs of literals with the propositional variable p in Γ. Let $N_\Gamma(p) = \{i_1, i_2, \ldots, i_{m(p)}\}$ and $i_1 \leq i_2 \leq \cdots \leq i_{m(p)}$ under the order $\leq$ associated with N. Now define $\Gamma'(p) = \{\overline{\uparrow i_{(j+1)}} : p \vee \uparrow i_j : p \mid 1 \leq j < m(p)\}$ for each propositional variable $p \in \Sigma$, and $\Gamma' = \bigcup_{p \in \Sigma} \Gamma'(p)$.

The clauses of Γ' are tautologies under many-valued semantics and capture the difference between classical and regular logic. It is obvious that if we add Γ' to Γ, then all regular inconsistencies can be classically detected.

Moreover, the size of Γ' is in $\mathcal{O}(|\Gamma|)$, although it may take $\mathcal{O}(|\Gamma| \log |\Gamma|)$ time to compute Γ'. The results for regular HORN-SAT and 2-CNF-SAT follow now from the observation that the clauses in Γ' are in the intersection of classical HORN-SAT and 2-CNF-SAT. Similar techniques can be used for non-linear orderings:

If N is a finite lattice, regular HORN-SAT is decidable in time linear in the length of the formula and polynomial in the cardinality of N via a reduction to classical HORN-SAT [10]. For distributive lattices, the more precise bound $|\Gamma| \cdot |N|^2$ can be shown [67]. The latter paper contains also some results on decidable first-order fragments of regular CNF formulas. A closer inspection of the proofs in the cited papers yields immediately that all regular HORN-SAT$_N$ versions (fixed size truth values) have linear complexity.

If the partial order of N is no lattice, then a natural notion of regular Horn formula can still be obtained by using signs that have the form $\uparrow S = \{i \in N \mid \text{there is } j \in S \text{ such that } i \geq j\}$, where $S \subseteq N$. This more general notion of regular HORN-SAT is still decidable in time linear in the length of the formula, but exponential in the cardinality of N provided that N possesses a maximal element [10].

If N is infinite, then regular HORN-SAT is decidable provided that N is a **locally finite lattice**, that is, every sub-lattice generated by a finite subset is finite [10].

A set of on-line (that is: incremental) algorithms for Horn formulas with numerical uncertainties has been proposed and studied in [4]. The resulting complexities are (at most cubic) polynomials, also in the infinite-valued case.

On the CNF level, some authors used many-valued semantics to approximate classical propositional consequence. In some cases this lead to polynomial decision procedures and results of this kind were used in knowledge representation [56]. An overview of related results is [17].

The NP-complete satisfiability problems for regular formulas and Regular-DPL exhibit the phase transition phenomena encountered in many decision procedures for NP-complete problems [51]: (i) there is a sharp increase (phase transition) of the percentage of unsatisfiable random CNF-SAT instances around a certain point when the ratio $\frac{c}{v}$ between the number c of clauses and the number v of variables is varied; (ii) there is an easy-hard-easy pattern in the computational difficulty of solving problem instances as $\frac{c}{v}$ is varied—the hard instances tend to be found near the crossover point [49, 15, 13].

Satisfiability in *propositional* infinite-valued Łukasiewicz logic with standard temporal operators of LTL [66] is undecidable [69], although $\mathrm{SAT}_{\mathcal{L}^0_{\text{Ł}}}$ is NP-complete and satisfiability of LTL alone is PSPACE-complete. An informal explanation for this surprising result is the following: Łukasiewicz sum $\oplus$ provides capability to add numbers, while temporal logic provides "registers" (worlds) for storing them and the temporal operators allow conditional jumps to different register contents. This is enough to simulate a suitable universal computation mechanism.

7 Representations

The following theorem gives a general representation of many-valued propositional signed formulas in terms of a Boolean combination of
their direct subformulas.

Theorem 5 ([35]). *Let $\mathcal{S}$ be one of the sets of signs (5)–(7) for an n-valued logic $\mathcal{L}^0$, $S \in \mathcal{S}$, and let $\varphi = S{:}\theta(\varphi_1, \ldots, \varphi_m)$ $(m \geq 1)$ be a signed $\mathbf{L}^0_\Sigma$-formula. $\mathbf{I}$ is an arbitrary n-valued Σ-interpretation.*

Then there are numbers $M_1, M_2 \leq n^m$, index sets $I_1, \ldots, I_{M_1}, J_1, \ldots, J_{M_2} \subseteq \{1, \ldots, m\}$, and signs $S_{rs}, S_{kl} \in \mathcal{S}$ with $1 \leq r \leq M_1$, $1 \leq k \leq M_2$ and $s \in I_r$, $l \in J_k$ such that φ is satisfiable by $\mathbf{I}$ iff

$$\bigvee_{r=1}^{M_1} \bigwedge_{s \in I_r} S_{rs}{:}\varphi_s \textit{ is satisfiable by } \mathbf{I}$$
$$\textit{iff}$$
$$\bigwedge_{k=1}^{M_2} \bigvee_{l \in J_k} S_{kl}{:}\varphi_l \textit{ is satisfiable by } \mathbf{I}.$$

The first characterization is called a ***signed DNF representation*** *of φ, the second a* ***signed CNF representation*** *of φ.*

The relevance of this theorem comes from the fact that a DNF representation of a formula $S{:}\theta(\varphi_1, \ldots, \varphi_m)$ gives rise to a generalized tableau rule:

$$\frac{S{:}\theta(\varphi_1, \ldots, \varphi_m)}{\begin{array}{c|c|c} \vdots & \vdots & \vdots \\ S_{1s}{:}\varphi_s & \cdots & S_{M_1 s}{:}\varphi_s \\ \vdots & \vdots & \vdots \end{array}}$$

It is then a routine task to build generic tableau-based deduction systems for generic finite-valued logics, which are at the same time decision procedures. CNF representations lead to sequent calculi and to signed CNF computation procedures.

Recall that the semantics of a first-order quantifier λ in many-valued logic is defined via a distribution function $Q_\lambda : \mathcal{P}^+(N) \to N$. Similar as in the propositional case, one may obtain a representation of a signed quantified formula in terms of certain signed instances. Informally, what one needs to do is to characterize the distributions (appearing as $(Q_\lambda)^{-1}(S)$ below) that are mapped to one of the truth values that occur in the sign S of a quantified formula.

Theorem 6 ([35]). *Let $\varphi = S{:}(\lambda x)\varphi$ be a signed $\mathbf{L}_\Sigma$-formula from an N-valued first-order logic $\mathcal{L}$ with finite N. Further, let $(Q_\lambda)^{-1}(S) = \{\emptyset \neq I \subseteq N \mid Q_\lambda(I) \in S\}$; then a signed quantified formula φ is satisfiable iff*

$$\bigvee_{I \in (Q_\lambda)^{-1}(S)} \Big(\bigwedge_{i \in I} \{i\}{:}\varphi\{x/c_i\} \wedge \bigwedge_{t \in Term^0_\Sigma} I{:}\varphi\{x/t\} \Big) \tag{10}$$

is satisfiable, where the c_i are new Skolem constants.

Each disjunct in this representation says that the distribution of φ at x is I: the first conjunction assures that *at least* the elements in I occur in the distribution, the second conjunction says that *at most* the elements in I occur. It must be stressed that monosigned first-order representations are *much* more complicated, even for simple quantifiers, see also [18].

A tableau rule is obtained from (10) in a standard way by replacing the second, infinite, conjunction with a non-deterministic guess for $I{:}\varphi(t)$, where $t \in \mathrm{Term}^0_\Sigma$ is an arbitrary ground term.

A CNF representation is obtained by duality: compute a DNF representation for $\overline{S}$ and replace "not $\bigvee\bigwedge \cdots S' \cdots$" with "$\bigwedge\bigvee \cdots \overline{S'} \cdots$" using generalized de Morgan rules.

The size of representations and tableau/sequent rules for many-valued operators is closely related to the deterministic complexity of decision problems, because the size of rules determines the size of sequent/tableau proofs, which in turn yield upper bounds of the complexity of VAL and SAT.

Representations (Theorems 5, 6) and expansions (Section 5) do also determine the size of various normal forms for many-valued logic formulas in an essential way.

In [60, 70, 29, 5] the following results on the maximal size of signed CNF/DNF representations
of finite-valued connectives θ are stated and proven:

$n = \|N\|,\ r = \alpha(\theta)$	DNF	CNF
monosigned	n^r	n^{r-1}
general	n^{r-1}	n^{r-1}

All bounds are tight. The r-ary Łukasiewicz sum $\oplus_{\text{Ł}}^r$ can be used to show this in all cases; it is defined as $\oplus_{\text{Ł}}^r(p_1, \ldots, p_r) \equiv p_1 \oplus_{\text{Ł}} \cdots \oplus_{\text{Ł}} p_r$.

These results can be used to prove that any signed formula φ of any finite-valued logic can be translated into a signed CNF with size in $\mathcal{O}(n^R|\varphi|)$, where $R = \max\{\alpha(\theta) \mid \theta \in \Theta,\ \theta \text{ occurs in } \varphi\}$ [31]. For logics defined by distributive lattices, up to exponentially better results are possible by using dual space representations. General complexity results for the latter and many concrete examples are in [68].

The branching factor of signed tableau rules for quantifiers (the number of disjuncts in Theorem 6) is at most $2^n - 2$. For the dual sequent rules the slightly better bound 2^{n-1} can be obtained [5], which is sharp as well.

A slightly different question is to ask how many different signs are needed in general to build a sound and complete signed tableau or sequent calculus for a given n-valued logic. It is shown in [6] for families of signs fulfilling a certain condition (including (5)–(7)) that this number is logarithmic in n (and the bound is tight).

One of the results relating to infinite-valued Łukasiewicz logic is that every $\mathbf{L}_{\text{Ł}}^0$-formula over just one variable can be polynomially translated into a regular signed atom with respect to the natural order on $N = [0, 1]$ [55]. The complexity of McNaughton functions in one variable is investigated in [1, 2]. In [53] it is shown that a $\mathbf{L}_{\text{Ł}}^0$-formula φ is valid in infinite-valued $\mathcal{L}_{\text{Ł}}^0$ iff it is valid in $(2^{(2|\varphi|)^2} + 1)$-valued $\mathcal{L}_{\text{Ł}}^0$. This bound was later improved to $(2^{|\varphi|} + 1)$ values [3, 19]. An analog result is established for logical consequence.

Space complexity of various kinds of MDDs (Section 5) is discussed, for example, in [62], where further pointers to the literature can be found. In fact, every known kind of MDD has exponential worst-case space complexity. Increased space complexity is frequently traded in for more efficient computation in practice.

The worst-case, best-case and relative space complexity of various kinds of polynomial representations of finite-valued functions is investigated in many papers, for example, [64].

Acknowledgements

I would like to thank Petr Hájek and Helmut Thiele for pointing out errors in an earlier version of this article. Thanks also to the anonymous referee for pointing out inaccuracies.

References

1. Aguzzoli, S.: The complexity of McNaughton functions of one variable. Advances in Applied Mathematics, 21, (1) (1998) 58–77
2. Aguzzoli, S.: Geometric and Proof Theoretic Issues in Łukasiewicz Propositional Logics. PhD thesis, University of Siena, Italy, Feb. (1999)
3. Aguzzoli, S. and Ciabattoni, A.: Finiteness in infinite-valued logic. Journal of Logic, Language and Information, 9, (1) (2000) 5–29
4. Ausiello, G. and Giaccio, R.: On-line algorithms for satisfiability problems with uncertainty. Theoretical Computer Science, 171 (1–2) Jan. (1997) 3–24
5. Baaz, M. and Fermüller, C.G.: Resolution-based theorem proving for many-valued logics. Journal of Symbolic Computation, 19, (4) Apr. (1995) 353–391
6. Baaz, M., Fermüller, C.G., Salzer, G. and Zach, R.: Labeled calculi and finite-valued logics. Studia Logica, 61, (1) (1998) 7–33
7. Baaz, M., Hájek, P., Montagna, F. and Veith, H.: Complexity of t-tautologies. Annals of Pure and Applied Logic, to appear (2001)
8. Baaz, M., Hájek, P., Svejda, D. and Krajíček, J.: Embedding logics into product logic. Studia Logica, 61, (1), July (1998) 35–47
9. Baaz, M., Leitsch, A. and Zach, R.: Incompleteness of a first-order Gödel logic and some temporal logics of programs. In Kleine Büning, H. editor, Selected Papers from Computer Science Logic, CSL'95, Paderborn, Germany, volume 1092 of LNCS. Springer-Verlag (1996) 1–15
10. Beckert, B., Hähnle, R. and Manyá, F.: Transformations between signed and classical clause logic. In Proc. 29th International Symposium on Multiple-Valued Logics, Freiburg, Germany. IEEE CS Press, Los Alamitos, May (1999) 248–255
11. Beckert, B., Hähnle, R. and Manyá, F.: The 2-SAT problem of regular signed CNF formulas. In Proc. 30th International Symposium on Multiple-Valued Logics, Portland/OR, USA. IEEE CS Press, Los Alamitos, May (2000) 331–336
12. Beckert, B., Hähnle, R. and Manyá, F.: The SAT problem of signed CNF formulas. In Basin, D., D'Agostino, M., Gabbay, D., Matthews, S. and Viganò, L. editors, Labelled Deduction, volume 17 of Applied Logic Series. Kluwer, Dordrecht, May (2000) 61–82
13. Béjar, R.: Systematic and Local Search Algorithms for Regular-SAT. PhD thesis, Facultat de Ciències, Universitat Autònoma de Barcelona (2000)
14. Béjar, R., Hähnle, R. and Manyá, F.: A modular reduction of regular logic to classical logic. In Proc. 31st International Symposium on Multiple-Valued Logics, Warsaw, Poland. IEEE CS Press, Los Alamitos, May (2001) 221–226
15. Béjar, R. and Manyà, F.: Phase transitions in the regular random 3-SAT problem. In Z. W. Raś and A. Skowron, editors, Proc. International Symposium on Methodologies for Intelligent Systems, ISMIS'99, Warsaw, Poland, number 1609 in LNCS. Springer-Verlag (1999) 292–300
16. Burch, J., Clarke, E., McMillan, K. and Dill, D.: Sequential circuit verification using symbolic model checking. In Proc. 27th ACM/IEEE Design Automation Conference (DAC). ACM Press, (1990) 46–51
17. Cadoli, M. and Schaerf, M.: On the complexity of entailment in propositional multivalued logics. Annals of Mathematics and Artificial Intelligence, 18, (1) (1996) 29–50

18. Carnielli, W.A.: On sequents and tableaux for many-valued logics. Journal of Non-Classical Logic, 8, (1), May (1991) 59–76
19. Ciabattoni, A.: Proof Theoretic Techniques in Many-Valued Logics. PhD thesis, University of Milan, Italy, Jan. (2000)
20. Cignoli, R.L.O., D'Ottaviano, I.M.L. and Mundici, D.: Algebraic Foundations of Many-Valued Reasoning, volume 7 of Trends in Logic. Kluwer, Dordrecht, Nov. (1999)
21. Clarke, E.M., Grumberg, O. and Peled, D.A.: Model Checking. The MIT Press, Cambridge, Massachusetts (1999)
22. Dowling, W. and Gallier, J.: Linear-time algorithms for testing the satisfiability of propositional Horn formulæ. Journal of Logic Programming, 3 (1984) 267–284
23. Dueck, G.W. and Butler, J.T.: Multiple-valued logic operations with universal literals. In Proc. 24th International Symposium on Multiple-Valued Logic, Boston/MA. IEEE CS Press, Los Alamitos, May (1994) 73–79
24. Escalada-Imaz, G. and Manyà, F.: The satisfiability problem for multiple-valued Horn formulæ. In Proc. International Symposium on Multiple Valued Logics, ISMVL'94, Boston/MA, USA. IEEE CS Press, Los Alamitos (1994) 250–256
25. Even, S., Itai, A. and Shamir, A.: On the complexity of timetable and multicommodity flow problems. SIAM Journal of Computing, 5, (4) (1976) 691–703
26. Gödel, K.: Zum intuitionistischen Aussagenkalkül. Anzeiger Akademie der Wissenschaften Wien, mathematisch-naturwiss. Klasse, 32 (1932) 65–66. Reprinted and translated in [27].
27. Gödel, K.: Collected Works : Publications 1929–1936, volume 1. Oxford University Press (1986). Edited by Solomon Feferman, John Dawson, and Stephen Kleene.
28. Gottwald, S.: A Treatise on Many-Valued Logics, volume 9 of Studies in Logic and Computation. Research Studies Press, Baldock, Nov. (2000)
29. Hähnle, R.: Automated Deduction in Multiple-Valued Logics, volume 10 of International Series of Monographs on Computer Science. Oxford University Press (1994)
30. Hähnle, R.: Many-valued logic and mixed integer programming. Annals of Mathematics and Artificial Intelligence, 12, (3,4), Dec. (1994) 231–264
31. Hähnle, R.: Short conjunctive normal forms in finitely-valued logics. Journal of Logic and Computation, 4, (6) (1994) 905–927
32. Hähnle, R.: Exploiting data dependencies in many-valued logics. Journal of Applied Non-Classical Logics, 6, (1) (1996) 49–69
33. Hähnle, R.: Proof theory of many-valued logic—linear optimization—logic design: Connections and interactions. Soft Computing—A Fusion of Foundations, Methodologies and Applications, 1, (3), Sept. (1997) 107–119
34. Hähnle, R.: Commodious axiomatization of quantifiers in multiple-valued logic. Studia Logica, Special Issue on Many-Valued Logics, their Proof Theory and Algebras, 61, (1) (1998) 101–121.
35. Hähnle, R.: Tableaux for many-valued logics. In D'Agostino, M., Gabbay, D., Hähnle, R. and Posegga, J. editors, Handbook of Tableau Methods. Kluwer, Dordrecht (1999) 529–580
36. Hähnle, R.: Advanced many-valued logics. In Gabbay, D.M. and Guenthner, F. editors, Handbook of Philosophical Logic, volume 2. Kluwer, Dordrecht, 2nd edition, Aug. (2001) 297–395

37. Hähnle, R.: Complexity of many-valued logics. In Proc. 31st International Symposium on Multiple-Valued Logics, Warsaw, Poland. Invited Tutorial. IEEE CS Press, Los Alamitos, May (2001) 137–146
38. Hájek, P.: Metamathematics of Fuzzy Logic, volume 4 of Trends in Logic: Studia Logica Library. Kluwer Academic Publishers, Dordrecht (1998)
39. Hájek, P., Godo, L. and Esteva, F.: A complete many-valued logic with product conjunction. Archive for Mathematical Logic, 35 (1996) 191–208
40. Hájek, P. and Tulipani, S.: Complexity of fuzzy probability logics. Fundamenta Informaticæ, 45, (3) (2001) 207–213
41. Hooker, J.N.: A quantitative approach to logical inference. Decision Support Systems, 4 (1988) 45–69
42. Jeroslow, R.G.: Logic-Based Decision Support. Mixed Integer Model Formulation. Elsevier, Amsterdam (1988)
43. Kirin, V.G.: Gentzen's method of the many-valued propositional calculi. Zeitschrift für mathematische Logik und Grundlagen der Mathematik, 12 (1966) 317–332
44. Łukasiewicz, J.: Jan Łukasiewicz, Selected Writings. Edited by L. Borowski. North-Holland (1970)
45. Łukasiewicz, J. and Tarski, A.: Untersuchungen über den Aussagenkalkül. Comptes Rendus des Séances de la Societé des Sciences et des Lettres de Varsovie, Classe III, 23 (1930) 1–21. Reprinted and translated in [44].
46. Manyà, F.: Proof Procedures for Multiple-Valued Propositional Logics. PhD thesis, Facultat de Ciències, Universitat Autònoma de Barcelona (1996). Published as [47]
47. Manyà, F.: Proof Procedures for Multiple-Valued Propositional Logics, volume 9 of Monografies de l'Institut d'Investigació en Intel·ligència Artificial. IIIA-CSIC, Bellaterra (Barcelona) (1999)
48. Manyà, F.: The 2-SAT problem in signed CNF formulas. Multiple-Valued Logic. An International Journal, 5 (2000)
49. Manyà, F., Béjar, R. and Escalada-Imaz, G.: The satisfiability problem in regular CNF-formulas. Soft Computing—A Fusion of Foundations, Methodologies and Applications, 2, (3), Sept. (1998) 116–123
50. McNaughton, R.: A theorem about infinite-valued sentential logic. Journal of Symbolic Logic, 16, (1) (1951) 1–13
51. Mitchell, D., Selman, B. and Levesque, H.: Hard and easy distributions of SAT problems. In Proc. of AAAI-92, San Jose, CA (1992) 459–465
52. Mostowski, A.: Proofs of non-deducibility in intuitionistic functional calculus. Journal of Symbolic Logic, 13, (4), Dec. (1948) 204–207
53. Mundici, D.: Satisfiability in many-valued sentential logic is NP-complete. Theoretical Computer Science, 52 (1987) 145–153
54. Mundici, D.: A constructive proof of McNaughton's Theorem in infinite-valued logic. Journal of Symbolic Logic, 59, (2), June (1994) 596–602
55. Mundici, D. and Olivetti, N.: Resolution and model building in the infinite-valued calculus of Łukasiewicz. Theoretical Computer Science, 200, (1–2) (1998) 335–366
56. Patel-Schneider, P. F.: A decidable first-order logic for knowledge representation. Journal of Automated Reasoning, 6 (1990) 361–388
57. Pogorzelski, W.A.: The deduction theorem for Łukasiewicz many-valued propositional calculi. Studia Logica, 15 (1964) 7–23

58. Ragaz, M.E.: Arithmetische Klassifikation von Formelmengen der unendlichwertigen Logik. PhD thesis, ETH Zürich (1981)
59. Rosser, J.B. and Turquette, A.R.: Many-Valued Logics. North-Holland, Amsterdam (1952)
60. Rousseau, G.: Sequents in many valued logic I. Fundamenta Mathematicæ, LX (1967) 23–33
61. Sasao, T.: Multiple-valued decomposition of generalized Boolean functions and the complexity of programmable logic arrays. IEEE Transactions on Computers, C-30, Sept. (1981) 635–643
62. Sasao, T.: Logic synthesis with EXOR gates. In Sasao, T. editor, Logic Synthesis and Optimization, chapter 12. Kluwer, Norwell/MA, USA (1993) 259–286
63. Sasao, T.: Switching Theory for Logic Synthesis. Kluwer, Norwell/MA, USA (1999)
64. Sasao, T. and Butler, J.T.: Comparison of the worst and best sum-of-products expressions for multiple-valued functions. In Proc. 27th International Symposium on Multiple-Valued Logic, Nova Scotia, Canada. IEEE CS Press, Los Alamitos, May (1997) 55–60
65. Scarpellini, B.: Die Nichtaxiomatisierbarkeit des unendlichwertigen Prädikatenkalküls von Łukasiewicz. Journal of Symbolic Logic, 27, (2), June (1962) 159–170
66. Sistla, A. and Clarke, E.: The complexity of propositional linear temporal logics. Journal of the ACM, 32, (3) (1985) 733–749
67. Sofronie-Stokkermans, V.: On translation of finitely-valued logics to classical first-order logic. In Prade, H. editor, Proc. 13th European Conference on Artificial Intelligence, Brighton. John Wiley & Sons (1998) 410–411
68. Sofronie-Stokkermans, V.: Automated theorem proving by resolution for finitely-valued logics based on distributive lattices with operators. Multiple-Valued Logic, 6, (3–4) (2001) 289–344
69. Wagner, H.: Nonaxiomatizability and undecidability of an infinite-valued temporal logic. Multiple-Valued Logic, 2, (1) (1997) 47–58
70. Zach, R.: Proof theory of finite-valued logics. Master's thesis, Institut für Algebra und Diskrete Mathematik, TU Wien, Sept. (1993). Also Technical Report TUW-E185.2-Z.1-93.

Part III

Fuzzy Logics and Their Applications

Chapter 10
Ternary Kleenean Non-additive Measures

Tomoyuki Araki[1], Masao Mukaidono[2], Fujio Yamamoto[3]

[1] Department of Electronics and Photonic Systems Engineering,
Hiroshima Institute of Technology 731-5193 Japan
araki@ieee.org
[2] Department of Computer Science,
Meiji University 214-8571 Japan
masao@cs.meiji.ac.jp
[3] Department of Information and Computer Sciences,
Kanagawa Institute of Technology 243-0292 Japan
yamamoto@ic.kanagawa-it.ac.jp

Abstract. The theory of non-additive measures and integral calculus with respect to them is rich and has been developed in many streams. In this paper, the authors focus on and extend one special non-additive measure, which is called fuzzy measure. Then they focus on and extend an item from the integral calculus, called the Sugeno integral. This expansion enables us to treat concepts such as "negation" and "unknown" in the field of fuzzy measures—these concepts have never been treated well in the field. To achieve it, the authors attempt to translate the vagueness of fuzzy set theory and fuzzy logic to the ambiguity of fuzzy measure by considering the fact that the Sugeno integral in fuzzy measure theory can be represented using fuzzy switching functions with constants in fuzzy logic. Then, we attempt to apply the facts clarified in fuzzy logic to fuzzy measure. Consequently, two new concepts: "Ternary Kleenean Non-additive Measure" and "Kleene-Sugeno integral" are proposed.

1 Introduction

The theory of non-additive measures has been developed by many researchers in different streams. It has enabled us to have rich results so far. We enumerate some of them: modified probabilities by Medolaghi from 1907, now known as Sugeno lambda-measures; theory of capacities by Choquet from 1953; maxitive measure by Shilkret from 1971; Dempster-Shafer theory from 60-ties and 70-ties; premeasures by Sipos from 1979; several submeasures and supermeasures; and so on. These are observed in [23, 10, 20, 6, 7, 4, 5, 22] and so on.

Fuzzy measure is among these; it was proposed by Sugeno [23] in 1974, and can treat ambiguities in fuzzy theory. Recently its usefulness was shown by

Yager [26]. He discussed its use as a unifying structure for modeling knowledge about uncertain (or ambiguous) variables, and showed that a large class of well-established types of uncertain (or ambiguous) representations could be modeled within fuzzy measure's framework, such as Dempster-Shafer belief structure, uncertainty (or ambiguity) representation corresponding to a set of possible fuzzy measures, and so on. Similarly, the usefulness of fuzzy measures and integrals with respect to them are discussed in many places, for instance, [23, 10, 20, 6, 8] and so on. Then defects of Sugeno integrals with respect to fuzzy measures are considered and methods to avoid them are also considered [19].

On the other hand, from the point of view of logic, many attempts have been made to treat intermediate states between true and false. The origin is Łukasiewicz's three-valued logic in 30-ties. Since his well-known studies many logic systems have been proposed and developed by many researchers. It is seen in the literature such as [9, 13, 11, 3, 21] and so on.

We discuss in this paper the relationship between Kleene's logic system, [12], and non-additive measures and integrals (especially, extensions of fuzzy measures and Sugeno integrals). From the point of view of many-valued logic, Kleene's logic system is a special case of Łukasiewicz's logic system. Furthermore, a special case of fuzzy logic systems, constructed using Zadeh's fuzzy set theory of type 1, [27], is equivalent to Kleene's infinite-valued logic system [15–17]. Then, logic functions in the fuzzy logic are essentially three-valued from algebraic point of view.

Futhermore, in fuzzy theory, the concept "undefinable," which includes "ambiguity," "vagueness," "imprecision," "uncertainty," and so on, has been studied actively by many researchers from the time Zadeh proposed fuzzy sets, [27], in 1965, until now. Fuzzy concepts such as "vagueness" and "ambiguity" are important in discussing the concept "undefinable." In the field of fuzzy theory, the fuzziness that is treated in fuzzy set theory and fuzzy logic is called "vagueness," and that which is treated in fuzzy measure theory is called "ambiguity." It is well known that they are both very different in concept, although each is treated in the framework of fuzzy theory.

We will first attempt to translate the vagueness of fuzzy set theory and fuzzy logic to the ambiguity of (monotonic and normalized) fuzzy measure by considering the fact that the Sugeno integral in fuzzy measure theory can be represented using fuzzy switching functions with constants (Fuzzy/C switching functions, for short) in fuzzy logic, [25, 14]. (Note that the converse, however, does not hold. Sugeno integrals can only represent positive and monotone Fuzzy/C switching functions.) Then, we will attempt to apply the facts clarified in fuzzy logic to fuzzy measures (see Fig. 1). We propose two new concepts: "Ternary Kleenean non-additive measure" and "Kleene-Sugeno integral." We also clarify the algebraic structure of ternary Kleenean non-additive measure.

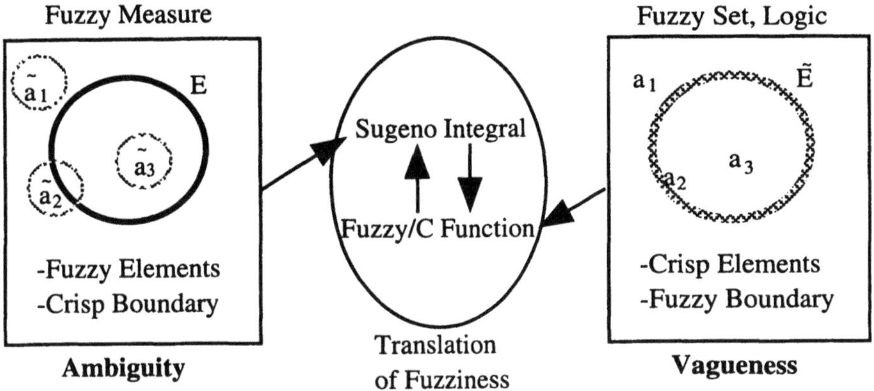

Fig. 1. Conceptual figure of mutual translation of ambiguity and vagueness

Based on our proposal, from the Sugeno integral's point of view the proposed framework can translate both vagueness to ambiguity and ambiguity to vagueness, transparently and mutually. Especially, translation from ambiguity to vagueness is important because vagueness is an easier fuzziness to handle than ambiguity when taking technological applications into consideration. This fact is evidence for clarifying the usefulness of the proposed framework. Furthermore, using the proposed framework, we can treat the concepts "negation" and "unknown," which are not only important concepts in the real world but also, from the Sugeno integral's point of view, there do not exist these concepts in fuzzy measures and integrals theory.

2 Preliminaries

2.1 Fuzzy measure and Sugeno integral

Definition 1. [23, 20] *Let X be the universe of discourse, which is a finite set. Let 2^X be the power set of X. Then* monotonic and normalized fuzzy measure *g is a mapping, such as $g : 2^X \to [0,1]$. It satisfies the following axioms:*

1. $g(\emptyset) = 0$, $g(X) = 1$.
2. $A, B \subseteq X$, $A \subset B$ *implies* $g(A) \leq g(B)$.

Hereafter, to simplify discussion, we use terminology "fuzzy measure" as "monotonic and normalized fuzzy measure" if confusion does not arise.

Definition 2. [23] *Let X be the universe of discourse, E be a subset of X, g be fuzzy measure over X, and h be a function, such as $h : E \to [0,1]$, to* *which integral calculus is applied. Then the* Sugeno integral *over E is defined as follows:*

$$(S)\int_{x\in E} h(x)\circ g = \bigvee_{A\subseteq E}((\bigwedge_{x\in A} h(x))\wedge g(A))$$

Note that notations $\wedge$ *and* $\bigwedge$ *mean minimum operations and the notation* $\bigvee$ *means a maximum operation.*

Example 1. Let $X = \{a_1, a_2\}$ and $E = X$. Let g be given as $g(\emptyset) = 0$, $g(a_1) = 0.3$, $g(a_2) = 0.6$, and $g(a_1, a_2) = 1$. h is given as $h(a_1) = 0.5$ and $h(a_2) = 0.6$. Then the Sugeno integral is obtained as $(S)\int_{x\in E} h(x)\circ g = 0.6$.

2.2 Fuzzy switching function with constants

Definitions
Hereafter, notations $V = [0,1]$, $V_3 = \{0, 0.5, 1\}$, and $V_2 = \{0,1\}$ will be used as sets of truth values. Then, a fuzzy switching function with constants is call a Fuzzy/C switching function, for short.

Definition 3. [1] *An* n-variable Fuzzy/C switching function *is a mapping from* V^n *to* V *and it is represented in special logic formulas defined as*

1. *Constants* $c_1, c_2, \cdots$ *and variables* $x_1, \cdots, x_n$ *over* V *are logic formulas,*
2. *If* F *and* G *are logic formulas,* $\sim F$, $F\wedge G$, $F\vee G$ *are logic formulas,*
3. *Only defined above are logic formulas.*

Note that logic connectives in the above special logic formulas, negation ($\sim$)*, conjunction* ($\wedge$)*, and disjunction* ($\vee$) *are defined as* $\sim x = 1 - x$, $x\wedge y = \min(x,y)$, *and* $x\vee y = \max(x,y)$, *respectively; we often omit writing the notation of a conjunction.*

Example 2. The following formulas represent Fuzzy/C switching functions:

$$f_1 = x_1x_2 \vee 0.7x_2 \vee 0.3{\sim}x_1$$
$$f_2 = x_1x_2 \vee 0.5{\sim}x_2 \vee x_1{\sim}x_1$$
$$f_3 = (x_1\vee x_2)(0.5\vee x_2)(x_1\vee {\sim}x_1)$$

Definition 4. (Partial ordering relation with respect to vagueness) *Let* $a, b \in V$. *Then,* $a \preceq_V b$ *hold if and only if either* $0\le a\le b\le 0.5$ *or* $0.5\le b\le a\le 1$.

Fundamental properties and representation of logic formulas of Fuzzy/C switching functions

Theorem 1. [1] (Monotonicity with respect to vagueness) *Let* f *be an* n*-variable Fuzzy/C switching function. Then,* $A, B \in V^n$ *and* $A \preceq_V B$ *imply* $F(A) \preceq_V F(B)$.

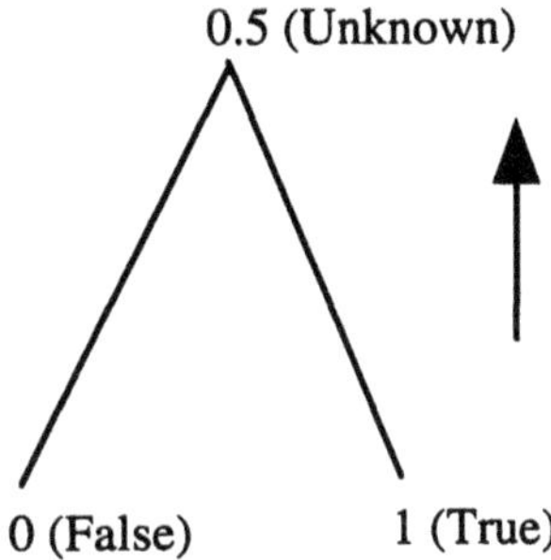

Fig. 2. Partial ordering relation with respect to vagueness($\preceq_V$)

Proof. Fuzzy/C switching functions are generalization of regular ternary logic functions in Ref. [18]. Fuzzy/C switching functions hold whole properties of regular ternary logic functions. Regularity defined in Ref. [18] is extension of Kleene's regularity [12], and it corresponds to the property of monotonicity with respect to vagueness. □

Theorem 2. [1] *Let f and g be n-variable Fuzzy/C switching functions. Then, $\forall A \in V^n(f(A) = g(A))$ is equivalent to $\forall B \in V_3^n(f(B) = g(B))$.*

Proof. As mentioned in Section 1, Fuzzy/C switching functions are logic functions based on Zadeh's set theory, and one of special cases of infinite-valued logic functions in terms of Kleene's logic system. Therefore, Fuzzy/C switching functions are essentially three-valued. □

According to Theorem 1 and Theorem 2, Fuzzy/C switching functions are monotonic with respect to vagueness and are essentially three-valued logic functions.

Definition 5. *A variable x_i and its negation $\sim x_i$ are called* literals. *A conjunction of some literals are called a* phrase. *A phrase which does not have both x_i and $\sim x_i$ for any i at the same time is called a* simple phrase. *If a simple phrase has all the variables, it is called a* minterm.

If a phrase has both x_i and $\sim x_i$ for at least one variable at the same time, then it is called a complementary phrase. *If a complementary phrase has all the variables, it is called a* complementary minterm.

Definition 6. *We assign an n-variable simple phrase $x = x_1^{t_1} \cdot \cdots \cdot x_n^{t_n}$ to an element $(t_1, t_2, \cdots, t_n) \in V_3^n$ as follows:*

$$x_i^{t_i} = \begin{cases} x_i & \text{if } t_i = 1 \\ 1 & \text{if } t_i = 0.5 \\ \sim x_i & \text{if } t_i = 0 \end{cases} \tag{1}$$

Then we assign an n-variable complementary minterm $x = x_1^{t_1} \cdot \cdots \cdot x_n^{t_n}$ *to an element* $(t_1, t_2, \cdots, t_n) \in V_3^n - V_2^n$ *as follows:*

$$x_i^{t_i} = \begin{cases} x_i & \text{if } t_i = 1 \\ x_i \sim x_i & \text{if } t_i = 0.5 \\ \sim x_i & \text{if } t_i = 0 \end{cases} \tag{2}$$

Theorem 3. [24] *Let* f *be a Fuzzy/C switching function. Then* f *can be represented in*

$$f(x) = \bigvee_{C \in V_3^n} \left\{ \bigwedge_{C \preceq_V C} (f(C) \wedge \alpha_C(\boldsymbol{x})) \vee (f(C) \wedge \beta_C(\boldsymbol{x})) \right\}.$$

$\alpha_C(\boldsymbol{x})$ *is a function represented in the simple phrase corresponding to* C *by Definition 6 (1). If* C *is included in* ${V_2}^n$, $\beta_C(x) = 0$, *otherwise* $\beta_C(x)$ *is a function represented in the complementary minterm corresponding to* C *by Definition 6 (2).*

Note that the notation $\bigwedge$ *means a minimum operation, and the notation* $\bigvee$ *means a maximum operation.*

Fuzzy/C switching functions can be represented in disjunctive and conjunctive forms because they are a model of Kleenean algebra and therefore satisfy the absorption law, the idempotent law, the double negation law, De Morgan's law, and so on.

3 The proposed measures and integrals

3.1 Representation of Sugeno integral in Fuzzy/C switching functions

The Sugeno integral is one of the most essential integral calculus with respect to fuzzy measure. Therefore, it is important that we investigate its extension for treating practical problems.

Takahagi and Araki [25] showed that the Sugeno integral with respect to fuzzy measure can be represented in positive and monotone Fuzzy/C switching functions (Property 1). Marichal [14] also showed independently that the Sugeno integral can be represented in aggregation functions which include Fuzzy/C switching functions as a special class.

Property 1. Let f be a Fuzzy/C switching function. Let $A = (a_1, \cdots, a_n)$ and $B = (b_1, \cdots, b_n)$ be elements in V^n. If f is *positive and monotone*, $A \leq B$ implies $f(A) \leq f(B)$, where $A \leq B$ means $\forall i (a_i \leq b_i)$.

We illustrate below how to correspond the Sugeno integral to a Fuzzy/C switching function [25]. Phrases in the third column of Table 1 are simple phrases, which are the conjunction of literals. Then the rule of making simple

Table 1. The Sugeno integral of Example 1

Subsets	$g(\bullet)$	Phrases	$h(x)$	$\bigwedge_{x\in A} h(x)$	$(\bigwedge_{x\in A} h(x)) \wedge g(\bullet)$	$(S)\int h(x) \circ g$
$\emptyset$	0	1	–	–	–	
$\{a_1\}$	0.3	x_1	$h(a_1) = 0.5$	0.5	0.3	
$\{a_2\}$	0.7	x_2	$h(a_2) = 0.6$	0.6	0.6	0.6
$\{a_1, a_2\}$	1	$x_1 \cdot x_2$	$h(a_1) = 0.5$ $h(a_2) = 0.6$	0.5	0.5	

phrases is that a literal x_i exists if a member a_i exists in the subset. We obtain a logic formula of Fuzzy/C switching function corresponding to the Sugeno integral by making the disjunction of all the conjunctions of values of fuzzy measure $g(\bullet)$ and simple phrases. In the case of Table 1, The obtained Fuzzy/C switching function is $f = 0.3X_1 \vee 0.7x_2 \vee x_1x_2$. Then, we can also obtain the value of the Sugeno integral by substitution such as $x_i = h(a_i)$. That is, $f(0.5, 0.6) = 0.6$.

3.2 Ternary Kleenean non-additive measures and Kleene-Sugeno integrals

In the real world, it is rare that we can decide whether an element is in a set or not. Therefore, it is necessary that we deal with the concept of the "unknown", that is, unknown whether an element is in or not. Furthermore, it is not rare that we need the concept "must not be in", that is, the existence of some elements harms the evaluation of others.

To satisfy these requirements, we propose a new concept: *ternary Kleenean non-additive measures.* Then we propose the *Kleene-Sugeno integral* as an integral calculus with respect to ternary Kleenean non-additive measures.

First, we consider properties of sets that can represent concepts such as "be in", "unknown", and "must not be in". Since there are three concepts, we need at least a kind of three-valued characteristic functions for the desired sets. Accordingly, we consider the following characteristic function.

Definition 7. *Let X be the universe of discourse and A be a subset of X. Then, we define the characteristic function of an element x in X as follows:*

$$\theta_A(x) = \begin{cases} \textit{1 if } x \textit{ is in } A. \\ \textit{0 if unknown whether } x \textit{ in } A \textit{ or not.} \\ \textit{-1 if } x \textit{ must not be in } A. \end{cases}$$

As shown in Definition 7, this characteristic function is based on a three-valued logic if we consider 1 to be "true" and -1 to be "false".

Secondly, we consider the meanings of the three values 1, 0, and -1, which are taken by the above characteristic function. It is valid and natural that there be a relationship among 1, 0, and -1, as shown in Fig. 3. That is, "be in (1)" and "unknown (0)" are comparable to each other. And similarly "must

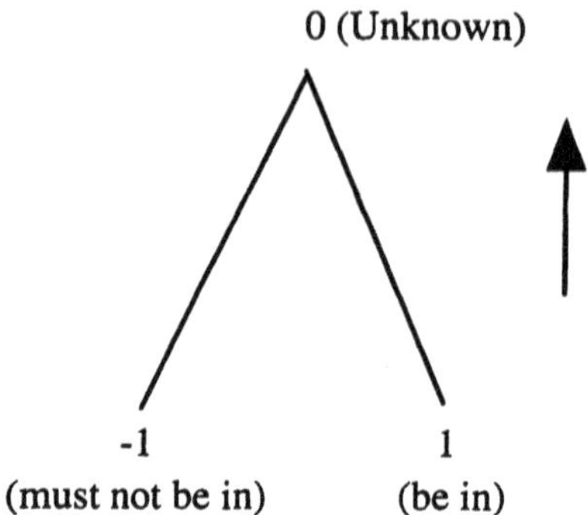

Fig. 3. Partial ordering relation with respect to ambiguity($\preceq_A$)

not be in (-1)" and "unknown (0)" are comparable to each other. However, both "be in (1)" and "must not be in (-1)" are fixed states, therefore they are not comparable to each other. Furthermore, "unknown (0)" is the most ambiguous state. The relationship shown in Fig. 3 is *partial ordering relation with respect to ambiguity* ($\preceq_A$). This relation is clearly isomorphic to the partial ordering relation with respect to vagueness ($\preceq_V$), as shown in Fig. 2. Furthermore, although this is a numerical relation, we apply and extend this relation to elements of the power set of the universe of discourse.

Definition 8. *Let X be the universe of discourse and $A, B \in 3^X$ (where 3^X means the power set of X, which is based on three-valued characteristic function in Definition 7). Then, $A \preceq_A B$ holds if and only if $\forall x \in X(\theta_A(x) \preceq_A \theta_B(x))$ holds.*

Definition 9. *Let X be the universe of discourse and $A, B \in 3^X$. Then, $A \subseteq B$ holds if and only if $\forall x \in X(\theta_A(x) \leq \theta_B(x))$ holds. Note that the definition of the relation $\in$ is the usual definition in terms of two-valued set theory.*

Next, we describe the notation of sets and elements.

- If $\theta_A(a) = 1$, then the element is represented as "a".
- If $\theta_A(a) = 0$, then the element is not represented, that is, " ".
- If $\theta_A(a) = -1$, then the element is represented as "$\overline{a}$".
- A set such as $\forall i(\theta_A(a_i) = 0)$ is represented as "$\emptyset_K$". This corresponds to the empty set in usual set theory.
- A set such as $\theta_A(a_j) = -1$ or 1 for $\forall j$ is represented in "M_i" ($i = 1, \cdots, 2^n$).
- The maximum element in M_i's ($i = 1, \cdots, 2^n$) with respect to $\subseteq$ is represented as "$\overline{M}$".
- The minimum element in M_i's ($i = 1, \cdots, 2^n$) with respect to $\subseteq$ is represented as "$\underline{M}$".

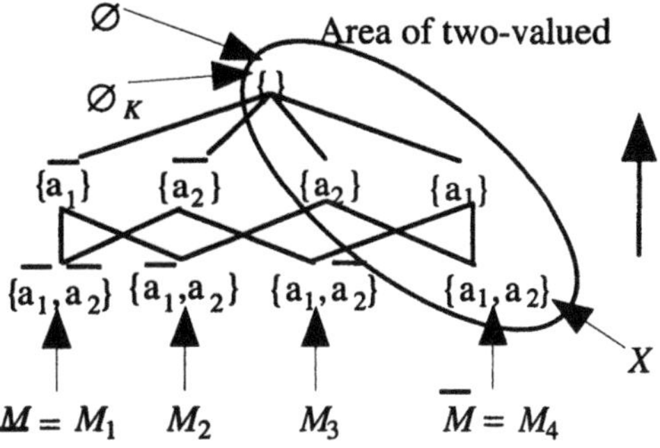

Fig. 4. Hasse diagram ($\preceq_A$)

Example 3. Let $\{a_1, a_2, a_3\}$ be the universe of discourse. Then, in its subset $A = \{a_1, \overline{a_3}\}$, the meaning of each element is as follows:

1. a_1 is in A.
2. a_2 is unknown whether it is in A or not.
3. a_3 must not be in A.

Example 4. Let $\{a_1, a_2\}$ be the universe of discourse. Then, $\{a_1, a_2\} \preceq_A \{a_1\}$, $\{a_1, \overline{a_2}\} \preceq_A \{a_1\}$, and $\{a_1\} \preceq_A \emptyset_K$ hold. $\{a_1, a_2\}$ and $\{a_1, \overline{a_2}\}$ are not comparable to each other.

Example 5. Let $X = \{a_1, a_2\}$ be the universe of discourse. Fig. 4 shows the Hasse diagram of partially ordered set $\langle 3^X, \preceq_A \rangle$.

Based on the definitions and notations, we propose the following definition of ternary Kleenean fuzzy measure:

Definition 10. (Ternary Kleenean non-additive measure) *Let X be the universe of discourse. Then,* ternary Kleenean non-additive measure *is a mapping $p : 3^X \to [-1, 1]$, which satisfies the following axioms:*

(B1) $p(\emptyset_K) = 0$, $p(\overline{M}) = 1$.
(B2) $p(\underline{M}) = -1$.
(M1) $A, B \in 3^X$, $A \preceq_A B$ *implies* $p(A) \preceq_A p(B)$.
(M2) *For* $M_i (i = 1, \cdots, 2^n)$, $M_i \subseteq M_j$ *implies* $p(M_i) \leq p(M_j)$.

If we interpret the concept "an element is not in" in usual two-valued sets as "unknown whether an element is in or not", we can consider usual two-valued power set to be one of the special subsets in three-valued power set. Then, usual fuzzy measure is a special case of ternary Kleenean non-additive measure.

First, we verify the boundary conditions. The empty set $\emptyset$ corresponds to $\emptyset_K$ and the universe of discourse X corresponds to $\overline{M}$ as shown in Fig. 4.

Poset $\langle 2^X, \preceq_A \rangle$ forms a lattice whose minimum element is $X(=\overline{M})$ and whose maximum element is $\emptyset(=\emptyset_K)$. Therefore, the boundary condition (B1) corresponds to Definition 1 1 (1).

Second, we verify the monotone conditions. In 2^X $(\subseteq 3^X)$, it is clear that the monotone condition (M1) corresponds to Definition 1 (2).

Consequently, if we interpret the concept "an element is not in" in usual two-valued sets as "unknown whether an element is in or not", usual two-valued power set is one of the special subsets in three-valued power set as shown in Fig. 4.

By adopting interpretation mentioned above, we define an integral calculus with respect to ternary Kleenean non-additive measures – the *Kleene-Sugeno integral* –

Before the definition of Kleene-Sugeno integral, we define KS-Fuzzy/C functions as Definition 11, which are isomorphic to Fuzzy/C switching functions in Definition 3 because we are to define Kleene-Sugeno integral in formulas of KS-Fuzzy/C functions.

Definition 11. *Let W be a closed interval* $[-1,1]$. *Then* n-variable KS-Fuzzy/C functions *are mappings from W^n to W, represented in special formulas defined as follows: (1) Constants $c_1, c_2, \cdots$ and variables $x_1, \cdots, x_n$ over W are formulas, (2) If F and G are formulas, $\neg F$, $F \wedge G$,$F \vee G$ are formulas, (3) Only defined above are formulas.*

Note that operators $\neg$, $\wedge$, and $\vee$ are defined as $\neg x = (-1) \cdot x$, $x \wedge y = \min(x, y)$, and $x \vee y = \max(x,)$, respectively; we often omit the notation $\wedge$.

Let V be $[0,1]$ and W be $[-1,1]$. Then by $\varphi(x) = 2 \cdot x - 1$, an element x in V is mapped onto an element in W (Clearly φ is one-to-one mapping.), and $\varphi(0) = -1$, $\varphi(0.5) = 0$, and $\varphi(1) = 1$. Therefore, clearly, KS-Fuzzy/C functions are isomorphic to Fuzzy/C switching functions, and the following properties are clearly hold by Theorem 1 and Theorem 2:

Property 2. (Monotonicity with respect to ambiguity) Let W be $[-1,1]$ and f be an n-variable KS-Fuzzy/C function. Then, $A, B \in W^n$ and $A \preceq_A B$ imply $f(A) \preceq_A f(B)$.

Note that the relation $\preceq_A$ is already defined by Figure 3.

Property 3. Let W be $[-1,1]$, W_3 be $\{-1,0,1\}$. And let f and g be KS-Fuzzy/C functions. Then, $\forall A \in W^n(f(A) = g(A))$ is equivalent to $\forall B \in W_3^n(f(B) = g(B))$.

The definition represented in KS-Fuzzy/C functions of Kleene-Sugeno Integrals are as follows:

Definition 12. (Kleene-Sugeno integral) *Let $X = \{a_1, \cdots, a_n\}$ be the universe of discourse. Then, the* Kleene-Sugeno integral *is an integral with respect to ternary Kleenean non-additive measure, which is represented with* *the following KS-Fuzzy/C function:*

$$(KS)\int_{x\in X} h\circ p = \bigvee_{C\in 3^X}\left\{\bigwedge_{C'\preceq_A C}(p(C')\wedge\alpha_C(\boldsymbol{x}))\vee(p(C)\wedge\beta_C(\boldsymbol{x}))\right\},$$

where the notation $\bigwedge$ means a minimum operation, and the notation $\bigvee$ means a maximum operation.

Note that $\alpha_C(\boldsymbol{x}) = x_1^{a_1}\cdot\dots\cdot x_n^{a_n}$ is a function represented in a sample phrase defined by the following correspondence:

$$x_i^{a_i} = \begin{cases} x_i & \text{if } a_i \text{ is in } C \\ 1 & \text{if } a_i \text{ is unknown whether in } C \text{ or not} \\ \neg x_i & \text{if } a_i \text{ must not be in } C, \end{cases}$$

and $\beta_C(\boldsymbol{x}) = x_1^{a_i}\cdot\dots\cdot x_n^{a_n}$ is a function represented in a complementary minterm defined by the following correspondence:

$$x_i^{a_i} = \begin{cases} x_i & \text{if } a_i \text{ is in } C \\ x_i\neg x_i & \text{if } a_i \text{ is unknown whether in } C \text{ or not} \\ \neg x_i & \text{if } a_i \text{ must not be in } C, \end{cases}$$

where if $C = M_i(i = 1,\cdots,2^n)$, $\beta_C(\boldsymbol{x}) = 0$. $h : X \to [-1,1]$ is a function to which integral calculus is applied. Thus, we can obtain the value of the Kleene-Segeno integral by substitutions $x_i = h(a_i)$ and $\neg x_j = h(\overline{a_j}) = (-1)\cdot h(a_j)$.

Example 6. Let $X = \{a_1, a_2\}$ be the universe of discourse and let the ternary Kleenean non-additive measure over X be given as $p(\emptyset_K) = 0$, $p(\{\overline{a_1}\}) = 0$, $p(\{\overline{a_2}\}) = 0$, $p(\{a_2\}) = 0$, $p(\{a_1\}) = 0.2$, $p(\{\overline{a_1},\overline{a_2}\}) = -1$, $p(\{\overline{a_1},a_2\}) = -1$, $p(\{a_1,\overline{a_2}\}) = 1$, and $p(\{a_1,a_2\}) = 1$. These assignments of measure are also shown in Fig. 5. Then, we can obtain the representation of this Kleene-Sugeno integral using Definition 12 as a KS-Fuzzy/C function $f = 1x_1x_2 \vee 1x_1\neg x_2 \vee 0.2x_1 \vee 0\neg x_1x_2\neg x_2$.

We describe the characteristic features of this formula as follows:

1. The first phrase $1x_1x_2$ means that "a_1 is in and a_2 is in" is valid with weight 1. The weight 1 means that this phrase is evaluated positively.
2. The second phrase $1x_1\neg x_2$ means that "a_1 is in and a_2 must not be in" is valid with weight 1.
3. The third phrase means that "a_1 is in" is valid with weight 0.2. This phrase is evaluated positively because 0.2 is greater than 0. However, the evaluation is low because 0.2 is close to 0 (unknown).
4. The final phrase $0\neg x_1x_2\neg x_2$ means that "a_1 must not be in and a_2 is in and furthermore a_2 must not be in" is valid. And "a_1 must not be in and a_2 is unknown" is not valid. The value of this phrase can never be greater than 0 (unknown). That is, this phrase can not be evaluated positively, and in almost all cases, is evaluated negatively.

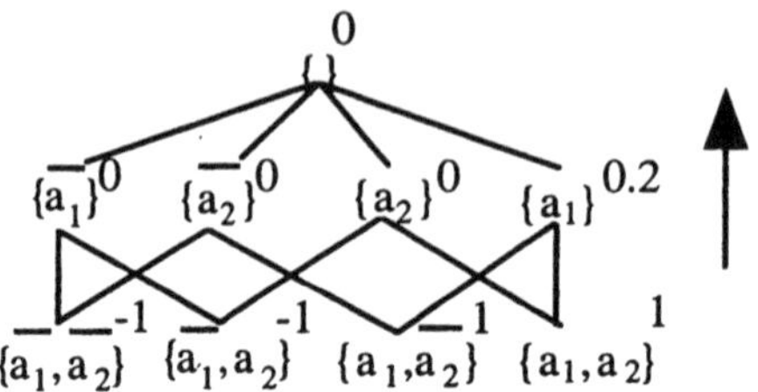

Fig. 5. Assignment of measure

We can interpret the semantics by using the formula representation of the Kleene-Sugeno integral. As shown in the above interpretations of phrases in the formula, especially phrases which include negation of variables, the concept "negation", that is, "must not be in" is represented. The value of Kleene-Sugeno integral can be obtained by substitution such that $x_i = h(a_i)$ and $\neg x_j = h(\overline{a_j}) = (-1) \cdot h(a_j)$. By obtaining the value of the Kleene-Sugeno integral, we can evaluate using the concept "negation", that is, "must not be in". Such representation and evaluation are impossible in usual fuzzy measure and integral theory.

4 Structure and meanings

Theorem 4. *The number of sets which are order isomorphic to usual two-valued power set and are included in three-valued power set is* 2^n.

Proof. Usual two-valued power set forms set-lattice $\langle 2^X, \subseteq \rangle$, whose the minimum element is $\emptyset_K$ and the maximum element is $\overline{M}$. There are 2^n subsets in 3^X, whose minimum element is $\emptyset_K$ and maximum element is $M_i (i = 1, \cdots, 2^n)$. They all are order isomorphic to $\langle 2^X, \subseteq \rangle$. □

Theorem 5. *Let the set-lattices (Boolean lattices) that correspond to each* $M_i (i = 1, \cdots, 2^n)$ *in Theorem 4 be* $B_i (i = 1, \cdots, 2^n)$ *and* $B = \{B_i | i = 1, \cdots, 2^n\}$. *And we define* $B_k \leq_B B_\ell$ *as* $M_k \subseteq M_\ell$. *Then Poset* $\langle B, \leq_B \rangle$ *forms a Boolean lattice.*

Proof. Let $M = \{M_i | i = 1, \cdots, 2^n\}$. Then, the Poset $\langle M, \subseteq \rangle$ forms a Boolean lattice. Therefore the theorem clearly holds. □

By Theorem 4 and Theorem 5, for a $\langle B_i, \subseteq \rangle$ which is order isomorphic to $\langle 2^X, \subseteq \rangle$, there always exists a complementary element B_i^C in $\{B_1, \cdots, B_{2^n}\} - \{B_i\}$ in terms of Boolean algebra. Therefore, a ternary Kleenean non-additve measure is an integration of complementary monotonic measures ($\langle B_i, \subseteq \rangle$s). Therefore, the Kleene-Sugeno integral is an integrated evaluation of complementary measures in terms of ternary Kleenean non-additive measures.

5 Powerfulness of representation

Based on facts clarified in this paper, we can easily estimate powerfulness of representation of the Sugeno integral and the Kleene-Sugeno integral in terms of Fuzzy/C switching functions. Let FC, KS, S be the set of all Fuzzy/C switching functions, the set of all Kleene-Sugeno integral represented in KS-Fuzzy/C functions, the set of all Sugeno integral represented in positive and monotone Fuzzy/C switching functions, respectively. Then, it is clear that $|S| < |KS| < |FC|$ holds, where $|\ |$ denotes cadinality of the set. $|KS| = |FC|$ does not hold because the conditions (B1), (B2), and (M2) in Definition10. $|S| < |KS|$ is necessarily holds because S is a set of positive and monotone Fuzzy/C switching functions, and functions in KS are not always positive and monotone. Thus, the Kleene-Sugeno integral is suitable for representing and evaluate objects to which integral calculus is applied more powerfully than the Sugeno integral.

6 Conclusions

In this paper, we applied facts clarified in fuzzy logic to fuzzy measure and defined ternary Kleenean non-additive measures and Kleene-Sugeno integrals. We demonstrated how ternary Kleenean non-additive measures and Kleene-Sugeno integrals can be used to evaluate negative estimation of elements in the universe of discourse for applying them to practical problems. We also clarified the algebraic structure and meaning of ternary Kleenean non-additive measures and Kleene-Sugeno integrals.

For usual Sugeno integrals, the range is the closed interval $[0,1]$. The values of the integrals mean relative estimation of objects, and not absolute estimation. This is because the usual Sugeno integrals are positive functions in terms of this paper. The range of Kleene-Sugeno integrals is the closed interval $[-1,1]$, and Kleene-Sugeno integrals are not positive functions in terms of this paper. Using similar way in the paper, we can easily expand the range $[-1,1]$ to $(-\infty,\infty)$ or $[-\infty,\infty]$. Therefore, it would be possible for us to use Kleene-Sugeno integrals as a tool for absolute estimation on real number.

References

1. Araki, T. and Mukaidono, M.: On the fuzzy switching function with constants (in Japanese), IEICE Trans. D-I, vol. J81-D-I, no.9 (1998)1037–1047
2. Araki, T., Mukaidono, M. and Yamamoto, F.: On a Kleenean extension of fuzzy measure, Proc. 31th International symposium on Multiple-valued logic, ISMVL2001, Poland (2001) 324–329
3. Bolc, L. and Borowik, P.: Many-valued logic I, Springer-Verlag (1992)
4. Choquet, G.: Theory of capacities, Annales de l'Institut Fourier 5 (1953) 131–295

5. Dempster, A.P.: Upper and lower probabilities included by a multi-valued mapping, Ann. Math. Stat., 38 (1967) 325–339
6. Denneberg, D.: Non-additive measure and integral, basic concepts and their role for applications, in [10] 42–69
7. Denneberg, D.: Non-additive measure and integral, Kluwer Academic Publishers (1994)
8. Dubois, D., Pradde, H. and Sabbadin, R.: Qualitative decision theory with Sugeno integrals, in [10] 314–332
9. Gottwald, S.: Many-valued logic and fuzzy set theory, in Hohle, U. and Rodabaugh, S.E. (eds.), Mathematics of fuzzy sets, Kluwer Academic Publishers (1999) 5–89
10. Grabisch, M., Murofushi, T. and Sugeno, M. (eds): Fuzzy measures and integrals, Physica-Verlag, Springer-Verlag (2000)
11. Hajek, P.: Metamathematics of fuzzy logic, Kluwer Academic Publishers (1998)
12. Kleene, S.C.: Introduction to metamathematics, North-Holland (1952)
13. Novak, V., Perfilieva, I. and Mockor, J.: Mathematical principles of fuzzy logic, Kluwer Academic Publishers (1999).
14. Marichal, J.: On Sugeno integral as an aggregation function, Fuzzy Sets and Systems, vol. 114, no. 3 (2000) 347–365
15. Mukaidono, M.: On some properties of a quantization in fuzzy logic, Proc. 7th International symposium on Multiple-valued logic, ISMVL1977 (1977) 103–106
16. Mukaidono, M.: A necessary and sufficient condition, Proc. Interna for fuzzy logic function, Proc. 9th international symposium on Multiple-valued logic, ISMVL1979 (1979) 159–166
17. Mukaidono, M.: A set of independent and complete axioms for fuzzy algebra (Kleene algebra), Proc. 11th international symposium on Multiple-valued logic, ISMVL1981 (1981) 27–34
18. Mukaidono, M.: Regular ternary logic functions – ternary logic functions suitable for treating ambiguity –, Proc. 13th international symposium on Multiple-valued logic, ISMVL1983 (1983) 286–291
19. Murofushi, T.: Lexicographic use of Sugeno integrals and monotonicity conditions, IEEE Trans. on Fuzzy Systems, vol. 9, no. 6 (2001) 785–794
20. Murofushi, T. and Sugeno, M.: Fuzzy measures and integrals, in [10] 3–41
21. Rescher, N.: Many-valued logic, McGraw-Hill (1969)
22. Sipos, J.: Integral with respect to a pre-measure, Math. Slovaca, 29 (1979) 141–155
23. Sugeno, M.: Theory of fuzzy integrals and its applications, Ph.D. Thesis, Tokyo Institute of Technology (1974)
24. Takagi, N. and Mukaidono, M.: Representation of Logic Formulas for Multiple-valued Kleenean Functions (in Japanese), IEICE Trans., D-I, vol. J75-D-I, no.2 (1992) 69–75
25. Takahagi, E. and Araki, T.: On fuzzy integral representation in fuzzy switching functions with constants, Proc. Vietnam-Japan Bilateral Symposium on Fuzzy Systems and Applications, VJFUZZY'98 (1998) 240–245
26. Yager, R.R.: Uncertain representation using fuzzy measure, IEEE Trans. on Systems, Man, and Cybernetics -Part B:Cybernetics, vol. 32, no. 1 (2002) 13–20
27. Zadeh, L.A.: Fuzzy sets, Inf. Control, 8 (1965) 338–353

Chapter 11
On the Hierarchy of t-norm Based Residuated Fuzzy Logics*

Francesc Esteva, Lluís Godo, and Àngel García-Cerdaña

Institut d'Investigació en Intel.ligència Artificial - CSIC
Campus Univ. Autònoma de Barcelona s/n
08193 Bellaterra, Spain
{esteva,godo,angel}@iiia.csic.es

Abstract. In this paper we overview recent results, both logical and algebraic, about [0, 1]-valued logical systems having a t-norm and its residuum as truth functions for conjunction and implication. We describe their axiomatic systems and algebraic varieties and show they can be suitably placed in a hierarchy of logics depending on their characteristic axioms. We stress that the most general variety generated by residuated structures in [0, 1], which are defined by left-continuous t-norms, is not the variety of residuated lattices but the variety of pre-linear residuated lattices, also known as MTL-algebras. Finally, we also relate t-norm based logics to substructural logics, in particular to Ono's hierarchy of extensions of the Full Lambek Calculus.

1 Introduction

Many-valued logics have become a subject of increasing interest as logics of vagueness; intermediate truth values are understood as partial degrees of truth of fuzzy propositions (see [29]). Many-valued semantics underlying fuzzy logic involve various binary operations on the unit real interval $[0,1]$, generalizing the classical Boolean truth functions on $\{0,1\}$. In this paper, we overview recent results, both logical and algebraic, about $[0,1]$-valued propositional logical systems having a t-norm and its residuum as truth functions for conjunction and implication. These systems have been called *t-norm based residuated fuzzy logics* and can be suitably placed in a hierarchy of logics depending on their characteristic axioms, all of them being extensions of the so-called Monoidal t-norm logic [5] (MTL, for short) and having classical logic as common extension.

* This is a revised and extended version of the paper "Complete residuated many-valued logics with t-norm conjunction" by Francesc Esteva and Lluís Godo which appears in Proc. of the 31st IEEE Int. Sypomsium on Multiple-Valued Logic, ISMVL-2001, Warsaw (Poland), 81-86.

Apart from MTL itself, which is the logic of left continuous t-norms, outstanding logics up in the hierarchy of extensions of MTL are Hájek's Basic logic [15] (BL for short), the logic of continuous t-norms, and Łukasiewicz, Gödel and Product logics, corresponding to the three main continuous t-norms (Łukasiewicz, minimum and product). The algebraic counterparts of all these logics are special subvarieties of residuated lattices, defined by some particular equations. As any residuated lattice, they contain multiplicative operations, a monoidal operation and its residuum (a t-norm and its residuum when the lattice is the unit interval [0,1]), and additive operations, corresponding to the lattice operations (min and max when in [0,1]).

In this paper we study a hierarchy of a class of subvarieties of residuated lattices which is obtained by the succesive and alternative addition of a set of equations related to characteristic properties of main MTL extensions. In this way, we consider a hierarchy of subvarieties, ranging from the variety of residuated lattices to Boolean algebras, where we emphasize that the most general variety generated by residuated structures in [0, 1] is not the variety of residuated lattices but the variety of pre-linear residuated lattices, also known as MTL-algebras. Moreover, taking into account the one-to-one correspondence between subvarieties of residuated lattices and axiomatic extensions of the logic of residuated lattices (equivalently defined by Hölhe in a Hilbert's style, called Monoidal logic, or as the extension of Full Lambek calculus with exchange and weakening, as noticed by Ono) we also consider the corresponding hierarchy of logical systems.

The structure of the paper is as follows. In Section 2 we focus on the main algebraic properties of the different t-norm based residuated fuzzy logics in contrast to residuated lattices. We stress that the most general variety generated by residuated structures in [0, 1], which are defined by left-continuous t-norms, is not the variety **RL** of residuated lattices but the variety **MTL** of pre-linear residuated lattices[1], also known as MTL-algebras. In Section 3 we describe some noticeable subvarieties of **MTL**, ranging from **MTL** itself to Boolean algebras. In Section 4 we make a logical reading of the varieties described in the previous section in terms of axiomatic systems of the corresponding logics. Finally, in Section 5 we show the strong relationship of the logical systems considered in Section 4 with Ono's hierarchy of extensions of the Full Lambek Calculus FL and described in [24], and we display a Hasse diagram containing all logical systems considered both in Ono's and in this papers.

[1] In this paper we keep the same name for a logical system and for its corresponding variety.

2 Residuated lattices versus t-norm based residuated lattices

As it is well known, a t-norm $*$ is a commutative and associative binary operation on $[0,1]$, non-decreasing in both variables and having 1 as neutral element (see [27] and [21]). The corresponding pseudo-inverse operation is defined as

$$x \Rightarrow_* y = \sup\{z \in [0,1] \mid z * x \leq y\}.$$

When the following condition holds

$$z \leq x \Rightarrow_* y \quad \text{iff} \quad x * z \leq y$$

we say that $\Rightarrow_*$ is the residuum (or residuated implication) of $*$ and then, as Pavelka observed (see [25]), $*$ and $\Rightarrow_*$ form an adjoint pair. A corresponding negation operation can be also defined by putting $n_*(x) = x \Rightarrow_* 0$.

The three outstanding examples of continuos t-norms are the Minimum, Product and Łukasiewicz t-norms, whose expressionns, together with their related residua and negation operations, are as follows:

$*$	Minimum	Product	Łukasiewicz
$x * y$	$\min(x,y)$	$x \cdot y$	$\max(0, x+y-1)$
$x \Rightarrow_* y$	$\begin{cases} 1, \text{ if } x \leq y, \\ y, \text{ otherwise} \end{cases}$	$\begin{cases} 1, & \text{if } x \leq y, \\ x/y, & \text{otherwise} \end{cases}$	$\min(1, 1-x+y)$
$n_*(x)$	$\begin{cases} 1, \text{ if } x = 0, \\ 0, \text{ otherwise} \end{cases}$	$\begin{cases} 1, \text{ if } x = 0, \\ 0, \text{ otherwise} \end{cases}$	$1-x$

Basic well-known results on t-norms and their residua are the following ones (see for instance [15, 21]):

(1) $*$ and $\Rightarrow_*$ form an adjoint pair iff $x \Rightarrow_* y = \max\{z \in [0,1] \mid z * x \leq y\}$.
(2) A t-norm has a residuum iff it is left continuous.
(3) A residuated implication satisfies the following ordering property: $x \leq y$ iff $x \Rightarrow y = 1$; this implies the so-called pre-linearity condition: $\max(x \Rightarrow y, y \Rightarrow x) = 1$.
(4) Given a left continuous t-norm $*$ and its residuum $\Rightarrow_*$, the following equation holds, $\max(x,y) = \min((x \Rightarrow_* y) \Rightarrow_* y, (y \Rightarrow_* x) \Rightarrow_* x)$.
(5) A t-norm is continuous iff it is an ordinal sum of copies of Łukasiewicz, Product and Minimum t-norms[2].

[2] Caution: in some cases there can exist "isolated" idempotent points not belonging to any component of the ordinal sum. Take, for example, Łukasiewicz in each interval $[\frac{1}{2n-1}, \frac{1}{2n}]$ and Product in each interval $[\frac{1}{2n}, \frac{1}{2n+1}]$ for n = 1,2,... and 0 is not a member of any component.

(6) A t-norm $*$ is continuous iff $*$ satisfies the following divisibility condition: for all $x \geq y$ there exists z such that $x * z = y$; and this holds iff $*$ and its residuum $\Rightarrow_*$ satisfy the equation $x * (x \Rightarrow_* y) = \min(x, y)$.
(7) The negation n_* defined by a continuous t-norm $*$ is involutive ($n_*(n_*(x)) = x$) iff the t-norm is isomorphic to the Łukasiewicz t-norm.
(8) A t-norm has no zero-divisors[3] iff the corresponding negation is Gödel negation, i.e., $n_*(x) = 0$ for all $x \neq 0$ and $n_*(0) = 1$, or equivalently iff the pseudo-complementation condition $\min(x, n_*(x)) = 0$ holds.

Throughout this paper we shall use *strict* t-norm to refer to a t-norm without zero-divisors[4].

On the other hand *bounded residuated lattices* are algebraic structures $\mathcal{RL} = (L, \wedge, \vee, *, \Rightarrow, 0, 1)$ such that:

(RL1) $(L, \wedge, \vee, 0, 1)$ is a bounded lattice,
(RL2) $(L, *, 1)$ is a commutative monoid with unit 1,
(RL3) the operations $*$ and $\Rightarrow$ form an adjoint pair.

On residuated lattices it is possible to define a negation operator $\neg$ as $\neg x = x \Rightarrow 0$, extending the usual negation in Boolean algebras. Residuated lattices on the real unit interval are exactly the structures defined by left continuous t-norms and their residua (see property (2) above). These (standard) structures do not generate however the whole variety of residuated lattices, since there are equations valid in these structures but not in the variety of residuated lattices. For example, consider the following properties:

Pre-linearity: $(x \Rightarrow y) \vee (y \Rightarrow x) = 1$ $\quad (Lin)$[5]

$\vee$-Definability: $x \vee y = ((x \Rightarrow y) \Rightarrow y) \wedge ((y \Rightarrow x) \Rightarrow x)$ $\quad$ ($\vee$-*Def*)

Distributivity between $\vee$ and $\wedge$ $\quad (Dist)$

These properties are satisfied by residuated lattices on [0,1] (because ([0, 1], max, min, 0, 1) is a distributive lattice and left continuous t-norms and their residua satisfy the above properties (3) and (4)), however, they are not always satisfied in the variety of residuated lattices as it is known and it will be shown in next examples. In these examples we also take into account three more interesting properties, namely:

Divisibility: $x * (x \Rightarrow y) = x \wedge y$ $\quad (Div)$

Pseudo-complementation: $x \wedge \neg x = 0$ $\quad (Pseudo)$

[3] $x > 0$ is a zero-divisor if there exists $y > 0$ such that $x * y = 0$.
[4] Notice that in the literature on t-norms, by a *strict* t-norm it is usually understood a continuous and strictly monotone t-norm, see e.g. [21].
[5] We use the notation "(Lin)" since this is the one used by Ono in [24].

Involution: $\neg\neg x = x$ (Inv)

which respectively correspond in [0, 1] to the properties of the t-norm of being continuous, of having no zero divisors and of having an associated involutive negation.

Example 1. Let $\mathcal{RL}_1 = (L_1, \wedge, \vee, \star, \Rightarrow_\star, 0, 1)$ where $(L_1, \wedge, \vee, 0, 1)$ is the distributive lattice given in Figure 1 and where $\star$ and $\Rightarrow_\star$ are the operations given in Table 1. One can check that $\mathcal{RL}_1$ is a distributive residuated lattice

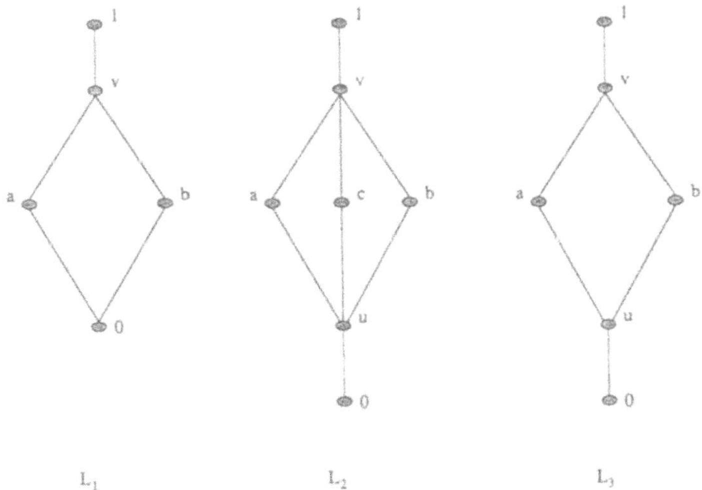

Fig. 1. Lattices L_1, L_2 and L_3.

$\star$	0	a	b	v	1
0	0	0	0	0	0
a	0	a	0	a	a
b	0	0	b	b	b
v	0	a	b	v	v
1	0	a	b	v	1

$\Rightarrow_\star$	0	a	b	v	1
0	1	1	1	1	1
a	b	1	b	1	1
b	a	a	1	1	1
v	0	a	b	1	1
1	0	a	b	v	1

Table 1. $\star$ and $\Rightarrow_\star$ operation on the lattice L_1.

in which the pre-linearity, $\vee$-definability and involution properties are not satisfied since we have

$$(a \Rightarrow_\star b) \vee (b \Rightarrow_\star a) = b \vee a = v \neq 1,$$

$$((a \Rightarrow_\star b) \Rightarrow_\star b) \wedge ((b \Rightarrow_\star a) \Rightarrow_\star a) = 1 \neq v = a \vee c,$$

$$\neg\neg v = 1 \neq v.$$

However, divisibility is satisfied and the corresponding negation $\neg x = x \Rightarrow_\star 0$ is a pseudo-complementation. □

Example 2. Let $\mathcal{RL}_2 = (L_2, \wedge, \vee, \circ, \Rightarrow_\circ, 0, 1)$ where $(L_2, \wedge, \vee, 0, 1)$ is the non-distributive lattice given in Figure 1 and where $\star$ and $\Rightarrow_\star$ are the operations given in Table 2.

$\circ$	0	u	a	b	c	v	1
0	0	0	0	0	0	0	0
u	0	0	0	0	0	0	u
a	0	0	u	0	u	u	a
b	0	0	0	u	u	u	b
c	0	0	u	u	0	u	c
v	0	0	u	u	u	u	v
1	0	u	a	b	c	v	1

$\Rightarrow_\circ$	0	u	a	b	c	v	1
0	1	1	1	1	1	1	1
u	v	1	1	1	1	1	1
a	b	v	1	v	v	1	1
b	a	v	v	1	v	1	1
c	c	v	v	v	1	1	1
v	u	v	v	v	v	1	1
1	0	u	a	b	c	v	1

Table 2. $\circ$ and $\Rightarrow_\circ$ operations on the lattice L_2.

An easy checking shows that these operations make $\mathcal{RL}_2$ a (non-distributive) residuated lattice where pre-linearity, divisibility and pseudo-complementation are not satisfied, since the reader can check that

$$(a \Rightarrow_\circ b) \vee (b \Rightarrow_\circ a) = v \vee v = v \neq 1,$$

$$v \circ (v \Rightarrow_\circ c) = v \circ v = u \neq c = v \wedge c,$$

$$u \wedge \neg u = u \wedge v = u \neq 0.$$

However this time $\vee$-definability and involution are satisfied. □

Example 3. On the same lattice L_2 used in Example 2, let $\bullet$ and $\Rightarrow_\bullet$ be the operations defined in Table 3.

Now, $\mathcal{RL}_3 = (L_2, \wedge, \vee, \bullet, \Rightarrow_\bullet, 0, 1)$ is a non-distributive residuated lattice in which pre-linearity, divisibility and involution are not satisfied, since we have for instance

$$(a \Rightarrow_\bullet b) \vee (b \Rightarrow_\bullet a) = v \vee v = v \neq 1,$$

$\bullet$	0	u	a	b	c	v	1
0	0	0	0	0	0	0	0
u	0	u	u	u	u	u	u
a	0	u	u	u	u	u	a
b	0	u	u	u	u	u	b
c	0	u	u	u	u	u	c
v	0	u	u	u	u	u	v
1	0	u	a	b	c	v	1

$\Rightarrow_\bullet$	0	u	a	b	c	v	1
0	1	1	1	1	1	1	1
u	0	1	1	1	1	1	1
a	0	v	1	v	v	1	1
b	0	v	v	1	v	1	1
c	0	v	v	v	1	1	1
v	0	v	v	v	v	v	1
1	0	u	a	b	c	v	1

Table 3. $\bullet$ and $\Rightarrow_\bullet$ operations on the lattice L_2.

$$v \bullet (v \Rightarrow_\bullet c) = v \bullet v = u \neq c = v \wedge c,$$
$$\neg\neg u = 1 \neq u.$$

Nevertheless, $\vee$-definability and pseudo-complementation are satisfied. □

Example 4.[6] Let L_3 be the distributive lattice given in Figure 1, sublattice of L_2, and let $\diamond$ and $\Rightarrow_\diamond$ be the operations defined in Table 4.

$\diamond$	0	u	a	b	v	1
0	0	0	0	0	0	0
u	0	0	0	0	0	u
a	0	0	a	0	a	a
b	0	0	0	b	b	b
v	0	0	a	b	v	v
1	0	u	a	b	v	1

$\Rightarrow_\diamond$	0	u	a	b	v	1
0	1	1	1	1	1	1
u	v	1	1	1	1	1
a	b	b	1	b	1	1
b	a	a	a	1	1	1
v	u	u	a	b	1	1
1	0	u	a	b	v	1

Table 4. $\diamond$ and $\Rightarrow_\diamond$ operations defined on the lattice L_4.

The structure $\mathcal{RL}_4 = (L_3, \wedge, \vee, \diamond, \Rightarrow_\diamond, 0, 1)$ is a distributive residuated lattice in which pre-linearity, divisibility, pseudo-complementation and $\vee$-definability are not satisfied since, in particular, we have

$$(a \Rightarrow_\diamond b) \vee (b \Rightarrow_\diamond a) = b \vee a = v \neq 1,$$
$$v \diamond (v \Rightarrow_\diamond u) = v \diamond u = 0 \neq u = v \wedge u,$$
$$a \wedge \neg a = a \wedge b = u \neq 0,$$
$$((a \Rightarrow_\diamond b) \Rightarrow_\diamond b) \wedge ((b \Rightarrow_\diamond a) \Rightarrow_\diamond a) = 1 \neq v = a \vee b.$$

In this case involution is however satisfied. □

[6] This example is taken from [24].

Example 5.[7] On the previous lattice L_3, let $\circledast$ and $\Rightarrow_{\circledast}$ be the operations defined in Table 5.

$\circledast$	0	u	a	b	v	1
0	0	0	0	0	0	0
u	0	u	u	u	u	u
a	0	u	a	u	a	a
b	0	u	u	u	u	b
v	0	u	a	u	a	v
1	0	u	a	b	v	1

$\Rightarrow_{\circledast}$	0	u	a	b	v	1
0	1	1	1	1	1	1
u	0	1	1	1	1	1
a	0	b	1	b	1	1
b	0	v	v	1	1	1
v	0	b	v	b	1	1
1	0	u	a	b	v	1

Table 5. $\circledast$ and $\Rightarrow_{\circledast}$ operations defined on the lattice L_3.

These operations make $\mathcal{RL}_5 = (L_3, \wedge, \vee, \circledast, \Rightarrow_{\circledast}, 0, 1)$ a distributive residuated lattice in which pre-linearity, divisibility and involution are not satisfied since, in particular, we have

$$(a \Rightarrow_{\circledast} b) \vee (b \Rightarrow_{\circledast} a) = b \vee v = v \neq 1,$$

$$v \circledast (v \Rightarrow_{\circledast} b) = v * b = u \neq b,$$

$$\neg\neg v = 1 \neq v.$$

On the contrary, pseudo-complementation and $\vee$-definability are satisfied in $\mathcal{RL}_5$. □

Example 6.[8] Finally, consider the operations $*$ and $\Rightarrow_*$ of Table 6 defined once more over the lattice L_3.

$\odot$	0	u	a	b	v	1
0	0	0	0	0	0	0
u	0	0	0	0	0	u
a	0	0	u	0	u	a
b	0	0	0	u	u	b
v	0	0	u	u	u	v
1	0	u	a	b	v	1

$\Rightarrow_{\odot}$	0	u	a	b	v	1
0	1	1	1	1	1	1
u	v	1	1	1	1	1
a	b	v	1	v	1	1
b	a	v	v	1	1	1
v	u	v	v	v	1	1
1	0	u	a	b	v	1

Table 6. $\odot$ and $\Rightarrow_{\odot}$ operations defined on the lattice L_3.

These operations define a residuated lattice $\mathcal{RL}_6 = (L_3, \wedge, \vee, \odot, \Rightarrow_{\odot}, 0, 1)$ in which this time pre-linearity, divisibility and pseudo-complementation are

[7] This example is taken from [20].
[8] This example is taken from [20].

not satisfied, since, in particular, we have

$$(a \Rightarrow_* b) \vee (b \Rightarrow_* a) = v \vee v = v \neq 1,$$

$$v * (v \Rightarrow_* a) = v * v = u \neq b,$$

$$v \wedge \neg v = u \neq 0.$$

However $\mathcal{RL}_6$ satisfies involution and $\vee$-definability. □

As a summary, Table 7 shows which properties are satisfied by the different residuated lattices of the previous examples.

	Pseudo	Inv	Dist	$\vee$-Def	Div	Lin
Example 1	yes	no	yes	no	yes	no
Example 2	no	yes	no	yes	no	no
Example 3	yes	no	no	yes	no	no
Example 4	no	yes	yes	no	no	no
Example 5	yes	no	yes	yes	no	no
Example 6	no	yes	yes	yes	no	no

Table 7. Summary of properties of the Examples.

The examples clearly show that the properties (Lin), ($\vee$-Def) and (Dist), satisfied for left continuous t-norms and their residua, are not universally verified in residuated lattices. Therefore, **RL**, the variety of residuated lattices, is not the variety generated by residuated lattices on the real unit interval defined by (left continuous) t-norms.

Moreover, the above table shows a number of negative results for residuated lattices like, for instance, the facts that distributivity does not imply $\vee$-definability (Examples 1 and 4), distributivity plus divisibility imply neither $\vee$-definability nor pre-linearity (Example 1), etc. These negative results will be used in the proof of the last proposition of the paper.

On contrast, there are indeed some positive relationships among the above properties, which will be also used in the last section of this paper. We collect them in the next lemma.

Lemma 1. *The following implications are valid in the framework of Residuated Lattices:*
(1) divisibility implies distributivity,
(2) pre-linearity implies distributivity and $\vee$-def,
(3) divisibility and Involution imply pre-linearity,
(4) divisibility plus v-definability imply pre-linearity.

Proof. The proofs of (1), (2) and (3) can be found in [18]. The first one is in Proposition 2.6, the second one (referring to distributivity) in lemma 2.4 and the third one in lemma 2.14. With respect to the fact that pre-linearity implies $\vee$-def one previously proves that pre-linearity implies that any algebra is a subdirect product of chains (see [18, theorem 4.8]) and then proving the implication for chains is an easy checking. Finally, here is a proof of (4) based on results of [9]. Using residuation and divisibility it is easy to prove that $((x \Rightarrow y) \Rightarrow x) \Rightarrow y \leq x \Rightarrow y \leq ((x \Rightarrow y) \Rightarrow x) \Rightarrow x$, which implies $(((x \Rightarrow y) \Rightarrow x) \Rightarrow y) \Rightarrow (((x \Rightarrow y) \Rightarrow x) \Rightarrow x) = 1$. On the other hand, by divisibility, $(a * (a \Rightarrow b)) \Rightarrow c = (b * (b \Rightarrow a)) \Rightarrow c$, which implies, using exchange, that $(a \Rightarrow b) \Rightarrow (a \Rightarrow c) \leq (b \Rightarrow a) \Rightarrow (b \Rightarrow c)$. Taking $a = (x \Rightarrow y) \Rightarrow x$, $b = y$ and $c = x$, we have $(y \Rightarrow ((x \Rightarrow y) \Rightarrow x)) \Rightarrow (y \Rightarrow x) = 1$, which is equivalent to $((x \Rightarrow y) \Rightarrow (y \Rightarrow x)) \Rightarrow (y \Rightarrow x) = 1$. Analogously, it is also true that $((y \Rightarrow x) \Rightarrow (x \Rightarrow y)) \Rightarrow (x \Rightarrow y) = 1$. Therefore $(x \Rightarrow y) \vee (y \Rightarrow x) = 1$ and pre-linearity is proved. □.

3 Varieties of t-norm based residuated lattices

As mentioned in the previous section, any residuated lattice on [0, 1] is defined by a left-continuous t-norm and satisfies the pre-linearity condition (*Lin*). Höhle proves in [18, Theorem 4.8, pag.76] that any pre-linear residuated lattice is a subdirect product of chains and this implies both the distributivity of the lattice and the definability of $\vee$ with respect to $\Rightarrow$ and $\wedge$. Thus, it seems reasonable to restrict ourselves to pre-linear residuated lattices as suitable algebraic structures for t-norm based residuated logics. This was done in [5] when defining the Monoidal t-norm based Logic (MTL for short), claimed to be the logic corresponding to left-continuous t-norms. This conjecture has been recently proved by Jenei and Montagna in [19]. In algebraic terms this result corresponds to the following proposition.

Proposition 1. *The variety generated by structures* $([0,1], *, \Rightarrow, \min, \max, 0, 1)$, *where* $*$ *is a left continuous t-norm and* $\Rightarrow$ *is its residuum, is the variety* **MTL** *of pre-linear residuated lattices. That is, an identity is valid in* **MTL** *iff it is valid in [0,1] for any left continuous t-norm and its residuum.*

Girard Monoids (see [18]) are residuated lattices in which the negation is involutive. But again, pre-linearity and distributivity of the lattice operations do not universally hold in the variety **GM** of Girard Monoids as Example 2 shows. Therefore **GM** is not generated by the corresponding class of structures in [0, 1], induced by involutive[9] left continuous t-norms. Actually, to obtain the variety generated by involutive (left continuous) t-norms and their residua we have to restrict ourselves to involutive pre-linear Residuated lattices (or pre-linear Girard monoids). These structures are used in

[9] By involutive t-norm we mean a (left continuous) t-norm $*$ such that $(x \Rightarrow_* 0) \Rightarrow_* 0 = x$.

[5] as semantics for the Involutive Monoidal t-norm based Logic (IMTL for short), trying to capture the logic of involutive, left-continuous t-norms, finally proved in [6]. This is stated algebraically in the following proposition.

Proposition 2. *The variety generated by structures* ([0, 1], $*$, $\Rightarrow$, min, max, 0, 1) *where* $*$ *is an involutive (left continuous) t-norm and* $\Rightarrow$ *is its residuum is the variety* **IMTL** *of involutive pre-linear residuated lattices.*

Some other genuine subvarieties of **MTL** have been considered in [5] and in [6], namely the following subvarieties:

SMTL: obtained by adding the equation (*Pseudo*),
ΠMTL: obtained by adding axioms (*Pseudo*) and (Π1),
WNM: obtained by adding the equation

$$((x * y) \Rightarrow 0) \vee ((x \wedge y) \Rightarrow (x * y)) = 1 \qquad (wnm)$$

NM: obtained by adding the equations (*wnm*) and (*Inv*)

All their corresponding logics, except for ΠMTL, have been proved to be standard complete (complete with respect to the algebras of the variety whose lattice reduct is [0,1]). Algebraically this can be expressed as follows.

Proposition 3. *The variety generated by structures* ([0, 1], $*$, $\Rightarrow$, min, max, 0, 1), *where the left-continuous t-norm* $*$ *and its residuum* $\Rightarrow$ *define an SMTL, WNM or NM algebra, respectively, is the full variety* **SMTL**, **WNM** *or* **NM**, *respectively.*

T-norms defining WNM algebras over the real unit interval are called *weak nilpotent minimum* t-norms and they were introduced in [5]. Given a weak negation function[10] n on [0, 1], the corresponding weak nilpotent minimum is defined as

$$x *_n y = \begin{cases} 0, & \text{if } x \leq n(y) \\ \min(x, y), & \text{otherwise} \end{cases}$$

and its corresponding residuated implication is given by

$$x \Rightarrow_n y = \begin{cases} 1, & \text{if } x \leq y \\ \max(n(x), y), & \text{otherwise} \end{cases}$$

It is straightforward to see that $x \Rightarrow_n 0 = n(x)$, so n is actually the negation corresponding to $*_n$. A particular subclass of these t-norms is the class of those t-norms defining NM-algebras, i.e. those WNM-algebras whose negation n is a strong (involutive) negation function, and called *nilpotent minimum* t-norms, introduced by Fodor in [10]. It is also proved in [5] that all nilpotent minimums in [0,1] are isomorphic, which implies that the variety **NM** is generated by only one NM-chain on [0, 1].

[10] A weak negation function is an order-reversing function $n : [0, 1] \rightarrow [0, 1]$ satisfying $n(n(x)) \geq x$ for all $x \in [0, 1]$ (See [8]).

With respect to the logic corresponding to ΠMTL, it has only been proved in [6] to be rational complete, in the following algebraic sense (t-norms defining a ΠMTL-algebra over [0,1] are called cancellative).

Proposition 4. *The variety generated by structures* $([0,1] \cap \mathbf{Q}, *, \Rightarrow, \min, \max, 0, 1)$ *where* $*$ *is a cancellative left continuous t-norm (on* $[0,1] \cap \mathbf{Q}$*) and* $\Rightarrow$ *its residuum, is the full variety* ΠMTL.

The question whether the variety generated by analogous structures on the real unit interval [0, 1] is the full variety ΠMTL is so far an open problem, obviously related to the standard completeness of the corresponding logic.

Finally, let us consider possibly the most prominent subvariety of **RL**, the variety generated by residuated lattices on the real unit interval defined by *continuous* t-norms and their residua. As already mentioned, the characteristic property of these structures is the satisfaction of the divisibility condition (see item (6) of section 2), which is equivalent in the setting of residuated lattices to the equation (See, for example [12]),

$$x * (x \Rightarrow y) = x \wedge y. \qquad (Div)$$

Taking this property in mind, Hájek defines in [15] the so-called Basic Fuzzy logic (BL for short), which has as corresponding algebraic structures prelinear residuated lattices satisfying equation (Div), called BL-algebras. In [15] it was conjectured that Basic Logic BL was the logic of continuous t-norms and this was indeed proved in [1], which amounts in algebraic terms to the following proposition.

Proposition 5. *The variety generated by structures* $([0,1], *, \Rightarrow, \min, \max, 0, 1)$ *where* $*$ *is a continuous t-norm and* $\Rightarrow$ *is its residuum, is the variety* **BL** *of BL-algebras.*

Let us briefly recall the varieties generated by the residuated chains defined on the real unit interval by each one of the three basic continuous t-norms and their residua: the variety **Ł** of MV-algebras (related to the Łukasiewicz t-norm), the variety **Π** of Product algebras (related to the product t-norm), and the variety **G** of Gödel algebras (related to the minimum t-norm) respectively. In fact, these varieties were (chronologically) the first ones which were studied from the logical point of view. The original completeness proofs of their corresponding logics can be found in [26], [17] and [4] respectively. Moreover, Hájek provides in [15] the equational characterization of these varieties as subvarieties of **BL**. Namely, the BL-subvariety of MV-algebras is characterized by the equation

$$(x \Rightarrow 0) \Rightarrow 0 = x \qquad (Inv)$$

the BL-subvariety of Product algebras by the equations [11]

[11] Cintula proves [2] that these two axioms can be simplified to a single axiom with only two propositional variables, and moreover there is not an equivalent axiom with just one variable.

$$\neg\neg z \Rightarrow (((x * z) \Rightarrow (y * z)) \Rightarrow (x \Rightarrow y)) = 1 \tag{$\Pi 1$}$$
$$x \wedge \neg x = 0 \tag{$Pseudo$}$$

and the BL-subvariety of Gödel algebras by the equation of contraction for $*$

$$x * x = x \tag{Con}$$

Another interesting subvariety of **BL**, that contains Π and **G**, is the subvariety **SBL** introduced in [7]. The **SBL** variety is defined by the equations of **BL** plus the equation

$$((x * y) \Rightarrow 0) \Rightarrow ((x \Rightarrow 0) \vee (y \Rightarrow 0)) = 1$$

or equivalently plus the above equation (*Pseudo*). Hence, SBL-algebras can also be called pseudo-complemented BL-algebras. In [1] it is proved that this variety is the variety generated by continuous t-norms without zero divisors. Finally let us also mention that in [3] the lattice of subvarieties of **BL** whose linearly ordered algebras are only single-component[12] BL-chains is fully described and equationally characterized.

A graph of all the logical systems corresponding to varieties described in this section is depicted in Figure 2. We could summarize this section stressing that t-norm based varieties of residuated lattices are all contained in the variety **MTL** of pre-linear residuated lattices.

4 Monoidal logic and t-norm based residuated logics

In this section we turn our attention to the logical systems corresponding to the algebraic varieties described in the previous section. Take into account that we will talk about completeness of these systems in three different ways: (1) completeness with respect to a whole variety (general completeness), (2) completeness with respect to the class of chains of a variety (chain completeness), and finally (3) completeness with respect to the class of [0,1] structures of a given variety (standard completeness).

First of all we consider Monoidal Logic (ML for short), introduced by Höhle in [18]. The language of propositional Monoidal logic is built in the usual way from a denumerable set of propositional symbols using the connectives $\vee, \wedge, \&, \rightarrow$ and the truth-constant 0. As definable connectives it has the negation and the double implication:

$$\neg\varphi \text{ is } \varphi \rightarrow 0$$
$$\varphi \leftrightarrow \psi \text{ is } (\varphi \rightarrow \psi)\&(\psi \rightarrow \varphi)$$

[12] By single-component BL-chains we mean BL-chains that are either MV, Product or Gödel chains.

The corresponding truth functions for these connectives are given by max, min, a t-norm and its corresponding residuum, negation and equivalence respectively. An axiomatic system of Monoidal Logic can be found in [18] and also in [12][13]. Höhle proved that this logic is complete w.r.t. the variety of Residuated Lattices. It is easy to prove that this kind of completeness extends to axiomatic extensions of ML. Thus, there is a one-to-one correspondence between axiomatic extensions of Monoidal logic and subvarieties of Residuated lattices, and consequently all logical systems presented in this section are in fact complete w.r.t. the corresponding subvariety of Residuated lattices. Moreover, if a variety satisfies the pre-linearity equation, the algebras of the variety are subdirect product of linearly ordered ones, and thus, in all extensions of Monoidal Logic where the formula

$$(\varphi \rightarrow \psi) \vee (\psi \rightarrow \varphi) \qquad (Lin)$$

is a theorem, we have chain completeness. For standard completeness the problem is more difficult, one needs to prove it case by case since standard completeness is not preserved by axiomatic extensions, as we shall show in Example 6 at the end of this section.

It is also well-known that the logical system corresponding to Girard Monoids is the propositional exponential-free multiplicative fragment of Girard's Linear Logic [11] assuming that the monoidal unit coincides with the greatest element [11], the so-called affine Multiplicative Additive Linear Logic, aMALL for short (see for example [23, 28]) . This logic is the axiomatic extension of Monoidal Logic with the axiom

$$\neg\neg\varphi \rightarrow \varphi \qquad (Inv)$$

requiring the negation to be involutive (see [12]). Nevertheless, as we explained in Section 2, these logics (Monoidal and aMALL) are neither chain nor standard complete.

The extension of Monoidal logic with the above pre-linearity axiom (*Lin*) is the so-called Monoidal t-norm based Logic (MTL for short) and it was defined and studied in [5]. In MTL, $\vee$ becomes definable (it is not a primitive connective any longer) as

$$\varphi \vee \psi \quad \text{is} \quad ((\varphi \rightarrow \psi) \rightarrow \psi) \wedge ((\psi \rightarrow \varphi) \rightarrow \varphi).$$

Axioms of MTL logic are:

[13] In this book there are axiomatic systems corresponding to many of the logics presented in this paper in a unified point of view, visualizing when one logic is an axiomatic extension of another. However, we want to warn the reader since there is an axiom missing in the axiomatization of Monoidal Logic. See [13] for a correction.

(A1) $(\varphi \to \psi) \to ((\psi \to \chi) \to (\varphi \to \chi))$
(A2) $(\varphi \& \psi) \to \varphi$
(A3) $(\varphi \& \psi) \to (\psi \& \varphi)$
(A4) $(\varphi \wedge \psi) \to \varphi$
(A5) $(\varphi \wedge \psi) \to (\psi \wedge \varphi)$
(A6) $(\varphi \& (\varphi \to \psi)) \to (\varphi \wedge \psi)$
(A7a) $(\varphi \to (\psi \to \chi)) \to ((\varphi \& \psi) \to \chi)$
(A7b) $((\varphi \& \psi) \to \chi) \to (\varphi \to (\psi \to \chi))$
(A8) $((\varphi \to \psi) \to \chi) \to (((\psi \to \varphi) \to \chi) \to \chi)$
(A9) $\bar{0} \to \varphi$

The rule of inference of MTL is *modus ponens*. Notice that this axiomatic system includes axiom (A8) instead of the pre-linearity axiom (*Lin*), but it is not difficult to prove that (A8) and (*Lin*) are indeed equivalent in the frame of Monoidal logic. Therefore, any axiomatic extension of MTL is chain complete. Furthermore, standard completeness for MTL has been recently proved by Jenei and Montagna in [19]. Therefore, since MTL-structures in the unit interval [0, 1] are defined by left-continuous t-norms, MTL can be properly called the logic of left continuous t-norm and their residua, hence MTL is the most general residuated fuzzy logic related to t-norms (recall that a t-norm has residuum iff it is left continuous).

The axiomatic extensions of MTL corresponding to the varieties studied in the previous section (i.e. IMTL, SMTL, ΠMTL, WNM, NM) have been studied in [6, 5] where issues of standard and finite strong standard completeness are analyzed. The lattice of these logics, together with the well-known extensions of BL (Product, Gödel, Łukasiewicz and SBL) is depicted in Figure 2 in the framework of Monoidal Logic. The axioms needed for obtaining these axiomatic extensions of MTL are the obvious logical translations of the equations defining the corresponding subvarieties of **MTL** and thus we shall not repeat them here, instead we think it is of worth stressing some interesting facts about these logics which we list below:

- The logics IMTL, SMTL and ΠMTL are to MTL what Łukasiewicz, SBL and Product logics are to BL, that is, they are obtained respectively as extensions of MTL by adding the same characteristic axioms of Łukasiewicz -(*Inv*)-, SBL -(*Pseudo*)- and Product -(*Pseudo*), (Π1)- as extensions of BL. All these systems are proved to be different (see [16] for ΠMTL and Product, [5] for IMTL and Łukasiewicz and [6] for SMTL and SBL), however when we add the contraction axiom (*Con*) to MTL the system obtained collapses with Gödel logic [16].
- An alternative axiomatization of IMTL not using residuation can be found in [14], where they use the so-called rotational property, $((\varphi \& \psi) \to \chi) \leftrightarrow ((\psi \& \neg \chi) \to \neg \varphi)$, and where the implication connective is taken as definable from the conjunction and negation, namely $(\varphi \to \psi)$ is $\neg(\varphi \& \neg \psi))$, definition which is also true in classical or Łukasiewicz logic.

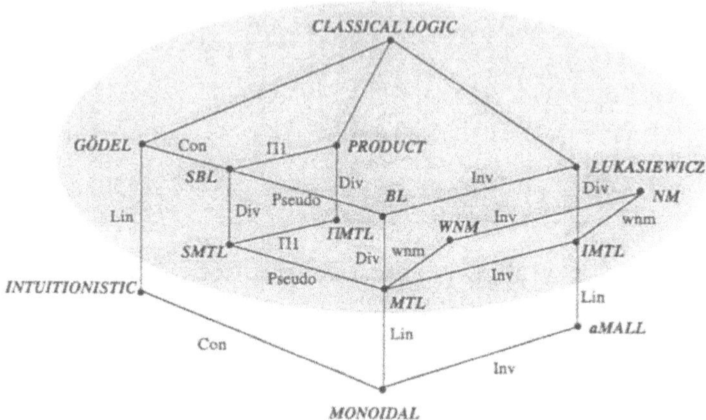

Fig. 2. Graph of main residuated many-valued logics with the shadowed part containing t-norm based logics.

- IMTL and Łukasiewicz logics are quite close. In both the negation is involutive and the implication can be defined as in the classical case from negation and conjunction as $\neg(\varphi \& \neg\psi)$. One of the relevant formulas that is a theorem for Łukasiewicz but not for IMTL is the equivalence $((\varphi \to \psi) \to \psi) \leftrightarrow ((\psi \to \varphi) \to \varphi)$, which gives us the usual definition of $\vee$ in Łukasiewicz logic as a simplification of the definition of $\vee$ in BL, MTL and IMTL.
- With respect to completeness results, all these logics, except ΠMTL, are proved to be finite strong standard complete and thus standard complete, i.e. complete with respect to the corresponding class of left continuous t-norm based [0,1] structures. Only NM, besides Łukasiewicz, Product and Gödel logics, is complete with respect to a single [0,1] structure. On the other hand, ΠMTL is only proved to be finite strong *rational* complete (see [6] for full details), in the sense of being complete with respect to ΠMTL-algebras whose lattice reduct is the rational interval $\mathbf{Q} \cap [0,1]$. Finite strong standard completeness and even standard completeness of ΠMTL remains as an open problem, as already mentioned in the previous section.
- Finally, let us remark that standard completeness does not universally extend to axiomatic extensions. Besides the very well-known cases of extensions of BL corresponding to varieties generated by Chang's MV chains or any finite BL chain, which are not standard complete, here

there is one example that could be interesting by itself. Let Π' be the extension of BL with the axiom

$$((\varphi \wedge \neg\varphi) \to 0) \wedge ((\varphi \to (\varphi \& \varphi)) \to ((\varphi \to 0) \vee \varphi)) \qquad (\Pi')$$

It is easy to check that BL-chains for which this axiom is a 1-tautology are SBL-chains which have no idempotent element different from 0 and 1. As a consequence we have: (i) the standard Product algebra is the only (up to isomorphism) BL-algebra on [0, 1] satisfying (Π'); but (ii) Product algebras are not the only BL-algebras satisfying (Π'). In fact, besides product chains, any subalgebra of an ordinal sum of product chains obtained by withdrawing idempotent elements different from 0 and 1 also satisfy (Π'), and these are not Product algebras. Therefore the logic Π' is not standard complete (otherwise it would be equivalent to Product logic) although it is complete with respect to the above class of BL-chains.

5 Relation to Ono's hierarchy of substructural logics

There is a strong connection between MTL and Ono's family of substructural logics, described in [24], which are different extensions of the Full Lambek calculus FL (see [22] for an introduction of Lambek calculus). In fact, in [24] Höhle's Monoidal logic is shown to be equivalent to the extension of FL with *exchange*, i.e. $(\varphi \to (\psi \to \chi)) \to (\psi \to (\varphi \to \chi))$, and *weakening*, i.e. $\varphi \to (\psi \to \varphi)$, denoted FL_{ew} in [24], whose algebraic semantics is the variety of residuated lattices. Hence, MTL is equivalent to the extension of FL_{ew} with the pre-linearity axiom (*Lin*), denoted in [24] as FL_{ew}[Lin]. Indeed FL_{ew} is considered in [24] since it is the logic whose algebraic counterpart is the variety of residuated lattices. Moreover, Ono studies different axiomatic extensions of FL_{ew} which can be successively obtained by the addition of some of the following axioms:

- Weak Contraction [WCon]: $(\varphi \to \neg\varphi) \to \neg\varphi$
- Contraction [Con]: $\varphi \to \varphi \& \varphi$
- Involution[14] [Inv]: $\neg\neg\varphi \to \varphi$
- Distributivity [Dist]: $\varphi \wedge (\psi \vee \chi) \to (\varphi \wedge \psi) \vee (\varphi \wedge \chi)$
- Pre-linearity [Lin]: $(\varphi \to \psi) \vee (\psi \to \varphi)$

Based on these properties Ono provides the full Hasse diagram of the upper sub-semilattice of the corresponding logics over FL_{ew}, represented in Figure 3, where we have added the equivalent names used in this paper on the right hand side. Moreover Ono proves that all the inclusions showed in the diagram are proper [24, Lemmas 1.8, 1.9] giving some examples and proving mainly that:

[14] Ono calls this property *double negation* and uses the abbreviation DN instead of Inv.

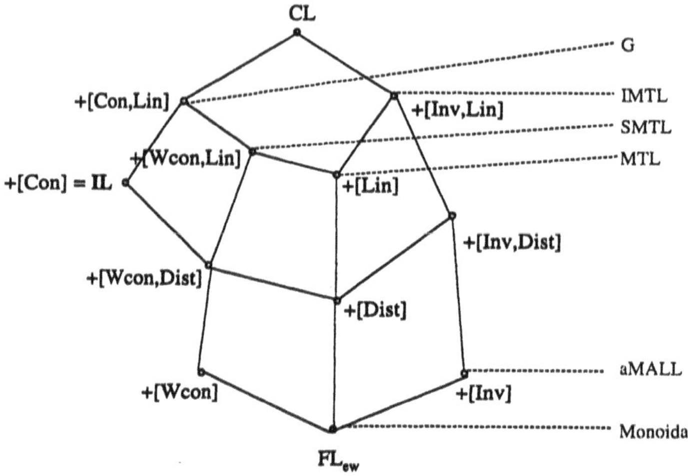

Fig. 3. Ono's diagram with the equivalent names used in this paper.

- FL_{ew}[WCon, Inv] (the extension of FL_{ew} with properties (*WCon*) and (*Inv*) in Ono's notation) is the Classical Logic (CL),
- FL_{ew}[Lin] proves the Distributivity axiom (*Dist*)[15], hence FL_{ew}[Lin, Dist] = FL_{ew}[Lin].

To understand the equivalences given in the diagram and to summarize them it is interesting to remark first that in the framework of Monoidal Logic [WCon] and (*Pseudo*) are equivalent (see [6] for a formal proof of this equivalence) and thus we will use only (*Pseudo*) for obvious simplification of the notation. Moreover the following equivalences hold:

- FL_{ew} is equivalent to Höhle's Monoidal Logic and its algebraic counterpart is the variety **RL** of residuated lattices.
- FL_{ew}[Lin] is equivalent to MTL and the Distributivity axiom (*Dist*) is provable in this logic, as already mentioned above.
- Obviously FL_{ew}[Pseudo,Lin] and FL_{ew}[Inv, Lin] coincide with SMTL and IMTL respectively. Moreover FL_{ew}[Inv] coincides with aMALL, FL_{ew}[Con] with Intuitionistic Logic IL, and thus FL_{ew}[Con, Lin] coincides with Gödel logic G as well.

The complementarity of this diagram with respect to the one presented in the previous section (Figure 2) is clear. In the following we complete it by taking into account results by Ono and results previously described in

[15] This result was already given by Höhle in [18]

this paper. Namely we add $\vee$-definability and divisibility conditions, ($\vee$-*def*) and (*Div*) respectively, into the picture and prove that all the inclusions are proper. Actually, from Lemma 1, we obtain the following equalities between

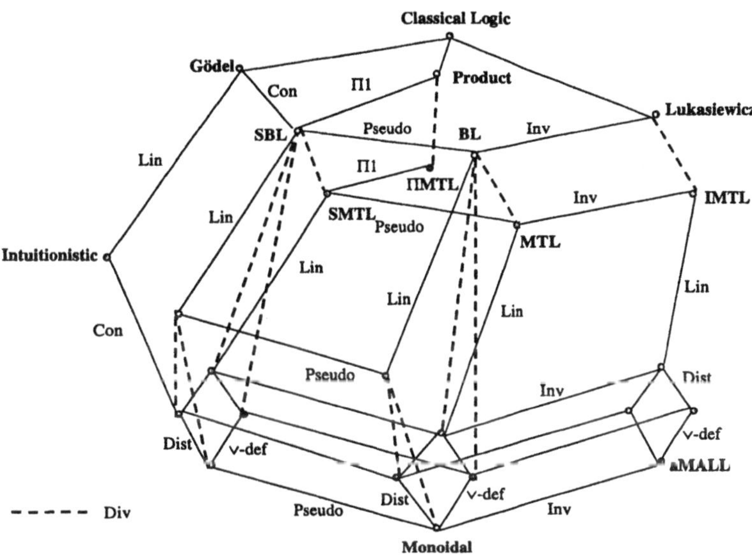

Fig. 4. Final diagram

FL_{ew} extensions:

- FL_{ew}[Dist, Div] = FL_{ew}[Div],
- FL_{ew}[Inv, Div] = Łukasiewicz,
- FL_{ew}[Div, $\vee$-def] = FL_{ew}[Div, Lin] = BL, and
- FL_{ew}[Pseudo, Div, $\vee$-def] = FL_{ew}[Pseudo, Div, Lin] = SBL.

Therefore the Hasse diagram is the one depicted in Fig. 4.

Proposition 6. *All inclusions shown in Figure 4 are proper.*

Proof. The proof is given by cases and using the one-to-one correspondence between logics and their counterpart as algebraic varieties. In the following we adopt Ono's notation and we write **RL**[...] to denote subvarieties of **RL**.

- It is clear that the varieties of Product and Gödel algebras are contained in **RL** but not in the variety of Girard's monoids **GM**. It is also true that the varieties of Product and Gödel algebras are contained in **RL**[$\vee$-def], **RL**[Dist], **RL**[Div] **RL**[$\vee$-def,Dist], **MTL** and **BL**, but they are not contained in the corresponding subvarieties obtained by adding involution. Thus all these inclusions are proper.

- The variety of MV-algebras is contained in **RL** and not in **RL**[Pseudo]. MV-algebras are also contained in **RL**[∨-def], **RL**[Dist], **RL**[Div], **RL**[∨-def,Dist], **MTL** and **BL**, but they are not contained in the corresponding subvarieties defined by adding pseudo-complementation.
- The algebra defined in Example 1 belongs to the variety **RL**[Pseudo] and not to **RL**[Pseudo,∨-def]. The algebra defined in Example 1 is also contained in **RL**, **RL**[Dist], **RL**[Div], **RL**[Pseudo,Dist] and **RL**[Pseudo,Div] and not in the corresponding subvarieties obtained by adding ∨-definability. Moreover, the algebra defined in Example 1 belongs to both **RL**[Div] and **RL**[Pseudo,Div] and not in their subvarieties obtained by adding pre-linearity (**BL** and **SBL** respectively).
- The algebra defined in Example 3 belongs to the variety **RL**[Pseudo] and not to **RL**[Pseudo,Dist]. It also belongs to **RL**, **RL**[∨-def], **RL**[Pseudo,∨-def] but not to the varieties obtained by adding distributivity.
- The algebra defined in Example 2 belongs to the variety **RL**[Inv] and not to **RL**[Inv,Dist]. The algebra defined in Example 3 is also contained in **RL**, **RL**[∨-def] and **RL**[Inv,∨-def] but not in the corresponding varieties obtained by adding distributivity.
- The algebra defined in Example 4 belongs to the variety **RL**[Inv] and not to **RL**[Inv,∨-def]. It also belongs to **RL**, **RL**[Dist] and **RL**[Inv,Dist] and neither to the varieties obtained by adding ∨-definability nor to the varieties adding divisibility.
- The algebra defined in Example 5 belongs both to the varieties **RL**[Pseudo, Dist,∨-def] and **RL**[Dist,∨-def] but not to the corresponding varieties obtained by adding pre-linearity, i.e. **SMTL** and **MTL** respectively. The algebra defined in Example 6 belongs to both the variety **RL**[Inv, Dist,∨-def] and **RL**[Dist,∨-def] but not to the correponding varieties obtained by adding pre-linearity, i.e. **IMTL** and **MTL** respectively.

Finally, we want to stress that we are still far away not only from a complete description of the lattice of subvarieties of **RL** corresponding to axiomatic extensions of Monoidal Logic but also from our goal of having the complete graph of the t-norm based logics (a subclass of the axiomatic extensions of the MTL). This paper is only a first step towards that direction and further and deeper work is needed to complete these hierarchies.

Acknowledgements

The authors are thankful to Ewa Orlowska for pointing them out the relationship between t-norm based logics and extensions of Full Lambek Calculus, and to Petr Hájek and Hiroakira Ono for their helpful remarks. The authors have been partially supported by the Spanish CYCIT project *LOGFAC*, TIC-2001-1577-C03-01.

References

1. Cignoli, R., Esteva, F., Godo, L., Torrens, A.: Basic Fuzzy Logic is the logic of continuous t-norms and their residua. Soft Computing 4 (2000) 106–112
2. Cintula, P.: About axiomatic system of the Product Fuzzy Logic. Soft Computing 5 (2001) 243–244
3. Di Nola, A., Esteva, F., Garcia, P., Godo, L. and Sessa, S.: Subvarieties of BL-algebras generated by single-component chains. Archive for Mathematical Logic, to appear
4. Dummett, M.: A propositional calculus with denumerable matrix. Journal of Symbolic Logic 24 (1959) 97–106
5. Esteva, F. and Godo, L.: Monoidal t-norm based Logic: Towards a logic for left continuous t-norms. Fuzzy Sets and Systems vol. 124, 3 (2001) 271–288
6. Esteva, F., Gispert, J., Godo, L. and Montagna, F.: On the standard and Rational Completeness of some Axiomatic extensions of Monoidal t-norm Based Logic. To appear in Studia Logica
7. Esteva, F., Godo, L., Hájek, P. and Navara, M.: Residuated fuzzy logics with an involutive negation. Archive for Mathematical Logic 39 (2000) 103–124
8. Esteva, F., Trillas, E. and Domingo, X.: Weak and strong negation functions for Fuzzy Set Theory. Proc. 11th Int. Symposium on Multiple-valued Logic ISMVL'81 (1981) 23–26
9. Ferreirim, I.M.: On varieties and quasivarieties of Hoops and their reducts. Phd. Thesis. University of Illinois at Chicago (1992)
10. Fodor, J.: Nilpotent minimum and related connectives for fuzzy logic. Proc. of FUZZ–IEEE'95 (1995) 2077–2082
11. Girard, J.Y.: Linear Logic. Theoretical Computer Science, 50 (1987) 1–102
12. Gottwald, S.: A Treatise on Many-valued Logics. Studies in Logic and Computation. Research Studies Press, Baldock (2001)
13. Gottwald, S., García-Cerdaña, A., Bou, F.: Axiomatizing Monoidal Logic – a Correction. Submmited
14. Gottwald, S. and Jenei, S.: On a new axiomatization for Involutive Monoidal t-norm based Logic. Fuzzy Sets and Systems, vol. 124, 3 (2001) 303–308
15. Hájek, P.: Metamathematics of Fuzzy Logic, Trends in Logic, vol. 4 Kluwer (1998)
16. Hájek, P.: Observations on the Monoidal t-norm Logic, Fuzzy Sets and Systems (to appear)
17. Hájek, P., Godo, L., Esteva, F.: A complete many-valued logic with product conjunction, Archive for Mathematical Logic 35 (1996) 191–208
18. Höhle, U.: Commutative, residuated l-monoids. In: Höhle, U. and Klement, E.P. eds., Non-Classical Logics and Their Applications to Fuzzy Subsets, Kluwer Acad. Publ., Dordrecht (1995) 53–106
19. Jenei, S. and Montagna, F.: A proof of standard completeness for Esteva and Godo's logic MTL. Studia Logica, to appear
20. Jipsen, P.: Some lists of finite structures at http://atlas.math.vanderbilt.edu/~pjipsen/gap/
21. Klement, E.P., Mesiar, R., Pap, E.: Triangular norms, Kluwer (2000)
22. Lambek, J.: Logic without structural rules (Another look at cut elimination). In: Schroeder-Heister, P. and Dosen, K. eds., Substructural Logics, Studies in Logic and Computation 2, Oxford Sciences Pub. (1993) 179–206

23. Okada, M.O. and Terui, K.: The finite model property for various fragments of intuitionistic linear logic Journal of Symbolic Logic 64 (2) (1999) 790–802
24. Ono, H.: Logic without contraction rule and residuated lattices I. To appear in the book for the Festschrift of Prof. R.K. Meyer, edited by E. Mares
25. Pavelka, J.: On Fuzzy Logic I, II, III. Z.Math. Logik Grundl. Math. 25 (1979) 45–52, 119–134, 447–464
26. Rose, P.A. and Rosser, J.B.: Fragments of many-valued statement calculi, Trans. A.M.S. 87 (1958) 1–53
27. Schweizer, B. and Sklar, A.: Probabilistic metric spaces, North Holland, Amsterdam (1983)
28. Troelstra, A.S.: Lectures in Linear Logic, CSLI Lecture Notes, vol. 29, Center for the Study of Language and Information, Stanford (1991)
29. Zadeh, L.A.: Fuzzy Logic, IEEE Comput. 1, 83 (1988)

Chapter 12
A Development of Set Theory in Fuzzy Logic

Petr Hájek, Zuzana Haniková

Institute of Computer Science
182 07 Praha 8
Czech Republic
hajek@cs.cas.cz, zuzana@cs.cas.cz

Abstract. This paper presents an axiomatic set theory FST ('Fuzzy Set Theory'), as a first-order theory within the framework of fuzzy logic in the style of [4]. In the classical ZFC, we use a construction similar to that of a Boolean-valued universe—over an algebra of truth values of the logic we use—to show the nontriviality of FST. We give the axioms of FST. Finally we show that FST interprets ZF.

1 Introduction

If anything comes to people's minds on the term 'fuzzy set theory' being used, it usually is the theory (or theories) handling fuzzy sets as real-valued functions on a fixed universe within the classical set-theoretic universe. However, as "fuzzy" or "many-valued" logic has been evolving into a formal axiomatic theory, a few exceptions from this general expectation have emerged: these are formal "set" theories *within* the respective many-valued logics, i.e., theories whose underlying logic is governed by a many-valued semantics.

We rely especially on those works which develop a theory in the language and style of the classical Zermelo-Fraenkel set theory (ZF). We have been inspired by a series of papers developing a theory generalizing ZF in a formally weaker logic—intuitionistic ([7], [2]) and later its strengthening commonly referred to as Gödel logic ([9], [10], [11]); some results and proofs carry over to our system. For an important example, the axiom of foundation, together with a very weak fragment of ZF, implies the law of the excluded middle, which yields the full classical logic (both in Gödel logic and in the logic we use in this paper), and thus the theory developed becomes crisp. For this reason we start with building a non-crisp universe in which we verify our axioms.

However, all these papers fall short in tackling one of the distinguishing traits of many-valued logic, which is the general non-idempotence of the conjunction (conjunction is idempotent in Gödel logic). The non-idempotence of conjunction affects the resulting theory considerably (cf. [3]); in coping with some of the difficulties we appreciated an elegant solution found in [8].

The author works over the so-called phase spaces as algebras of truth values and builds, using an analogy of the construction of a Boolean-valued universe (over a phase-space instead of a Boolean algebra), a class, with class operations evaluating formulas in the language $\{\in, \subseteq, =\}$, in which he verifies the chosen axioms of his set theory. Having observed that the standard Łukasiewicz algebra enriched with the Δ operator is a particular phase space, we have studied this paper thinking of a more general approach employing (linearly ordered) BL-algebras with Δ; this should form a common generalization of the approach of [9] and [10] and that of [8].

This paper is an extension of [6]; it brings in a simplified definition of the initial universe, and an inner model of ZF in FST. Regarding the nature of this paper we omit most of the logical background, which is to be found mainly in [4], as well as some technical details and some proofs.

The authors acknowledge gratefully a partial support from the grant IAA 103 0004, Grant Agency of the Academy of Sciences, Czech Republic.

2 Prerequisites

Our logic is a variant of the basic fuzzy predicate logic as developed in [4]. We present its formal definition and comment on its semantics.

Definition 1. BL$\forall\Delta$ *is a formal logical system with basic connectives* $\&$, $\to$, $\overline{0}$ *and* Δ, *defined connectives* $\neg$, $\wedge$, $\vee$ *and* $\equiv$*—where* $\neg\varphi$ *is* $\varphi \to \overline{0}$, $\varphi \wedge \psi$ *is* $\varphi\&(\varphi \to \psi)$, $\varphi \vee \psi$ *is* $((\varphi \to \psi) \to \psi) \wedge ((\psi \to \varphi) \to \varphi)$, *and* $\varphi \equiv \psi$ *is* $(\varphi \to \psi)\&(\psi \to \varphi)$*—and with two quantifiers* $\forall$ *and* $\exists$. *The axioms are as follows:*

(A1) $(\varphi \to \psi) \to ((\psi \to \chi) \to (\varphi \to \chi))$

(A2) $(\varphi\&\psi) \to \varphi$

(A3) $(\varphi\&\psi) \to (\psi\&\varphi)$

(A4) $(\varphi\&(\varphi \to \psi)) \to (\psi\&(\psi \to \varphi))$

(A5a) $(\varphi \to (\psi \to \chi)) \to ((\varphi\&\psi) \to \chi)$

(A5b) $((\varphi\&\psi) \to \chi) \to (\varphi \to (\psi \to \chi))$

(A6) $((\varphi \to \psi) \to \chi) \to (((\psi \to \varphi) \to \chi) \to \chi)$

(A7) $\overline{0} \to \varphi$

(Δ1) $\Delta\varphi \vee \neg\Delta\varphi$

(Δ2) $\Delta(\varphi \vee \psi) \to (\Delta\varphi \vee \Delta\psi)$

(Δ3) $\Delta\varphi \to \varphi$

(Δ4) $\Delta\varphi \to \Delta\Delta\varphi$

(Δ5) $\Delta(\varphi \to \psi) \to (\Delta\varphi \to \Delta\psi)$

($\forall$1) $\forall x\varphi(x) \to \varphi(t)$ (t substitutable for x in φ)

($\exists$1) $\varphi(t) \to \exists x\varphi(x)$ (t substitutable for x in φ)

($\forall$2) $\forall x(\chi \to \varphi) \to (\chi \to \forall x\varphi)$ (x not free in χ)

($\exists$2) $\forall x(\varphi \to \chi) \to (\exists x\varphi \to \chi)$ (x not free in χ)

($\forall$3) $\forall x(\varphi \vee \chi) \to (\forall x\varphi \vee \chi)$ (x not free in χ).

The deduction rules of BL$\forall\Delta$ *are modus ponens, generalization, and* $\{\varphi/\Delta\varphi\}$.

The standard semantics is many-valued with the real interval $[0,1]$ as the set of truth values; continuous t-norms are possible truth functions of the conjunction & and, given a continuous t-norm, its residuum is taken to be the corresponding truth function of the implication $\rightarrow$. Note that for any such choice the truth function of $\wedge$ is the minimum and of $\vee$ is the maximum. Recall the three important continuous t-norms and their residua (continuous t-norms are denoted with an $*$, their residua with a $\Rightarrow$):

Łukasiewicz: $x * y = \min(x, y)$,
Gödel: $x * y = \min(x, y)$,
product: $x * y = x - y$

For each continuous t-norm $x \Rightarrow y = 1$ iff $x \leq y$; otherwise:

Łukasiewicz: $x \Rightarrow y = 1 - x + y$,
Gödel: $x \Rightarrow y = y$,
product: $x \Rightarrow y = y/x$.

The semantics of Δ over $[0,1]$ is as follows: $\Delta(1) = 1, \Delta(x) = 0$ for $x < 1$.

The *general* semantics of BL is given by BL-algebras (BLΔ-algebras). A BL-algebra is a residuated lattice $\mathbf{L} = (L, \cap, \cup, *, \Rightarrow, 0, 1)$ satisfying the following for each x, y :

$x \cap y = x * (x \Rightarrow y)$ (divisibility)
$(x \Rightarrow y) \cup (y \Rightarrow x) = 1.$ (prelinearity)

A BLΔ-algebra is an expansion of a BL algebra with a unary operation Δ satisfying identities corresponding to (Δ1)–(Δ5). For completeness theorem of the propositional logic BL having the axioms (A1)–(A7) with respect to BL-algebras (and BLΔ having the axioms (A1)–(A7), (Δ1)–(Δ5)) see [4].

A *schematic extension* of the logic is given by adding schemata of axioms (e.g. for Łukasiewicz logic one adds the double negation axiom $\neg\neg\varphi \rightarrow \varphi$).

The semantics of the predicate logic BL$\forall$ and BL$\forall\Delta$ is given by $\mathbf{L}$-interpretations of the predicate language, $\mathbf{L}$ being a BL-algebra (BLΔ-algebra). Such an interpretation $\mathbf{M} = (M, (r_P)_{P\,\text{predicate}}(m_c)_{c\,\text{constant}})$ has a domain $M \neq \emptyset$, for each n-ary predicate P a $\mathbf{L}$-fuzzy n-ary relation $r_P : M^n \rightarrow \mathbf{L}$ and for each constant c an element $m_c \in M$. A Tarski-style definition of the value $\|\varphi\|^{\mathbf{L}}_{\mathbf{M},v}$ of a formula φ given by the interpretation $\mathbf{M}$, algebra $\mathbf{L}$ and evaluation v of variables by elements of M offers itself; in particular $\|(\forall x)\varphi\|^{\mathbf{L}}_{\mathbf{M},v} = \inf_{v'} \|\varphi\|^{\mathbf{L}}_{\mathbf{M},v'}$ where v' runs over all evaluations identical with v except possibly for the value of x. Since $\mathbf{L}$ need not be complete lattice, this value may be undefined; the $\mathbf{L}$-interpretation $\mathbf{M}$ is *safe* if $\|\varphi\|^{\mathbf{L}}_{\mathbf{M},v}$ is defined for each φ and v.

Definition 2. *Let C be an arbitrary schematic extension of* BL$\forall\Delta$. *A C-algebra is a BLΔ-algebra L in which all the axioms of C are L-tautologies.*

Theorem 1. (Strong completeness) *Let C be a schematic extension of* BL$\forall\Delta$, *let T be a theory over C, and let φ be a formula of the language of T. Then T proves φ iff φ holds in any safe model M of T over any linearly ordered C-algebra.*

3 The initial universe

Consider the classical ZFC. Fix $\mathcal{C}$ as a schematic extension of BL$\forall\Delta$, and fix a constant L for an arbitrary linearly ordered complete $\mathcal{C}$-algebra; write $L = (L, *, \Rightarrow, \wedge, \vee, 0, 1)$ (as usual, L denotes both the algebra and its support). In ZFC let us make the following construction: in analogy to the construction of a Boolean-valued universe over a complete Boolean algebra, we build the class V^L by ordinal induction. Define $L^+ = L - \{0\}$.

$$V_0^L = \{\emptyset\}$$

$$V_{\alpha+1}^L = \{f : \mathrm{Fnc}(f) \,\&\, D(f) \subseteq V_\alpha^L \,\&\, R(f) \subseteq L^+\}$$

for any ordinal α, and for limit ordinals λ

$$V_\lambda^L = \bigcup_{\alpha<\lambda} V_\alpha^L$$

Here $\mathrm{Fnc}(x)$ is a unary predicate stating that x is a function, and $D(x)$ and $R(x)$ are unary functions assigning to x its domain and range, respectively.

Note that functions taking the value 0 on any element of their domain are not considered as elements of the universe.

Finally we put

$$V^L = \bigcup_{\alpha\in\mathrm{On}} V_\alpha^L$$

Observe that for $\alpha \leq \beta$, $V_\alpha^L \subseteq V_\beta^L$.

We define two binary functions from V^L into L, assigning to any tuple $x, y \in V^L$ the values $||x \in y||$ and $||x = y||$ (representing the "truth values" of the two predicates $\in$ and $=$):
$||x \in y|| = y(x)$ if $x \in D(y)$, otherwise 0,
$||x = y|| = 1$ if $x = y$, otherwise 0.

We now use induction on the complexity of formulas to define for any formula $\varphi(x_1, \ldots, x_n)$—the free variables in φ being $x_1, \ldots, x_n$—a corresponding n-ary function from $(V^L)^n$ into L. The induction steps admit the following cases (we just write φ for short):
φ is $\overline{0}$: then $||\varphi|| = 0$;
φ is $\psi \,\&\, \chi$; then $||\varphi|| = ||\psi|| * ||\chi||$;
φ is $\psi \to \chi$: then $||\varphi|| = ||\psi|| \Rightarrow ||\chi||$;
φ is $\psi \wedge \chi$: then $||\varphi|| = ||\psi|| \wedge ||\chi||$;
φ is $\psi \vee \chi$: then $||\varphi|| = ||\psi|| \vee ||\chi||$;
φ is $\Delta\psi$; then $||\varphi|| = \Delta||\psi||$:
φ is $\forall x\psi$: then $||\varphi|| = \bigwedge_{u\in V^L} \psi(x/u)$;
φ is $\exists x\psi$: then $||\varphi|| = \bigvee_{u\in V^L} \psi(x/u)$.
Here $\psi(x/u)$ is the result of substituting u for the variable x in ψ. We use the symbols $\wedge$, $\vee$, Δ for both logical connectives and operations in L.

Definition 3. *Let φ be a closed formula. We say that φ is valid in V^L iff $||\varphi = 1||$ is provable in ZFC.*

(As usual, we take closures of formulas containing free variables when considering their validity.)

Lemma 1. *(1) ZFC proves for $u \in V^L$:*
$u(x) = ||x \in u||$ for $x \in D(u)$, $||u = u|| = 1$
(2) ZFC proves for $u, v, w \in V^L$:
(i)$||u = v|| * ||v = w|| \leq ||u = w||$
(ii)$||u \in v|| * ||v = w|| \leq ||u \in w||$
(iii)$||u = v|| * ||v \in w|| \leq ||u \in w||$

Proof. (1) immediate,
(2) (i) immediate,
(ii) if $v = w$, then $||v = w|| = 1$, so $||u \in v|| = ||u \in v|| * ||v = w|| = ||u \in w|| * ||v = w|| = ||u \in w||$; otherwise $||v = w|| = 0$, in which case the statement holds trivially.
(iii) analogously. □

Lemma 2. (Substitution) *For any formula φ, ZFC proves: $\forall u, v \in V^L$, $||u = v|| * ||\varphi(u)|| \leq ||\varphi(v)||$.*

Lemma 3. (Bounded quantifiers) *ZFC proves $\forall x \in V^L$:*
(i) $||\exists y \in x \varphi(y)|| = ||\exists y (y \in x \,\&\, \varphi(y))|| = \bigvee_{y \in D(x)} (x(y) * ||\varphi(y)||)$
(ii) $||\forall y \in x \varphi(y)|| = ||\forall y (y \in x \rightarrow \varphi(y))|| = \bigwedge_{y \in D(x)} (x(y) \Rightarrow ||\varphi(y)||)$

Proof. (i) $||\exists y (y \in x \,\&\, \varphi(y))|| = \bigvee_{y \in V^L} (||y \in x|| * ||\varphi(y)||) = \bigvee_{y \in D(x)} (x(y) * ||\varphi(y)||)$ since $||y \in x||$ is nonzero only if $y \in D(x)$, and in that case it is $x(y)$.
(ii) analogously. □

Corollary 1. $||x \subseteq y|| = ||\forall u \in x (u \in y)|| = ||\forall u (u \in x \rightarrow u \in y)||$.

4 The theory FST

We introduce the theory FST in the language $\{\in\}$; we take $=$ to be a logical symbol with the usual axioms (imposing reflexivity, symmetry, transitivity and congruence w.r.t $\in$ on the corresponding relation). The underlying logic of FST is a schematic extension $\mathcal{C}$ (possibly void) of BL$\forall\Delta$; when proving theorems within FST, we only rely on the logical axioms of BL$\forall\Delta$ (thus *any* schematic extension will, for example, interpret ZF). On the other hand, for a given extension $\mathcal{C}$, the universe V^L—for L a $\mathcal{C}$-algebra—will yield an interpretation of FST over $\mathcal{C}$.

The reader will have noticed that ours is a crisp equality. This was imposed by the following fact:

Lemma 4. *A theory with comprehension (for open formulas) and pairing (or singletons) over a logic which proves the propositional formula* $(\varphi \rightarrow \varphi \& \varphi) \rightarrow (\varphi \vee \neg\varphi)$ *proves* $\forall x, y(x = y \vee \neg(x = y))$.

Proof. Given x, y, let z be s. t. $u \in z \equiv (u \in \{x\} \& u = x)$, i. e., $u \in z \equiv (u = x)^2$. Since $(x = x)^2$, we have $x \in z$. If $y = x$ then $y \in z$ by congruence, but then $(y = x)^2$; thus we have proved $y = x \rightarrow (y = x)^2$, thus (by assumption on the logic) $(x = y \vee \neg(x = y))$. □

Thus e.g. in Łukasiewicz logic and in product logic we get a crisp $=$. Also, under the usual formulation of extensionality, crispness of $=$ implies crispness of $\in$ in a theory with pairings and unions (cf. [3]). We borrow the elegant solution from [8]: a modification of extensionality uses the Δ operator, which invalidates the proof of crispness of $\in$ from the crispness of $=$.

Definition 4. *FST is a first order theory in the language* $\{\in\}$, *with the following axioms:*
(i) (extensionality) $\forall x \forall y(x = y \equiv (\Delta(x \subseteq y) \& \Delta(y \subseteq x)))$
(ii) (empty set) $\exists x \Delta \forall y \neg(y \in x)$
(iii) (pair) $\forall x \forall y \exists z \Delta \forall u(u \in z \equiv (u = x \vee u = y))$
(iv) (union) $\forall x \exists z \Delta \forall u(u \in z \equiv \exists y(u \in y \& y \in x))$
(v) (weak power) $\forall x \exists z \Delta \forall u(u \in z \equiv \Delta(u \subseteq x))$
(vi) (infinity)$\exists z \Delta(\emptyset \in z \& \forall x \in z(x \cup \{x\} \in z))$
(vii) (separation) $\forall x \exists z \Delta \forall u(u \in z \equiv (u \in x \& \varphi(u, x))$, for any formula not containing z as a free variable
(viii) (collection) $\forall x \exists z \Delta[\forall u \in x \exists v \varphi(u, v) \rightarrow \forall u \in x \exists v \in z \varphi(u, v))]$ for any formula not containing z as a free variable
(ix) ($\in$-induction) $\Delta \forall x(\forall y \in x \varphi(y) \rightarrow \varphi(x)) \rightarrow \Delta \forall x \varphi(x)$ for any formula φ
(x) (support) $\forall x \exists z(\mathrm{Crisp}(z) \& \Delta(x \subseteq z)))$

The Δ's in the axioms of weak power and $\in$-induction are introduced to weaken the statements; the Δ's after an existential quantifiers is used to guarantee that an element with the postulated property exists in *every* model of the theory (which does not follow from the semantics of the $\exists$ quantifier).

As usual, in the formulation of some axioms we use the functions of empty set, singleton, and union, which can be introduced using the appropriate axioms as in classical ZF, and also the usual definition of $\subseteq$. For a detailed treatment of introducing functions in BL$\forall$, see [5]. The function introduced by the weak power set axiom will be denoted $WP(x)$ for a set x. We use the following definition of crispness of a set:

Definition 5. $\mathrm{Crisp}(x) \equiv \forall u(\Delta(u \in x) \vee \Delta\neg(u \in x))$

Theorem 2. *Let* φ *be a closed formula provable in FST. Then* ZFC *proves* $||\varphi|| = 1$ *in* V^L.

Having proved this theorem we shall take the liberty to call V^L a *model* of FST. (More pedantically, we have constructed an interpretation of FST (over $\mathcal{C}$) in ZFC.) We omit all validity proofs for logical axioms and inference rules. The validity of congruence axioms for $\in$ has been verified in Lemma 1. It remains to verify the set-theoretic axioms.

Lemma 5. *The set-theoretical axioms (i)-(x) hold in V^L.*

Proof. **(extensionality)** For fixed $x, y \in V^L$, either $x = y$ and then the axiom holds, or $x \neq y$ and $||x = y|| = 0$; then either $D(x) = D(y)$ and w. l. o. g. there is a $z \in D(x)$ s. t. $x(z) < y(z)$, so $||\Delta y \subseteq x|| = 0$, or w. l. o. g. there is a $z \in D(x)$ s. t. $z \notin D(y)$, and then $||z \in x||$ is nonzero while $||z \in y||$ is zero thus $||\Delta x \subseteq y|| = 0$.

(empty set) There is only one candidate for the role of an "empty set in V^L", and this is the $\emptyset$ in V_0^L. Indeed, for an arbitrary x we get $||x \in \emptyset|| = 0$ since no x can be in the domain of $\emptyset$ (taken as a function).

(pair) For fixed $x, y \in V^L$, there is a $z \in V^L$ such that $D(z) = \{x, y\}$ and $z(x) = z(y) = 1$. The set z has the desired properties: for arbitrary $u \in V^L$, either $u \in D(z)$, then either $u = x$ or $u = y$ and $||u \in z|| = z(u) = 1$, or $u \notin D(z)$, and then $||u \in z|| = ||u = x|| = ||u = y|| = 0$.

(union) For a fixed $x \in V^L$, define (auxiliary) $D^2(x) = \bigcup\{D(v) : v \in D(x)\}$. Define z s. t. $D(z) = \{u \in D^2(x) : \bigvee_{v \in D(x)}(v(u) * x(v)) > 0\}$ (with a nilpotent t-norm, the union of a nonempty set may well be empty), and for $u \in D(z)$ set $z(u) = \bigvee_{v \in D(x)}(v(u) * x(v))$. Then for an arbitrary $u \in V^L$, if $u \in D(z)$ then $||u \in z|| = z(u) = \bigvee_{v \in D(x)}(v(u) * x(v)) = ||\exists y(u \in y \in x)||$. If $u \notin D(z)$, then $||u \in z|| = 0$, and also $||\exists y \in x(u \in y)|| = 0$ by definition of $D(z)$.

(weak power) For a fixed $x \in V^L$, define z s.t. $D(z) = \{u \in V^L : D(u) \subseteq D(x) \,\&\, u(v) \leq x(v)$ for $v \in D(u)\}$, and $z(u) = 1$ for $u \in D(z)$. For $u \in V^L$, either $u \in D(z)$ and then $||u \in z|| = z(u) = 1$, and also (by definition of $D(z)$) $||u \subseteq x|| = 1 = ||\Delta u \subseteq x||$, or $u \notin D(z)$, thus $||u \in z|| = 0$, and (by definition of $D(z)$) either $D(u) \not\subseteq D(x)$, or for some $v \in D(u)$, $u(v) > x(v)$, and in either case $||\Delta \forall v \in u(v \in x)|| = \Delta \bigvee_{v \in D(u)}(u(v) \Rightarrow ||v \in x||) = 0$.

(infinity) Define a function z with $D(z) = V_\omega^L$ and $z(u) = 1$ for $u \in D(z)$. Then $||\emptyset \in z|| = 1$ and, since for $x \in V_\alpha^L$ we have $x \cup \{x\} \in V_{\alpha+1}^L$, also $||\forall x \in z(x \cup \{x\} \in z)|| = 1$.

(separation) For a fixed $x \in V^L$, and a given φ, define z s. t. $D(z) = \{u \in D(x) : x(u) * ||\varphi(u)|| > 0\}$ and for $u \in D(z)$ set $z(u) = x(u) * ||\varphi(u)||$. Obviously this definition of z demonstrates the validity of separation in V^L.

(collection) in [8]

($\in$-induction) Fix a formula φ and suppose the axiom does not hold in V^L. Then (since Δ is two-valued) it must be the case that $||\Delta \forall x(\forall y \in x\varphi(y) \rightarrow \varphi(x))|| = 1$ and $||\Delta \forall x \varphi(x)|| = 0$, thus there is a successor ordinal α s. t. $\exists x \in V_\alpha^L(\varphi(x) < 1) \,\&\, \forall \beta < \alpha \forall y \in V_\beta^L(\varphi(y) = 1)$. Suppose first $\alpha = 0$; but $||\forall y \in$

$\emptyset\varphi(y) \to \varphi(\emptyset)|| = 1 \Rightarrow \varphi(\emptyset) = \varphi(\emptyset) < 1$, thus the antecedent would be 0. Suppose now $\alpha > 0$, and $x \in V^L_\alpha$ is s. t. $\varphi(x) < 1$. From the condition that $||\forall y \in x\varphi(y) \to \varphi(x)|| = 1$, and since $||\forall y \in x\varphi(y)|| = 1$, we get $||\varphi(x)|| = 1$, a contradiction.

(support) For a fixed $x \in V^L$ take z such that $D(z) = D(x)$ and $\forall u \in D(z)$ $z(u) = 1$. Then $||\mathrm{Crisp}(z)|| = 1$ and $\bigwedge_{u \in D(x)}(x(u) \Rightarrow ||u \in z||) = 1$. □

5 An interpretation of ZF in FST

Within FST, we shall define a class (i.e., we shall give a formula with one free variable, in the language of FST) of hereditarily crisp sets, which will be proved an inner model of ZF in FST.

We start with a bunch of technical statements.

Lemma 6. $\mathrm{BL}\forall\Delta \vdash \forall x\Delta\varphi \equiv \Delta\forall x\varphi$.

Proof. The right-to-left implication is easy. We give a BL$\forall\Delta$-proof of the converse one (an analogy to the proof of the Barcan formula in S5). Let $\Diamond\varphi$ stand for $\neg\Delta\neg\varphi$.
(i) $\varphi \to \neg\Delta\neg\varphi$.
In BLΔ, $\Delta\neg\varphi \to \neg\varphi$, thus $(\Delta\neg\varphi\&\varphi) \to \overline{0}$, thus $\varphi \to (\Delta\neg\varphi \to \overline{0})$.
(ii) $\neg\Delta\varphi \to \Delta\neg\Delta\varphi$.
In BLΔ, $\Delta\varphi \vee \neg\Delta\varphi$, thus $\Delta\Delta\varphi \vee \Delta\neg\Delta\varphi$. This gives $\neg\Delta\varphi \to \Delta\neg\Delta\varphi$.
Thus the following two are provable in BLΔ:
(iii) $\varphi \to \Delta\Diamond\varphi$
(iv) $\Diamond\Delta\varphi \to \Delta\varphi$.
Next, (v) $\Delta(\varphi \to \psi) \to (\Diamond\varphi \to \Diamond\psi)$,
since $\Delta(\varphi \to \psi) \to \Delta(\neg\psi \to \neg\varphi) \to (\Delta\neg\psi \to \Delta\neg\varphi) \to (\Diamond\varphi \to \Diamond\psi)$.
(vi) $\Diamond\forall x\varphi \to \forall x\Diamond\varphi$ is a consequence of (v).
Finally, $\forall x\Delta\varphi \to \Delta\Diamond\forall x\Delta\varphi \to \Delta\forall x\Diamond\Delta\varphi \to \Delta\forall x\Delta\varphi \to \Delta\forall x\varphi$. □

Lemma 7. $\mathrm{BL}\forall\Delta \vdash \Delta(\varphi \vee \neg\varphi) \equiv \Delta(\varphi \to \Delta\varphi)$.

Proof. Takes place inside BL$\forall\Delta$. Implication left-to-right: $\Delta(\varphi \vee \neg\varphi) \to (\Delta\varphi \vee \Delta\neg\varphi) \to [\varphi \to (\varphi\&(\Delta\varphi \vee \Delta\neg\varphi))] \to [\varphi \to (\Delta\varphi \vee (\varphi\&\Delta\neg\varphi))]$. In the last formula in the chain, $\varphi\&\Delta\neg\varphi \to \overline{0}$, thus the last formula implies $\varphi \to \Delta\varphi$; we get $\Delta(\varphi \vee \neg\varphi) \to (\varphi \to \Delta\varphi)$, which Δ-generalizes to the desired implication.

Conversely, $\Delta\varphi \to \Delta(\varphi \vee \neg\varphi)$, thus also (i) $\Delta\varphi \to [\Delta(\varphi \to \Delta\varphi) \to \Delta(\varphi \vee \neg\varphi)]$. Moreover $\Delta(\varphi \to \Delta\varphi) \to (\Delta\neg\Delta\varphi \to \Delta\neg\varphi) \to [\Delta\neg\Delta\varphi \to \Delta(\varphi \vee \neg\varphi)]$, thus (ii) $\Delta\neg\Delta\varphi \to [\Delta(\varphi \to \Delta\varphi) \to \Delta(\varphi \vee \neg\varphi)]$. Since BL$\forall\Delta$ proves $\Delta\varphi \vee \Delta\neg\Delta\varphi$, we get the right-to-left implication by putting together (i) and (ii).

□

Recall the definition of a crisp set. We get

Corollary 2. $\forall x(\mathrm{Crisp}(x) \equiv \forall u \Delta(u \in x \to \Delta(u \in x))$.

Note that $\mathrm{Crisp}(x) \equiv \Delta \forall u(u \in x \to \Delta(u \in x)) \to \Delta\Delta\forall u\ (u \in x \to \Delta(u \in x))$, so crispness is a crisp property:

Lemma 8. $\mathrm{Crisp}(x) \to \Delta\mathrm{Crisp}(x)$.

We write $\bowtie \varphi$ for $\Delta(\varphi \vee \neg\varphi)$.

Lemma 9. $\mathrm{BL}\Delta \vdash (\bowtie \varphi \& (\varphi \to \Delta\psi)) \to \Delta(\varphi \to \psi)$.

Proof. $[\Delta\varphi\&(\varphi \to \Delta\psi)] \to \Delta\psi \to \Delta(\varphi \to \psi)$, and $\Delta\neg\varphi \to \Delta(\varphi \to \psi)$, so $\bowtie \varphi \equiv (\Delta\varphi \vee \neg\Delta\varphi) \to ((\varphi \to \Delta\psi) \to \Delta(\varphi \to \psi))$. □

Lemma 10. $\mathrm{BL}\forall\Delta \vdash \forall xy((\mathrm{Crisp}(x)\&\mathrm{Crisp}(y)\&x \subseteq y) \to \Delta(x \subseteq y))$.

Proof. $(\mathrm{Crisp}(x)\&\mathrm{Crisp}(y)\&(u \in x \to u \in y)) \to$
$(\bowtie (u \in x)\&(u \in x \to \Delta u \in y)) \to \Delta(u \in x \to u \in y)$. □

Definition 6. (Hereditarily crisp transitive set) $HCT(x) \equiv \mathrm{Crisp}(x)\ \&$ $\&\ \forall u \in x(\mathrm{Crisp}(u)\&u \subseteq x)$

The formula $HCT(x)$ defines a class, and we adopt the habit of writing $x \in HCT$ instead of $HCT(x)$ and approaching classes in a similar way we approach sets. A "crisp class" C is a class for which $\forall x\Delta(x \in C \vee \neg(x \in C))$, or equivalently, $\forall x \Delta(x \in C \to \Delta(x \in C))$.

Lemma 11. (Crispness of HCT) $x \in HCT \to \Delta(x \in HCT)$.

Proof. $x \in HCT \equiv [\mathrm{Crisp}(x)\&\forall u(u \in x \to (\mathrm{Crisp}(u)\&u \subseteq x))] \to$
$[\Delta\mathrm{Crisp}(x)\&\forall u(\bowtie (u \in x)\&(u \in x \to \Delta(\mathrm{Crisp}(u)\&u \subseteq x)))] \to$
$[\Delta\mathrm{Crisp}(x)\&\forall u\Delta(u \in x \to (\mathrm{Crisp}(u)\&u \subseteq x))] \to$
$\Delta(x \in HCT)$. □

Definition 7. (Hereditarily crisp set) $H(x) \equiv \mathrm{Crisp}(x)\ \&\ \exists x' \in HCT(x \subseteq x')$.

Lemma 12. (Crispness of H) $x \in H \to \Delta(x \in H)$.

Proof. By definition of H, we are to prove
$(\mathrm{Crisp}(x)\&\exists y \in HCT(x \subseteq y)) \to \Delta(\mathrm{Crisp}(x)\&\exists y \in HCT(x \subseteq y))$.
Since $\mathrm{Crisp}(x) \to \Delta\mathrm{Crisp}(x)$, it suffices to prove
$(\mathrm{Crisp}(x)\&\exists y \in HCT(x \subseteq y)) \to \Delta(\exists y \in HCT(x \subseteq y))$
(we may use the presumption $\mathrm{Crisp}(x)$ in each of the two implications since it is idempotent).
First, $(\mathrm{Crisp}(x)\&y \in HCT\&x \subseteq y) \to \Delta(y \in HCT\&x \subseteq y)$ by Lemma 11 and Lemma 10. Now generalize: $\forall y((\mathrm{Crisp}(x)\&y \in HCT\&x \subseteq y) \to \Delta(y \in HCT\&x \subseteq y))$; hence $(\mathrm{Crisp}(x)\&\exists y(y \in HCT\&x \subseteq y)) \to \exists y\Delta(y \in HCT\&x \subseteq y)$; and the succedent implies $\Delta\exists y(y \in HCT\&x \subseteq y)$. □

We show that FST proves H to be an inner model of ZF. In more detail, for φ a formula in the language of ZF, define φ^H inductively as follows:
$\varphi^H = \varphi$ for φ atomic;
$\varphi^H = \varphi$ for $\varphi = \overline{0}$;
$\varphi^H = \psi^H \& \chi^H$ for $\varphi = \psi \& \chi$;
$\varphi^H = \psi^H \to \chi^H$ for $\varphi = \psi \to \chi$;
$\varphi^H = (\forall x \in H)\psi^H$ for $\varphi = (\forall x)\psi$;
$\varphi^H = (\exists x \in H)\psi^H$ for $\varphi = (\exists x)\psi$.

Theorem 3. *Let φ be a theorem of ZF. Then* FST $\vdash \varphi^H$.

To prove this theorem, we first show that the law of the excluded middle (LEM) holds in H; since BL$\forall$ together with LEM yield classical logic, we will have proved that all logical axioms of ZF are provable when relativized to H. Then we prove the H-relativized versions of all the axioms of the ZF set theory.

Lemma 13. *Let $\varphi(x_1, \ldots, x_n)$ be a ZF-formula whose free variables are among $x_1, \ldots, x_n$. Then FST proves*

$$\forall x_1 \in H \ldots \forall x_n \in H(\varphi^H(x_1, \ldots, x_n) \vee \neg\varphi^H(x_1, \ldots, x_n))$$

Proof. We consider a formula φ with (at most) one free variable x, the modification for multiple free variables being easy. The formula to be proved in FST is $\forall x(x \in H \to (\varphi^H(x) \vee \neg\varphi^H(x))$; it suffices to prove $\forall x \Delta(x \in H \to (\varphi^H(x) \vee \neg\varphi^H(x))$, and by Lemma 9 and Lemma 7, it suffices to prove $\forall x(x \in H \to \Delta(\varphi^H(x) \to \Delta\varphi^H(x))$.

The proof proceeds by induction on the complexity of φ.

Atomic subformulas: $=$ is a crisp predicate, for $\in$ we have to prove $x \in y \to \Delta x \in y$ assuming $x, y \in H$. In fact $y \in H$ implies Crisp(y), which entails $\forall x(x \in y \to \Delta x \in y)$.

Conjunction: for a subformula $\psi_1(x,y) \& \psi_2(x,z)$ of φ (we assume one free variable in common, and one distinct free variable for each subformula, and we henceforth omit their explicit listings, for legibility's sake) assume $x, y \in H \to (\psi_1^H \to \Delta\psi_1^H)$ and $x, z \in H \to (\psi_2^H \to \Delta\psi_2^H)$. Then $(x \in H)^2 \& (y, z \in H) \to ((\psi_1^H \& \psi_2^H) \to \Delta(\psi_1^H \& \psi_2^H))$, and since $x \in H$ is idempotent, this completes the induction step for conjunction.

Implication: for a subformula $\psi_1(x,y) \to \psi_2(x,z)$ of φ, assume $x, y \in H \to (\psi_1^H \to \Delta\psi_1^H)$ and $x, z \in H \to (\psi_2^H \to \Delta\psi_2^H)$. Thus $(\psi_1^H \to \psi_2^H) \& (x, z \in H) \to (\psi_1^H \to \Delta\psi_2^H)$, and since $x, y \in H$ implies crispness of ψ_1^H, we get $x, y \in H \to ((\psi_1^H \to \Delta\psi_2^H) \to \Delta(\psi_1^H \to \psi_2^H))$. Thus $x, y, z \in H \to ((\psi_1^H \to \psi_2^H) \to \Delta(\psi_1^H \to \psi_2^H))$.

The universal quantifier: for a subformula $\forall y \psi(x,y)$ of φ, the induction hypothesis is $x, y \in H \to (\psi^H(x,y) \to \Delta\psi^H(x,y))$. Generalize in y: $x \in H \to \forall y(y \in H \to (\psi^H(x,y) \to \Delta\psi^H(x,y))$; now since for a crisp α, BL proves $(\alpha \to (\beta \to \gamma)) \to ((\alpha \to \beta) \to (\alpha \to \gamma))$, we may modify the succedent to

$\forall y((y \in H \to \psi^H(x,y)) \to (y \in H \to \Delta\psi^H(x,y))$, and distributing $\forall y$, we have proved: $x \in H \to (\forall y \in H\psi^H(x,y) \to \forall y \in H\Delta\psi^H(x,y))$. To flip the Δ and the $\forall y \in H$ in the succedent, use Lemma 9.

The existential quantifier: for a subformula $\forall y\psi(x,y)$ of φ, the induction hypothesis is $x, y \in H \to (\psi^H(x,y) \to \Delta\psi^H(x,y))$. Generalize in y: $x \in H \to \forall y(y \in H \to (\psi^H(x,y) \to \Delta\psi^H(x,y))$; now $(y \in H \to (\psi^H(x,y) \to \Delta\psi^H(x,y))) \to (y \in H \to (y \in H\&\psi^H(x,y) \to y \in H\&\Delta\psi^H(x,y))) \to (y \in H\&\psi^H(x,y) \to y \in H\&\Delta\psi^H(x,y))$. We get $x \in H \to (\exists y(y \in H\&\psi^H(x,y)) \to \exists y\Delta(y \in H\&\psi^H(x,y)))$, and the last succedent implies $\Delta\exists y(y \in H\&\psi^H(x,y))$. □

Definition 8. $WPC(x) = \{u \in WP(x); \text{Crisp}(u)\}$.

Lemma 14. $\forall x(x \subseteq H\&\text{Crisp}(x) \to x \in H)$.

Proof. Fix an x. $x \subseteq H$ is by definition $\forall u \in x\exists u'(u' \in HCT\&u \subseteq u')$. Thus $\exists v_0\forall u \in x\exists u' \in v_0(u' \in HCT\&u \subseteq u')$. Fix v_0 and separate: $v = \{u \in v_0 : \Delta u \in v_0\&u \in HCT\&\exists u_0 \in x(u_0 \subseteq u)\}$. Note that $\forall u \in x\exists u' \in v(u' \in HCT\&u \subseteq u')$, because x is crisp. v is a crisp set and all its elements are crisp (all of them are in HCT); hence $\bigcup v$ is crisp; its elements are crisp since $a \in \bigcup v$ implies $\exists b \in HCT(a \in b \in v)$, thus a is crisp. $\bigcup v$ is also transitive: $b \in a \in \bigcup v$ implies $\exists y(b \in a \in y \in v)$, and since $y \in HCT$ is transitive, $b \in y \in v$ and $b \in \bigcup v$. Now consider $WPC(\bigcup v)$ (the crisp elements of the weak power set of $\bigcup v$); this is a crisp set, and it is a subset of $WPC(\bigcup v) \cup \bigcup v$, which is a crisp transitive set of crisp elements, thus in HCT, thus $WPC(\bigcup v) \in H$. Since $v \subseteq WPC(\bigcup v)$, we get $\forall u \in x\exists u' \in WPC(\bigcup v)(u \subseteq u')$, thus $\forall u \in x(u \in WPC(\bigcup v)$; this means $x \subseteq WPC(\bigcup v) \in HCT$, so $x \in H$. □

We consider ZF with the following axioms: empty set, pair, union, power set, infinity, separation, collection, extensionality, $\in$-induction. The exact spelling of these axioms is given separately when proving in FST their H-ed versions.

Theorem 4. *For φ being any of the abovementioned axioms of ZF, FST proves φ^H.*

Proof. **(empty set)** $\exists z\forall u\neg(u \in z)$.
The H-translation, which reads $\exists z \in H\forall u \in H\neg u \in z$, is provable since $0 \in HCT$.
(pair) $\forall x, y\exists z\forall u(u \in z \equiv (u = x \vee u = y))$.
The H-translation $\forall x, y \in H\exists z \in H\forall u \in H(u \in z \equiv (u = x) \vee (u = y))$ is absolute: the set $\{x, y\}$ is crisp (since $=$ is crisp); to show that it is a subset of a set which is in HCT, consider $x' \in HCT$ a witness for $x \in H$ and $y' \in HCT$ a witness for $y \in H$. Then $\{x, y\} \cup x' \cup y'$ is a crisp transitive set with crisp elements, hence in HCT, and thus $\{x, y\}$ is in H.
(union) $\forall x\exists z\forall u(u \in z \equiv \exists y(u \in y \in x))$.

The H-translation $\forall x \in H \exists z \in H \forall u \in H(u \in z \equiv \exists y \in H(u \in y \in x))$ is absolute: let $x' \in HCT$ witness $x \in H$. Then $\bigcup x$ is a crisp set with crisp elements (since $x \subseteq x'$), and $\bigcup x \subseteq \bigcup x' \in HCT$, which witnesses $\bigcup x \in H$.
(power set) $\forall x \exists z \forall u(u \in z \equiv u \subseteq x)$
The H-translation reads $\forall x \in H \exists z \in H \forall u \in H(u \in z \equiv \forall y \in H(y \in u \rightarrow y \in x))$. Let x' be a witness for $x \in H$. then for $x \in H$ it holds that $\forall u \in H(u \in WPC(x) \equiv \Delta(u \subseteq x)^H \equiv (u \subseteq x)^H)$, $WPC(x)$ is crisp, and $WPC(x) \subseteq WPC(x') \cup x'$, which is a transitive crisp set with crisp elements, thus in HCT and a witness for $WPC(x) \in H$.
(separation) $\forall x \exists z \forall u(u \in z \equiv (u \in x \& \varphi(u))$ for a ZF-formula not containing z freely.
The H-translation is absolute: let $x' \in HCT$ witness $x \in H$ and set $z = \{u \in x; \varphi^H(u)\}$, then z is a crisp set and $z \subseteq x \subseteq x' \in HCT$ (i.e., x' is a witness for $z \in H$).
(infinity) $\exists z(0 \in z \& \forall u \in z\ (u \cup \{u\} \in z))$.
It suffices to prove that there is a set $z \in H$ s.t. $0 \in z \& \forall u \in z(u \cup \{u\} \in z)$. Let z_0 be any set satisfying the axiom of infinity in FST and define $z_1 = \{x \in z_0 : \Delta(x \in z_0) \& \mathrm{Crisp}(x)\}$. Then z_1 is a subset of z_0 and $0 \in z_1$ and $u \in z_1 \rightarrow u \cup \{u\} \in z_1$. Now let $z = \{x \in z_1 : x \subseteq z_1\}$, i.e., a transitive subset of z_1. Obviously $0 \in z$, let us prove $x \in z \rightarrow (x \cup \{x\}) \in z$, that is by definition of z, $[x \in z_1 \& \forall y \in x(y \in z_1)] \rightarrow [x \cup \{x\} \in z_1 \& \forall a(a \in x \vee a = x \rightarrow a \in z_1)]$. We know $x \in z_1 \rightarrow (x \cup \{x\} \in z_1)$. Also, $x \in z_1 \& (y \in x \rightarrow y \in z_1) \rightarrow (y \in x \vee y = x \rightarrow y \in z_1)$. (Note that we made multiple use of the presumption $x \in z_1$; we may do that because the formula is crisp, thus $x \in z_1 \equiv (x \in z_1)^2$.) Finally, z is a crisp transitive set with crisp elements, so $z \in HCT$.
(extensionality) $\forall xy(x = y \equiv \forall z(z \in x \equiv z \in y))$.
The H-translation $\forall x, y \in H(x = y \equiv \forall z \in H(z \in x \equiv z \in y))$ follows from extensionality in FST by H being transitive and by the crispness of its elements (the Δ's may be left out).
(collection) $\forall u(\forall x \in u \exists y \varphi(x, y) \rightarrow \exists v \forall x \in u \exists y \in v \varphi(x, y))$ for φ not containing v freely.
The H-translation reads $\forall u \in H(\forall x \in H(x \in u \rightarrow \exists y \in H \varphi^H(x, y)) \rightarrow \exists v \in H \forall x \in H(x \in u \rightarrow \exists y \in H(y \in v \& \varphi^H(x, y))))$. Fix $u \in H$ and a ZF-formula φ; we want to find a corresponding $v \in H$ s.t. the above is true. Define v_0 by collection in FST for u and the formula $y \in H \& \varphi^H(x, y)$; separate $v = \{x \in v_0 : \Delta(x \in v_0) \& x \in H\}$. Then $v \subseteq H$ is a crisp set and, since u is crisp, satisfies the collection axiom. By Lemma 14, $v \in H$.
($\in$-induction) $\forall x(\forall y \in x \varphi(y) \rightarrow \varphi(x)) \rightarrow \forall x \varphi(x)$ for any ZF-formula φ.
The H-translation is $\forall x \in H(\forall y \in H(y \in x \rightarrow \varphi^H(y)) \rightarrow \varphi^H(x)) \rightarrow \forall x \in H \varphi^H(x)$. Fix a ZF-formula φ, and consider the instance of $\in$-induction in FST for the formula $x \in H \rightarrow \varphi^H(x)$:
$\Delta \forall x(\forall y \in x(y \in H \rightarrow \varphi^H(y)) \rightarrow (x \in H \rightarrow \varphi^H(x))) \rightarrow \Delta \forall x(x \in H \rightarrow \varphi^H(x))$. This formula is our aim except for the Δ's. Let us denote A the antecedent and S the succedent in the implication, with the Δ's omitted.

Since $\Delta S \to S$, it remains to be proved that $A \to \Delta A$. Let us further denote $\forall y \in H(y \in x \to \varphi^H(y))$ with I. First, it is obvious that
$x \in H \to \forall y(y \in x \to\bowtie \varphi^H(y))$, and
$x \in H \to \forall y \bowtie (y \in x)$. Thus,
$x \in H \to \forall y(\bowtie (y \in x) \& (y \in x \to\bowtie \varphi^H(y)))$, hence
$x \in H \to \forall y(\bowtie (y \in x \to \varphi^H(y)))$, and
$x \in H \to\bowtie \forall y(y \in x \to \varphi^H(y))$, thus $x \in H \to\bowtie I$. Since $x \in H \equiv\bowtie (x \in H)$, we get
$A \to \forall x \in H(\bowtie I \& (I \to \Delta\varphi^H(x)))$, hence by use of Lemma 9
$A \to \forall x \in H\Delta(I \to \varphi^H(x))$, so
$A \to \Delta\forall x \in H(I \to \varphi^H(x))$, which is $A \to \Delta A$. □

References

1. Cignoli, R., Esteva, F., Godo, L., Torrens, A.: Basic Fuzzy Logic is the logic of continuous t-norms and their residua, Soft Computing 4 (2000) 106–112
2. Grayson, R.J.: Heyting-valued models for intuitionistic set theory, Application of Sheaves: Lect. Notes Math. 753 (1979) 402–414
3. Grishin, V.N.: O vese aksiomy svertyvania v teorii, osnovannoj na logike bez sokrashchenij, Matematicheskije zametki 66, no.5 (1999) 643–652
4. Hájek, P.: Metamathematics of Fuzzy Logic, Kluwer Academic Publishers (1998)
5. Hájek, P.: Function symbols in fuzzy predicate logic, Proc. East-West Fuzzy Colloq. Zittau (2000)
6. Hájek, P., Haniková, Z.: A set theory within fuzzy logic, Proc. 31st IEEE ISMVL Warsaw (2001) 319–324
7. Powell, W. C.: Extending Gödel's negative interpretation to ZF, J. Symb. Logic 40, no. 2 (1975) 221–229
8. Shirahata, M.: Phase-valued models of linear set theory, preprint (1999)
9. Takeuti, G., Titani, S.: Intuitionistic fuzzy logic and intuitionistic fuzzy set theory, J. Symb. Logic 49 (1984) 851–866
10. Takeuti, G., Titani, S.: Fuzzy logic and fuzzy set theory, Arch. Math. Logic 32 (1992) 1–32
11. Titani, S.: A lattice-valued set theory, Arch. Math. Logic 38 (1999) 395–421

Chapter 13
A Fuzzy Generalisation of Information Relations

Anna Maria Radzikowska[1], Etienne E. Kerre[2]

[1] Faculty of Mathematics and Information Science, Warsaw University of Technology
Plac Politechniki 1, 00–661 Warsaw, Poland
annrad@mini.pw.edu.pl

[2] Department of Applied Mathematics and Computer Science, Ghent University
Krijgslaan 281 (S9), B-9000 Gent, Belgium.
Etienne.Kerre@rug.ac.be

Abstract. In this paper we consider a fuzzy generalisation of some information relations. Basic properties of these relations are provided. We give characterisations of these relations formalised by means of fuzzy information operators. For particular classes of fuzzy information relations the corresponding classes of fuzzy information logics are defined and briefly discussed.

1 Introduction

In real–life problems we usually deal with incomplete information. Generally speaking, there are two reasons for incompleteness of information. First, we often have only partial data about a domain the problem under consideration is related to. Second, the acquired information, if available, is often imprecise (e.g. when expressed by means of linguistic terms like "quite good" or "rather tall"). As a consequence, the knowledge derived from the incomplete information is uncertain.

In many application domains the available information has the form of a collection of objects together with descriptions of these objects, usually expressed in terms of properties of objects. Such systems of data are referred to as *information systems*. Formal methods for representing and analysing information in information systems have been extensively developed within the theory of rough sets.[3] In these approaches descriptions of objects, viewed as the explicit information contained in the system, are traditionally assumed to be precise. Relationships among objects, which constitute implicit information, are derived from properties of objects. Formally, they are binary

[3] For the extended survey of rough set-style analysis of information, we refer to [16].

relations on a domain of objects and are referred to as *information relations.* These relations represent either *indistinguishabilities* of objects (characterising some kinds of "sameness" of objects) or *distinguishabilities* (reflecting some types of differences among objects). Logical systems capable to derive information about relations among objects represented in information systems are called *information logics.* During the last decade information logics were intensively investigated in the literature (see [1],[3],[4],[7],[13],[16],[17],[15],[28]).

When an imprecise information is admitted, it is clear that it cannot be adequately represented by means of standard methods based on classical two–valued structures. A natural solution seems to be fuzzy generalisations of the respective methods. Recently, fuzzy rough sets have been extensively investigated ([5],[21],[18],[22],[23],[26]).

Orłowska ([15]) has recently proposed a general framework for generalising information logics for the multi–valued case. In [19] and [20] we have provided preliminary extensions of this approach. In the present paper we continue these investigations. More specifically, we generalise the notion of information system for the case where properties of objects are assumed to be imprecise. Formally, they are represented by fuzzy sets in the respective domains. We analyse several binary relations between fuzzy sets. These relations, in turn, give the basis for defining *fuzzy information relations.* Basic properties of fuzzy information relations are given. We define several *fuzzy information operators* that are used to provide operator–oriented characterisations of some binary fuzzy relations. Finally, we consider some classes of fuzzy information logics and give characteristic tautologies of these systems.

2 Information relations

In this section we recall the definitions of main classes of information relations and give their basic properties.

Definition 1. *An* ***information system*** *is a tuple* $\Sigma=(OB, AT, \{V_a : a\in AT\})$*, where* OB *is a nonempty set of objects,* AT *is a nonempty and finite set of attributes and for each* $a\in AT$*,* V_a *is a nonempty set of values of attribute* a*. Every* $a\in AT$ *is a mapping* $a: OB \rightarrow \wp(V_a)$*.* □

Intuitively, for any object $x\in OB$ and any attribute $a\in AT$, $a(x)\subseteq V_a$ is the set of values x has on the attribute a.

Example 1 Consider an information system $\Sigma=(OB, AT, \{V_a : a\in AT\})$ given in Table 1, where

$$OB = \{P_1,\ldots,P_6\}$$
$$AT = \{\text{PR}\,(\textit{profession}), \text{LAN}\,(\textit{Languages spoken})\}$$
$$V_{Pr} = \{\text{e}\,(\textit{engineer}), \text{r}\,(\textit{researcher}), \text{t}\,(\textit{teacher}), \text{md}\,(\textit{medical doctor}).\}$$
$$V_{Lan} = \{\text{E}\,(\textit{English}), \text{S}\,(\textit{Spanish}), \text{F}\,(\textit{French}), \text{D}\,(\textit{Dutch}), \text{I}\,(\textit{Italian})\}.$$

OB	PR	LAN
P_1	{r}	{F,I}
P_2	{t}	{E,D,I}
P_3	{md,r}	{D,E,F}
P_4	{e,md}	{S,F}
P_5	{e,r}	{F,I}

Table 1. Information system

For P_1, we know that he/she is a researcher who speaks French and Italian; P_4 is a medical doctor and an engineer speaking Spanish and French. □

Definition 2. *For an information system $\Sigma = (OB, AT, \{V_a : a \in AT\})$ and any $\mathcal{A} \subseteq AT$, let us define the following two classes of binary relations on OB, called **information relations in Σ induced by $\mathcal{A}$**: for all $x, y \in OB$,*

1. INDISTINGUISHABILITY RELATIONS:
   - *a strong (weak) indiscernibility relation $ind^s(\mathcal{A})$ (resp. $ind^w(\mathcal{A})$):*
     $(x,y) \in ind^s(\mathcal{A})$ $(ind^w(\mathcal{A})$ *iff* $a(x) = a(y)$ *for all (some)* $a \in \mathcal{A}$
   - *a strong (weak) similarity relation $sim^s(\mathcal{A})$ $(sim^w(\mathcal{A}))$:*
     $(x,y) \in sim^s(\mathcal{A})$ $(sim^w(\mathcal{A}))$ *iff* $a(x) \cap a(y) \neq \emptyset$ *for all (some)* $a \in \mathcal{A}$
   - *a strong (weak) incomplementarity relation*[4] $icm^s(\mathcal{A})$ $(icm^w(\mathcal{A}))$:
     $(x,y) \in icm^s(\mathcal{A})$ $(icm^w(\mathcal{A}))$ *iff* $a(x) \neq co\, a(y)$ *for all (some)* $a \in \mathcal{A}$
2. DISTINGUISHABILITY RELATIONS:
   - *a strong (weak) diversity relation $div^s(\mathcal{A})$ $(div^w(\mathcal{A}))$:*
     $(x,y) \in div^s(\mathcal{A})$ $(div^w(\mathcal{A}))$ *iff* $a(x) \neq a(y)$ *for all (some)* $a \in \mathcal{A}$
   - *a strong (weak) orthogonality relation $ort^s(\mathcal{A})$ $(ort^w(\mathcal{A}))$:*
     $(x,y) \in ort^s(\mathcal{A})$ $(ort^w(\mathcal{A}))$ *iff* $a(x) \cap a(y) = \emptyset$ *for all (some)* $a \in \mathcal{A}$
   - *a strong (weak) complementarity relation $cmp^s(\mathcal{A})$ $(cmp^w(\mathcal{A}))$:*
     $(x,y) \in cmp^s(\mathcal{A})(cmp^w(\mathcal{A}))$ *iff* $a(x) = co\, a(y)$ *for all (some)* $a \in \mathcal{A}$.□

Remark 1 Note that for every information system $\Sigma = (OB, AT, \{V_a : a \in AT\})$, for every attribute $a \in AT$ and for every information relation R defined above, it holds $R^s(\{a\}) = R^w(\{a\})$.

Example 1 (cont.) Note that P_1 and P_5 are weakly indiscernible wrt both attributes (they both speak only French and Italian), whereas P_4 and P_5 are strongly similar wrt AT (they are both engineers and speak French). Also, $(P_2, P_4) \in cmp^w(AT)$, since $\text{LAN}(P_2) = \{\text{E}, \text{D}, \text{I}\} = V_{Lan} \setminus \text{LAN}(P_4)$. □

[4] For any domain $\mathfrak{X}$ and $A \subseteq \mathfrak{X}$, $co\, A$ stands for $\mathfrak{X} \setminus A$.

Proposition 1. [16] *For any information system* $\Sigma = (OB, AT, \{V_a : a \in AT\})$, *any information relation* $R \in \{ind, sim, icm, div, ort, cmp\}$ *in* Σ *and any sets* $\mathcal{A}, \mathcal{B} \subseteq AT$ *of attributes,*

$$R^s(\mathcal{A} \cup \mathcal{B}) = R^s(\mathcal{A}) \cap R^s(\mathcal{B}) \qquad R^s(\emptyset) = OB \times OB$$
$$R^w(\mathcal{A} \cup \mathcal{B}) = R^w(\mathcal{A}) \cup R^w(\mathcal{B}) \qquad R^w(\emptyset) = \emptyset. \quad \blacksquare$$

We say that a binary relation R on OB is

- *quasi–reflexive* iff for all $x, y \in OB$, $(x, y) \in R$ implies $(x, x) \in R$
- *weakly irreflexive* iff for all $x, y \in OB$, $(x, y) \notin R$ implies $(x, x) \notin R$
- *n–transitive*, $n \geq 1$, iff $(x, x_1) \in R, \ldots, (x_n, y) \in R$ implies $(x, y) \in R$ for all $x, x_1, \ldots, x_n, y \in OB$
- *n–cotransitive* (resp. irreflexive) iff $co\,R$ is n–transitive (resp. reflexive).

In other words, R is quasi–reflexive (resp. weakly irreflexive) iff x is R–related (resp. not R–related) with itself provided that some element $y \in OB$ is R–related with x (resp. not R–related with x). Clearly, all reflexive (resp. irreflexive) relations are quasi–reflexive (resp. weakly irreflexive), but not conversely. Transitive (resp. cotransitive) relations are just 1-transitive (resp. 1–cotransitive).

In view of results presented in [16] and [17] we have:

Theorem 1. *For every information system* $\Sigma = (OB, AT, \{V_a : a \in AT\})$ *and every* $\mathcal{A} \subseteq AT$,

1. $ind^s(\mathcal{A})$ *is an equivalence relation;*
   $ind^w(\mathcal{A})$ *is a tolerance relation (for* $\mathcal{A} \neq \emptyset$*) such that for every* $a \in AT$, $ind^w(\{a\})$ *is an equivalence relation.*
2. $sim^s(\mathcal{A})$ *and* $sim^w(\mathcal{A})$ *are quasi–reflexive and symmetric*
3. $icm^w(\mathcal{A})$ *(for* $\mathcal{A} \neq \emptyset$*) and* $icm^s(\mathcal{A})$ *are reflexive and symmetric;*
   $icm^w(\mathcal{A})$ *is 2–cotransitive,* $icm^s(\{a\})$ *is 2–cotransitive for every* $a \in AT$
4. $div^s(\mathcal{A})$ *(for* $\mathcal{A} \neq \emptyset$*) and* $div^w(\mathcal{A})$ *are irreflexive and symmetric;*
   $div^w(\mathcal{A})$ *is cotransitive,* $div^s(\{a\})$ *is cotransitive for every* $a \in AT$
5. $ort^s(\mathcal{A})$ *and* $ort^w(\mathcal{A})$ *are weakly irreflexive and symmetric*
6. $cmp^s(\mathcal{A})$ *(for* $\mathcal{A} \neq \emptyset$*) and* $cmp^w(\mathcal{A})$ *are irreflexive and symmetric;*
   $cmp^s(\{a\})$ *is 2–transitive,* $cmp^w(\{a\})$ *is 2–transitive for every* $a \in AT$. ■

3 Fuzzy logical operators

In this section we recall the definitions and basic properties of fuzzy logical operators, which are fuzzy generalisations of logical connectives of classical logic. In particular, triangular norms generalise classical conjunction.

Definition 3. [25] *A **triangular norm** (t–norm, for short) is an increasing, associative and commutative mapping $\mathcal{T} : [0,1]^2 \to [0,1]$ satisfying the boundary condition $\mathcal{T}(x,1)=x$ for every $x \in [0,1]$.* □

The most popular continuous t–norms are ([12]):

- the standard t–norm $\mathcal{T}_M(x,y) = \min\{x,y\}$
- the product operation $\mathcal{T}_P(x,y) = x * y$
- the Łukasiewicz t–norm $\mathcal{T}_L(x,y) = \max\{0, x+y-1\}$.

A t–norm is called *left–continuous* iff it has left–continuous partial mappings. A point $x_0 \in]0,1[$ is called a *zero divisor of $\mathcal{T}$* iff $\mathcal{T}(x_0,y_0)=0$ for some $y_0 \in]0,1[$. We will write T^+ to denote the class of all t–norms without zero divisors. Note that $\mathcal{T}_M$, $\mathcal{T}_P \in T^+$, but $\mathcal{T}_L \notin T^+$.

Definition 4. [25] *A **triangular conorm** (t–conorm, for short) is an increasing, associative and commutative mapping $\mathcal{S} : [0,1]^2 \to [0,1]$ satisfying the boundary condition $\mathcal{S}(x,0)=x$ for every $x \in [0,1]$.* □

Three well-known continuous t–conorms are ([12]):

- the standard t–conorm $\mathcal{S}_M(x,y) = \max\{x,y\}$
- the probabilistic sum $\mathcal{S}_P(x,y) = x+y-x*y$
- the Łukasiewicz t–conorm $\mathcal{S}_L(x,y) = \min\{1, x+y\}$.

Due to the associativity and commutativity of t–norms and t–conorms, these operators can be generalised for n–ary case, $n \geq 0$, in the following way: for any $A = \{x_1, \ldots, x_n\} \subseteq [0,1]$,

$$\mathcal{T}_{x\in A}(x) = \begin{cases} 1 & \text{iff } A=\emptyset \\ \mathcal{T}(\mathcal{T}(x_1,\ldots,x_{n-1}),x_n) & \text{otherwise} \end{cases} \tag{1}$$

$$\mathcal{S}_{x\in A}(x) = \begin{cases} 0 & \text{iff } A=\emptyset \\ \mathcal{S}(\mathcal{S}(x_1,\ldots,x_{n-1}),x_n) & \text{otherwise.} \end{cases} \tag{2}$$

Definition 5. *A **negator** is a decreasing mapping $\mathcal{N} : [0,1] \to [0,1]$ such that $\mathcal{N}(0)=1$ and $\mathcal{N}(1)=0$.* □

The standard negator $\mathcal{N}_s$ is defined by: $\mathcal{N}_s(x) = 1-x$, $x \in [0,1]$.
A negator $\mathcal{N}$ is called *involutive* (resp. *semi-involutive*) iff $\mathcal{N}(\mathcal{N}(x)) = x$ (resp. $\mathcal{N}(\mathcal{N}(x)) \geqslant x$) for every $x \in [0,1]$.

Definition 6. [12] *An **implicator** is any mapping $\mathcal{I} : [0,1]^2 \to [0,1]$ with decreasing first and increasing second partial mappings such that $\mathcal{I}(1,0)=0$ and $\mathcal{I}(0,0)=\mathcal{I}(0,1)=\mathcal{I}(1,1)=1$.* □

For a left–continuous t–norm $\mathcal{T}$, the *residual implicator based on $\mathcal{T}$* (the *residuum of $\mathcal{T}$*) is a mapping $\mathcal{I}_\mathcal{T}$ defined by:

$$\mathcal{I}_\mathcal{T}(x,y) = \sup\{\lambda \in [0,1] : \mathcal{T}(x,\lambda) \leqslant y\} \quad \text{for all } x,y \in [0,1]. \tag{3}$$

The following are the most popular examples of residual implicators ([12]:

- the Gödel implicator based on $\mathcal{T}_M$: $\mathcal{I}_G(x,y)=1$ for $x \leq y$ and $\mathcal{I}_G(x,y)=y$ elsewhere
- the Łukasiewicz implicator based on $\mathcal{T}_L$: $\mathcal{I}_L(x,y) = \min\{1, 1-x+y\}$
- the Gaines implicator based on $\mathcal{T}_P$: $\mathcal{I}_g(x,y)=1$ for $x \leq y$ and $\mathcal{I}_g(x,y) = \frac{y}{x}$ otherwise.

Given an implicator $\mathcal{I}$, the mapping $\mathcal{N}_\mathcal{I}(x)=\mathcal{I}(x,0)$, $x \in [0,1]$, is a negator, called a *negator induced by* $\mathcal{I}$. If $\mathcal{I}$ is the residuum of a left–continuous t–norm $\mathcal{T}$ then the negator induced by $\mathcal{I}$ will be denoted by $\mathcal{N}_\mathcal{T}$. By T^{inv} we will denote the class of all left–continuous t–norms such that $\mathcal{N}_\mathcal{T}$ are involutive.

Remark 2 In the literature some other classes of implicators are broadly studied ([11], [12]). In particular, given a t–conorm $\mathcal{S}$ and a negator $\mathcal{N}$, the mapping $\mathcal{I}(x,y)=\mathcal{S}(\mathcal{N}(x),y)$, $x,y \in [0,1]$, is called an *S–implicator based on* $\mathcal{S}$ *and* $\mathcal{N}$. The Kleene–Dienes implicator $\mathcal{I}_{KD}(x,y) = \max\{1-x, y\}$ for all $x,y \in [0,1]$, is the example of an S–implicator.

Also, it is worth noting that $\mathcal{N}_s$ is the nagator induced by the Łukasiewcz implicator $\mathcal{I}_L$, whereas the mapping $\mathcal{N}(0)=1$ and $\mathcal{N}(x)=0$ for every $x \in]0,1]$ is the negator induced both by $\mathcal{I}_G$ and $\mathcal{I}_g$. □

Let us recall several basic properties of residual implicators.

Proposition 2. *Let* $\mathcal{T}$ *be a left–continuous t–norm,* $\mathcal{I}$ *be its residuum and let* $\mathcal{N}=\mathcal{N}_\mathcal{T}$. *Then for all* $x,y,z \in [0,1]$,

(i) $\mathcal{T}(x,\mathcal{I}(x,y)) \leqslant y$
(ii) $x \leqslant y \Leftrightarrow \mathcal{I}(x,y)=1$
(iii) $\mathcal{I}(1,x)=x$
(iv) $\mathcal{N}$ *is semi-involutive*
(v) $\mathcal{N}(\mathcal{N}(\mathcal{N}(x))) = \mathcal{N}(x)$
(vi) $\mathcal{N}(\sup_{i\in I} x_i) = \inf_{i\in I} \mathcal{N}(x_i)$ *for any indexed family* $\{x_i\}_{i\in I} \subseteq [0,1]$
(vii) $\mathcal{I}(x,\mathcal{N}(y)) = \mathcal{N}(\mathcal{T}(x,y))$
(viii) $\mathcal{T}(\mathcal{I}(x,y),\mathcal{I}(y,z)) \leqslant \mathcal{I}(x,z)$
(ix) $\mathcal{I}(x,y) \leqslant \mathcal{I}(\mathcal{N}(y),\mathcal{N}(y))$
(x) $\mathcal{I}(\mathcal{T}(x,y),z) = \mathcal{I}(x,\mathcal{I}(y,z))$
(xi) *if* $\mathcal{T} \in T^+$ *then* $\mathcal{N}(x)=0$ *for every* $x \in]0,1]$.

Proof. **(i)–(x)** are proved in [9], [27]. To show **(xi)**, assume that $\mathcal{T} \in T^+$ is a left–continuous t–norm and $\mathcal{N}_\mathcal{T}(x_0)>0$ for some $x_0>0$. From the definition of $\mathcal{N}_\mathcal{T}$ we have $\mathcal{I}_\mathcal{T}(x_0,0)>0$. Put $\lambda=\mathcal{I}_\mathcal{T}(x_0,0)$. By (3) and left–continuity of $\mathcal{T}$ it gives $\mathcal{T}(x_0,\lambda)=0$, so $\mathcal{T}$ has zero divisors – a contradiction. ■

For a t–norm $\mathcal{T}$, a $\mathcal{T}$*-equivalence* is a mapping $\mathcal{E}_\mathcal{T} : [0,1]^2 \to [0,1]$ such that $\mathcal{E}_\mathcal{T}(x,x)=1$, $\mathcal{E}_\mathcal{T}(x,y)=\mathcal{E}_\mathcal{T}(y,x)$ and $\mathcal{T}(\mathcal{E}_\mathcal{T}(x,y),\mathcal{E}_\mathcal{T}(y,z)) \leqslant \mathcal{E}_\mathcal{T}(x,z)$ for all $x,y,z \in [0,1]$.

Proposition 3. [2] *For every left-continuous t-norm $\mathcal{T}$ and its residuum $\mathcal{I}$, the mapping $\mathcal{E}_\mathcal{T}(x,y)=\mathcal{T}(\mathcal{I}(x,y),\mathcal{I}(y,x))$, $x,y\in[0,1]$, is a $\mathcal{T}$-equivalence.* ■

Finally, the unary operator $\mathcal{M}:[0,1]\to[0,1]$ is defined by:[5]

$$\mathcal{M}(x)=\begin{cases}0 & \text{if } x=0\\ 1 & \text{if } x>0.\end{cases} \tag{4}$$

It can be shown that for every $\mathcal{T}\in T^+$, $\mathcal{M}(x)=\mathcal{N}_\mathcal{T}(\mathcal{N}_\mathcal{T}(x))$ for any $x\in[0,1]$.

4 Fuzzy sets and binary fuzzy relations

Let $\mathfrak{X}$ be a nonempty universe. A *fuzzy set A in $\mathfrak{X}$* is any mapping $A:\mathfrak{X}\to[0,1]$. By $\mathcal{F}(\mathfrak{X})$ we will denote the family of all fuzzy sets in $\mathfrak{X}$. For any $A\in\mathcal{F}(\mathfrak{X})$ and any $x\in\mathfrak{X}$, $A(x)$ is the *membership degree* to which x belongs to A.

Basic operations on sets in classical set theory are generalised in the following way ([11]). Let $\mathcal{T}$ and $\mathcal{S}$ be a t–norm and a t–conorm, respectively. By the $\mathcal{T}$–intersection and the $\mathcal{S}$–union of $A\in\mathcal{F}(\mathfrak{X})$ and $B\in\mathcal{F}(\mathfrak{X})$ we mean the fuzzy sets $A\cap_\mathcal{T}B$ and $A\cup_\mathcal{S}B$ in $\mathfrak{X}$ defined by: for every $x\in\mathfrak{X}$,

$$\begin{aligned}(A\cap_\mathcal{T}B)(x) &= \mathcal{T}(A(x),B(x))\\ (A\cup_\mathcal{S}B)(x) &= \mathcal{S}(A(x),B(x)).\end{aligned}$$

For an implicator $\mathcal{I}$, a $\mathcal{T}$–equivalence operator $\mathcal{E}_\mathcal{T}$ and $A,B\in\mathcal{F}(\mathfrak{X})$, we write $A\Rightarrow_\mathcal{I}B$ and $A\Leftrightarrow_{\mathcal{E}_\mathcal{T}}B$ to denote the fuzzy sets in $\mathfrak{X}$ given by: for every $x\in\mathfrak{X}$,

$$\begin{aligned}&(A\Rightarrow_\mathcal{I}B)(x)=\mathcal{I}(A(x),B(x)),\\ &(A\Leftrightarrow_{\mathcal{E}_\mathcal{T}}B)(x)=\mathcal{E}_\mathcal{T}(A(x),B(x)).\end{aligned}$$

For a negator $\mathcal{N}$, the *$\mathcal{N}$–complement of* $A\in\mathfrak{X}$ is the fuzzy set in $\mathfrak{X}$ given by:

$$(co_\mathcal{N}A)(x)=\mathcal{N}(A(x)) \quad \text{for every } x\in\mathfrak{X}.$$

For $A,B\in\mathcal{F}(\mathfrak{X})$, we will write $A\subseteq B$ iff $A(x)\leqslant B(x)$ for every $x\in\mathfrak{X}$.

A *binary fuzzy relation on $\mathfrak{X}$* is any mapping $R:\mathfrak{X}\times\mathfrak{X}\to[0,1]$, i.e, it is a fuzzy set on $\mathfrak{X}\times\mathfrak{X}$. The family of all binary fuzzy relations on $\mathfrak{X}$ will be denoted by $\mathcal{R}(\mathfrak{X})$. We will write $\mathbf{1}$ to denote the fuzzy universal relation, i.e. $\mathbf{1}(x,y)=1$ for all $x,y\in\mathfrak{X}$.

Let us recall some basic classes of binary fuzzy relations. Let $\mathcal{T}$ and $\mathcal{N}$ be a t–norm and a negator, respectively, and let $n\geq 1$ be an integer. A binary fuzzy relation $R\in\mathcal{R}(\mathfrak{X})$ is called

- *symmetric* iff $R(x,y)=R(y,x)$ for all $x,y\in\mathfrak{X}$

[5] In [9] this operator was denoted by ∇.

- *reflexive* iff $R(x,x)=1$ for every $x \in \mathfrak{X}$
  pseudo–reflexive iff $R(x,x)>0$ for every $x \in \mathfrak{X}$
  quasi–reflexive iff for every $x \in \mathfrak{X}$, $\sup_{y\in\mathfrak{X}} R(x,y)>0$ implies $R(x,x)>0$
  irreflexive iff $R(x,x)=0$ for every $x \in \mathfrak{X}$
  weakly irreflexive iff $R(x,x)=\inf_{y\in\mathfrak{X}} R(x,y)$ for every $x \in \mathfrak{X}$
- $\mathcal{T}^n$*–transitive* iff for all $x, x_1, \ldots, x_n, y \in \mathfrak{X}$,

$$\mathcal{T}\left(R(x,x_1), R(x_1,x_2), \ldots, R(x_n,y)\right) \leqslant R(x,y)$$

- $(\mathcal{T}^n,\mathcal{N})$*–cotransitive* iff $co_{\mathcal{N}}R$ is $\mathcal{T}^n$–transitive.

If R is $\mathcal{T}^1$–transitive then it is called $\mathcal{T}$–transitive. For a left–continuous t–norm $\mathcal{T}$, a $(\mathcal{T}^n,\mathcal{N}_{\mathcal{T}})$–cotransitive binary fuzzy relation R will be shortly called $\mathcal{T}^n$–cotransitive.

Observe that reflexivity implies pseudo–reflexivity which, in turn, implies quasi–reflexivity (the converse implications do not hold in general). Similarly, every irreflexive relation is weakly irreflexive (but not conversely).

A binary fuzzy relation R on $\mathfrak{X}$ is called a *tolerance* relation iff it is reflexive and symmetric. For a t–norm $\mathcal{T}$, a $\mathcal{T}$–transitive tolerance relation is called a $\mathcal{T}$*–equivalence* relation. A $\mathcal{T}$–equivalence relation R on $\mathfrak{X}$ is called a $\mathcal{T}$*–equality on* $\mathfrak{X}$ iff it satisfies the *separation property*:

$$R(x,y)=1 \text{ iff } x=y \text{ for all } x,y \in \mathfrak{X}.$$

4.1 Some binary fuzzy relations on $\mathcal{F}(\mathfrak{X})$

In this section we investigate some fuzzy relations between two fuzzy sets. More precisely, we will pay particular attention to fuzzy relations measuring degrees to which two fuzzy sets $A, B \in \mathcal{F}(\mathfrak{X})$ are either *compatible* or *incompatible.* For simplicity, we will use the same terminology as it was introduced for information relations in Section 2.

Given an implicator $\mathcal{I}$, the following binary fuzzy relation on $\mathcal{F}(\mathfrak{X})$ is called a *fuzzy inclusion*:

$$Inc_{\mathcal{I}}(A,B) = \inf_{x\in\mathfrak{X}} \mathcal{I}(A(x),B(x)) \quad \text{for all } A,B \in \mathcal{F}(\mathfrak{X}). \tag{5}$$

For $A, B \in \mathcal{F}(\mathfrak{X})$, $Inc_{\mathcal{I}}(A,B)$ is the degree to which A is included in B.
If $\mathcal{I}$ is the residuum of a left–continuous t–norm $\mathcal{T}$ then (5) will be referred to as a $\mathcal{T}$***–inclusion***.

Lemma 1. *If $\mathcal{I}$ is the residuum of a left–continuous t–norm $\mathcal{T}$ then $Inc_{\mathcal{I}}$ is reflexive and $\mathcal{T}$–transitive.*

Proof. Reflexivity of $Inc_{\mathcal{T}}$ results from Proposition 2(ii). To show that it is $\mathcal{T}$–transitive, note that for all $A, B, C \in \mathcal{F}(\mathfrak{X})$,

$$\begin{aligned} \mathcal{T}(Inc_{\mathcal{T}}(A,B), Inc_{\mathcal{T}}(B,C)) &= \mathcal{T}\Big(\inf_{x\in\mathfrak{X}} \mathcal{I}(A(x),B(x)), \inf_{x\in\mathfrak{X}} \mathcal{I}(B(x),C(x))\Big) \\ &\leqslant \inf_{x\in\mathfrak{X}} \mathcal{T}(\mathcal{I}(A(x),B(x)), \mathcal{I}(B(x),C(x))) \\ &\leqslant \inf_{x\in\mathfrak{X}} \mathcal{I}(A(x),C(x)) = Inc_{\mathcal{T}}(A,C) \end{aligned}$$

by Proposition 2(viii). ∎

By a *compatibility* relation we mean a binary fuzzy relation measuring the degree to which two fuzzy sets $A, B \in \mathcal{F}(\mathfrak{X})$ overlap. In the fuzzy literature a very popular measure of compatibility of two fuzzy sets is the following relation:

$$Com_{\mathcal{T}}(A,B) = \sup_{x\in\mathfrak{X}} \mathcal{T}(A(x),B(x)) \quad \text{for all } A, B \in \mathcal{F}(\mathfrak{X}),$$

where $\mathcal{T}$ is a t–norm. This relation is called a ***$\mathcal{T}$–compatibility relation***.

Lemma 2. *For every t–norm $\mathcal{T}$, $Com_{\mathcal{T}}$ is symmetric and, if $\mathcal{T} \in T^+$, quasi–reflexive.*

Proof. Symmetry of $Com_{\mathcal{T}}$ results from commutativity of t–norms. To show that $Com_{\mathcal{T}}$ is quasi–reflexive for $\mathcal{T} \in T^+$, consider an arbitrary $A \in \mathcal{F}(\mathfrak{X})$ and assume that $\sup_{B\in\mathcal{F}(\mathfrak{X})} Com_{\mathcal{T}}(A,B) > 0$. Then $Com_{\mathcal{T}}(A,B) > 0$ for some $B \in \mathcal{F}(\mathfrak{X})$, so $\mathcal{T}(A(x_0), B(x_0)) > 0$ for some $x_0 \in \mathfrak{X}$. Hence $A(x_0) > 0$. Since $\mathcal{T} \in T^+$, $\mathcal{T}(A(x_0), A(x_0)) > 0$, so $\sup_{x\in\mathfrak{X}} \mathcal{T}(A(x), A(x)) = Com_{\mathcal{T}}(A,A) > 0$. ∎

Next, let us consider a binary fuzzy relation measuring disjointness (also called *orthogonality*) of two fuzzy sets. Recall that two crisp sets $A \subseteq \mathfrak{X}$ and $B \subseteq \mathfrak{X}$ are orthogonal iff $A \subseteq co\, B$. In the fuzzy literature this very property is often used to measure the degree to which two fuzzy sets are disjoint (cf. [11]). Specifically, given an implicator $\mathcal{I}$ and a negator $\mathcal{N}$, let us consider the following binary fuzzy relation:

$$Ort_{\mathcal{I},\mathcal{N}}(A,B) = Inc_{\mathcal{I}}(A, co_{\mathcal{N}} B) \quad \text{for all } A, B \in \mathcal{F}(\mathfrak{X}).$$

For a left–continuous t–norm $\mathcal{T}$, its residuum $\mathcal{I}$ and $\mathcal{N} = \mathcal{N}_{\mathcal{T}}$, $Ort_{\mathcal{I},\mathcal{N}}$ is called a ***$\mathcal{T}$–orthogonality relation*** and denoted simply by $Ort_{\mathcal{T}}$.

Lemma 3.

(i) *$Ort_{\mathcal{T}}$ is symmetric*
(ii) *if $\mathcal{T} \in T^+$ then $Ort_{\mathcal{T}} = co_{\mathcal{N}} Com_{\mathcal{T}}$*
(iii) *if $\mathcal{T} \in T^+$ then $Ort_{\mathcal{T}}$ is weakly irreflexive.*

Proof. By way of example we show **(iii)**. Assume that $\mathcal{T} \in T^+$, but $Ort_\mathcal{T}$ is not weakly irreflexive, i.e.

$$\inf_{B \in \mathcal{F}(\mathfrak{X})} Ort_\mathcal{T}(A, B) < Ort_\mathcal{T}(A, A) \quad \text{for some } A \in \mathcal{F}(\mathfrak{X}).$$

Then $Ort_\mathcal{T}(A, B) < Ort_\mathcal{T}(A, A)$ for some $B \in \mathcal{F}(\mathfrak{X})$. By **(ii)**, this is equivalent to $\mathcal{N}(Com_\mathcal{T}(A, B)) < \mathcal{N}(Com_\mathcal{T}(A, A))$. Since $\mathcal{N}$ is induced by the residuum of $\mathcal{T}$, this means that $\mathcal{T}(Com_\mathcal{T}(A, B), \mathcal{N}(Com_\mathcal{T}(A, A))) > 0$. Next, since $\mathcal{T} \in T^+$, we have **(a)** $Com_\mathcal{T}(A, B) > 0$ and **(b)** $\mathcal{N}(Com_\mathcal{T}(A, A)) > 0$. From **(b)** we get $Com_\mathcal{T}(A, A) = 0$ by Proposition 2**(xi)**. Whence $A = \emptyset$, which contradicts **(a)**. ■

Remark 3 Note that in **(iii)** of Lemma 3 the assumption $\mathcal{T} \in T^+$ is essential. For the Łukasiewicz implicator $\mathcal{I}_L$ (based on $\mathcal{T} = \mathcal{T}_L$ with zero divisors) and $A \in \mathcal{F}(\mathfrak{X})$ such that $A(x) = \frac{1}{2}$ for every $x \in \mathfrak{X}$, by simple calculations we get $Ort_\mathcal{T}(A, A) = 1$. Next, $Ort_\mathcal{T}(A, \mathfrak{X}) = \frac{1}{2}$. Then $\inf_{B \in \mathcal{F}(\mathfrak{X})} Ort_\mathcal{T}(A, B) < Ort_\mathcal{T}(A, A)$, so $Ort_\mathcal{T}$ is not weakly irreflexive.

Observe also that for any t–norm $\mathcal{T} \in T^+$ and any fuzzy sets $A, B \in \mathcal{F}(\mathfrak{X})$, $Ort_\mathcal{T}(A, B) \in \{0, 1\}$, so $Ort_\mathcal{T}$ is actually a crisp relation. Indeed, for any $x \in \mathfrak{X}$ such that $B(x) = 0$ we have $\mathcal{N}(B(x)) = 1$ and $\mathcal{I}_\mathcal{T}(A(x), \mathcal{N}(B(x))) = 1$ by Proposition 2**(ii)**. Also, if $B(x) > 0$ then $\mathcal{N}(B(x)) = 0$ by Proposition 2**(xi)**. Therefore, $\mathcal{I}(A(x), \mathcal{N}(B(x))) = \mathcal{I}(A(x), 0) = \mathcal{N}(A(x))$ which is either 0 (if $A(x) > 0$) or 1 (if $A(x) = 0$). □

Consider now a fuzzy relation measuring degrees to which two fuzzy sets are equal. To this end, for a left–continuous t–norm $\mathcal{T}$ and its residuum $\mathcal{I}$, define the following binary fuzzy relation on $\mathcal{F}(\mathfrak{X})$:

$$Eq_\mathcal{T}(A, B) = \mathcal{T}(Inc_\mathcal{I}(A, B), Inc_\mathcal{I}(B, A)) \quad \text{for all } A, B \in \mathcal{F}(\mathfrak{X}). \tag{6}$$

Proposition 4. [2] *$Eq_\mathcal{T}$ is a $\mathcal{T}$-equality on $\mathcal{F}(\mathfrak{X})$.* ■

The relation $Eq_\mathcal{T}$ will be referred to as a ***$\mathcal{T}$–indiscernibility relation***.

Let us consider now a binary fuzzy relation representing degrees to which two fuzzy sets differ. Note first that for any crisp sets A and B:

$$A \neq B \Leftrightarrow (\exists x)(x \in A \wedge x \notin B) \vee (\exists x)(x \notin A \wedge x \in B). \tag{7}$$

The following binary fuzzy relation on $\mathcal{F}(\mathfrak{X})$ is a fuzzy generalisation of (7): for all $A, B \in \mathcal{F}(\mathfrak{X})$,

$$Div_{\mathcal{T},\mathcal{S},\mathcal{N}}(A, B) = \mathcal{S}\big(\sup_{x \in \mathfrak{X}} \mathcal{T}(A(x), \mathcal{N}(B(x))), \sup_{x \in \mathfrak{X}} \mathcal{T}(\mathcal{N}(A(x)), B(x))\big),$$

where $\mathcal{T}$, $\mathcal{S}$ and $\mathcal{N}$ is a t–norm, a t–conorm and a negator, respectively, or equivalently,

$$Div_{\mathcal{T},\mathcal{S},\mathcal{N}}(A, B) = \mathcal{S}(Com_\mathcal{T}(A, co_\mathcal{N} B), Com_\mathcal{T}(co_\mathcal{N} A, B)).$$

In particular, for a left–continuous t–norm $\mathcal{T}$, $\mathcal{N} = \mathcal{N}_{\mathcal{T}}$ and a t–conorm $\mathcal{S}$ we have a ***(T, S)–diversity relation***: for all $A, B \in \mathcal{F}(\mathfrak{X})$,

$$Div_{\mathcal{T},\mathcal{S}}(A, B) = \mathcal{S}(Com_{\mathcal{T}}(A, co_{\mathcal{N}}B), Com_{\mathcal{T}}(co_{\mathcal{N}}A, B)).$$

Lemma 4. *$Div_{\mathcal{T},\mathcal{S}}$ is irreflexive, symmetric and for $\mathcal{T} \in T^+$, $\mathcal{T}$–cotransitive.*

Proof. By way of example we show irreflexivity of $Div_{\mathcal{T},\mathcal{S}}$. Notice that for any $A \in \mathcal{F}(\mathfrak{X})$,

$$Com_{\mathcal{T}}(A, co_{\mathcal{N}}A) = \sup_{y \in \mathfrak{X}} \mathcal{T}(A(x), \mathcal{N}(A(x))) = \sup_{y \in \mathfrak{X}} \mathcal{T}(A(x), \mathcal{I}(A(x), 0)) = 0$$

by Proposition 2(**i**). Next, by symmetry of $Com_{\mathcal{T}}$,

$$Com_{\mathcal{T}}(co_{\mathcal{N}}A, A) = Com_{\mathcal{T}}(A, co_{\mathcal{N}}A) = 0.$$

Therefore we get $Div_{\mathcal{T},\mathcal{S}}(A, A) = \mathcal{S}(0, 0) = 0$. ∎

Two crisp sets $A \subseteq \mathfrak{X}$ and $B \subseteq \mathfrak{X}$ are called *complementary* iff $A = co\, B$. Note that this relation is irreflexive, symmetric and 2–transitive.

Given a t–norm $\mathcal{T}$ and a $\mathcal{T}$–equality $\mathcal{E}_{\mathcal{T}}$, a fuzzy complementarity relation can be defined in a straightforward way. In particular, for a left–continuous t–norm $\mathcal{T}$ and $\mathcal{N} = \mathcal{N}_{\mathcal{T}}$, we define a ***T–complementarity relation*** by:

$$Cmp_{\mathcal{T}}(A, B) = Eq_{\mathcal{T}}(A, co_{\mathcal{N}}B) \quad \text{for all } A, B \in \mathcal{F}(\mathfrak{X}),$$

where $Eq_{\mathcal{T}}$ is defined by (6).

Lemma 5.

(**i**) *$Cmp_{\mathcal{T}}$ is T^2–transitive*

(**ii**) *if $\mathcal{T} \in T^{inv}$ then $Cmp_{\mathcal{T}}$ is symmetric*

(**iii**) *if $\mathcal{T} \in T^+$ then $Cmp_{\mathcal{T}}$ is irreflexive.*

Proof. By way of example we show (**iii**). Assume that this is not the case, i.e. $Cmp_{\mathcal{T}}(A, A) > 0$ for some $A \in \mathcal{F}(\mathfrak{X})$. Then

$$Cmp_{\mathcal{T}}(A, A) = Eq_{\mathcal{T}}(A, co_{\mathcal{N}}A) = \mathcal{T}(Inc_{\mathcal{I}}(A, co_{\mathcal{N}}A), Inc_{\mathcal{I}}(co_{\mathcal{N}}A, A)) > 0,$$

where $\mathcal{I}$ is the residuum of $\mathcal{T}$. This means that (**a**) $Inc_{\mathcal{T}}(A, co_{\mathcal{N}}A) > 0$, and (**b**) $Inc_{\mathcal{T}}(co_{\mathcal{N}}A, A) > 0$. By Lemma 3(**ii**), from (**a**) we get $Inc_{\mathcal{T}}(A, co_{\mathcal{N}}A) = Ort_{\mathcal{T}}(A, A) = co_{\mathcal{N}}Com_{\mathcal{T}}(A, A) > 0$, or equivalently, $\mathcal{N}(Com_{\mathcal{T}}(A, A)) > 0$. Hence, since $\mathcal{T} \in T^+$, by Proposition 2(**xi**), $Com_{\mathcal{T}}(A, A) = 0$, so $A = \emptyset$. Then we get $Inc_{\mathcal{I}}(co_{\mathcal{N}}A, A) = Inc_{\mathcal{I}}(\mathfrak{X}, \emptyset) = 0$, which contradicts (**b**). ∎

Two crisp sets $A \subseteq \mathfrak{X}$ and $B \subseteq \mathfrak{X}$ are said to be *incomplementary* iff $A \neq co\, B$. Consider the following fuzzification of this relation: for a left-continuous t-norm $\mathcal{T}$, $\mathcal{N} = \mathcal{N}_\mathcal{T}$ and a t-conorm $\mathcal{S}$: for all $A, B \in \mathcal{F}(\mathfrak{X})$,

$$Icm_{\mathcal{T},\mathcal{S}}(A, B) = \mathcal{S}(Com_\mathcal{T}(A, B), Com_\mathcal{T}(co_\mathcal{N} A, co_\mathcal{N} B)).$$

This relation is called a ***$(\mathcal{T}, \mathcal{S})$–incomplementarity relation***. Its formulation is inspired by the following equivalence: for crisp sets $A, B \subseteq \mathfrak{X}$ it holds:

$$A \neq co\, B \Leftrightarrow (A \cap B \neq \emptyset) \vee (co\, A \cap co\, B \neq \emptyset).$$

Lemma 6. *$Icm_{\mathcal{T},\mathcal{S}}$ is symmetric and for $\mathcal{T} \in T^+$, pseudo-reflexive and $\mathcal{T}^2$-cotransitive.*

Proof. By way of example we show pseudo-reflexivity of $Icm_{\mathcal{T},\mathcal{S}}$ for $\mathcal{T} \in T^+$. Assume that it is not the case, i.e. for some $A \in \mathcal{F}(\mathfrak{X})$, $Icm_{\mathcal{T},\mathcal{S}}(A, A) = 0$. This means that $\mathcal{S}(Com_\mathcal{T}(A, A), Com_\mathcal{T}(co_\mathcal{N} A,\ co_\mathcal{N} A)) = 0$. Hence **(a)** $Com_\mathcal{T}(A, A) = 0$, and **(b)** $Com_\mathcal{T}(co_\mathcal{N} A, co_\mathcal{N} A) = 0$. Since $\mathcal{T} \in T^+$, **(a)** implies $A = \emptyset$. Then we get $Com_\mathcal{T}(co_\mathcal{N} A, co_\mathcal{N} A) = 1$ – a contradiction with **(b)**. ∎

Example 2 To illustrate relations discussed above, let us consider two fuzzy sets A and B on the set $\mathfrak{X} = \{a_1, \ldots, a_6\}$ given by:

$$\begin{aligned} A &= (0.9\ \ 1\ \ 0\ \ 0.5\ \ 0.8\ \ 0.3) \\ B &= (0.6\ \ 0.4\ \ 0.2\ \ 1\ \ 0.6\ \ 0). \end{aligned}$$

The table below shows values of particular fuzzy relations between A and B for the popular fuzzy logical operators.

	$\mathcal{T} = \mathcal{T}_L$, $\mathcal{S} = \mathcal{S}_L$	$\mathcal{T} = \mathcal{T}_P$, $\mathcal{S} = \mathcal{S}_P$	$\mathcal{T} = \mathcal{T}_Z$, $\mathcal{S} = \mathcal{S}_Z$
$Eq_\mathcal{T}(A, B)$	0	0	0
$Div_{\mathcal{T},\mathcal{S}}(A, B)$	1	0.2	0.2
$Com_\mathcal{T}(A, B)$	0.5	0.54	0.6
$Ort_\mathcal{T}(A, B)$	0.5	0	0
$Cmp_\mathcal{T}(A, B)$	0.2	0	0
$Icm_\mathcal{T}(A, B)$	0.8	0.54	0.6

5 Fuzzy information relations

In this section we will consider information systems with fuzzy data. We will define several fuzzy relations derived from such systems and give basic properties of these relations.

To begin with, let us introduce a fuzzy generalisation of the notion of an information system.

Definition 7. *A* ***fuzzy information system*** *is a tuple* $\Sigma = (OB, AT, \{V_a : a \in AT\})$*, where* OB *is a nonempty set of objects,* AT *is a nonempty and finite set of attributes and for every* $a \in AT$*,* V_a *is a nonempty set. Every* $a \in AT$ *is a mapping* $a : OB \to \mathcal{F}(V_a)$. □

Intuitively, for every object $x \in OB$, every attribute $a \in AT$ and every value $v \in V_a$, $a(x)(v)$ is the degree to which the object x has the value v on the attribute a.

Example 3 Consider a set $OB = \{P_1, \ldots, P_7\}$ of seven people characterised by languages they speak (LAN) and their personal interests (INT). We have then the fuzzy information system $\Sigma = (OB, AT, \{V_a : a \in AT\})$, where

$$AT = \{\text{LAN}, \text{INT}\}$$
$$V_{\text{INT}} = \{\text{Fine art}, \text{Music}, \text{Sport}, \text{Travelling}, \}$$
$$V_{\text{LAN}} = \{\text{English}, \text{French}, \text{Dutch}, \text{Spanish}, \text{Italian}\}$$

and the values of attributes are given in Table 2 below.

OB	*Languages* (LAN)	*Interests* (INT)
P_1	(0.8 0.6 1 0.1 0)	(0.2 1 0 0.8)
P_2	(0.9 0.6 1 0.2 0)	(0.2 0.9 0.1 0.8)
P_3	(0 0.4 0.1 0.9 0.8)	(0.3 0 0.1 0)
P_4	(0.1 0.4 0 0.8 0.9)	(0.8 0.1 1 0.2)
P_5	(0 0.7 0 0.6 0.9)	(0.2 0.6 0 0.7)
P_6	(0 1 0 1 0)	(1 1 0 1)
P_7	(0 1 0 0 1)	(0 0 1 0)

Table 2: Fuzzy information system

For $a \in AT$, $x \in OB$ and $V_a = \{v_1, \ldots, v_n\}$, $(\delta_1, \ldots, \delta_n)$ stands for the membership function $a(x)(v_i) = \delta_i$, $i = 1, \ldots, n$.

We know that P_1 speaks English fluently (to the degree 0.8), quite well French (to the degree 0.6), perfectly Dutch (to the degree 1), but his Spanish is very weak (to the degree 0.1) and he cannot speak Italian at all. Also, he particularly interests in music (to the degree 1), is very fond of travelling (to the degree 0.8), does not like fine art much (to the degree 0.2) and is not interested in sport at all. Note also that P_6 and P_7 have crisp characteristics: all we know of them is which language they can speak (resp. do not speak) and what they are (resp. are not) interest in. □

Definition 8. *Let* $\mathcal{T}$ *be a left–continuous t–norm and let* $\mathcal{S}$ *be a t–conorm. For any fuzzy information system* $\Sigma = (OB, AT, \{V_a : a \in AT\})$ *and any* $\mathcal{A} \subseteq AT$*, let us define the following classes of binary fuzzy relations on* $\mathcal{F}(OB)$ *in* Σ*, called* ***$(\mathcal{T}, \mathcal{S})$–information relations in Σ induced by $\mathcal{A}$****: for all* $x, y \in OB$,

A. FUZZY INDISTINGUISHABILITY RELATIONS:

(1) *strong and weak fuzzy indiscernibility relations:*

$$\mathsf{ind}^{s}_{\mathcal{T}}(\mathcal{A})(x,y) = \mathcal{T}_{a\in\mathcal{A}}\big(Eq_{\mathcal{T}}(a(x),a(y))\big)$$
$$\mathsf{ind}^{w}_{\mathcal{T},\mathcal{S}}(\mathcal{A})(x,y) = \mathcal{S}_{a\in\mathcal{A}}\big(Eq_{\mathcal{T}}(a(x),a(y))\big)$$

(2) *strong and weak fuzzy compatibility relations:*

$$\mathsf{com}^{s}_{\mathcal{T}}(\mathcal{A})(x,y) = \mathcal{T}_{a\in\mathcal{A}}\big(Com_{\mathcal{T}}(a(x),a(y))\big)$$
$$\mathsf{com}^{w}_{\mathcal{T},\mathcal{S}}(\mathcal{A})(x,y) = \mathcal{S}_{a\in\mathcal{A}}\big(Com_{\mathcal{T}}(a(x),a(y))\big)$$

(3) *strong and weak fuzzy incomplementarity relations:*

$$\mathsf{icm}^{s}_{\mathcal{T},\mathcal{S}}(\mathcal{A})(x,y) = \mathcal{T}_{a\in\mathcal{A}}\big(Icm_{\mathcal{T},\mathcal{S}}(a(x),a(y))\big)$$
$$\mathsf{icm}^{w}_{\mathcal{T},\mathcal{S}}(\mathcal{A})(x,y) = \mathcal{S}_{a\in\mathcal{A}}\big(Icm_{\mathcal{T},\mathcal{S}}(a(x),a(y))\big)$$

B. FUZZY DISTINGUISHABILITY RELATIONS:

(4) *strong and weak fuzzy diversity relations:*

$$\mathsf{div}^{s}_{\mathcal{T},\mathcal{S}}(\mathcal{A})(x,y) = \mathcal{T}_{a\in\mathcal{A}}\big(Div_{\mathcal{T},\mathcal{S}}(a(x),a(y))\big)$$
$$\mathsf{div}^{w}_{\mathcal{T},\mathcal{S}}(\mathcal{A})(x,y) = \mathcal{S}_{a\in\mathcal{A}}\big(Div_{\mathcal{T},\mathcal{S}}(a(x),a(y))\big)$$

(5) *strong and weak fuzzy orthogonality relations:*

$$\mathsf{ort}^{s}_{\mathcal{T}}(\mathcal{A})(x,y) = \mathcal{T}_{a\in\mathcal{A}}\big(Ort_{\mathcal{T}}(a(x),a(y))\big)$$
$$\mathsf{ort}^{w}_{\mathcal{T},\mathcal{S}}(\mathcal{A})(x,y) = \mathcal{S}_{a\in\mathcal{A}}\big(Ort_{\mathcal{T}}(a(x),a(y))\big)$$

(6) *strong and weak fuzzy complementarity relations:*

$$\mathsf{cmp}^{s}_{\mathcal{T}}(\mathcal{A})(x,y) = \mathcal{T}_{a\in\mathcal{A}}\big(Cmp_{\mathcal{T}}(a(x),a(y))\big)$$
$$\mathsf{cmp}^{w}_{\mathcal{T},\mathcal{S}}(\mathcal{A})(x,y) = \mathcal{S}_{a\in\mathcal{A}}\big(Cmp_{\mathcal{T}}(a(x),a(y))\big). \qquad \square$$

We will write $\mathsf{INF}_{\mathcal{T},\mathcal{S}}(\Sigma)$ ($\mathsf{INF}^{s}_{\mathcal{T},\mathcal{S}}(\Sigma)$, $\mathsf{INF}^{w}_{\mathcal{T},\mathcal{S}}(\Sigma)$) to denote the class of all (strong, weak) $(\mathcal{T},\mathcal{S})$–information relations in Σ; $\mathsf{INF}_{\mathcal{T},\mathcal{S}}$ will stand for the class of all $(\mathcal{T},\mathcal{S})$–information relations in every Σ.

Example 3 (cont.) Taking the Łukasiewicz logical operators $\mathcal{T}_L$, $\mathcal{I}_L$, $\mathcal{S}_L$ and $\mathcal{N}_s$, by simple calculations one can easily check that $Eq_{\mathcal{T}_L}(\text{LAN}(P_1),\text{LAN}(P_2)) = 0.9$ and $Eq_{\mathcal{T}_L}(\text{INT}(P_1),\text{INT}(P_2)) = 0.8$, whence P_1 and P_2 are strongly $\mathcal{T}_L$–indiscernible wrt both attributes to the degree $\mathcal{T}_L(0.9,0.8) = 0.7$.
Moreover, since $Ort_{\mathcal{T}_L}(\text{LAN}(P_2),\text{LAN}(P_3)) = 0.9$ and $Ort_{\mathcal{T}_L}(\text{INT}(P_2),\text{INT}(P_3)) = 1$, P_2 and P_3 are strongly $\mathcal{T}_L$–orthogonal wrt AT to the degree $\mathcal{T}_L(0.9,1) = 0.9$.
We also have $Cmp_{\mathcal{T}_L}(\text{LAN}(P_2),\text{LAN}(P_4)) = Cmp_{\mathcal{T}_L}(\text{INT}(P_2),\text{INT}(P_4)) = 0.9$, so P_2 and P_4 are strongly $\mathcal{T}_L$–complementary wrt AT to the degree $\mathcal{T}_L(0.9,0.9) = 0.8$.

Next, P_4 and P_5 are weakly $(\mathcal{T}_L, \mathcal{S}_L)$–compatible wrt AT to the degree 0.9 (since $Com_{\mathcal{T}_L}(\text{LAN}(P_4), \text{LAN}(P_5)) = 0.8$ and $Com_{\mathcal{T}_L}(\text{INT}(P_4), \text{INT}(P_5)) = 0.1$). Finally, since P_6 and and P_7 have crisp characteristics, for any t-norm $\mathcal{T}$ and any t–conorm $\mathcal{S}$, they are totally weakly $(\mathcal{T}, \mathcal{S})$–compatible wrt AT (i.e. to the degree 1), since they both speak French. □

Remark 4 Note that a $(\mathcal{T}, \mathcal{S})$–information relation $\mathsf{inf} \in \mathsf{INF}_{\mathcal{T},\mathcal{S}}(\Sigma)$ might be viewed as a mapping $\mathsf{inf} : \wp(AT) \to \mathcal{R}(OB)$. Also, for every $a \in AT$ and for every $\mathsf{inf} \in \mathsf{INF}_{\mathcal{T},\mathcal{S}}(\Sigma)$, it holds $\mathsf{inf}^s(\{a\}) = \mathsf{inf}^w(\{a\})$. □

Proposition 5. *For every fuzzy information system Σ and for every strong $(\mathcal{T}, \mathcal{S})$–information relation we have $\mathsf{inf}^s_{\mathcal{T},\mathcal{S}}(\emptyset) = 1$, and for every weak $(\mathcal{T}, \mathcal{S})$–information relation it holds $\mathsf{inf}^w_{\mathcal{T},\mathcal{S}}(\emptyset) = \emptyset$.*

Proof. Follows from (1) and (2). ■

The following theorem provides basic properties of $(\mathcal{T}, \mathcal{S})$–information relations.

Theorem 2. *Let $\mathcal{T}$ be a left–continuous t–norm and let $\mathcal{S}$ be a t–conorm. Then for every fuzzy information system $\Sigma = (OB, AT, \{V_a : a \in AT\})$ and for every $\mathcal{A} \subseteq AT$,*

1. *$\mathsf{ind}^s_{\mathcal{T}}(\mathcal{A})$ is a $\mathcal{T}$–equivalence relation*
   $\mathsf{ind}^w_{\mathcal{T},\mathcal{S}}(\mathcal{A})$ is a fuzzy tolerance relation for $\mathcal{A} \neq \emptyset$, and for any $a \in AT$, $\mathsf{ind}^w_{\mathcal{T},\mathcal{S}}(\{a\})$ is a $\mathcal{T}$–equivalence relation
2. *$\mathsf{com}^s_{\mathcal{T}}(\mathcal{A})$ and $\mathsf{com}^w_{\mathcal{T},\mathcal{S}}(\mathcal{A})$ are symmetric and for $\mathcal{T} \in T^+$, quasi–reflexive*
3. *$\mathsf{icm}^s_{\mathcal{T},\mathcal{S}}(\mathcal{A})$ and $\mathsf{icm}^w_{\mathcal{T},\mathcal{S}}(\mathcal{A})$ are symmetric*
   for $\mathcal{T} \in T^+$,
   $\mathsf{icm}^s_{\mathcal{T},\mathcal{S}}(\mathcal{A})$ (resp. $\mathsf{icm}^w_{\mathcal{T},\mathcal{S}}(\mathcal{A})$ for $\mathcal{A} \neq \emptyset$) is pseudo–reflexive
   $\mathsf{icm}^w_{\mathcal{T},\mathcal{S}}(\mathcal{A})$ (resp. $\mathsf{icm}^s_{\mathcal{T},\mathcal{S}}(\{a\})$ for any $a \in AT$) is $\mathcal{T}^2$–cotransitive
4. *$\mathsf{div}^s_{\mathcal{T},\mathcal{S}}(\mathcal{A})$ is symmetric and for $\mathcal{A} \neq \emptyset$, irreflexive*
   $\mathsf{div}^w_{\mathcal{T},\mathcal{S}}(\mathcal{A})$ is symmetric and irreflexive
   for $\mathcal{T} \in T^+$, $\mathsf{div}^w_{\mathcal{T},\mathcal{S}}(\mathcal{A})$ (resp. $\mathsf{div}^s_{\mathcal{T},\mathcal{S}}(\{a\})$ for any $a \in AT$) is $\mathcal{T}$-cotransitive
5. *$\mathsf{ort}^s_{\mathcal{T}}(\mathcal{A})$ and $\mathsf{ort}^w_{\mathcal{T},\mathcal{S}}(\mathcal{A})$ are symmetric and for $\mathcal{T} \in T^+$, weakly irreflexive*
6. *$\mathsf{cmp}^s_{\mathcal{T}}(\mathcal{A})$ (resp. $\mathsf{cmp}^w_{\mathcal{T}}(\{a\})$ for any $a \in AT$) is $\mathcal{T}^2$–transitive*
   for $\mathcal{T} \in T^{inv}$, $\mathsf{cmp}^s_{\mathcal{T}}(\mathcal{A})$ and $\mathsf{cmp}^w_{\mathcal{T},\mathcal{S}}(\mathcal{A})$ are symmetric
   for $\mathcal{T} \in T^+$, $\mathsf{cmp}^w_{\mathcal{T},\mathcal{S}}(\mathcal{A})$ (resp. $\mathsf{cmp}^s_{\mathcal{T}}(\mathcal{A})$ for $\mathcal{A} \neq \emptyset$) is irreflexive.

Proof. Let us show (i). By Proposition 4, $Eq_{\mathcal{T}}$ is a $\mathcal{T}$–equivalence relation, so clearly $\mathsf{ind}^s_{\mathcal{T}}(\mathcal{A})(x, y)$ is reflexive and symmetric. To show that it is also

$\mathcal{T}$–transitive, for any $\mathcal{A} \subseteq AT$ and any $x, y, z \in OB$, let us consider

$$\begin{aligned}
&\mathcal{T}(\mathsf{ind}^s_{\mathcal{T}}(\mathcal{A})(x,y), \mathsf{ind}^s_{\mathcal{T}}(\mathcal{A})(y,z)) \\
&\quad = \mathcal{T}\big(\mathcal{T}_{a\in\mathcal{A}}(Eq_{\mathcal{T}}(a(x),a(y))), \mathcal{T}_{a\in\mathcal{A}}(Eq_{\mathcal{T}}(a(y),a(z)))\big) \\
&\quad = \mathcal{T}_{a\in\mathcal{A}}\big(\mathcal{T}(Eq_{\mathcal{T}}(a(x),a(y)), Eq_{\mathcal{T}}(a(y),a(z)))\big)
\end{aligned}$$

by associativity of t–norms. Hence, since $Eq_{\mathcal{T}}$ is $\mathcal{T}$–transitive, we get

$$\mathcal{T}(\mathsf{ind}^s_{\mathcal{T}}(\mathcal{A})(x,y), \mathsf{ind}^s_{\mathcal{T}}(\mathcal{A})(y,z)) \leqslant \mathcal{T}_{a\in\mathcal{A}}(Eq_{\mathcal{T}}(a(x),a(z))),$$

so $\mathsf{ind}^s_{\mathcal{T}}(\mathcal{A})$ is also $\mathcal{T}$–transitive for every $\mathcal{A} \subseteq AT$.
By Proposition 5, $\mathsf{ind}^w(\emptyset)$ is irreflexive. For every $\mathcal{A} \neq \emptyset$, $\mathsf{ind}^w_{\mathcal{T},\mathcal{S}}(\mathcal{A})$ is a tolerance relation by Proposition 4. Moreover, by Remark 4, for every $a \in AT$, $\mathsf{ind}^w_{\mathcal{T},\mathcal{S}}(\{a\}) = \mathsf{ind}^s_{\mathcal{T}}(\{a\})$, so it is a $\mathcal{T}$–equivalence relation.

The remaining properties can be proved in the analogous way. ■

6 Fuzzy information operators

In this section we define several fuzzy information operators and provide characterisations of some binary fuzzy relations using these operators.

Definition 9. *Any mapping $\Omega : \mathcal{R}(\mathfrak{X}) \times \mathcal{F}(\mathfrak{X}) \rightarrow \mathcal{F}(\mathfrak{X})$ is called a **fuzzy information operator**.* □

In the following we will consider specific classes of fuzzy information operators, each one is determined by a left–continuous t–norm $\mathcal{T}$.

Definition 10. *Let $\mathcal{T}$ be a left–continuous t–norm, $\mathcal{I}$ be the residuum of $\mathcal{T}$ and let $\mathcal{N} = \mathcal{N}_{\mathcal{T}}$. The following fuzzy information operators, called **$\mathcal{T}$–information operators**, are defined as follows: for any $R \in \mathcal{R}(\mathfrak{X})$, any $A \in \mathcal{F}(\mathfrak{X})$ and any $x \in \mathfrak{X}$,*

- **(O.1)** $([R]_{\mathcal{T}} A)(x) = \inf_{y\in\mathfrak{X}} \mathcal{I}_{\mathcal{T}}(R(x,y), A(y))$
- **(O.2)** $(\langle R\rangle_{\mathcal{T}} A)(x) = \sup_{y\in\mathfrak{X}} \mathcal{T}(R(x,y), A(y))$
- **(O.3)** $([\![R]\!]_{\mathcal{T}} A)(x) = \inf_{y\in\mathfrak{X}} \mathcal{I}_{\mathcal{T}}(A(y), R(x,y))$
- **(O.4)** $(\langle\!\langle R\rangle\!\rangle_{\mathcal{T}} A)(x) = \sup_{y\in\mathfrak{X}} \mathcal{T}(\mathcal{N}(R(x,y)), \mathcal{N}(A(y)))$
- **(O.5)** $(\mathbf{[}R]_{\mathcal{T}} A)(x) = \inf_{y\in\mathfrak{X}} \mathcal{I}_{\mathcal{T}}(\mathcal{M}(R(x,y)), A(y))$
- **(O.6)** $(\boldsymbol{\langle} R\rangle_{\mathcal{T}} A)(x) = \sup_{y\in\mathfrak{X}} \mathcal{T}(\mathcal{M}(R(x,y)), A(y))$
- **(O.7)** $([\![R]_{\mathcal{T}} A)(x) = \inf_{y\in\mathfrak{X}} \mathcal{I}(\mathcal{N}(R(x,y)), A(y))$
- **(O.8)** $(\langle\!\langle R\rangle_{\mathcal{T}} A)(x) = \sup_{y\in\mathfrak{X}} \mathcal{T}(\mathcal{N}(R(x,y)), A(y))$. □

Notice that the operators **(O.1)** and **(O.2)** are fuzzy counterparts of the necessity and the possibility operators[6], whereas **(O.3)** and **(O.4)** are fuzzy generalisations of the sufficiency and the impossibility operators introduced in [10].

In crisp case it is well–known (see e.g. [16]) that information operators $[\,]$ and $\langle\rangle$ (resp. $[\![\,]\!]$ and $\langle\!\langle\,\rangle\!\rangle$) are dual in the sense that for every binary relation R on $\mathfrak{X}$ and every $A \subseteq \mathfrak{X}$, $[R]A = co\,\langle R\rangle co\,A$ (resp. $[\![R]\!] = co\,\langle\!\langle R\rangle\!\rangle co\,A$). In fuzzy case, however, duality does not hold in general. We have, however, the following weaker property. First, let us introduce the following notion.

Definition 11. *Let $\mathcal{N}$ be a negator and let Ω_1 and Ω_2 be fuzzy information operators. We say that Ω_1 and Ω_2 are $\mathcal{N}$–**semidual** iff for every $R \in \mathcal{R}(\mathfrak{X})$ and every $A \in \mathcal{F}(\mathfrak{X})$, $\Omega_i(R)(A) \subseteq co_{\mathcal{N}}\,\Omega_j(R)(co_{\mathcal{N}}A)$, $i,j = 1,2$, $i \neq j$.* □

Proposition 6. *The following pairs of $\mathcal{T}$–information operators are semidual wrt $\mathcal{N}_{\mathcal{T}}$: $[\,]_{\mathcal{T}}$ and $\langle\rangle_{\mathcal{T}}$, $[\![\,]\!]_{\mathcal{T}}$ and $\langle\!\langle\,\rangle\!\rangle_{\mathcal{T}}$, $[\![\,]_{\mathcal{T}}$ and $\langle\!\langle\,\rangle_{\mathcal{T}}$, $\mathbf{[\,]}_{\mathcal{T}}$ and $\mathbf{\langle\,\rangle}_{\mathcal{T}}$.*

Proof. By way of example we show that $[\![\,]\!]_{\mathcal{T}}$ and $\langle\!\langle\,\rangle\!\rangle_{\mathcal{T}}$ are $\mathcal{N}$–semidual. For every $R \in \mathcal{R}(\mathfrak{X})$, every $A \in \mathcal{F}(\mathfrak{X})$ and every $x \in \mathfrak{X}$ we have:

$$\begin{aligned} co_{\mathcal{N}}(\langle\!\langle R\rangle\!\rangle_{\mathcal{T}}\, co_{\mathcal{N}}A)(x) &= \mathcal{N}\Big(\sup_{y\in\mathfrak{X}} \mathcal{T}(\mathcal{N}(R(x,y)), \mathcal{N}(\mathcal{N}(A(y))))\Big) \\ &= \inf_{y\in\mathfrak{X}} \mathcal{N}\Big(\mathcal{T}(\mathcal{N}(R(x,y)), \mathcal{N}(\mathcal{N}(A(y))))\Big) \end{aligned}$$

by Proposition 2**(vi)**. Next, by Proposition 2**(vii)**

$$\begin{aligned} &\inf_{y\in\mathfrak{X}} \mathcal{N}\Big(\mathcal{T}(\mathcal{N}(R(x,y)), \mathcal{N}(\mathcal{N}(A(y))))\Big) \\ &\qquad = \inf_{y\in\mathfrak{X}} \mathcal{I}\Big(\mathcal{N}(R(x,y)), \mathcal{N}(\mathcal{N}(\mathcal{N}(A(y))))\Big). \end{aligned}$$

Furthermore, by Proposition 2**(v)**,

$$\inf_{y\in\mathfrak{X}} \mathcal{I}(\mathcal{N}(R(x,y)), \mathcal{N}(\mathcal{N}(\mathcal{N}(A(y))))) = \inf_{y\in\mathfrak{X}} \mathcal{I}(\mathcal{N}(R(x,y)), \mathcal{N}(A(y))).$$

Finally, by Proposition 2**(x)**,

$$\inf_{y\in\mathfrak{X}} \mathcal{I}(\mathcal{N}(R(x,y)), \mathcal{N}(A(y))) \geqslant \inf_{y\in\mathfrak{X}} \mathcal{I}(A(y), R(x,y)) = ([\![R]\!]_{\mathcal{T}} A)(x).$$

Therefore, $[\![R]\!]_{\mathcal{T}} A \subseteq co_{\mathcal{N}}(\langle\!\langle R\rangle\!\rangle_{\mathcal{T}} A)$. Proceeding in the similar way one can show that $\langle\!\langle R\rangle\!\rangle_{\mathcal{T}} A \subseteq co_{\mathcal{N}}[\![R]\!]_{\mathcal{T}} A$. Hence $[\![\,]\!]_{\mathcal{T}}$ and $\langle\!\langle\,\rangle\!\rangle_{\mathcal{T}}$ are $\mathcal{N}$–semidual. ■

Fuzzy information operators are useful to characterise properties of binary fuzzy relations. The next theorem provides characterisations of particular fuzzy relations in terms of $\mathcal{T}$–information operators.

---

[6] In fuzzy modal logics **(O.1)** and **(O.2)** are used to interprete necessity □ and possibility ◇ operators, respectively (see, for example, [8]).

The following auxiliary notation will be used. For $R \in \mathcal{R}(\mathfrak{X})$, an integer $n \geqslant 1$ and a fuzzy information operator $\Omega(R)$, we will write $\Omega^n(R)(A)$ to denote $\Omega(R)(A)$ iff $n=1$ and $\Omega(R)(\Omega^{n-1}(R)(A))$ for $n>1$.

Theorem 3. *Let $\mathcal{T}$ and $\mathcal{I}$ be a left–continuous t–norm and its residuum, respectively, and let $R \in \mathcal{R}(\mathfrak{X})$. For any $A \in \mathcal{F}(\mathfrak{X})$ and any integer $n \geqslant 1$,*

(i)	*R is reflexive*	*iff*	$[R]_{\mathcal{T}} A \subseteq A$
(ii)	*R is pseudo–reflexive*	*iff*	$[R\,]_{\mathcal{T}} A \subseteq A$
(iii)	*R is quasi–reflexive*	*iff*	$A \cap_{\mathcal{T}} [R\,]_{\mathcal{T}} \mathfrak{X} \subseteq [R\,]_{\mathcal{T}} A$
(iv)	*R is irreflexive*	*iff*	$[\![R]\!]_{\mathcal{T}} A \subseteq co_{\mathcal{N}} A$
(v)	*R is weakly irreflexive*	*iff*	$A \cap_{\mathcal{T}} [\![R]\!]_{\mathcal{T}} A \subseteq [\![R]\!]_{\mathcal{T}} \mathfrak{X}$
(vi)	*R is symmetric*	*iff*	$A \subseteq [R]_{\mathcal{T}} \langle R \rangle_{\mathcal{T}} A$
(vii)	*R is $\mathcal{T}^n$–transitive*	*iff*	$[R]^n_{\mathcal{T}} A \subseteq [R]^{n+1}_{\mathcal{T}} A.$
(viii)	*R is $\mathcal{T}^n$–cotransitive*	*iff*	$[\![R]\!]^n_{\mathcal{T}} A \subseteq [\![R]\!]^{n+1}_{\mathcal{T}} A.$

Proof. (i) and (iv)–(viii) are proved in [24].

(ii) ($\Rightarrow$) Assume that for some $A \in \mathcal{F}(\mathfrak{X})$ it holds $A \cap_{\mathcal{T}} \langle R \rangle_{\mathcal{T}} \mathfrak{X} \not\subseteq \langle R \rangle_{\mathcal{T}} A$, i.e. there is $x_0 \in \mathcal{F}(\mathfrak{X})$ such that

$$\mathcal{T}\big(A(x_0), \sup_{y \in \mathfrak{X}} \mathcal{T}(\mathcal{M}(R(x_0,y)), 1)\big) > \sup_{y \in \mathfrak{X}} \mathcal{T}(\mathcal{M}(R(x_0,y)), A(y)). \tag{8}$$

We show that R is not quasi–reflexive, i.e. for some $x_0 \in \mathfrak{X}$,

$$\sup\nolimits_{y \in \mathfrak{X}} R(x_0,y) > 0 \tag{9}$$
$$R(x_0,x_0) = 0. \tag{10}$$

From (8), $\mathcal{T}\big(A(x_0), \sup_{y \in \mathfrak{X}} \mathcal{M}(R(x_0,y))\big) > \mathcal{T}(\mathcal{M}(R(x_0,x_0)), A(x_0))$, so

$$\sup_{y \in \mathfrak{X}} \mathcal{M}(R(x_0,y)) > \mathcal{M}(R(x_0,x_0)).$$

Hence $\mathcal{M}(R(x_0,y_0)) > \mathcal{M}(R(x_0,x_0))$ for some $y_0 \in \mathfrak{X}$. By the definition (4) this is equivalent to

$$R(x_0,y_0) > 0 \tag{11}$$
$$R(x_0,x_0) = 0. \tag{12}$$

(11) implies that $\sup_{y \in \mathfrak{X}} R(x_0,y) > 0$, which together with (12) means that R is not quasi–reflexive.

($\Leftarrow$) Assume that R is not quasi–reflexive, i.e. (9)–(10) hold. Put $A = \{x_0\}$. Then we have:

$$\begin{aligned}(A \cap_{\mathcal{T}} \langle R \rangle_{\mathcal{T}} \mathfrak{X})(x_0) &= \mathcal{T}\big(A(x_0), \sup_{y \in \mathfrak{X}} \mathcal{T}(\mathcal{M}(R(x_0,y)), 1)\big) \\ &= \sup_{y \in \mathfrak{X}} \mathcal{M}(R(x_0,y)) = 1.\end{aligned}$$

Also, one can easily note that

$$(\langle R\rangle_{\mathcal{T}} A)(x_0) = \sup_{y\in\mathfrak{X}} \mathcal{T}(\mathcal{M}(R(x_0,y)), A(y)) = \mathcal{M}(R(x_0,x_0)),$$

which in view of (10) gives $(\langle R\rangle_{\mathcal{T}} A)(x_0)=0$. Therefore we get

$$(A \cap_{\mathcal{T}} \langle R\rangle_{\mathcal{T}} \mathfrak{X})(x_0) > (\langle R\rangle_{\mathcal{T}} A)(x_0).$$

Obviously, $A \cap_{\mathcal{T}} \langle R\rangle_{\mathcal{T}} \mathfrak{X} \not\subseteq \langle R\rangle_{\mathcal{T}} A$.

(iii) ($\Rightarrow$) Assume that R is pseudo–reflexive, i.e. for every $x\in\mathfrak{X}$, $R(x,x)>0$. From (4) we get $\mathcal{M}(R(x,x))=1$. For every $A\in\mathcal{F}(\mathfrak{X})$ and every $x\in\mathfrak{X}$ we have:

$$\begin{aligned}([R]_{\mathcal{T}} A)(x) &= \inf_{y\in\mathfrak{X}} \mathcal{I}(\mathcal{M}(R(x,y)), A(y)) \leqslant \mathcal{I}(\mathcal{M}(R(x,x)), A(x))\\ &= \mathcal{I}(1, A(x)) = A(x)\end{aligned}$$

by Proposition 2(iii). Hence $[R]_{\mathcal{T}} A \subseteq A$.

($\Leftarrow$) Assume that R is not pseudo–reflexive, that is $R(x_0,x_0)=0$ for some $x_0\in\mathfrak{X}$. Then $\mathcal{M}(R(x_0,x_0))=0$. Consider $A=\mathfrak{X}\setminus\{x_0\}$. Notice that

$$\begin{aligned}([R]_{\mathcal{T}} A)(x_0) &= \inf_{y\in\mathfrak{X}} \mathcal{I}(\mathcal{M}(R(x_0,y)), A(y)) = \mathcal{I}(\mathcal{M}(R(x_0,x_0)), A(x_0))\\ &= \mathcal{I}(0,1) = 1 > 0 = A(x_0),\end{aligned}$$

so $[R]_{\mathcal{T}} A \not\subseteq A$. ■

7 Fuzzy frames

In this section we will consider relational systems with fuzzy relations. We show that fuzzy information systems determine specific classes of such systems.

To begin with, let us introduce some basic notions.

Definition 12. *A* ***fuzzy frame*** *is a system $\mathfrak{F}=(\mathfrak{X},\Pi,R)$, where $\mathfrak{X}$ is a nonempty universe, Π is a finite set of* parameters *and $R:\wp(\Pi)\to\mathcal{R}(\mathfrak{X})$, i.e. for every $P\subseteq\Pi$, $R(P)$ is a binary fuzzy relation on $\mathfrak{X}$.* □

Given a fuzzy frame $\mathfrak{F}=(\mathfrak{X},\Pi,R)$, an *atom of* $\mathfrak{F}$ is a fuzzy relation $R(\{\pi\})$, where $\pi\in\Pi$.

Definition 13. *Let $\mathcal{T}$ and $\mathcal{S}$ be a t–norm and a t–conorm, respectively. We say that a fuzzy frame $\mathfrak{F}=(\mathfrak{X},\Pi,R)$ is*

- *a* ***strong $\mathcal{T}$–frame*** *iff for every $P\subseteq\Pi$ and for all $x,y\in\mathfrak{X}$,*

$$R(P)(x,y) = \mathcal{T}_{p\in P}\big(R(\{p\})(x,y)\big)$$

- *a* ***weak*** $\mathcal{S}$***-frame*** *iff for every* $P \subseteq \Pi$ *and for all* $x, y \in \mathfrak{X}$,
$$R(P)(x,y) = \mathcal{S}_{p\in P}\big(R(\{p\})(x,y)\big). \quad \square$$

Remark 5 The above definition and (1)–(2) imply that

- if $\mathfrak{F} = (\mathfrak{X}, \Pi, R)$ is a strong $\mathcal{T}$–frame then $R(P_1 \cup P_2) \supseteq R(P_1) \cap_{\mathcal{T}} R(P_2)$ for all $P_1, P_2 \subseteq \Pi$ and $R(\emptyset) = \mathbf{1}$,
- if $\mathfrak{F} = (\mathfrak{X}, \Pi, R)$ is a weak $\mathcal{S}$–frame then $R(P_1 \cup P_2) \subseteq R(P_1) \cup_{\mathcal{S}} R(P_2)$ for all $P_1, P_2 \in \Pi$ and $R(\emptyset) = \emptyset$. $\square$

Proposition 7. *Let* $\mathfrak{F} = (\mathfrak{X}, \Pi, R)$ *be a fuzzy frame,* $\mathcal{T}$ *be a left-continuous t-norm,* $\mathcal{N} = \mathcal{N}_{\mathcal{T}}$ *and* $\mathcal{S}$ *be a t-conorm. For every* $A \in \mathcal{F}(\mathfrak{X})$ *we have:*

(i) *if* $\mathfrak{F}$ *is a strong* $\mathcal{T}$*-frame then* $co_{\mathcal{N}} \langle\langle R(\emptyset) \rangle\rangle_{\mathcal{T}} A = \mathfrak{X}$.
(ii) *if* $\mathfrak{F}$ *is a weak* $\mathcal{S}$*-frame then* $co_{\mathcal{N}} \langle R(\emptyset) \rangle_{\mathcal{T}} A = \mathfrak{X}$.

Proof. (i) Let $\mathfrak{F}$ be a strong $\mathcal{T}$–frame. By Remark 5, $R(\emptyset) = \mathbf{1}$. Then, for every $A \in \mathcal{F}(\mathfrak{X})$ and every $x \in \mathfrak{X}$ we have:

$$\begin{aligned}(co_{\mathcal{N}} \langle\langle R(\emptyset) \rangle\rangle_{\mathcal{T}} A)(x) &= \mathcal{N}\big(\sup_{y\in\mathfrak{X}} \mathcal{T}(\mathcal{N}(R(\emptyset)(x,y)), \mathcal{N}(A(y)))\big) \\ &= \mathcal{N}\big(\sup_{y\in\mathfrak{X}} \mathcal{T}(0, \mathcal{N}(A(y)))\big) = \mathcal{N}(0) = 1.\end{aligned}$$

(ii) can be proved in the analogous way. ■

7.1 Fuzzy information frames

It is easily noted that every fuzzy information system Σ determines fuzzy frames with $(\mathcal{T}, \mathcal{S})$–information relations.

Definition 14. *Let* $\mathcal{T}$ *and* $\mathcal{S}$ *be a left-continuous t-norm and a t-conorm, respectively. Moreover, let* $\Sigma = (OB, AT, \{V_a : a \in AT\})$ *be a fuzzy information system and let* $\mathsf{inf} \in \mathsf{INF}_{\mathcal{T},\mathcal{S}}(\Sigma)$*. A* ***$(\mathcal{T}, \mathcal{S})$–information frame determined by Σ wrt*** inf *is a fuzzy frame of the form* $\mathfrak{F}_{\Sigma}^{\mathcal{T},\mathcal{S}}(\mathsf{inf}) = (OB, AT, \mathsf{inf})$. $\square$

We will write $\mathfrak{F}_{\Sigma}^{\mathcal{T},\mathcal{S}}(\mathsf{inf}_+)$ to denote a $(\mathcal{T}, \mathcal{S})$–information frame determined by Σ wrt $\mathsf{inf}_+ \in \mathsf{INF}_{\mathcal{T},\mathcal{S}}(\Sigma)$, where ind_+ is induced by a nonempty set of attributes.
Note that for any Σ, if $\mathsf{inf} \in \{\mathsf{ind}^s_{\mathcal{T}}, \mathsf{com}^s_{\mathcal{T}}, \mathsf{ort}^s_{\mathcal{T}}, \mathsf{cmp}^s_{\mathcal{T}}\}$ then the fuzzy frame (OB, AT, inf) is actually the $\mathcal{T}$–information frame (since the respective relations are determined by a t-norm $\mathcal{T}$ only) and will be denoted by $\mathfrak{F}_{\Sigma}^{\mathcal{T}}(\mathsf{inf})$.

Proposition 8. *Let* $\mathcal{T}$ *be a left-continuous t-norm and let* $\mathcal{S}$ *be a t-conorm. Then for every fuzzy information system* Σ*, every* $\mathsf{inf}^s \in \mathsf{INF}^s_{\mathcal{T},\mathcal{S}}(\Sigma)$ *and every* $\mathsf{inf}^w \in \mathsf{INF}^w_{\mathcal{T},\mathcal{S}}(\Sigma)$, $\mathfrak{F}_{\Sigma}^{\mathcal{T},\mathcal{S}}(\mathsf{inf}^s)$ *is a strong* $\mathcal{T}$*-frame and* $\mathfrak{F}_{\Sigma}^{\mathcal{T},\mathcal{S}}(\mathsf{inf}^w)$ *is a weak* $\mathcal{S}$*-frame.*

Proof. Directly results from Definitions 8, 13 and 14. ∎

Depending on a choice of $(\mathcal{T},\mathcal{S})$–information relations, we obtain specific classes of $(\mathcal{T},\mathcal{S})$–information frames. The following proposition gives the basic properties of particular classes of $(\mathcal{T},\mathcal{S})$–information frames.

Proposition 9. *For every fuzzy information system Σ and for every left–continuous t–norm $\mathcal{T}$,*

1. *all relations in $\mathfrak{F}^{\mathcal{T}}_{\Sigma}(\mathsf{ind}^s)$ are $\mathcal{T}$–equivalence relations;*
   all relations in $\mathfrak{F}^{\mathcal{T},\mathcal{S}}_{\Sigma}(\mathsf{ind}^w_+)$ are tolerance relations and all its atoms are $\mathcal{T}$–equivalence relations
2. *all relations in $\mathfrak{F}^{\mathcal{T}}_{\Sigma}(\mathsf{com}^s)$ and $\mathfrak{F}^{\mathcal{T},\mathcal{S}}_{\Sigma}(\mathsf{com}^w)$ are symmetric, for $\mathcal{T}\in T^+$ they are also quasi–reflexive*
3. *all relations in $\mathfrak{F}^{\mathcal{T},\mathcal{S}}_{\Sigma}(\mathsf{icm}^s)$ and $\mathfrak{F}^{\mathcal{T}}_{\Sigma}(\mathsf{icm}^w)$ are symmetric*
   for $\mathcal{T}\in T^+$, all relations in $\mathfrak{F}^{\mathcal{T},\mathcal{S}}_{\Sigma}(\mathsf{icm}^s)$ and $\mathfrak{F}^{\mathcal{T},\mathcal{S}}_{\Sigma}(\mathsf{icm}^w_+)$ are reflexive and all relations in $\mathfrak{F}^{\mathcal{T},\mathcal{S}}_{\Sigma}(\mathsf{icm}^w)$ and all atoms of $\mathfrak{F}^{\mathcal{T},\mathcal{S}}_{\Sigma}(\mathsf{icm}^s)$ are $\mathcal{T}^2$–cotransitive
4. *all relations in $\mathfrak{F}^{\mathcal{T},\mathcal{S}}_{\Sigma}(\mathsf{div}^s_+)$ and $\mathfrak{F}^{\mathcal{T},\mathcal{S}}_{\Sigma}(\mathsf{div}^w)$ are irreflexive and symmetric for $\mathcal{T}\in T^+$, all relations in $\mathfrak{F}^{\mathcal{T},\mathcal{S}}_{\Sigma}(\mathsf{div}^w)$ and all atoms of $\mathfrak{F}^{\mathcal{T},\mathcal{S}}_{\Sigma}(\mathsf{div}^s)$ are $\mathcal{T}$–cotransitive*
5. *all relations in $\mathfrak{F}^{\mathcal{T}}_{\Sigma}(\mathsf{ort}^s)$ and $\mathfrak{F}^{\mathcal{T},\mathcal{S}}_{\Sigma}(\mathsf{ort}^w)$ are symmetric and for $\mathcal{T}\in T^+$, weakly irreflexive*
6. *all relations in $\mathfrak{F}^{\mathcal{T}}_{\Sigma}(\mathsf{cmp}^s)$ and all atoms of $\mathfrak{F}^{\mathcal{T},\mathcal{S}}_{\Sigma}(\mathsf{cmp}^w)$ are $\mathcal{T}^2$–transitive*
   for $\mathcal{T}\in T^+$, all relations in $\mathfrak{F}^{\mathcal{T}}_{\Sigma}(\mathsf{cmp}^s_+)$ and $\mathfrak{F}^{\mathcal{T},\mathcal{S}}_{\Sigma}(\mathsf{cmp}^w)$ are irreflexive
   for $\mathcal{T}\in T_{inv}$, all relations in $\mathfrak{F}^{\mathcal{T}}_{\Sigma}(\mathsf{cmp}^s)$ and $\mathfrak{F}^{\mathcal{T},\mathcal{S}}_{\Sigma}(\mathsf{cmp}^w)$ are symmetric.

Proof. Follows from Theorem 2. ∎

8 Fuzzy information logics

Information logics are modal logics with relative accessibility relations. More precisely, in these systems modal operators are determined by accessibility relations which are information relations. Since information relations are determined by sets of attributes of objects, the respective accessibility relations are actually binary relations parameterised by sets of these attributes. Accordingly, modal operators are defined for parameters corresponding to respective sets of attributes. It is well-known in the literature ([3],[4],[14],[16],[17],[15],[28]) that relational systems with parameterised binary relations (frames) give Kripke–style models for information logics. Analogously, fuzzy frames might be viewed as Kripke models for fuzzy information logics. In this section we define a general class of *fuzzy information logics* FIL. In these logics basic connectives $\wedge$ and $\vee$ are interpreted by means of an arbitrary left–continuous t–norm and a t–conorm, respectively.

8.1 Syntax

The language of any logic FIL is determined by two pairwise disjoint denumerable sets: Var of propositional variables and $Apar$ of *atomic parameters*, a parameter constructor $\sqcup$, two truth constants $\perp$ and $\top$, logical connectives: $\wedge$, $\vee$, $\rightarrow$ and modal operators: $[\,]$, $\langle\,\rangle$, $[\![\,]\!]$, $\langle\!\langle\,\rangle\!\rangle$, $[\![\,]$, $\langle\!\langle\,\rangle$, $\pmb{[}\,]$, $\pmb{\langle}\,\rangle$.[7] The remaining logical connectives are defined by:

$$\neg\varphi \stackrel{\text{def}}{=} \varphi \rightarrow \perp$$

$$\varphi \equiv \psi \stackrel{\text{def}}{=} (\varphi \rightarrow \psi) \wedge (\psi \rightarrow \varphi).$$

The set Par of *parameters* is the smallest set satisfying the following conditions:

(a) $Apar \subset Par$
(b) if $P_1, P_2 \in Par$ then so is $P_1 \sqcup P_2$.

Intuitively, every parameter corresponds to a set of attributes.

The set $\mathfrak{L}$ of all formulae (language) of FIL is the smallest set satisfying:

(a) $Var \subseteq \mathfrak{L}$
(b) $\perp, \top \in \mathfrak{L}$
(c) if $\varphi, \psi \in \mathfrak{L}$ and $P \in Par$ then $\varphi \wedge \psi$, $\varphi \rightarrow \psi$, $\varphi \vee \psi$, $[P]\varphi$, $\langle P\rangle\varphi$, $[\![P]\!]\varphi$, $\langle\!\langle P\rangle\varphi$, $[\![P]\varphi$, $\langle\!\langle P\rangle\!\rangle\varphi$, $\pmb{[}P\,]\varphi$, $\pmb{\langle}P\rangle\varphi \in \mathfrak{L}$.

8.2 Semantics

For an arbitrary left–continuous t–norm $\mathcal{T}$ and a t–conorm $\mathcal{S}$, logical connectives $\wedge$, $\rightarrow$, $\neg$, $\vee$ and $\equiv$ are interpreted by means of $\mathcal{T}$, its residuum $\mathcal{I}$, $\mathcal{N} = \mathcal{N}_{\mathcal{T}}$, $\mathcal{S}$ and $\mathcal{E} = \mathcal{E}_{\mathcal{T}}$ (defined in Proposition 3), respectively. Also, the fuzzy operator $\mathcal{M}$ will be used to define the meaning of two specific fuzzy modal operators $\pmb{[}\,]$ and $\pmb{\langle}\,\rangle$.

Definition 15. *Let $\mathfrak{L}$ be a language of FIL and let $\mathfrak{F} = (W, \Pi, R)$ be a fuzzy frame. An* **interpretation of $\mathfrak{L}$ based on $\mathfrak{F}$** *is a pair $\mathfrak{I} = (\mathfrak{F}, m)$, where m is a meaning function such that*

(a) $m(\perp) = 0$, $m(\top) = 1$
(b) *for every $p \in Var$, $m(p) \in \mathcal{F}(W)$*
(c) *for every $a \in Apar$, $m(a) \in \Pi$.* □

In other words, to each propositional variable $p \in Var$ and each $w \in W$ (called a *possible world*), m assigns a real number $d \in [0, 1]$ viewed as a degree to which p is true in w. Also, every atomic parameter $a \in Apar$ is interpreted as a single parameter $P \in \Pi$.

[7] For the sake of simplicity we will use the same notation for fuzzy modal operators in a language of FIL and fuzzy information operators **(O.1)**–**(O.8)**.

For an interpretation $\mathfrak{I}$ of a language $\mathfrak{L}$ of FIL, a *valuation in* $\mathfrak{I}$ is the extension of m to the set of all parameters and the set of all formulae, respectively. For any $P \in Par$ we write $val_{\mathfrak{I}}(P)$ to denote the valuation of P in $\mathfrak{I}$ inductively defined by: (i) $val_{\mathfrak{I}}(a) = m(a)$ for every $a \in Apar$ (ii) $val_{\mathfrak{I}}(P_1 \sqcup P_2) = val_{\mathfrak{I}}(P_1) \cup val_{\mathfrak{I}}(P_2)$ for all $P_1, P_2 \in Par$. The truth value of a formula $\varphi \in \mathfrak{L}$ is defined by:

- $val_{\mathfrak{I}}(p) = m(p)$ for all $p \in Var$
- $val_{\mathfrak{I}}(\varphi \wedge \psi) = val_{\mathfrak{I}}(\varphi) \cap_{\mathcal{T}} val_{\mathfrak{I}}(\psi)$
- $val_{\mathfrak{I}}(\varphi \to \psi) = val_{\mathfrak{I}}(\varphi) \Rightarrow_{\mathcal{I}} val_{\mathfrak{I}}(\psi)$
- $val_{\mathfrak{I}}(\neg\varphi) = co_{\mathcal{N}} val_{\mathfrak{I}}(\varphi)$
- $val_{\mathfrak{I}}(\varphi \vee \psi) = val_{\mathfrak{I}}(\varphi) \cup_{\mathcal{S}} val_{\mathfrak{I}}(\psi)$
- $val_{\mathfrak{I}}(\varphi \equiv \psi) = val_{\mathfrak{I}}(\varphi) \Leftrightarrow_{\mathcal{E}} val_{\mathfrak{I}}(\psi)$
- for every $P \in Par$ and every $\Omega \in \{[\,], \langle\,\rangle, [\![\,]\!], \langle\!\langle\,\rangle\!\rangle, [\![\,], \langle\!\langle\,\rangle, \mathbf{[}\,\mathbf{]}, \mathbf{\langle}\,\mathbf{\rangle}\}$,
  $val_{\mathfrak{I}}(\Omega(P)\varphi) = \Omega_{\mathcal{T}}(R(val_{\mathfrak{I}}(P)))(val_{\mathfrak{I}}(\varphi))$.

Let $\mathfrak{I}$ be an interpretation of FIL based on a fuzzy frame $\mathfrak{F} = (\mathfrak{X}, \Pi, R)$. A formula φ of FIL is said to be *true in* $\mathfrak{I}$, in symbols $\mathfrak{I} \models \varphi$, iff $val_{\mathfrak{I}}(\varphi) = \mathfrak{X}$. For a fuzzy frame $\mathfrak{F}$, if $\mathfrak{I} \models \varphi$ for all $\mathfrak{I}$ based on $\mathfrak{F}$, then φ it is called *true in* $\mathfrak{F}$, in symbols $\mathfrak{F} \models \varphi$. For a class $\mathcal{K}$ of fuzzy frames, φ is called a $\mathcal{K}$*–tautology*, written $\models_{\mathcal{K}} \varphi$, iff $\mathfrak{F} \models \varphi$ for every $\mathfrak{F} \in \mathcal{K}$; if $\models_{\mathcal{K}} \varphi$ for all classes $\mathcal{K}$ of frames then φ is called a *tautology*.
For a class $\mathcal{K}$ of fuzzy frames, by a logic $\mathcal{L}(\mathcal{K})$ we mean the set of all $\mathcal{K}$–tautologies. Particular logics $\mathcal{L}(\mathcal{K})$ can be characterised by properties of fuzzy relations in $\mathfrak{F}$. As usual, these properties are expressed by means of fuzzy modal operators. Then each logic $\mathcal{L}(\mathcal{K})$ is characterised by a corresponding set of $\mathcal{K}$–tautologies representing properties of fuzzy relations in $\mathfrak{F}$.

For a property Γ of binary fuzzy relations, a class of frame $\mathcal{K}$ will be called a *class with* Γ iff for every fuzzy frame $\mathfrak{F} \in \mathcal{K}$, all fuzzy relations in $\mathfrak{F}$ have the property Γ.

Proposition 10. *Let* $\mathcal{K}$ *be a class of fuzzy frames and* $\mathfrak{L}$ *be a language of the logic* $\mathcal{L}(\mathcal{K})$ *with the set Par of parameters. For every* $P \in Par$*, every* $\varphi, \psi \in \mathfrak{L}$ *and every integer* $n \geq 1$,

(T.1)	*for* $\mathcal{K}$ *with reflexivity:*	$\models_{\mathcal{K}} [P]\varphi \to \varphi$
(T.2)	*for* $\mathcal{K}$ *with pseudo–reflexivity:*	$\models_{\mathcal{K}} \mathbf{[}P\mathbf{]}\varphi \to \varphi$
(T.3)	*for* $\mathcal{K}$ *with quasi–reflexivity:*	$\models_{\mathcal{K}} (\varphi \wedge \mathbf{\langle}P\mathbf{\rangle}\top) \to \mathbf{\langle}P\mathbf{\rangle}\varphi)$
(T.4)	*for* $\mathcal{K}$ *with irreflexivity:*	$\models_{\mathcal{K}} [\![P]\!]\varphi \to \neg\varphi$
(T.5)	*for* $\mathcal{K}$ *with weak irreflexivity:*	$\models_{\mathcal{K}} (\varphi \wedge [\![P]\!]\varphi) \to [\![P]\!]\top$
(T.6)	*for* $\mathcal{K}$ *with symmetry:*	$\models_{\mathcal{K}} \varphi \to [P]\langle P\rangle\varphi$
(T.7)	*for* $\mathcal{K}$ *with* $\mathcal{T}^n$*–transitivity:*	$\models_{\mathcal{K}} [P]^n\varphi \to [P]^{n+1}\varphi$
(T.8)	*for* $\mathcal{K}$ *with* $\mathcal{T}^n$*–cotransitivity:*	$\models_{\mathcal{K}} [\![P]\!]^n\varphi \to [\![P]\!]^{n+1}\varphi.$

Proof. Note first that for every two formulae φ and ψ in $\mathfrak{L}$ of FIL and any interpretation $\mathfrak{I}$ of $\mathfrak{L}$, by Proposition 2(ii) we have that the condition $val_{\mathfrak{I}}(\varphi) \subseteq val_{\mathfrak{I}}(\psi)$ implies $\mathfrak{I} \models \varphi \rightarrow \psi$. Therefore, if $\mathfrak{I} \models \varphi \rightarrow \psi$ holds for all interpretations $\mathfrak{I}$ based on any fuzzy frame $\mathfrak{F} \in \mathcal{K}$ then $\varphi \rightarrow \psi$ is a $\mathcal{K}$–tautology. Analogously, if $val_{\mathfrak{I}}(\varphi) \cap_{\mathcal{T}} val_{\mathfrak{I}}(\psi) \subseteq val_{\mathfrak{I}}(\varphi)$ for all $\varphi, \psi, \varphi \in \mathfrak{L}$ and for every $\mathfrak{I} \in \mathcal{K}$ then $(\varphi \wedge \psi) \rightarrow \varphi$ is a $\mathcal{K}$–tautology. Hence, by Theorem 3, we get the results. ∎

Furthermore, by Proposition 7 we have:

Corollary 1. *Let $\mathcal{T}$ be a left-continuous t-norm, $\mathcal{S}$ be a t-conorm, $\mathcal{K}$ be a class of fuzzy frames and $\mathfrak{L}$ be a language of the logic $\mathcal{L}(\mathcal{K})$. For any $\varphi \in \mathfrak{L}$,*

- *if $\mathcal{K}$ is the class of strong $\mathcal{T}$-frames then*

$$\textbf{(Fs)} \quad \models_{\mathcal{K}} \neg \langle\!\langle \emptyset \rangle\!\rangle \varphi$$

- *if $\mathcal{K}$ is the class of weak $\mathcal{S}$-frames then*

$$\textbf{(Fw)} \quad \models_{\mathcal{K}} \neg \langle \emptyset \rangle \varphi. \quad ∎$$

Let $\mathsf{inf} \in \mathsf{INF}_{\mathcal{T},\mathcal{S}}$ and let $\mathcal{K}$-inf stand for the class of fuzzy frames which relations have all properties characterising inf. Theorem 2, Proposition 10 and Corollary 1 imply sets of tautologies of particular logics $\mathcal{L}(\mathcal{K}\text{-}\mathsf{inf})$. For example,

$$Taut(\mathsf{ind}^s) = \{\textbf{(T.1)}, \textbf{(T.6)}, \textbf{(T.7)}\ (n=1), \textbf{(Fs)}\} \in \mathcal{L}(\mathcal{K}\text{-}\mathsf{ind}^s)$$
$$Taut(\mathsf{div}^w) = \{\textbf{(T.4)}, \textbf{(T.6)}, \textbf{(T.8)}(n=1), \textbf{(Fw)}\} \in \mathcal{L}(\mathcal{K}\text{-}\mathsf{div}^w), \quad \text{if } \mathcal{T} \in T^+.$$

$Taut(\mathsf{ind}^s)$ are tautologies of $\mathcal{L}(\mathcal{K}\text{-}\mathsf{ind}^s)$ for an arbitrary left–continuous t–norm $\mathcal{T}$ taken as an underlying semantical fuzzy operator. In particular, it is the case for the Łukasiewicz t–norm $\mathcal{T}_L$. Yet since $\mathcal{T}_L \notin T^+$, $Taut(\mathsf{div}^w)$ is not the set of tautologies of $\mathcal{L}(\mathcal{K}\text{-}\mathsf{div}^w)$, if $\mathcal{T}_L$ is taken as a semantical logical operator.

9 Conclusions

In this paper we have studied a fuzzy generalisation of information relations. We have proposed definitions of these relations based on an arbitrary left–continuous t–norm and an arbitrary t–conorm. Basic properties of these relations have been provided. This gave us the basic for defining several classes of fuzzy information frames which, in turn, are Kripke–style models for fuzzy information logics. We have given basic tautologies characterising particular classes of fuzzy information logics.

This paper is the starting point for further studies. The obtained results give rise to at least two open problems. The first one concerns deduction systems for particular classes of logics considered here. Another problem is related to so-called *information representability* ([3],[6]) of fuzzy frames. Our discussion implies that a class $\mathcal{K}(\mathsf{inf})$ of fuzzy information frames determined by a particular fuzzy information relation inf is a subclass of $\mathcal{K}$-inf. The question is whether this is a proper subclass or not, i.e. whether for any fuzzy frame $\mathfrak{F} \in \mathcal{K}$-$\mathsf{inf}$ there is a fuzzy information system Σ such that $\mathfrak{F}_{\Sigma}^{\mathcal{T},\mathcal{S}}(\mathsf{inf}) = \mathfrak{F}$. These problems will be discussed in our future work.

Acknowledgements

A. Radzikowska was partially supported by the KBN Grant No 8T11C01617. The work was carried out in the framework of COST Action 274 on *Theory and Applications of Relational Structures as Knowledge Instruments.*

References

1. Balbiani, P., Orłowska, E.: A hierarchy of modal logics with relative accessibility relations. Journal of Applied Non-Classical Logics 9, no 2-3 (1999) 303–348
2. De Baets, B., Mesiar, R.: Pseudo-metrics and $\mathcal{T}$–equivalences. Journal of Fuzzy Mathematics 5 (2) (1997) 471–481
3. Demri, S., Orłowska, E., Vakarelov, D.: Indiscernibility and complementarity relations in information systems. J. Gerbrandy, M. Marx, M. de Rijke and Y. Venema (eds) JFAK, Essays Dedicated to Johan van Benthem on the Occasion of his 50th Birthday. Amsterdam University Press (1999)
4. Demri, S. and Orłowska, E. Incomplete Information: Structure, Inference, Complexity. EATCS Monographs in Theoretical Computer Science, Springer (2002)
5. Dubois, D., Prade, H.: Putting fuzzy sets and rough sets together. Intelligent Decision Support, Słowiński, R. (ed.), Kluwer Academic (1992) 203–232
6. Düntsch, I., Orłowska, E.: Beyond modalities: Sufficiency and Mixed Algebras. Relational Methods for Computer Science Applications, Orłowska, E. and Szałas, A. (eds), Physica–Verlag, Heidelberg (2000) 263–283
7. Düntsch, I. and Orłowska, E.: Logics of complementarity in information systems. Mathematical Logic Quarterly 46 (2000) 267–288
8. Godo, L., Rodriquez, R.O.: Fuzzy Modal Logic for Similarity Reasoning. Kluwer Academic Publishers 6 (1999) 33–48
9. Hájek, P.: Metamathematics of Fuzzy Logic. Kluwer Academic Publishers, Dordrecht (1998)
10. Humberstone, I.: Inaccessible words. Notre Dame Journal of Formal Logic 24 (1983) 346–352
11. Kerre, E.E. (ed.): Introduction to the Basic Principles of Fuzzy Set Theory and Some of Its Applications, Communication & Cognition, Gent (1993)
12. Klir, G.J., Yuan, B.: Fuzzy Logic: Theory and Applications. Prentice–Hall, Englewood Cliffs, NJ (1995)
13. Konikowska, B.: A logic for reasoning about relative similarity. Studia Logica 58 (1997) 185–226

14. Orłowska, E.: Kripke models with relative accessibility and their applications to inference from incomplete information. Mathematical Problems in Computation Theory, Mirkowska, G. & Rasiowa, H. (eds.), Banach Center Publications 21 (1988) 329–339
15. Orłowska, E.: Many–valudness and uncertainty. Proceedings of the 27th International Symposium on Multiple-Valued Logic ISMVL–97, Antigonish, Canada (1997) 153–160. Also in Multiple Valued Logics 4 (1999) 207–227
16. Orłowska, E. (ed.): Incomplete Information – Rough Set Analysis. Studies in Fuzziness and Soft Computing, Springer–Verlag (1998)
17. Orłowska, E.: Studying incompleteness of information: a class of information logics. in Kijania–Placek, K. and Woleński, J. (eds.), The Lvov–Warsaw School and Contempotary Philosophy, Kluwer Academic Press (1998) 283–300
18. Radzikowska, A.M., Kerre, E.E.: Fuzzy rough sets revisited. Proceedings of European Congress of Intelligent Techniques and Soft Computing EUFIT–99, Aachen, Germany. Published on CD-ROM (1999)
19. Radzikowska, A.M., Kerre, E.E.: On some classes of fuzzy information relations. Proceedings of the 31st International Symposium of Multiple Valued Logics ISMVL–2001, Warsaw, Poland (2001) 75–80
20. Radzikowska, A.M., Kerre, E.E.: Towards studying of fuzzy information relations. Proceedings of International Conference in Fuzzy Logic and Technology EUSFLAT-2001, Leicester, UK (2001) 365–368
21. Radzikowska, A.M., Kerre, E.E.: A comparative study on fuzzy rough sets. Fuzzy Sets and Systems 126 (2) (2002) 137–155
22. Radzikowska, A.M., Kerre, E.E.: A general calculus of fuzzy rough sets. Submitted (2002)
23. Radzikowska, A.M., Kerre, E.E.: Fuzzy rough sets based on residuated lattices. Submitted (2002)
24. Radzikowska, A.M., Kerre, E.E.: Characterisations of main classes of fuzzy relations using information operators. Submitted (2002)
25. Schweizer, B., Sklar, A.: Probabilistic Metric Spaces. North Holland, Amsterdam (1983)
26. Thiele, H.: Fuzzy Rough Sets versus Rough Fuzzy Sets – An Interpretation and a Comparative Study using Concepts of Modal Logics. Proceedings of European Congress of Intelligent Techniques and Soft Computing EUFIT–97, Aachen, Germany (1997) 159–167
27. Turunen, E.: Mathematics Behind Fuzzy Logics. Physica–Verlag (1999)
28. Vakarelov, D.: A modal logic for similarity relations in Pawlak knowledge representation systems. Fundamenta Informaticae 15 (1991) 61–79

Part IV

Multiple-valued Logics for Control Theory and Rational Belief

Chapter 14
Weierstrass Approximation Theorem and Łukasiewicz Formulas with one Quantified Variable

Stefano Aguzzoli, Daniele Mundici

Department of Computer Science, University of Milan
Via Comelico 39-41, 20135 Milan, Italy
{aguzzoli,mundici}@dsi.unimi.it

Abstract. The logic $\exists Ł$ of continuous piecewise linear functions with rational coefficients has enough expressive power to formalize Weierstrass approximation theorem. Thus, up to any prescribed error, every continuous (control) function can be approximated by a formula of $\exists Ł$. As shown in this work, $\exists Ł$ is just infinite-valued Łukasiewicz propositional logic with one quantified propositional variable. We evaluate the computational complexity of the decision problem for $\exists Ł$. Enough background material is provided for all readers wishing to acquire a deeper understanding of the rapidly growing literature on Łukasiewicz propositional logic and its applications.

1 Łukasiewicz logic and nonclassical control

After several decades of relative quiescency, the infinite-valued propositional calculus of Łukasiewicz (denoted $Ł_\infty$ throughout this work) and its associated algebras, Chang's MV-algebras, have acquired an important role in various areas, ranging from Berlekamp's fault-tolerant communication with feedback [6], [21], [26], to ordered groups, algebras of operators of quantum spin systems, and toric desingularization. We refer to [10] for a comprehensive account. A short survey can be found in [32]. See [9] for further historical information.

An *MV-algebra* is a set equipped with an associative-commutative operation $\oplus$, with a neutral element 0, and with an operation $\neg$ such that $\neg\neg x = x$, $x \oplus \neg 0 = \neg 0$ and, finally,

$$\neg(\neg x \oplus y) \oplus y = \neg(\neg y \oplus x) \oplus x. \tag{1}$$

These equations are meant to capture *some* properties of the real unit interval $[0,1]$ equipped with negation $\neg x = 1 - x$ and truncated addition $x \oplus y = \min(1, x+y)$. For instance, direct inspection shows that the left hand member of equation (1) coincides with the maximum of x and y, whence the equation expresses the commutativity of the max operation. Conversely, Chang's completeness theorem, [7], [8], [35], [10, 2.5] states that the above equations do capture *all* properties of $[0,1]$, in the sense that an equation follows from the above axioms iff it is valid in $[0,1]$. This result is a far reaching generalization of the well known fact that the two-element set $\{0,1\}$ equipped with the operations $1-x$ and $\max(x,y)$ generates the variety of boolean algebras.

As a second important property, there is a categorical equivalence Γ between abelian lattice-ordered groups with strong unit, and MV-algebras [19], [10, Section 7]. For any such group G with strong unit u, Γ equips the unit interval $[0,u] \subseteq G$ with the operation $u-x$ and with truncated addition $u \wedge (x+y)$. Further, for any unit-preserving lattice-preserving homomorphism $\psi\colon (G,u) \to (G',u')$, the functor Γ restricts ψ to the unit interval $[0,u] \subseteq G$. As a consequence, in any MV-algebra one has precisely one genuine group-theoretical addition. Equivalently, *formulas* $\psi(X_1,\ldots,X_n)$ *of* Ł$_\infty$ *can express Ruspini's condition*, stating that the X_i "sum up to one", thus forming a (non-boolean) partition. See [19, Proposition 3.3], [26], [27, 3.1].

As a third major result, *McNaughton's representation theorem* states that, up to logical equivalence in Ł$_\infty$, formulas $\varphi=\varphi(X_1,\ldots,X_n)$ precisely yield continuous $[0,1]$-valued piecewise linear functions f over $[0,1]^n$, where each piece of f is a linear polynomial with integer coefficients. Again, this is a far reaching generalization of the well known fact that boolean formulas can represent all boolean functions. Intuitively, *the Łukasiewicz calculus* Ł$_\infty$ *is the logic of piecewise linear functions with integer coefficients.* McNaughton [16] proved this theorem in 1951 using a non-constructive argument. In [22] one can find a direct geometric proof, stressing the role of desingularization: the latter is not only a basic notion in the theory of toric varieties, but, as we shall see, it is also a key tool for the proof of our main representation theorem. We refer to the Appendix for further information.

* * * * *

It is often the case that a complex mechanical system S is modelled by a set $\mathcal{S}$ of highly nonlinear differential equations, [37], [4]. While the solution of $\mathcal{S}$ is hopeless, in some cases one is able to get useful information from the hamiltonian of S, and to approximate the desired control function $\mathbf{x} \mapsto y$ for S, by means of a *disjunction* D of "rules" (or, more appropriately, of "cases") of the form

- either $\mathbf{x}$ satisfies hypothesis H_1 and then y must obey condition C_1
- or $\mathbf{x}$ satisfies hypothesis H_2 and then y must obey condition C_2
- ...
- or $\mathbf{x}$ satisfies hypothesis H_m and then y must obey condition C_m

Sometimes, the above disjunction D may be the result of interrogating a human expert, rather than the hamiltonian of a system. In any case, from D one can conveniently try to construct/compute an efficient controller for S.

Typically, the hypotheses $H_1, \ldots, H_m$ form a non-boolean partition—a set of pairwise incompatible, exhaustive conditions satisfying Ruspini's law. As remarked above, this can be formalized in $Ł_\infty$. Conditions $C_1, \ldots, C_m$ usually have the form of $[0,1]$-valued functions over some compact Hausdorff space X, which, after a suitable normalization, can be safely identified with some n-cube $[0,1]^n$ (the case $n = 1$ being most often encountered). The problem arises to approximate each C_i by a suitable formula, and to express the "and then" connective in the above disjunction.

2 Many-values, product, and the rationals

While non-hamiltonian control has many aspects [37], the approximation of continuous functions by disjunctions of rules is an important part of it. One is interested in algorithms to approximate a continuous (control) function $f\colon [0,1]^n \to [0,1]$ by a logical formula. To this purpose, one must firstly predetermine a logic which is powerful enough to uniformly approximate all $[0,1]$-valued continuous functions on $[0,1]^n$, and whose deductive apparatus allows useful manipulations of these formulas.

As authoritatively shown by [14], (also see [27], [24] and [4]) suitably expressive many-valued logical systems can offer flexible tools to represent, explain, refine, simplify, manipulate sets of rules.

Recent research on "generalized conjunction connectives" (alias, triangular norms) [15], [14], [34], [29] shows that a logic incorporating *polynomials with rational coefficients* would have a decisive role for the above approximation problems. One must then investigate logics adding to $Ł_\infty$ the *product* connective and the rational numbers (see for instance, [18] [11]). [1]

In the paper [4] one can find an algorithm for the automatic generation of a logical formula of the above form D approximating any (normalized) continuous function. This formula belongs to the Esteva-Godo-Montagna logic $ŁΠ\frac{1}{2}$. This logic is obtained by adding to $Ł_\infty$ a second conjunction connective (natural multiplication of real numbers in the unit interval) together with "truncated" division and a constant symbol for the element 1/2. Suitable axioms are added to take care of these new connectives. The logic $ŁΠ\frac{1}{2}$ contains all the most important propositional fuzzy logics: $Ł_\infty$, Product and Gödel, to name a few (see [11]). The authors of [4] use the following variant of Weierstrass Theorem:

Theorem 1. *Suppose the function $f : [0,1]^n \to [0,1]$ is continuous and let $0 < \epsilon \in \mathbb{R}$. Then there is a continuous piecewise linear function $g : [0,1]^n \to [0,1]$, such that each piece of g has rational coefficients, and $|f(\mathbf{x}) - g(\mathbf{x})| < e$ for every $\mathbf{x} \in [0,1]^n$.*

[1] Alternatively, in [25] the author only uses tensor products of semisimple MV-algebras, the bare minimum needed to write down disjunctions like D above.

Given an effective representation of the function f, say by a Turing machine $\mathcal{M}$ that computes the approximate value $f(\mathbf{x})$ for any rational n-tuple $\mathbf{x} \in [0,1]^n$, the algorithm of [4] outputs, for any rational ϵ, a formula in Ł$\Pi\frac{1}{2}$ logic yielding an ϵ-approximation of the information supplied by $\mathcal{M}$. The desired Weierstrass-like approximation of f is now obtainable, on condition that the partial derivatives of f are suitably bounded.

Since the deductive machinery of Ł$\Pi\frac{1}{2}$ should be much more complex than for Ł$_\infty$, one might try to obtain Weierstrass approximation by adding to Ł$_\infty$ as little as possible. In this respect, already the authors of [4] and [30] note that Pavelka's rational propositional logic [33] (i.e., Ł$_\infty$ plus constant connectives for all rational truth-values in $[0,1]$)
satisfies Weierstrass approximation theorem.

In this work we shall describe another product-free logic, denoted ∃Ł, where rational numbers are introduced in a natural way (by quantifying over a single propositional variable). We shall see that ∃Ł coincides with quantified propositional Łukasiewicz logic, and that its expressive power precisely yields continuous piecewise linear functions with rational coefficients.

3 Łukasiewicz propositional logic

Let $V = \{X_1, X_2, \ldots\}$ be a denumerable set of symbols, called *(propositional) variables*. The set *Form* of formulas of Łukasiewicz propositional logic is inductively defined as follows:

- Each variable $X \in V$ is a formula.
- If φ is a formula then $\neg\varphi$ is a formula.
- If ψ and φ are formulas, then $(\psi \oplus \varphi)$ is a formula.

Definition 1. *Each formula $\varphi(X_1, \ldots, X_n)$ is canonically associated with the function $f_\varphi\colon [0,1]^n \to [0,1]$ defined by the following inductive stipulation:*

- $f_{X_i}(x_1, \ldots, x_n) = x_i$ = *the ith projection.*
- $f_{\neg\varphi} = 1 - f_\varphi$.
- $f_{(\varphi\oplus\psi)} = \min(1, f_\varphi + f_\psi)$.

The function f_φ is said to be represented by (or, f_φ is associated with) φ. Two formulas φ and ψ are logically equivalent iff $f_\varphi = f_\psi$.

A formula φ is *valid* in Ł$_\infty$ (or, φ is a *tautology* in Ł$_\infty$) iff f_φ is identically 1 over $[0,1]^n$.

Ł$_\infty$ is axiomatized by a few axioms, essentially amounting to a reformulation of the equations defining MV-algebras, together with modus ponens. In [20] it is proved that the tautology problem for Ł$_\infty$ is co-NP-complete, thus having the same complexity as its boolean counterpart.

In Ł$_\infty$ one usually introduces the following derived connectives (where outer parentheses are omitted for the sake of readability): $\varphi\odot\psi = \neg(\neg\varphi\oplus\neg\psi)$,

$\varphi \to \psi = \neg\varphi \oplus \psi$, $\varphi \vee \psi = (\varphi \to \psi) \to \psi$, $\varphi \wedge \psi = \neg(\neg\varphi \vee \neg\psi)$. As one can readily check: $f_{\varphi\odot\psi} = \max(0, f_\varphi + f_\psi - 1)$, $f_{\varphi\to\psi} = \min(1, 1 - f_\varphi + f_\psi)$, $f_{\varphi\vee\psi} = \max(f_\varphi, f_\psi)$, $f_{\varphi\wedge\psi} = \min(f_\varphi, f_\psi)$.

For each integer $k > 0$ we shall write $k\varphi$ for the formula $\varphi \oplus \varphi \oplus \cdots \oplus \varphi$, where φ occurs k many times.

Definition 2. *We say that* $f : [0,1]^n \to [0,1]$ *is a* McNaughton function *iff it satisfies the following conditions:*

1. f *is continuous.*
2. f *is piecewise linear, that is, there are finitely many linear polynomials* $p_1, \ldots, p_u$, *of the form* $p_i(\mathbf{x}) = a_{i1}x_1 + \cdots + a_{in}x_n + b_i$, *such that for every* $\mathbf{x} \in [0,1]^n$ *there is* $j \in \{1, \ldots, u\}$ *for which* $f(\mathbf{x}) = p_j(\mathbf{x})$.
3. *Each piece of* f *has integer coefficients, that is, for every* $i \in \{1, \ldots, u\}$ *we have* $a_{i1}, a_{i2}, \ldots, a_{in}, b_i \in \mathbb{Z}$.

It is easy to see [10, 3.3] that every McNaughton function f determines a polyhedral complex C_f with rational vertices, over the n-cube, in such a way that over every n-dimensional (closed, convex) polyhedron of $P \in C_f$, f is linear, and in fact coincides with a uniquely determined polynomial p_i. We say that any such P is a *linear domain* of f, and that p_i is its corresponding *linear piece* of f.

A straightforward induction argument shows that for every formula ψ the function represented by ψ is a McNaughton function. The converse is a deep result due to McNaughton [16]. We then have the following representation theorem:

Theorem 2. *Fix* $n = 1, 2, \ldots$. *Then the set of McNaughton functions over* $[0,1]^n$ *is equal to the set of functions represented by formulas of Łukasiewicz logic with* n *propositional variables.*

Owing to its importance for the main results of this work, a constructive proof of Theorem 2 shall be sketched in the Appendix. See [22] or [10, 9.1] for the complete proof.

It is easy to see that Weierstrass theorem does not hold for McNaughton functions. The latter, however, can approximate any continuous function in a weaker topology, as follows:

Theorem 3. *Let* $f : [0,1]^n \to [0,1]$ *be continuous and* $0 < \epsilon \in \mathbb{R}$. *Then there is a McNaughton function* $g : [0,1]^n \to [0,1]$ *such that* $\int_{[0,1]^n} |f(\mathbf{x}) - g(\mathbf{x})| d\mathbf{x} < \epsilon$.

To obtain Weierstrass approximation theorem, we shall enrich the Łukasiewicz propositional calculus by allowing quantification over propositional variables as follows:

Definition 3. *The set $QForm \supseteq Form$ of* quantified propositional formulas *is defined by the following inductive stipulation: If φ and ψ are formulas in $QForm$, then so are $\neg\varphi$ and $(\varphi \oplus \psi)$; further, for every propositional variable X, both $\exists X\, \varphi$ and $\forall X\, \varphi$ are in $QForm$.*

Definition 4. *For every formula $\varphi \in QForm$ its associated function f_φ is inductively defined as in Definition 1, with the following extra proviso:*

$$f_{\exists X_i \varphi} = \max_{x_i \in [0,1]} f_\varphi \quad \text{and} \quad f_{\forall X_i \varphi} = \min_{x_i \in [0,1]} f_\varphi .$$

The definitions of validity and logical equivalence for this logic, denoted QŁ, are, mutatis mutandis, the same as for $Ł_\infty$.

We denote by ∃Ł the fragment of QŁ given by all formulas of the form $\exists X \varphi$, where φ is a formula of $Ł_\infty$ and X is a variable. One easily sees that all functions represented by formulas in ∃Ł are piecewise linear with *rational* coefficients. The problem whether *all* such functions are representable in ∃Ł will be solved in the next section.

4 Main representation theorem

The definition of a $[0,1]$-*valued continuous piecewise linear function f over $[0,1]^n$ with rational coefficients* coincides with Definition 2, where Condition 3 is weakened as follows:

3′. Each piece of f has rational coefficients, that is, for every linear piece $p_i(\mathbf{x}) = \sum_{j=1}^{n} a_{ij}x_j + b_i$, we have $a_{i1}, a_{i2}, \ldots, a_{in}, b_i \in \mathbb{Q}$.

Definition 5. *Any function $f : [0,1]^n \to [0,1]$ satisfying 1., 2. and 3′. will be called a* rational McNaughton function.

Theorem 4. *Let $f : [0,1]^n \to [0,1]$ be a rational McNaughton function. Then there is a McNaughton function $g : [0,1]^{n+1} \to [0,1]$ such that, for every $\mathbf{x} = (x_1, \ldots, x_n) \in [0,1]^n$,*

$$f(\mathbf{x}) \quad = \max_{x_{n+1} \in [0,1]} g(\mathbf{x}, x_{n+1}).$$

Proof. Define the two subsets f^+ and f^- of the $(n+1)$-cube $[0,1]^{n+1}$ by

$$f^+ = \{(\mathbf{x}, x_{n+1}) \in [0,1]^{n+1} \mid x_{n+1} \geq f(\mathbf{x})\},$$

$$f^- = \{(\mathbf{x}, x_{n+1}) \in [0,1]^{n+1} \mid x_{n+1} \leq f(\mathbf{x})\}.$$

Identifying (the graph of) f with a set of $(n+1)$-tuples, f then coincides with $f^+ \cap f^-$. We shall also identify the domain $[0,1]^n$ of f with the *floor* $[0,1]^n \times \{0\}$ of the $(n+1)$-cube, and we shall call $[0,1]^n \times \{1\}$ its *ceiling*.

Further, a function $k\colon [0,1]^{n+1} \to [0,1]$ will be said to be *f-amenable* iff is it continuous, $k(\mathbf{x}, x_{n+1}) = x_{n+1}$ whenever $(\mathbf{x}, x_{n+1}) \in f^-$, and k is nonincreasing with respect to x_{n+1} whenever $(\mathbf{x}, x_{n+1}) \in f^+$. In other words, for all $\mathbf{x} \in [0,1]^n$ and $y, z \in [0,1]$, if $y \leq z$ and $(\mathbf{x}, y) \in f^+$ then $k(\mathbf{x}, y) \geq k(\mathbf{x}, z)$. If k is f-amenable then for every $\mathbf{x} \in [0,1]^n$, $f(\mathbf{x}) = \max_{x_{n+1} \in [0,1]} k(\mathbf{x}, x_{n+1})$.

As a preliminary step for the construction of the desired McNaughton function g, we shall construct an f-amenable piecewise linear function h. To ensure the piecewise linearity of h we shall suitably triangulate the $(n+1)$-cube.

By our assumptions about f there exists a triangulation $\mathcal{T}_0$ of the floor of the $(n+1)$-cube such that [2]

(i) each simplex of $\mathcal{T}_0$ has rational vertices,
(ii) f is linear over each simplex of $\mathcal{T}_0$.

For every n-simplex $T \in \mathcal{T}_0$ let us agree to call the set $P_T = T \times [0,1]$ the *prism* in $[0,1]^{n+1}$ with basis T. P_T is an $(n+1)$-dimensional (compact convex) polyhedron whose $2(n+1)$ many vertices can be displayed as

$$(\mathbf{w}_0^T, 0), \ldots, (\mathbf{w}_n^T, 0), \quad (\mathbf{w}_0^T, 1), \ldots, (\mathbf{w}_n^T, 1), \tag{2}$$

where the first $n+1$ vertices are precisely those of T. For each $i = 0, \ldots, n$, the edge joining the two vertices $(\mathbf{w}_i^T, 0)$ and $(\mathbf{w}_i^T, 1)$ will be said to be *vertical*. Thus, every edge (= one-dimensional face) of P_T is either vertical, or else it lies in the floor, or in the ceiling of the $(n+1)$-cube. The intersection of any two prisms $P_{T'}, P_{T''}$ is either empty or is a common face of both prisms.

By (ii) above, for each n-simplex $T \in \mathcal{T}_0$ the graph of f splits P_T into two (compact convex) polyhedra with rational vertices, namely $P_T^+ = P_T \cap f^+$ and $P_T^- = P_T \cap f^-$. Recalling (2), the vertices of P_T^+ can be listed [3] as

$$(\mathbf{w}_0^T, f(\mathbf{w}_0^T)), \ldots, (\mathbf{w}_n^T, f(\mathbf{w}_n^T)), \quad (\mathbf{w}_0^T, 1), \ldots, (\mathbf{w}_n^T, 1). \tag{3}$$

The n-simplex $P_T^+ \cap P_T^- = f \cap P_T$ is a common face of both P_T^+ and P_T^-. In case f constantly equals 1 over T, $P_T^- = P_T$ and P_T^+ is the intersection of P_T with the ceiling of the $(n+1)$-cube. In case f constantly vanishes over T, $P_T^+ = P_T$ and P_T^- coincides with T. Otherwise, both polyhedra P_T^+ and P_T^- are $(n+1)$-dimensional.

Let $\mathcal{F}^+ = \{P_T^+ \mid T \text{ an } n\text{-dimensional simplex in } \mathcal{T}_0\}$. Then $\bigcup \mathcal{F}^+ = f^+$, and the intersection of any two polyhedra in $\mathcal{F}^+$ is either empty or is a common face of both polyhedra. A routine argument [4] yields a triangulation $\mathcal{T}^+$ of f^+ *refining* $\mathcal{F}^+$ and without new vertices. In other words, $\mathcal{T}^+$ is a simplicial complex over f^+ and each polyhedron $P_T^+ \in \mathcal{F}^+$ is the set-theoretic

[2] All polyhedra, and in particular, all simplexes in this work are closed and convex.
[3] In this list there are repeated vertices iff $f(\mathbf{w}) = 1$ for some vertex $\mathbf{w}$ of T.
[4] See, e.g., [12].

union of simplexes of $\mathcal{T}^+$; further, the vertices of all these simplexes are already vertices of polyhedra in $\mathcal{F}^+$.

Claim 1. Each $(n+1)$-simplex $S \in \mathcal{T}^+$ has a *vertical* edge, i.e., a one-dimensional face η_S joining a ceiling vertex $(\mathbf{x}, 1)$ of S with another vertex of S of the form $(\mathbf{x}, f(\mathbf{x}))$.

Indeed, since $\mathcal{T}^+$ refines $\mathcal{F}^+$, for each n-simplex $T \in \mathcal{T}_0$ the set of simplexes of $\mathcal{T}^+$ which are contained in the prism P_T^+ is a triangulation $\mathcal{T}_T^+$ of P_T^+. Thus in particular, S is a member of $\mathcal{T}_U^+$ for a unique n-simplex $U \in \mathcal{T}_0$. At most $n+1$ of the $n+2$ vertices of S can lie on the ceiling. Since $\mathcal{T}^+$ has no new vertices, the vertices of $\mathcal{T}_U^+$ are among the vertices of P_U^+. By (2)-(3), the vertices of S not lying on the ceiling must belong to the set $\{(\mathbf{w}_0^U, f(\mathbf{w}_0^U)), \ldots, (\mathbf{w}_n^U, f(\mathbf{w}_n^U))\}$. Therefore, for some $j = 0, \ldots, n$, S has two distinct vertices $(\mathbf{w}_j^U, f(\mathbf{w}_j^U))$ and $(\mathbf{w}_j^U, 1)$. The edge η_S joining these two vertices satisfies our claim.

We now define the function $h \colon [0,1]^{n+1} \to [0,1]$ as follows:

(a) h is linear over each $(n+1)$-simplex of $\mathcal{T}^+$,
(b) $h(\mathbf{x}, x_{n+1}) = x_{n+1}$ whenever $(\mathbf{x}, x_{n+1}) \in f^-$,
(c) $h(\mathbf{x}, x_{n+1}) = 0$ whenever $(\mathbf{x}, x_{n+1})$ is a vertex of $\mathcal{T}^+$ not lying in f^-. [5]

The function h is continuous and piecewise linear, because on the one hand, $\mathcal{T}^+$ is a simplicial complex over f^+ without new vertices besides those of the prisms P_T^+ and, on the other hand, the intersection between any simplex of $\mathcal{T}^+$ and any polyhedron of P_T^- is either empty or else is a common face of both.

Claim 2. h is f-amenable.

As a matter of fact, by Claim 1 each $(n+1)$-simplex $S \in \mathcal{T}^+$ has a vertical edge η_S joining its two vertices, say, $(\mathbf{w}, f(\mathbf{w}))$ and $(\mathbf{w}, 1)$, with $f(\mathbf{w}) \neq 1$. By definition, $h((\mathbf{w}, f(\mathbf{w}))) = f(\mathbf{w})$ and $h((\mathbf{w}, 1)) = 0$, because $(\mathbf{w}, 1)$ does not lie in f^-. Since h is linear along η_S, its directional derivative $\partial h / \partial x_{n+1}$ is ≤ 0 at each point $(\mathbf{w}, t)$ in the relative interior of η_S. Since h is linear over all of S, $\partial h / \partial x_{n+1} \leq 0$ at each point in the interior of S. Since h is continuous, then h is nonincreasing with respect to x_{n+1} over all of f^+. Thus h is f-amenable and our second claim is settled.

As already noted, from the f-amenability of h it follows that, for each $(\mathbf{x}, x_{n+1}) \in [0,1]^{n+1}$,

$$f(\mathbf{x}) = \max_{x_{n+1} \in [0,1]} h(\mathbf{x}, x_{n+1}). \tag{4}$$

Let now $\mathcal{F}^- = \{P_T^- \mid T \text{ an } n\text{-dimensional simplex in } \mathcal{T}_0\}$. Let $\mathcal{T}^-$ be a triangulation of f^- refining $\mathcal{F}^-$ with no new vertices besides those of the

[5] There may be vertices $(\mathbf{x}, f(\mathbf{x}))$ of $\mathcal{T}^+$ lying on the ceiling of the $(n+1)$-cube and also belonging to f^-. Then $h(\mathbf{x}, f(\mathbf{x})) = f(\mathbf{x}) = 1$.

prisms P_T^-. Since P_T^- intersects P_T^+ in an n-simplex of $\mathcal{T}^+$, then $\mathcal{T}^- \cup \mathcal{T}^+$ is a triangulation of the $(n+1)$-cube; further, each simplex of the form $P_T^+ \cap P_T^-$ contained in the graph $f^+ \cap f^-$ is also an n-simplex of $\mathcal{T}^- \cap \mathcal{T}^+$. All vertices of $\mathcal{T}^- \cup \mathcal{T}^+$ are rational.

Following [22, p.598], [10, 9.1], we now construct a *unimodular* triangulation $\mathcal{T}$ of the $(n+1)$-cube, refining $\mathcal{T}^- \cup \mathcal{T}^+$. In other words,

— each simplex in $\mathcal{T}^- \cup \mathcal{T}^+$ is the set-theoretic union of simplexes in $\mathcal{T}$,
— each $(n+1)$-simplex $W \in \mathcal{T}$ has rational vertices $\mathbf{w}_0^W, \ldots, \mathbf{w}_{n+1}^W$, where for each $j = 0, \ldots, n+1$, $\mathbf{w}_j^W = (r_{j1}, \ldots, r_{jn+1}) = (a_{j1}/d_j, \ldots, a_{jn+1}/d_j)$, with $0 < d_j$ being the least common denominator of the rational coordinates r_{jt} of $\mathbf{w}_j^W$ $(t = 1, \ldots, n+1)$
— the $(n+2) \times (n+2)$ matrix whose jth line is given by the integers $d_j, a_{j1}, \ldots, a_{jn+1}$ has its determinant equal to ± 1.

(See the Appendix for a sketch of this basic construction in many-valued logic.)

We now define the continuous function $g\colon [0,1]^{n+1} \to [0,1]$ as follows:

(I) g is linear over each simplex of $\mathcal{T}$,
(II) $g(\mathbf{x}, x_{n+1}) = x_{n+1}$ whenever $(\mathbf{x}, x_{n+1})$ is a vertex of $\mathcal{T}$ lying in f^-,
(III) $g(\mathbf{x}, x_{n+1}) = 0$ whenever $(\mathbf{x}, x_{n+1})$ is a vertex of $\mathcal{T}$ lying in $f^+ \setminus f^-$.

In the light of (I) we can rephrase (II) in the equivalent form

(II') $g(\mathbf{x}, x_{n+1}) = x_{n+1}$ whenever $(\mathbf{x}, x_{n+1}) \in f^-$.

Since $\mathcal{T}$ refines $\mathcal{T}^+ \cup \mathcal{T}^-$, one can equivalently define g as the sum

$$g = \sum_{\mathbf{w} \in \text{vertices of } \mathcal{T} \text{ lying in } f^-} a_{\mathbf{w}} H_{\mathbf{w}}^{\mathcal{T}},$$

where for each vertex $\mathbf{w} = (a_1/d, \ldots, a_{n+1}/d)$ of $\mathcal{T}$ lying in f^-, the function $H_{\mathbf{w}}^{\mathcal{T}}\colon [0,1]^{n+1} \to [0,1]$ is the Schauder hat with apex $\mathbf{w}$ with respect to the unimodular triangulation $\mathcal{T}$ as defined in the appendix, and $a_{\mathbf{w}} = a_{n+1}$. The unimodularity of $\mathcal{T}$ ensures that all coefficients of g are integers. Since the value of g is always ≤ 1 over its domain $[0,1]^{n+1}$, we immediately see that g is a McNaughton function.

Claim 3. $g \leq h$.

As a matter of fact, by (II'), g coincides with h over each simplex of $\mathcal{T}$ contained in f^-. Further, $g = 0 \leq h$ over each simplex of $\mathcal{T}$ which is contained in f^+ and is disjoint from f^-. Finally, let U be a simplex of $\mathcal{T}$ which is contained in f^+ and has some vertex in f^-. Then U is contained in some simplex of $\mathcal{T}^+$, because $\mathcal{T}$ refines $\mathcal{T}^+ \cup \mathcal{T}^-$. Since both g and h are linear over U, it is then sufficient to prove the inequality $g(\mathbf{w}) \leq h(\mathbf{w})$ for each vertex $\mathbf{w}$ of U. In case $\mathbf{w} \in f$ we have $g(\mathbf{w}) = h(\mathbf{w})$. In case $\mathbf{w} \in f^+ \setminus f$ we have $0 = g(\mathbf{w}) \leq h(\mathbf{w})$. The claim is settled.

Recalling (4), since h and g coincide over f^-, from Claim 3 we get the desired conclusion $f(\mathbf{x}) = \max_{x_{n+1} \in [0,1]} g(\mathbf{x}, x_{n+1})$.

Corollary 1. *Let $f\colon [0,1]^n \to [0,1]$ be a rational McNaughton function. Then there is a formula φ in $n+1$ variables of the infinite-valued Łukasiewicz propositional logic $Ł_\infty$ such that, for every $\mathbf{x} = (x_1, \ldots, x_n) \in [0,1]^n$,*

$$f(\mathbf{x}) = f_{\exists X_{n+1}\varphi(X_1,\ldots,X_n,X_{n+1})}(\mathbf{x}).$$

Thus, formulas in $\exists Ł$ represent all piecewise linear functions with rational coefficients.

From Theorems 1 and 4 we immediately get the following variant of Weierstrass theorem:

Corollary 2. *Let $f\colon [0,1]^n \to [0,1]$ be a continuous function, and $0 < \epsilon \in \mathbb{R}$. Then there is a McNaughton function $g\colon [0,1]^{n+1} \to [0,1]$ such that*

$$|f - \max_{x_{n+1}\in[0,1]} g| < \epsilon.$$

4.1 Another representation theorem

Formulas in the logic $\exists Ł$ are certain simple *prenex* formulas of QŁ. One can then interpret Corollary 1 as giving prenex forms for rational McNaughton functions. If we allow occurrences of quantifiers at arbitrary levels of nesting in the body of formulas, then we can give an alternative representation theorem, having a much simpler proof.

Given a formula $\varphi = \varphi(X_1, \ldots, X_n)$ in $Ł_\infty$, an integer $k > 1$ and a variable Y not occurring in φ, we define the formula $\Psi(\varphi, k)$ of $Ł_\infty$ by

$$\Psi(\varphi, k) = (\varphi \odot \neg(k-1)Y) \wedge Y. \tag{5}$$

It follows that, for every $\mathbf{x} \in [0,1]^n$,

$$f_{\exists Y \Psi(\varphi,k)}(\mathbf{x}) = \frac{f_\varphi(\mathbf{x})}{k}.$$

By direct inspection, $f_{\exists Y\Psi(\varphi,k)}(\mathbf{x}) = \max_y f_{\Psi(\varphi,k)}(\mathbf{x}, y) = f_{\Psi(Z,k)}(z, z/k) = z/k$, where $z = f_\varphi(\mathbf{x})$ and Z is a new propositional variable.

This formula will be used to prove Theorem 5 below. [6]

We assume the reader to be familiar with the notion of free and bound occurrence of a variable within a quantified formula.

Theorem 5. *Let $f\colon [0,1]^n \to [0,1]$ be a rational McNaughton function. Then there is a quantified propositional Łukasiewicz formula φ in which the variables $X_1, \ldots, X_n, Y$ occur, such that all quantifiers in φ have the form $\exists Y$, the variable Y never occurs free in φ, and for every $\mathbf{x} = (x_1, \ldots, x_n) \in [0,1]^n$, $f(\mathbf{x}) = f_\varphi(\mathbf{x})$.*

[6] A similar result was proved in [5], using *division connectives* $\square_k(x) = x/k$, for each prime k. As the authors remark, no finite subset of the set of division connectives can capture the full expressive power of the infinite set.

Proof. Let $\mathcal{U}$ be a unimodular triangulation of $[0,1]^n$ such that f is linear over each simplex of $\mathcal{U}$. Let $\mathbf{w} = (a_1/d, \ldots, a_n/d)$ be a vertex of $\mathcal{U}$ and $H^{\mathcal{U}}_{\mathbf{w}}$ be the Schauder hat in $\mathcal{U}$ with apex $\mathbf{w}$ as defined in the Appendix. Let Δ be an n-simplex of $\mathcal{U}$ having $\mathbf{w}$ among its vertices. Also let $p: [0,1]^n \to [0,1]$ be the linear polynomial

$$p(\mathbf{x}) = \frac{b_0}{c_0} + \sum_{i=1}^{n} \frac{b_i}{c_i} x_i \qquad (b_0, b_1, \ldots, b_n, c_0, c_1, \ldots, c_n \in \mathbb{Z})$$

such that f equals p over Δ. For some $h_{\mathbf{w}}, k_{\mathbf{w}} \in \mathbb{Z}$ we can write

$$f(\mathbf{w}) = \frac{b_0}{c_0} + \sum_{i=1}^{n} \frac{b_i a_i}{c_i d} = \frac{h_{\mathbf{w}}}{k_{\mathbf{w}} d}.$$

Let

$\varphi_{\mathbf{w}}$ be a formula of Ł_∞ such that $f_{\varphi_{\mathbf{w}}} = H^{\mathcal{U}}_{\mathbf{w}}$ over $[0,1]^n$. Then $h_{\mathbf{w}} H^{\mathcal{U}}_{\mathbf{w}}(\mathbf{w}) = h_{\mathbf{w}}/d$, and the continuous function $g_{\mathbf{w}} = h_{\mathbf{w}} f_{\exists Y \Psi(\varphi_{\mathbf{w}}, k_{\mathbf{w}})}$ is linear over each simplex of $\mathcal{U}$, Moreover, $g_{\mathbf{w}}(\mathbf{v}) = 0$ for every vertex of $\mathcal{U}$ other than $\mathbf{w}$, while $g_{\mathbf{w}}(\mathbf{w}) = h_{\mathbf{w}}/(k_{\mathbf{w}} d)$. In conclusion,

$$f = \sum_{\mathbf{w} \in \text{ vertices of } \mathcal{U}} g_{\mathbf{w}} = f_{\oplus_{\mathbf{w}} h_{\mathbf{w}} \exists Y \Psi(\varphi_{\mathbf{w}}, k_{\mathbf{w}})}. \tag{6}$$

The unimodular triangulation $\mathcal{U}$ built in the proof of Theorem 5 covers $[0,1]^n$, while the unimodular triangulation $\mathcal{T}$ of Theorem 4 covers $[0,1]^{n+1}$. Since the complexity of the algorithmic construction of such triangulations rapidly increases with dimension ([3],[12]) the construction in Theorem 5 is significantly simpler than its counterpart in Theorem 4.

While only the variable Y is quantified in formula (6), quantifications may occur several times within the body of the formula. In general, the formula $\exists Y \varphi \oplus \exists Y \psi$ is not logically equivalent to $\exists Y(\varphi \oplus \psi)$.

As a dual counterpart of Theorem 5, by taking minima instead of maxima, and replacing formula (5) by [7]

$$\Phi(\varphi, k) = (\varphi \to kY) \to Y, \tag{7}$$

one has the following

Theorem 6. *Let $f: [0,1]^n \to [0,1]$ be a rational McNaughton function. Then there is a quantified propositional Łukasiewicz formula φ in which the variables $X_1, \ldots, X_n, Y$ occur, such that all quantifiers in φ have the form $\forall Y$, the variable Y never occurs free in φ, and for every $\mathbf{x} = (x_1, \ldots, x_n) \in [0,1]^n$, $f(\mathbf{x}) = f_\varphi(\mathbf{x})$.*

[7] Formula (7) was suggested by Petr Cintula.

5 Computations in ∃Ł

Since rational piecewise linear functions are closed under taking maxima and minima, as an effect of Theorem 4, one can naturally limit attention to formulas in ∃Ł, rather than QŁ, and consider the computational complexity of the following problem:

Problem 1 (TAUT-∃Ł).

> INSTANCE: A formula $\varphi = \varphi(X_1, \ldots, X_n, Y)$ in $Ł_\infty$.
> QUESTION: Is the formula $\exists Y \varphi$ valid in ∃Ł ? Equivalently: Is the function $k(\mathbf{x}) = \max_{y \in [0,1]} f_\varphi(\mathbf{x}, y)$ equal to 1 for all $\mathbf{x} \in [0,1]^n$?

In Theorem 7 below we shall prove that this problem is in Π_2^p, relative to the Meyer-Stockmeyer polynomial hierarchy [17]. As the reader will recall, for each integer $i \geq 0$, the complexity classes Σ_i^p and Π_i^p are defined by

$$\Sigma_0^p = \Pi_0^p = \mathrm{P}, \qquad \Sigma_{i+1}^p = \mathrm{NP}^{\Sigma_i^p}, \qquad \Pi_{i+1}^p = \text{co}-\Sigma_{i+1}^p.$$

Here, as usual, P is the class of problems which are decidable in deterministic polynomial-time; further, for every complexity class C, the class NP^C is the set of problems L such that there is a problem $L' \in C$ and a polynomial-time non-deterministic Turing reduction from L to L', [13].

Turning back to our problem TAUT-∃Ł, in order to *disprove* that $\exists Y \varphi$ is a tautology, one must

> *Guess* an n-tuple $\mathbf{x}^* = (x_1^*, \ldots, x_n^*) \in [0,1]^n$, where the function $k(\mathbf{x}) = \max_y f_\varphi(\mathbf{x}, y)$ has a local minimum (in the sense that for all $\mathbf{x}$ in an open neighbourhood of $\mathbf{x}^*$ in $[0,1]^n$, $k(\mathbf{x}) \geq k(\mathbf{x}^*)$), and then
>
> *Check* that $\max_{y \in [0,1]} f_\varphi(\mathbf{x}^*, y) < 1$.

Recalling Definition 5, k is a rational McNaughton function. Let us fix, once and for all, a polyhedral complex $\mathcal{K}$ over $[0,1]^n$ with rational vertices such that, over each polyhedron of $\mathcal{K}$, the map k is linear. In other words, each n-dimensional (closed, convex) polyhedron D in $\mathcal{K}$ is a linear domain of k. There is a unique linear polynomial k_D such that $k = k_D$ over D.

Lemma 1. *The function k attains its absolute minimum value at some vertex $\mathbf{x}^*$ of $\mathcal{K}$. Among all vertices $\mathbf{x}^*$ of $\mathcal{K}$ where k attains its absolute minimum, at least one satisfies the following condition: Let $e(1) < e(2) < \cdots < e(u)$ be the complete list of those indexes $e(i)$ of the coordinates of $\mathbf{x}^*$ such that $x_{e(i)}^*$ is either equal to 0 or to 1 (say, $x_{e(i)}^* = \beta_i \in \{0,1\}$.) Then there are $d = n+1-u$ linear domains $D_1, \ldots, D_d$ of k in $\mathcal{K}$ with $\mathbf{x}^*$ as their common* *vertex, and with $k_1, \ldots, k_d$ their corresponding linear pieces of k, such that the system of linear equations*

$$\begin{cases} k_1(\mathbf{x}) &= k_2(\mathbf{x}) \\ k_2(\mathbf{x}) &= k_3(\mathbf{x}) \\ \ldots & \ldots\ldots \\ k_{d-1}(\mathbf{x}) &= k_d(\mathbf{x}) \\ x_{e(1)} &= \beta_1 \\ \ldots & \ldots\ldots \\ x_{e(u)} &= \beta_u \end{cases} \tag{8}$$

in the n-tuple of unknowns $\mathbf{x} = (x_1, \ldots, x_n)$, has $\mathbf{x}^*$ as its unique solution.

Proof. Let z^* be the absolute minimum value of k, and let $M = k^{-1}(z^*)$. The piecewise linearity of k ensures that M is the set-theoretic union of a sub-complex $\mathcal{K}'$ of $\mathcal{K}$, whence M contains at least one vertex of $\mathcal{K}$. In case M contains an isolated point $\mathbf{x}^*$, then automatically $\mathbf{x}^*$ is a vertex of $\mathcal{K}$. The integer u equals the number of coordinates of $\mathbf{x}^*$ which are either equal to 0 or to 1. The remaining coordinates of $\mathbf{x}^*$ can be uniquely determined via $n + 1 - u$ linear pieces of k as in (8) above. Note in particular that if $\mathbf{x}^*$ lies in the interior of $[0, 1]^n$ then $u = 0$, and $\mathbf{x}^*$ arises as the unique solution of a system of n equations where $n + 1$ pieces of k are set to be equal. In case M does not contain any isolated point, for each vertex $\mathbf{x}$ of $\mathcal{K}'$ one defines rank($\mathbf{x}$) as the maximum rank of a linear system like (8) above, which is satisfied by $\mathbf{x}$. Whenever a vertex $\mathbf{x}$ is of rank $< n$ there is a path of 1-dimensional edges in $\mathcal{K}'$ connecting $\mathbf{x}$ to some vertex $\mathbf{x}'$ with rank($\mathbf{x}'$) $>$ rank($\mathbf{x}$) .

We now choose, once and for all, a point $\mathbf{x}^*$ where k attains its absolute minimum value, with $\mathbf{x}^*$ satisfying the conditions of the above Lemma. Upon fixing an index $j \in \{1, 2, \ldots, d\}$, we pick a point $\mathbf{x}$ in the interior of the linear domain D_j, and we define the map $g_{\mathbf{x}} \colon [0, 1] \to [0, 1]$ by

$$g_{\mathbf{x}}(y) = f_{\varphi}(\mathbf{x}, y). \tag{9}$$

The set $\nabla_{\mathbf{x}} \subseteq [0, 1]$ of nondifferentiability points of $g_{\mathbf{x}}$ is finite and, by definition, $0, 1 \in \nabla_{\mathbf{x}}$. Further, for each relative maximum value z of $g_{\mathbf{x}}$ there is $y \in \nabla_{\mathbf{x}}$ such that $g_{\mathbf{x}}(y) = z$. Let $L_{\mathbf{x}}$ denote the set of those points $y \in \nabla_{\mathbf{x}}$ where $g_{\mathbf{x}}$ attains its *absolute* maximum value, in symbols,

$$L_{\mathbf{x}} = \{y \in \nabla_{\mathbf{x}} \mid g_{\mathbf{x}}(y) = k(\mathbf{x}) = k_j(\mathbf{x})\}. \tag{10}$$

Note that $L_{\mathbf{x}}$ is finite and non-empty. The largest element $y_{\mathbf{x}}$ of $L_{\mathbf{x}}$ shall be called the *witness of k at $\mathbf{x}$*, in symbols,

$$f_{\varphi}(\mathbf{x}, y_{\mathbf{x}}) = g_{\mathbf{x}}(y_{\mathbf{x}}) = k(\mathbf{x}) = k_j(\mathbf{x}), \quad y_{\mathbf{x}} = \max L_{\mathbf{x}}. \tag{11}$$

Fix now a polyhedral complex $\mathcal{C}$ over $[0,1]^{n+1}$, with rational vertices, such that over every polyhedron of $\mathcal{C}$ the function f_φ is linear. In other words, each $(n+1)$-dimensional polyhedron T of $\mathcal{C}$ is a linear domain of f_φ. Let p_T be the uniquely determined linear piece of f_φ over T. For definiteness assume $y_{\mathbf{x}} \neq 0,1$ (the case $y_{\mathbf{x}} = 0,1$ only requiring trivial notational modifications). Then the non-differentiability of $g_{\mathbf{x}}$ at $y_{\mathbf{x}}$ ensures that the point $(\mathbf{x}, y_{\mathbf{x}})$ lies on the intersection $F_{\mathbf{x}}$ of at least two linear domains $T_{1\mathbf{x}}$ and $T_{2\mathbf{x}}$ of f_φ. Let $p_{1\mathbf{x}}$ and $p_{2\mathbf{x}}$ be their corresponding linear pieces of f_φ. Then for all $y \geq y_{\mathbf{x}}$ sufficiently close to $y_{\mathbf{x}}$ we have

$$g_{\mathbf{x}}(y) = p_{1\mathbf{x}}(y). \tag{12}$$

Similarly, for all $y \leq y_{\mathbf{x}}$ close to $y_{\mathbf{x}}$ we have

$$g_{\mathbf{x}}(y) = p_{2\mathbf{x}}(y). \tag{13}$$

For each $i = 1,2$ let us display $p_{i\mathbf{x}}$ as follows

$$p_{i\mathbf{x}}(x_1, \ldots, x_n, y) = a_{i\mathbf{x}1}x_1 + \cdots + a_{i\mathbf{x}n}x_n + b_{i\mathbf{x}}y + c_{i\mathbf{x}}. \tag{14}$$

Since f_φ is a McNaughton function, all the above coefficients a, b, c are integers. The non-differentiability of $g_{\mathbf{x}}$ at $y_{\mathbf{x}}$ ensures that $b_{1\mathbf{x}} \neq b_{2\mathbf{x}}$, whence

$$y_{\mathbf{x}} = \frac{(a_{1\mathbf{x}1}x_1 + \cdots + a_{1\mathbf{x}n}x_n + c_{1\mathbf{x}}) - (a_{2\mathbf{x}1}x_1 + \cdots + a_{2\mathbf{x}n}x_n + c_{2\mathbf{x}})}{b_{2\mathbf{x}} - b_{1\mathbf{x}}}. \tag{15}$$

Letting now $\mathbf{x}$ range over the interior of D_j, we define $\mathcal{X} \subseteq [0,1]^n$ to be the set of those points $\mathbf{x} = (x_1, \ldots, x_n) \in D_j$ such that the witness $y_{\mathbf{x}}$ lies on the intersection of *at least three* linear domains of f_φ. Since $\mathcal{X}$ is the union of finitely many sets whose dimension is $< n$, then there exists a non-empty open set $U \subseteq D_j$ such that

> for every $\mathbf{x} \in U$, the witness $y_{\mathbf{x}}$ lies on the intersection of *exactly two* linear domains of f_φ, say $T_{1\mathbf{x}}$ and $T_{2\mathbf{x}}$, whose corresponding linear pieces $p_{1\mathbf{x}}$ and $p_{2\mathbf{x}}$ satisfy (14)-(15).

For each pair (p_1, p_2) of *distinct* linear pieces of f_φ, let $m = m(p_1, p_2)$ be the maximum number of affinely independent points $\mathbf{x}_1, \ldots, \mathbf{x}_m \in U$ such that $p_1 = p_{1\mathbf{x}_1} = \cdots = p_{1\mathbf{x}_m}$ and $p_2 = p_{2\mathbf{x}_1} = \cdots = p_{2\mathbf{x}_m}$. Since U is n-dimensional and f_φ has only finitely many linear pieces, there exists a pair $(p_1^{(j)}, p_2^{(j)})$ of distinct linear pieces of f_φ with $m(p_1^{(j)}, p_2^{(j)}) = n+1$. Therefore we have $n+1$ affinely independent points $\mathbf{x}_1, \ldots, \mathbf{x}_{n+1} \in U$ such that

$$p_1^{(j)} = p_{1\mathbf{x}_1} = \cdots = p_{1\mathbf{x}_{n+1}} \text{ and } p_2^{(j)} = p_{2\mathbf{x}_1} = \cdots = p_{2\mathbf{x}_{n+1}}. \tag{16}$$

For each $i = 1,2$ let us display $p_i^{(j)}$ as follows

$$p_i^{(j)}(x_1, \ldots, x_n, y) = a_{i1}^{(j)}x_1 + \cdots + a_{in}^{(j)}x_n + b_i^{(j)}y + c_i^{(j)}. \tag{17}$$

Let the function $w^{(j)} \colon [0,1]^n \to [0,1]$ be defined by

$$w^{(j)}(\mathbf{x}) = \frac{(a_{11}^{(j)}x_1 + \cdots + a_{1n}^{(j)}x_n + c_1^{(j)}) - (a_{21}^{(j)}x_1 + \cdots + a_{2n}^{(j)}x_n + c_2^{(j)})}{b_2^{(j)} - b_1^{(j)}}. \tag{18}$$

Then by (15) for each $\mathbf{x}_t$, $(t = 1, \ldots, n+1)$ we have

$$y_{\mathbf{x}_t} = w^{(j)}(\mathbf{x}_t). \tag{19}$$

By (10)-(13) and (16)-(19) we can write for each $t = 1, \ldots, n+1$,

$$\begin{aligned} f_\varphi(\mathbf{x}_t, y_{\mathbf{x}_t}) = f_\varphi(\mathbf{x}_t, w^{(j)}(\mathbf{x}_t)) &= p_1^{(j)}(\mathbf{x}_t, w^{(j)}(\mathbf{x}_t)) = p_2^{(j)}(\mathbf{x}_t, w^{(j)}(\mathbf{x}_t)) \\ &= k(\mathbf{x}_t) = k_j(\mathbf{x}_t). \end{aligned}$$

Since $k = k_j$ over D_j, and k_j is linear, knowledge of its values at points $\mathbf{x}_1, \ldots, \mathbf{x}_{n+1}$ uniquely determines k_j. Since the functions $p_1^{(j)}$, $p_2^{(j)}$, and $w^{(j)}$ are linear, we conclude that, for all $\mathbf{x} \subset D_j$,

$$k(\mathbf{x}) = k_j(\mathbf{x}) = p_1^{(j)}(\mathbf{x}, w^{(j)}(\mathbf{x})) = p_2^{(j)}(\mathbf{x}, w^{(j)}(\mathbf{x})).$$

We have shown

Lemma 2. *Adopt the above notation. Let, in particular, $\mathbf{x}^*$ and $D_1, \ldots, D_d$ be as given by Lemma 1. For each $j = 1, \ldots, d$ there are two distinct linear pieces $p_1^{(j)}$ and $p_2^{(j)}$ of f_φ such that, letting for each $i = 1, 2$*

$$p_i^{(j)}(x_1, \ldots, x_n, y) = a_{i1}^{(j)}x_1 + \cdots + a_{in}^{(j)}x_n + b_i^{(j)}y + c_i^{(j)}, \tag{20}$$

we have $b_2^{(j)} \neq b_1^{(j)}$, and defining the map $w^{(j)} \colon [0,1]^n \to [0,1]$ by

$$w^{(j)}(\mathbf{x}) = \frac{(a_{11}^{(j)}x_1 + \cdots + a_{1n}^{(j)}x_n + c_1^{(j)}) - (a_{21}^{(j)}x_1 + \cdots + a_{2n}^{(j)}x_n + c_2^{(j)})}{b_2^{(j)} - b_1^{(j)}}, \tag{21}$$

it follows that, for all $\mathbf{x} \in D_j$,

$$k(\mathbf{x}) = k_j(\mathbf{x}) = p_1^{(j)}(\mathbf{x}, w^{(j)}(\mathbf{x})) = p_2^{(j)}(\mathbf{x}, w^{(j)}(\mathbf{x})). \tag{22}$$

In particular, the above identities hold for $\mathbf{x} = \mathbf{x}^*$.

Lemma 3. *Let $\mathbf{x}^*$ be as above. Let $\bar{z}$ be a value of relative maximum for the function $g_{\mathbf{x}^*} \colon [0,1] \to [0,1]$ defined by $g_{\mathbf{x}^*}(y) = f_\varphi(\mathbf{x}^*, y)$. Then there is a rational point $\bar{y}$ in the set $\nabla_{\mathbf{x}^*}$ of non-differentiability points of $g_{\mathbf{x}^*}$, together with a pair of linear domains D_1 and D_2 of f_φ with corresponding linear pieces p_1 and p_2 such that, letting $p_i(x_1, \ldots, x_n, y) = a_{i1}x_1 + \cdots + a_{in}x_n + b_iy + c_i$ for each $i = 1, 2$, we have*

(i) $g_{\mathbf{x}^*}(\bar{y}) = \bar{z}$;
(ii) $\bar{y}$ *lies on the intersection of* D_1 *and* D_2;
(iii) $\bar{y} = ((a_{11}x_1 + \cdots + a_{1n}x_n + c_1) - (a_{21}x_1 + \cdots + a_{2n}x_n + c_2))/(b_2 - b_1)$.

Proof. We use a variant of the initial argument leading to Lemma 2. For each relative maximum value $\bar{z}$ of $g_{\mathbf{x}^*}$ there is $y \in \nabla_{\mathbf{x}^*}$ such that $g_{\mathbf{x}^*}(y) = \bar{z}$. Let $N(\bar{z})$ be the set of those points $y \in \nabla_{\mathbf{x}^*}$ where $g_{\mathbf{x}^*}$ attains value $\bar{z}$. The rationality of $\mathbf{x}^*$ ensures that all elements of $N(\bar{z})$ are rational. Note that $N(\bar{z})$ is finite and non-empty. The largest element $\bar{y}$ of $N(\bar{z})$ satisfies (i).

Assume for definiteness [8] $\bar{y} \neq 0, 1$. Then the non-differentiability of $g_{\mathbf{x}^*}$ at $\bar{y}$ ensures that the point $(\mathbf{x}^*, \bar{y})$ lies on the intersection F of at least two linear domains D_1 and D_2 of f_φ in C, with their corresponding linear pieces p_1 and p_2. This settles (ii).

To conclude the proof, for all $y \geq \bar{y}$ sufficiently close to $\bar{y}$ we have $g_{\mathbf{x}^*}(y) = p_1(y)$. Similarly, for all $y \leq \bar{y}$ close to $\bar{y}$ we have $g_{\mathbf{x}^*}(y) = p_2(y)$. From the non-differentiability of $g_{\mathbf{x}^*}$ at $\bar{y}$ we get $b_1 \neq b_2$, whence the fact that $\bar{y}$ lies on F can be reformulated as

$$\bar{y} = \frac{(a_{11}x_1 + \cdots + a_{1n}x_n + c_1) - (a_{21}x_1 + \cdots + a_{2n}x_n + c_2)}{b_2 - b_1}$$

as required to settle (iii).

We are now in a position to evaluate the computational complexity of TAUT-∃Ł. To this purpose we fix some notation and terminology.

Let $\mathbf{x} = (h_1/k_1, \ldots, h_n/k_n) \in ([0,1] \cap \mathbb{Q})^n$, be a rational point in the n-cube. Assume for each $i \in \{1, \ldots, n\}$ the integers h_i and k_i to be relatively prime. Let $\mathrm{den}(\mathbf{x})$ denote the least common multiple of $k_1, \ldots, k_n$.

Let further $\#(X, \varphi)$ denote the total number of occurrences of the variable X in φ; also, let $\#\varphi$ be the total number of occurrences of variables in φ. For any n-tuple $\mathbf{x} \in ([0,1] \cap \mathbb{Q})^n$, we define the *size* of $\mathbf{x}$, to be the logarithm to the base two of $\mathrm{den}(\mathbf{x})$, in symbols, $\mathrm{size}(\mathbf{x}) = \log_2 \mathrm{den}(\mathbf{x})$.

Lemma 4. *Assume* $\varphi = \varphi(X_1, \ldots, X_n, Y)$ *is an arbitrary formula of* Ł_∞. *Let* k, $\mathbf{x}^*$, $D_1, \ldots, D_d$ *and* $k_1, \ldots, k_d$ *be as given by Lemma 1. Let* $z^* = k(\mathbf{x}^*)$, *and, as above, let* $g_{\mathbf{x}^*}(y) = f_\varphi(\mathbf{x}^*, y)$.

Then there exists a polynomial q *(independent of* φ*) having the following property:*

(i) $\mathrm{size}(\mathbf{x}^*) < q(\#\varphi)$;
(ii) each $y \in g_{\mathbf{x}^*}^{-1}(z^*) \cap \nabla_{\mathbf{x}^*}$ *is rational, and* $\mathrm{size}(y) < q(\#\varphi)$.

[8] It is possible that $\bar{y} \in \{0, 1\}$; the modifications in the present lemma are trivial.

Proof. In the light of Lemma 2, system (8) of Lemma 1 can be rewritten as

$$\begin{cases} p_1^{(1)}(\mathbf{x}, w^{(1)}(\mathbf{x})) & = p_1^{(2)}(\mathbf{x}, w^{(2)}(\mathbf{x})) \\ p_1^{(2)}(\mathbf{x}, w^{(2)}(\mathbf{x})) & = p_1^{(3)}(\mathbf{x}, w^{(3)}(\mathbf{x})) \\ \dots & \dots\dots \\ p_1^{(d-1)}(\mathbf{x}, w^{(d-1)}(\mathbf{x})) & = p_1^{(d)}(\mathbf{x}, w^{(d)}(\mathbf{x})) \\ x_{e(1)} & = \beta_1 \\ \dots & \dots\dots \\ x_{e(u)} & = \beta_u \end{cases} \tag{23}$$

The n-tuple $\mathbf{x}^*$ is the unique solution of the above system. Following (20), for every $i = 1, 2$ let us display each $p_i^{(j)}$ $j = 1, \dots, d$ as follows:

$$p_i^{(j)}(\mathbf{x}, w^{(j)}(\mathbf{x})) = a_{i1}^{(j)} x_1 + \dots + a_{in}^{(j)} x_n + b_i^{(j)} w^{(j)}(\mathbf{x}) + c_i^{(j)},$$

where $w^{(j)}(\mathbf{x})$ is as in (21). A standard induction argument (see, e.g., [2, Lemma 12]) shows that for each $i = 1, 2$ and $s = 1, \dots, n$,

$$|a_{is}^{(j)}| \leq \#(X_s, \varphi), \text{ and } |b_i^{(j)}| \leq \#(Y, \varphi). \tag{24}$$

Letting $d^{(j)} = b_2^{(j)} - b_1^{(j)}$, it follows that, for every $l = 1, \dots, d-1$, the lth line of the above system (23) can be written as

$$\frac{\sum_{s=1}^{n} \left(d^{(l+1)}(a_{1s}^{(l)} b_2^{(l)} - a_{2s}^{(l)} b_1^{(l)}) - d^{(l)}(a_{1s}^{(l+1)} b_2^{(l+1)} - a_{2s}^{(l+1)} b_1^{(l+1)})\right) x_s}{d^{(l)} d^{(l+1)}} = t^{(l)},$$

where

$$t^{(l)} = \frac{d^{(l)}(c_1^{(l+1)} b_2^{(l+1)} - c_2^{(l+1)} b_1^{(l+1)}) - d^{(l+1)}(c_1^{(l)} b_2^{(l)} - c_2^{(l)} b_1^{(l)})}{d^{(l)} d^{(l+1)}}.$$

Multiplying both sides in line l by the quantity $d^{(l)} d^{(l+1)}$, we obtain from (23) a linear system $\mathcal{M}\mathbf{x} = \mathbf{t}$ where, by (24), all entries m_{rs} of the $n \times n$ matrix $\mathcal{M}$ are integers satisfying the inequality $|m_{rs}| \leq 2 \cdot 2\#(Y, \varphi) \cdot 2\#(X_s, \varphi) \cdot \#(Y, \varphi)$. Also the vector $\mathbf{t}$ of constant terms of the above system has integer coordinates. It follows that $\mathrm{den}(\mathbf{x}^*) \leq |\det(\mathcal{M})|$, and a straightforward computation yields $\mathrm{den}(\mathbf{x}^*) \leq (2\#\varphi)^{3(d-1)} \leq (2\#\varphi)^{3n} \leq (2\#\varphi)^{3\#\varphi}$, whence $\mathrm{size}(\mathbf{x}^*) \leq 3\#\varphi \log_2(2\#\varphi) \leq 3(\#\varphi)^2$, which settles (i).

The proof of (ii) is now an immediate consequence of Lemma 3(iii).

We are now ready to prove the main theorem of this section.

Theorem 7. TAUT-$\exists$Ł $\in \Pi_2^p$.

Proof. Adopt the above notation and terminology. Let $\varphi = \varphi(X_1, \dots, X_n, Y)$ be a formula in $Ł_\infty$. With the intent of disproving that $\exists Y\, \varphi$ is a tautology, if there is an $\mathbf{x} \in [0,1]^n$ such that $k(\mathbf{x}) = \max_{y\in[0,1]} f_\varphi(\mathbf{x}, y) < 1$ then, by the above Lemmas 1 and 4 there is a polynomial q and a rational n-tuple $\mathbf{x}^*$ such that $k(\mathbf{x}^*) < 1$, and the size of $\mathbf{x}^*$ is $< q(\#\varphi)$. Again let the function $g_{\mathbf{x}^*}: [0,1] \to [0,1]$ be defined by $g_{\mathbf{x}^*}(y) = f_\varphi(\mathbf{x}^*, y)$. After guessing such $\mathbf{x}^*$ via a non-deterministic polynomial time procedure, we must only check that for all $y \in [0,1]$,

$$k(\mathbf{x}^*) = \max_{y\in[0,1]} g_{\mathbf{x}^*}(y) < 1.$$

Since $\mathbf{x}^*$ is rational, Lemma 3 is to the effect that for each relative maximum value $\bar{z}$ of $g_{\mathbf{x}^*}$ there is a rational point $\bar{y} \in \nabla_{\mathbf{x}^*}$, such that $g_{\mathbf{x}^*}(\bar{y}) = \bar{z}$, and in addition, by Lemma 4(ii), the size of $\bar{y}$ is $< q(\#\varphi)$. Thus for our final check we may safely restrict to computing the value of $g_{\mathbf{x}^*}$ for the set A of all rationals $y \in [0,1]$ whose size is $< q(\#\varphi)$. One immediately sees that each computation $y \in A \mapsto z = g_{\mathbf{x}^*}(y) \in [0,1] \cap \mathbb{Q}$ can be performed in deterministic polynomial time with respect to $\#\varphi$. Therefore, this final check has co-NP complexity, and our theorem is proved.

Appendix: Normal forms and desingularizations

In many-valued logic, as well as in boolean logic, disjunctive normal forms (DNF) play a fundamental role, both for automated deduction, and for a deeper understanding of the algebra [28], [1]. McNaughton's theorem yields DNF reductions of formulas $\psi = \psi(X_1, \ldots, X_n)$ in Ł_∞, by representing their associated functions $p = f_\psi: [0,1]^n \to [0,1]$ in a special way.

To see this, following [22], [23] one proceeds in several steps as follows.

FIRST STEP. The n-cube $[0,1]^n$ is partitioned into a complex $\mathcal{C}$ of convex polyhedra with rational vertices such that, over any such polyhedron, the function p is linear. Generalizing the familiar construction for 2-dimensional polyhedra (where one adds a maximal set of diagonals) we now subdivide $\mathcal{C}$ into a simplicial complex $\mathcal{S}$ without adding new vertices. Using Minkowski's convex body theorem, we can further subdivide $\mathcal{S}$ into a *unimodular* simplicial complex $\mathcal{T}$: in other words, for each n-simplex Δ in $\mathcal{T}$ writing in homogeneous integer coordinates the vertices of Δ, we obtain an $(n+1) \times (n+1)$ integer valued matrix M_Δ whose determinant is equal to ± 1.

SECOND STEP. One then constructs the family $H(\mathcal{T}) = \{H^{\mathcal{T}}_{\mathbf{w}}\}_{\mathbf{w}}$ of Schauder hats of $\mathcal{T}$. By the *Schauder hat* $H^{\mathcal{T}}_{\mathbf{w}}$ of $\mathcal{T}$ at vertex $\mathbf{w}$ (the *apex* of the hat) we mean the uniquely determined piecewise linear continuous function such that $H^{\mathcal{T}}_{\mathbf{w}}(\mathbf{w}) = 1/d$ (where d is the least common denominator of the homogeneous integer coordinates of $\mathbf{w}$), $H^{\mathcal{T}}_{\mathbf{w}}(\mathbf{v}) = 0$ for every vertex $\mathbf{v}$ of $\mathcal{T}$ other than $\mathbf{w}$, and $H^{\mathcal{T}}_{\mathbf{w}}$ coincides with a linear function h_Δ over each n-simplex Δ of $\mathcal{T}$. As an equivalent reformulation of the unimodularity property, the coefficients of the linear polynomial representing h_Δ are integers. An easy lemma, first proved by McNaughton, and then simplified by Rose and Rosser ([16],[35]), yields a formula φ_Δ representing h_Δ over Δ. A routine min-max argument now shows that all Schauder hat functions are representable by formulas, and therefore p can be expressed as a disjunction (sum) of Schauder hat formulas. We naturally regard the formulas representing the functions in $H(\mathcal{T})$ as the basic constituents of a DNF reduction of p.

THIRD STEP. To grasp the connection between the above DNF construction and toric varieties [12], upon writing in homogeneous integer coordinates the vertices of each simplex Δ in $\mathcal{T}$, we obtain a complex Ψ of simplicial cones, also known as a *fan*. As one more equivalent reformulation of the unimodularity of $\mathcal{T}$ we can say that Ψ is "regular". By the well known vocabulary of toric geometry [12, p. 329-330], (smooth) toric varieties correspond to (regular) fans, and hence to (unimodular) simplicial complexes of the above kind. It follows [23] that desingularizing a toric variety amounts to subdividing a complex into a unimodular simplicial complex, precisely as is done to compute DNF reductions of McNaughton functions, or to prove our main representation theorem. Our desingularization algorithms, arising from DNF reduction algorithms in infinite-valued logic, yield tight estimates of the Euler characteristic of desingularizations of low-dimensional toric varieties [3]. Further, Panti [31] gives a geometric proof of Chang's completeness theorem using the De Concini-Procesi theorem on elimination of points of indeterminacy [12, p. 252].

Acknowledgments

We are grateful to Matthias Baaz, Brunella Gerla and Helmut Veith for many discussions about rational McNaughton functions.

References

1. Aguzzoli, S.: The complexity of McNaughton functions of one variable, Advances in Applied Mathematics, 21 (1998) 58–77
2. Aguzzoli, S., Ciabattoni, A.: Finiteness in infinite-valued Łukasiewicz logic. Journal of Logic, Language and Information. Special issue on Logics of Uncertainty Mundici, D. (Ed.), 9 (2000) 5–29
3. Aguzzoli, S., Mundici, D.: An algorithmic desingularization of 3–dimensional toric varieties. Tôhoku Mathematical Journal, 46 (1994) 557–572
4. Amato, P., Porto, M.: An algorithm for the automatic generation of a logical formula representing a control law. Neural Network World, 10 (2000) 777–786
5. Baaz, M., Veith, H.: Interpolation in fuzzy logic. Archive for Mathematical Logic, 38 (1999) 461–489
6. Berlekamp, E.R.: Block coding for the binary symmetric channel with noiseless, delayless feedback. In: Error-correcting Codes. Mann, H.B. (Ed.) Wiley, New York (1968) 330–335
7. Chang, C.C.: Algebraic analysis of many-valued logics. Transactions of the American Mathematical Society, 88 (1958) 467–490
8. Chang, C.C.: A new proof of the completeness of the Łukasiewicz axioms. Transactions of the American Mathematical Society, 93 (1959) 74–90

9. Chang, C.C.: The writing of the MV-algebras, Studia Logica, special issue on Many-valued Logics, Mundici, D. (Ed.) 61 (1998) 3–6
10. Cignoli, R., D'Ottaviano, I.M.L., Mundici, D. Algebraic Foundations of Many-valued Reasoning (Trends in Logic, Studia Logica Library), Kluwer, Dordrecht (2000)
11. Esteva, F., Godo, L., Montagna, F.: The ŁΠ and Ł$\Pi 1/2$ logics: two complete fuzzy systems joining Łukasiewicz and product logic, Archive for Mathematical Logic. 40 (2001) 39–67
12. Ewald, G.: Combinatorial convexity and algebraic geometry (Graduate Texts in Mathematics, vol. 168). Springer-Verlag, Berlin, Heidelberg, New York (1996)
13. Garey, M.R., Johnson, D.S.: Computers and intractability. A guide to the Theory of NP-completeness. W.H. Freeman and Company, San Francisco (1979)
14. Hájek, P.: Metamathematics of fuzzy logic. Kluwer, Dordrecht. (Trends in Logic, Studia Logica Library) (1998)
15. Klement, E.P., Mesiar, R., Pap, E.: Triangular Norms (Trends in Logic, Studia Logica Library).
Kluwer, Dordrecht (2000)
16. McNaughton, R.: A theorem about infinite-valued sentential logic. The Journal of Symbolic Logic, 16 (1951) 1–13
17. Meyer, A.R., Stockmeyer, L.J.: The equivalence problem for regular expressions with squaring requires exponential time. Proc. 13th Ann. Symp. on Switching and Automata Theory, IEEE Computer Society (1972) 125–129
18. Montagna, F.: An algebraic approach to propositional fuzzy logic. Journal of Logic, Language and Information. Special issue on Logics of Uncertainty Mundici, D. (Ed.), 9 (2000) 91–124
19. Mundici, D.: Interpretation of AF C^*-algebras in Łukasiewicz sentential calculus. Journal of Functional Analysis, 65 (1986) 15–63
20. Mundici, D.: Satisfiability in many-valued sentential logic is NP-complete. Theoretical Computer Science, 52 (1987) 145–153
21. Mundici, D.: The logic of Ulam's game with lies. In: Knowledge, Belief and Strategic Interaction. (Cambridge Studies in Probability, Induction and Decision Theory) Bicchieri, C., Dalla Chiara, M.L. (Eds.) Cambridge University Press (1992) 275–284
22. Mundici, D.: A constructive proof of McNaughton's Theorem in infinite-valued logic. The Journal of Symbolic Logic, 59 (1994) 596–602
23. Mundici, D.: Łukasiewicz normal forms and toric desingularizations. In: Proceedings of Logic Colloquium 1993, Keele, England. Hodges, W. et al. (Eds.) Oxford University
Press (1996) 401–423
24. Mundici, D.: Nonboolean partitions and their logic, In: First Springer-Verlag Forum on Soft Computing. Prague, August 1997, Soft Computing 2 (1998) 18–22
25. Mundici, D.: Tensor products and the Loomis-Sikorski theorem for MV-algebras. Advances in Applied Mathematics, 22 (1999) 227–248
26. Mundici, D.: Ulam game, the logic of Maxsat, and many-valued partitions. In: Fuzzy Sets, Logics, and Reasoning about Knowledge, Dubois, D., Prade, H., Klement, E.P. (Eds.), Kluwer, Dordrecht (2000) 121–137
27. Mundici, D.: Reasoning on imprecisely defined functions, In: Discovering the World with Fuzzy Logic, Novak, V., Perfilieva, I. (Eds.), Physica-Verlag, Springer, Heidelberg, New York (2000) 331–366

28. Mundici, D., Olivetti, N.: Resolution and model building in the infinite-valued calculus of Łukasiewicz. Theoretical Computer Science, 200 (1998) 335–366
29. Mundici, D., Riečan, B.: Probability on MV-algebras, In: Handbook of Measure Theory, Pap, E. (ed.), North-Holland, Amsterdam, to appear (200?)
30. Novák, V., Perfilieva, I., Močkoř, J.: Mathematical Principles of Fuzzy Logic. Kluwer, Dordrecht (1999)
31. Panti, G.: A geometric proof of the completeness of the Łukasiewicz calculus. Journal of Symbolic Logic, 60 (1995) 563–578
32. Panti, G.: Multi-valued Logics. In: Quantified Representation of Uncertainty and Imprecision, vol. 1, Smets, P. (Ed.), Kluwer, Dordrecht (1998) 25–74
33. Pavelka, J.: On Fuzzy Logic I,II,III. Zeitschrift für math. Logik und Grundlagen der Mathematik, 25 (1979) 45–52, 119–134, 447–464
34. Riečan, B., Neubrunn, T.: Integral, Measure, and Ordering. Kluwer, Dordrecht (1997)
35. Rose, A., Rosser, J.B.: Fragments of many-valued statement calculi. Transactions of the American Mathematical Society, 87 (1958) 1–53
36. Tarski, A.: Logic, Semantics, Metamathematics. Clarendon Press, Oxford (1956). Reprinted, Hackett, Indianapolis (1983).
37. Verbruggen, H.B., Bruijn, P.M.: Fuzzy control and conventional control: What is (and can be) the real contribution of Fuzzy Systems?, Fuzzy Sets and Systems, 90 (1997) 151–160

Chapter 15
A Łukasiewicz–style Many-valued Similarity Reasoning. Review

Esko Turunen

Tampere University of Technology,
P.O. Box 692, FIN-33101 Tampere, Finland
esko.turunen@tut.fi

Abstract. A Łukasiewicz many-valued similarity based fuzzy control algorithm is introduced, and tested in realistic control systems. The results are compared to fuzzy control systems where the inference is based on Mamdani–style system or Sugeno–style system. Simulation tests show that the new algorithm is comparable with the standard ones and gives, in some extreme cases, even statistical significant better results.
Keywords: Non-classical logics, fuzzy control.

1 Introduction

The content of this review paper has been published in [1], [4] [5] and [8], thus many details are omitted.

Our aim of is twofold. In the mathematical part we consider an Algorithm to construct fuzzy IF-THEN inference systems. The Algorithm utilizes Łukasiewicz many-valued equivalence, and its design has a special feature that whenever the output would not be unique the final decision should be left to human experts. Thus, as much as possible the intelligence relies on a real controller, and technical defuzzification methods are not needed. In the application part we show how the Algorithm can be utilized in control; we consider traffic signal control problems and water resources management.

Fuzzy logic [10] has been introduced and successfully applied to a wide range of automatic control tasks. The main benefit of fuzzy logic is the opportunity to model the ambiguity and the uncertainty of decision-making. The point in utilizing fuzzy logic in control theory is to model control based on human expert knowledge, rather than to model the process itself. Indeed, fuzzy control has proven to be successful in problems where exact mathematical modeling is hard or impossible but an experienced human operator can control process.

At present, there is a multitude of inference systems based on fuzzy technique. Most of them, however, suffer (logically) ill–defined foundations; even if they are mostly performing better that classical mathematical methods, they still contain black boxes, e.g. defuzzification, which are very difficult to justify mathematically or logically. For example, fuzzy IF – THEN rules, which are in the core of fuzzy inference systems, are often reported to be generalizations of classical Modus Ponens rule of inference, but literally this not the case; the relation between these rules and any known many–valued logic is complicated and artificial. Moreover, the performance of an expert system should be equivalent to that of human expert: it should give the same results that the expert gives, but warn when the control situation is so vague that an expert is not sure about the right action. The existing fuzzy expert systems very seldom fulfill this latter condition. Many researches observe that fuzzy inference is based on *similarity*. Kosko [6], for example, writes *'Fuzzy membership...represents similarities of objects to imprecisely defined properties'.*

We study systematically many–valued equivalence, or fuzzy similarity, the original notion by Zadeh [11], a genaralization of equivalence relation; a binary fuzzy relation that is reflexive, symmetric and weakly transitive. Later many other authors have developed Zadeh's ideas, see e.g. [9], [3] and [2]. Dubois and Prade write *'...The evaluation of similarity between two multi–feature descriptions of objects may be specially of interest in analogical reasoning. If we assume that each feature is associated with an attribute domain equipped with similarity relation modelling approximate equality on this domain, the problem is then to aggregate the degrees of similarity between the objects pertaining to each feature into a global similarity index. This means that the resulting index should still have properties like reflexivity, symmetry and max$\odot$-transitivity. ... . Moreover, we may think of a weighted aggregation if we consider that we are dealing with a fuzzy set of features having different levels of importance.'*

We give a solution to the problem stated by Dubois & Prade above. It turns out that, starting from the Łukasiewicz well–defined many–valued logic, we are able to construct a method performing fuzzy reasoning such that the inference relies only on experts knowledge and on well–defined logical concepts.

Our basic observation is that any fuzzy set generates a fuzzy similarity, and that these similarities can be combined to a fuzzy relation which turns out to a fuzzy similarity, too. We call this induced fuzzy relation *Total Fuzzy Similarity*. Fuzzy IF – THEN inference systems are, in fact, problems of choice: compare each IF–part of the rule base with an actual input value, find *the most similar* case and fire the corresponding THEN–part; if it is not unique, use a criteria given by an expert to proceed. Thus, the main distinction between our approach and other fuzzy inference systems is that only one IF–THEN rule determines the output. Based on the Łukasiewicz well–defined many–valued logic, we show how this method can be carried out formally. Thus, the aim is to give a mathematical approach to fuzzy reasoning, a method that is based on many–valued equivalence rather that many–valued implication. For missing details, see [1], [4], [5] and [8].

2 Mathematical preliminaries

Recall a binary operation $\odot : [0,1]^2 \searrow [0,1]$ is called *t-norm* if, for all elements $x, y, z \in [0,1]$, (i) if $x \leq y$, then $x \odot z \leq y \odot z$, (ii) $x \odot y = y \odot x$, (iii) $x \odot 1 = x$, (iv) $x \odot (y \odot z) = (x \odot y) \odot z$, in particular, *continuous* t-norms $\odot$ and their residua $\rightarrow$, defined by (v) $x \odot y \leq z$ iff $x \leq y \rightarrow z$ play a fundamental role in fuzzy logic. The most frequently used continuos t-norms in various fuzzy inference systems generate the following algebraic structures
Gödel algebra:

$$x \odot y = \min\{x, y\},\ x \rightarrow y = \begin{cases} 1 & \text{if } x \leq y \\ y & \text{otherwise.} \end{cases}$$

Product t-algebra:

$$x \odot y = xy,\ x \rightarrow y = \begin{cases} 1 & \text{if } x \leq y \\ \frac{y}{x} & \text{otherwise.} \end{cases}$$

Łukasiewicz algebra:

$$x \odot y = \max\{0, x + y - 1\},\ x \rightarrow y = \begin{cases} 1 & \text{if } x \leq y \\ 1 - x + y & \text{otherwise.} \end{cases}$$

These three examples are fundamental since, in a certain sense, they characterize all possible continuos t-norms. They are the generators of all *BL-algebras* (cf. [8]) of the real unit interval, too; by fixing a continuous t-norm we fix a *Basic Logic*, a well-defined many-valued logic modeling mathematically fuzzy reasoning. The operations $\odot$ and $\rightarrow$ are the algebraic counterparts of the logical connectives conjunction and implication, respectively. In particular, the complement x^* of an element $x \in [0,1]$ defined by $x^* = x \rightarrow 0$, is the algebraic counterpart of negation, while many-valued equivalence is interpreted algebraically by *bi-residuum* defined, for all $x, y \in [0,1]$, via

$$x \leftrightarrow y = \min\{(x \rightarrow y), (y \rightarrow x)\}.$$

In any BL-algebra, a bi-residuum $\leftrightarrow$ has the following properties (cf. [8])

$$x \leftrightarrow x = 1, \tag{1}$$

$$x \leftrightarrow y = y \leftrightarrow x, \tag{2}$$

$$(x \leftrightarrow y) \odot (y \leftrightarrow z) \leq x \leftrightarrow z, \tag{3}$$

$$x \leftrightarrow 1 = x. \tag{4}$$

It is easy to see that, in Łukasiewicz algebra, the $\leftrightarrow$ bi-residuum is calculated via

$$x \leftrightarrow y = 1 - \max\{x, y\} + \min\{x, y\}.$$

Castro & Klawonn [2] among others set the following important

Definition 1 *Let A be a non-void set and $\odot$ a continuos t-norm. Then a* fuzzy similarity *S on A is such a binary fuzzy relation that, for each $x, y, z \in A$,*
(i) $S\langle x, x\rangle = 1$ (everything is similar to itself),
(ii) $S\langle x, y\rangle = S\langle y, x\rangle$ (fuzzy similarity is symmetric),
(iii) $S\langle x, y\rangle \odot S\langle y, z\rangle \leq S\langle x, z\rangle$ (fuzzy similarity is weakly transitive).

Trivially, fuzzy similarity is a generalization of classical equivalence relation, thus called *many-valued equivalence*, too. Recall a *fuzzy set X* is an ordered couple (A, μ_X), where the *reference set A* is a non-void set and the *membership function* $\mu_X : A \searrow [0, 1]$ tells the degree to which an element $a \in A$ belongs to the fuzzy set X.

Theorem 1 *Any fuzzy set (A, μ_X) on a reference set A generates a fuzzy similarity S on A, defined by*

$$S(x, y) = \mu_X(x) \leftrightarrow \mu_X(y), \text{ where } x, y \text{ are elements of } A. .$$

Moreover,

$$\text{if } \mu_X(y) = 1 \text{ then } S(x, y) = \mu_X(x).$$

Theorem 2 *For each $x, z \in A$, condition (iii) is equivalent to*
(iv) $\bigvee_{y \in A}(S\langle x, y\rangle \odot S\langle y, z\rangle) \leq S\langle x, z\rangle$

Notice that the original definition of fuzzy similarity given by Zadeh in [11] is (i), (ii) and (iv). Moreover, it is worth noting that, in Łukasiewicz algebra, 'the negation of equivalence is distance'. Indeed, for all $a, b \in [0, 1]$,

$$(a \leftrightarrow b)^* = 1 - [1 - \max\{a, b\} + \min\{a, b\}] = |a - b| = \sqrt{(a - b)^2},$$

the one dimensional Euclidean distance between a and b.

Theorem 3 *Consider n Łukasiewicz valued fuzzy similarities S_i, $i = 1, \cdots, n$ on a set X. Then*

$$S\langle x, y\rangle = \tfrac{1}{n}\Sigma_{i=1}^{n} S_i\langle x, y\rangle$$

is a Łukasiewicz valued fuzzy similarity on X. More generally, the weighted mean

$$S\langle x, y\rangle = \tfrac{1}{M}\Sigma_{i=1}^{n} m_i \cdot S_i\langle x, y\rangle,$$

where $M = \Sigma_{i=1}^{n} m_i$ and m_i are natural numbers, is again a Łukasiewicz valued fuzzy similarity on X, and called Total Fuzzy Similarity.

Notece that Theorem 3 does not hold for other BL-algebras than Łukasiewicz algebra. Indeed, in general, weak transitivity does hold.

3 Algorithm to construct fuzzy inference systems

Consider an fuzzy IF–THEN inference system S, the input universe of discourse is X, the IF–parts, and an output universe of discourse is Y, the THEN–parts. We assume there are n input variables and one output variable. The dynamics of S are characterized by a finite collection of IF–THEN–rules; e.g.

Rule 1 IF x is A_1 and y is B_1 and z is C_1 THEN w is D_1
Rule 2 IF x is A_2 and y is B_2 and z is C_2 THEN w is D_2

$\vdots$ $\qquad\qquad\qquad\qquad\qquad\qquad\qquad\qquad$ $\vdots$

Rule k IF x is A_k and y is B_k and z is C_k THEN w is D_k

where $A_1, \cdots, D_k$ are fuzzy sets of height 1, that is, in each fuzzy set there is at least one element that obtains the membership degree 1. Generally, the output fuzzy sets $D_1, \cdots, D_k$ should obtain all the same values $\in [0,1]$ the input fuzzy sets $A_1, \cdots, C_k$ do, however, the outputs can be crisp actions, too. All these fuzzy sets are to be specified by the fuzzy control engineer. We avoid disjunction between the rules by allowing some of the output fuzzy sets D_i and $D_j, i \neq j$, be possibly equal . Thus, a fixed THEN–part can be followed by various IF–parts. Some of the input fuzzy sets may be equal, too (e.g. $B_i = B_j$ for some $i \neq j$). However, the rule base should be consistent; a fixed IF–part precedes a fixed THEN–part. Moreover, the rule base can be incomplete; if an expert is not able to define the THEN-part of some combination 'IF x is A_i and y is B_i and z is C_i' then the rule should be skipped.

Now we are in the position to formulate an algorithm a fuzzy control engineer has to perform to construct a total fuzzy similarity based inference system.

Step 1. Create the dynamics of S, i.e. define the IF–THEN rules, give the shapes of the input fuzzy sets (e.g. $A_1, \cdots, C_k$) and the shapes of the output fuzzy sets (e.g. $D_1, \cdots, D_k$).

Step 2. Give weights to various parts of the input fuzzy sets (e.g. to A_i.s, B_i.s and C_i.s) to emphasize the mutual importance of the corresponding input variables.

Step 3. Put the IF–THEN–rules in a linear order with respect to their mutual importance, or give some criteria on how this can be done when necessary; i.e. give a criteria on how to distinguish inputs causing equal degree of total fuzzy similarity in different IF–parts.

Step 4. For each THEN–part i, give a criteria on how to distinguish outputs with equal degree on membership (e.g. w_0 and v_0 such that $\mu_{D_i}(w_0) = \mu_{D_i}(v_0)$, $w_0 \neq v_0$).

A general framework for the inference system is now ready. Assume then that we have actual input values, e.g. (x_0, y_0, z_0). The corresponding output value w_0 is found in the following way.

Step 5. Consider each IF–part of the rule base as a crisp case, and compare the actual input values separately with each IF–part, in other words, count total fuzzy similarities between the actual inputs and each IF–part of the rule base; by the above Theorems, this is equivalent to counting weighted means, e.g.

$$\begin{aligned} m_1\mu_{A_1}(x_0) + m_2\mu_{B_1}(y_0) + m_3\mu_{C_1}(z_0) &= \text{Similarity(actual,Rule 1)} \\ m_1\mu_{A_2}(x_0) + m_2\mu_{B_2}(y_0) + m_3\mu_{C_2}(z_0) &= \text{Similarity(actual,Rule 2)} \\ \vdots \qquad\qquad & \qquad \vdots \\ m_1\mu_{A_k}(x_0) + m_2\mu_{B_k}(y_0) + m_3\mu_{C_k}(z_0) &= \text{Similarity(actual,Rule } k) \end{aligned}$$

where m_1, m_2 and m_3 are the weights given in Step 2.

Step 6. Fire an output value w_0 such that

$$\mu_{D_i}(w_0) = \text{Similarity(actual,Rule } i)$$

corresponding to the maximal Total Fuzzy Similarity Similarity(actual,Rule i), if such Rule i is not unique, use the mutual order given in Step 3, and if there are several such output values w_0 utilize the criteria given in Step 4.

Notice that, in case we drop the assumption that the fuzzy sets under consideration are of height 1 the computations will be more complicated (but still possible). Notice, too, that Step 6 can be view as an instance of Generalized Modus Ponens in the sense of Pavelka's fuzzy logic (cf. [7]), i.e.

$$R_{GMP} \quad : \quad \frac{\alpha, (\alpha \texttt{ imp } \beta)}{\beta}, \frac{a, b}{a \odot b}$$

where α corresponds to the IF–part, β corresponds to the THEN–part, a is the degree of maximal total fuzzy similarity and $b = 1$. In general, this gives a many–valued logic based theoretical justification to Step 6. In case the outputs are concrete actions that can either be taken or not we define 'take the action only in case $a \in [c, 1]$, where c is a suitable value'.

Of course, we can specify our algorithm by putting extra demands, for example, in some cases the degree of total fuzzy similarity of the best alternative should be greater than some fixed value $\alpha \in [0, 1]$, sometimes all the alternatives possessing the highest fuzzy similarity should be indicated, or the difference between the best candidate and second one should be larger than a fixed value $\beta \in [0, 1]$. All this depends on an expert's choice.

4 Application 1: Signalized isolated pedestrian crossing

For simplicity, the first application of the Algorithm controls a signalized isolated pedestrian crossing (see Fig. 1), for more complicated traffic signal control applications, cf. [4], [5]. Normally, the signals of isolated pedestrian crossings in Finland are working as traffic actuated, and the rest phase is

vehicle green (important safety aspect). Two detectors are located per each lane, one at the stop line and the other 60 meters from the stop line. Pedestrian green time is constant (10 seconds) or even actuated (6...14 seconds) in specific conditions if children or elderly people are numerous. The main goal of fuzzy control is to give pedestrians an opportunity to cross the street safely, and with minimum waiting time, but also that the risk of rear-end collisions is minimized (minimize the number of approaching vehicles at the termination moment). It is also important that control does not encourage pedestrians to cross the street during the pedestrian red phase. Controlling the timing of a traffic signal means making the following evaluation constantly: either to terminate the current phase and to change it to the next phase, or to continue the current phase. In other words, a controller incrementally evaluates these two options and takes the most appropriate option. This means the output is the decision about the termination (T) or the extension (E) of vehicle signal group (crisp value). The input parameters in use are
∘ pedestrian waiting time in seconds, PWT, corresponding to three fuzzy sets; *short, long, very long,*
∘ maximum number of approaching vehicles/lane, NoAV, corresponding to three fuzzy sets *none, some, many,*
∘ discharging queue indicator, gap between vehicles at stop line GAP, corresponding two fuzzy sets; *low, high.*

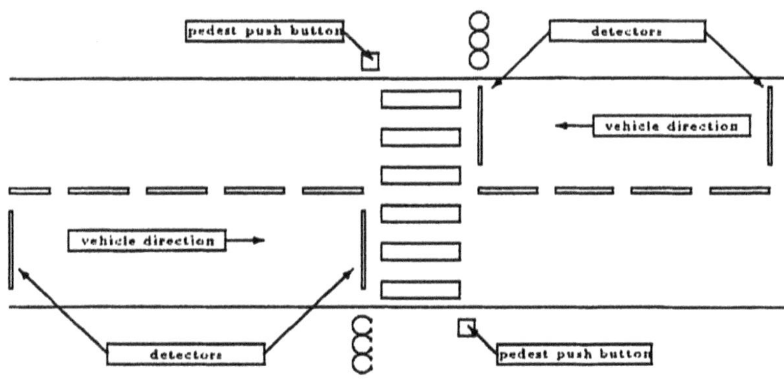

Fig. 1. Layout of signalized isolated pedestrian crossing

In this fuzzy IF-THEN inference system, general rule formulation is the following

IF PWT is short/long/verylong AND
NoAV is none/some/many AND
GAP is low/high
THEN Terminate/Extend

The rule base is complete, indeed, total number of rules is 18 ($= 3 \times 3 \times 2$). There are 9 rules for the extension and 9 rules for the termination decisions. We have now settled Step 1 and Step 2 of the Algorithm.

According to an experienced traffic signal designer, in fifty-fifty situation the decision is Extension. This corresponding to Step 3 of the Algorithm. Clearly, Step 4 is empty. Step 5 and Step 6 is straightforward.

To illustrate the performance of the control system, assume a pedestrian has been waiting 13 seconds, there are 2 vehicles approaching and their gap is 1.5 seconds. Such a situation is the most similar to the case PVT is long, $\mu_{long}(13) = 0.6$, NoAV is some, $\mu_{some}(2) = 1$, and GAP is small, $\mu_{small}(1.5) = 0.8$. The degree of total fuzzy similarity to this IF-part is 0.8, and the corresponding THEN-part is 'Extend vehicles green signal'.

4.1 Simulation results

A previous version of this control system was a Mamdani style fuzzy inference machine. Simulations made by HUTSIM, a traffic flow simulation package developed at the Laboratory of Transportation Engineering of the Helsinki University of Technology, Finland, showed [4] that even the previous version performed better or at least as well as traditional isolated pedestrian signals do. Therefore the performance of the Łukasiewicz based control system was compared to this Mamdani style inference machine[1]. The performances of the three control systems Mamdani-style, Total Fuzzy Similarity based with weights 1,1,1 and Total Fuzzy Similarity based with weights 3,2,1 were compared with respect to the two measures: average pedestrian and vehicle delays for three pedestrian volumes: 15 pedestrian per hour, 50 pedestrian per hour and 150 pedestrian per hour. A statistical hypothesis 'The average delays are equal in each case' was tested by approximate t-test on the risk level $\alpha = 0.01$. According to the results, there is no difference if pedestrian volume is low (15 ped/h). The pedestrian delays seems to be slightly smaller if control strategy is Mamdani. However, this difference is not statistically significant. If traffic volume is high and, especially, if pedestrian volume is high too, then the results of fuzzy similarity are better. This difference has statistical significance.

5 Application 2: Water level control

Real-time reservoir operation is a continuous decision-making process on the water level of a reservoir and release from it. The operation is always based on

[1] The rule base and the fuzzy sets were same

operating policy and rules defined and decided upon in strategic planning. The complexity of the real-time release decision, considering all the time-dependent information, warrants the importance of real-time operation.

A fuzzy rule-based control model for multipurpose real-time reservoir operation was constructed by Total Fuzzy Similarity method, and compared with the more traditional Sugeno–style method. Specifically the seasonal variation in both hydrological variables and operational targets were examined. This was done by considering the inputs as season-dependent relative values, instead of using absolute values. The inference drawn in several stages allowed a simple, accessible model structure. The control model was illustrated using Lake Paijanne, a regulated lake in Finland. The model was calibrated to simulate the actual operation, but also to better fulfill the new multipurpose operational objectives determined by experts.

Corresponding to Step 1 of the Algorithm, in Realease Model, the only rule is the following

IF Relative Water Level is very low/low/$\cdots$/very high AND
Relative Inflow is very small/$\cdots$/very large
THEN Release is exceptional low/ $\cdots$ / exceptional large

Corresponding to Step 2, all fuzzy sets are of triangular shape. Since there is only one rule, Step 3 is trivial. Corresponding to Step 4, it was decided to fire the left-hand value of the most similar output triangular fuzzy set.

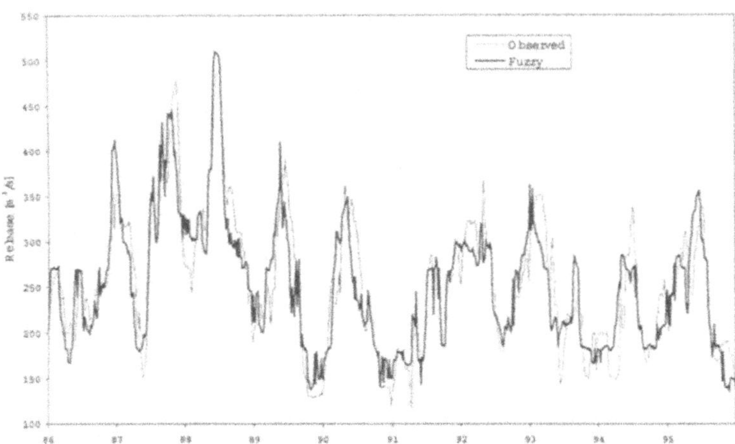

Fig. 2. Observed water release and water release ruled by Total Fuzzy Similarity Control

The model was tested using data from years 1985-1996. In the test phase the global optimization mode was used. Fig. 2 shows the release, when Total

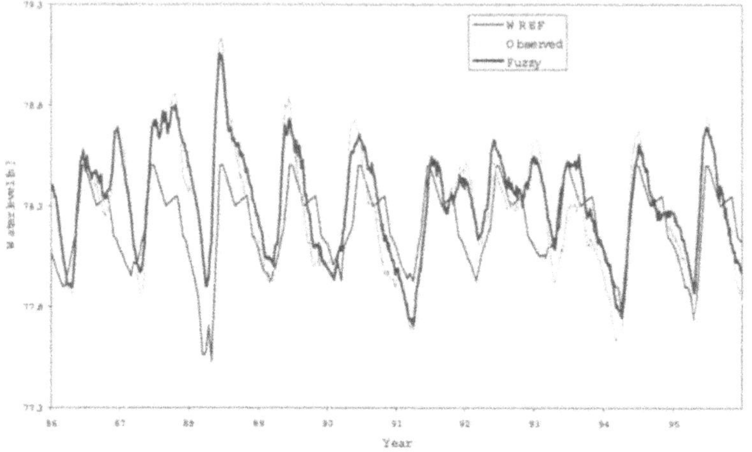

Fig. 3. Water reference level, observed water level and water level obtained by Total Fuzzy Similarity Control

Fuzzy Similarity is used. Correspondingly, the simulated water level is presented in Fig. 3. Sugeno method was chosen for comparison between different fuzzy inference methods, due to its simplicity and popularity. In both methods the system was kept as much as possible the same. To apply the Sugeno method the peak values of the triangular output membership functions were used as crisp values. Combination of the premises was implemented by taking a minimum and defuzzification was made using weighted average.

The performances of the two methods were almost indistinguishable. With the Total Fuzzy Similarity the water level targets in summer were sometimes better fulfilled, but the release tended to be somewhat more fluctuating, and the limitation of change of release was more relevant.

6 Conclusion

We have discussed an Algorithm based on Łukasiewicz logic to construct fuzzy inference. Relatively similar results with our Total Fuzzy Similarity method and Mamdani or Sugeno method indicate that the underlying concept in fuzzy control is related to many-valued similarity. The strong mathematical background of Total Fuzzy Similarity method puts fuzzy reasoning on a solid foundation.

References

1. Dubrovin, T., Jolma, A., Turunen, E.: Fuzzy model for real-time reservoir operation. Journal of Water Resources Planning and Management Vol. 128, No 1 (2002) 66–73
2. Castro, J.L., Klawonn, F.: Similarity in Fuzzy Reasoning, Using Fuzzy Logic. Proc. FUZZ-IEEE'99 (1371-1376). Mathware and Soft Computing 2 (1995) 197–228
3. Dubois, D. and Prade, H.: Similarity-Based Approximate Reasoning. In: Computational Intelligence Imitating Life (Proc. of the IEEE Symp., Orlando, FL, June 27- Julu 1st, 1994) Zurada, J.M., Marks II, R.J., Robinson, X.C.J. (eds.) IEEE Press, New York (1994) 69–80
4. Niittymäki, J., Kikuchi, S.: Application of Fuzzy Logic of a Pedestrian Crossing Signal, Transportation Research Record Nuishability in the setting of Fuzzy Logic. In: Management Decission Support Systems using Fuzzy Set and Possibility Theory, Yager, R.R. and Kacprzyk, J. (eds.) Verlag TUV, (1985) 198–212. Fuzzy Sets and Systems 16 (1985) 103–113
5. Niittymäki, J., Turunen, E.: Traffic Signal Control on Similarity Logic. To be published in Fuzzy Sets and Systems (2002)
6. Kosko, B.: The probability monopoly, IEEE Transactions of Fuzzy Systems, 2 (1994) 32–33
7. Pavelka, J.: On fuzzy logic, I,II,III Zeitsch. f. Math. Logik. 25 (1979) 45–52, 119–134, 447–464
8. Turunen, E.: Mathematics behind Fuzzy Logic, Advances in Soft Computing, Physica-Verlag, Heidelberg (1999)
9. Trillas, E., Valvelde, L.: On Implication and Indistinguishability in the setting of Fuzzy Logic. In: Management Decission Support Systems using Fuzzy Set and Possibility Theory, Yager, R.R. and Kacprzyk, J. (Eds.) Verlag TUV (1985) 198–212. Fuzzy Sets and Systems 16 (1985) 103–113
10. Zadeh, L.: Fuzzy Sets, Information and Control 8 (1965) 338–353
11. Zadeh, L.A.: Similarity Relations and Fuzzy Orderings, Information Sciences 3 (1971) 177–200

Chapter 16
Two Values, Three Values, Many Values, No Values

Charles G. Morgan

Department of Philosophy, University of Victoria,
Victoria, B.C. V8W 3P4, Canada
cmorgan@uvic.ca

Abstract. Classical formal semantics is based on bivalence and truth-functionality. Historically, problems with these two notions have motivated the move from two values to three values and from three values to many values. We reflect on ordinary reasoning and systems of rational belief to motivate numerically based probabilistic semantics, which abandons truth-functionality. But numerically based probability theory is overly specific compared to real belief systems. So we develop a formal semantics based on comparative probability structures. Thus we develop a formal semantics which is not truth-functional and which does not use any values at all. The semantics is shown to be universal for any extension, with or without quantifiers, of classical sentence logic. We briefly discuss some areas for additional research.

1 Some background

At its most basic, we may think of a logic as being a characterization of a *consequence relation.* Given a set of sentences, the premises, we wish to know whether or not it is legitimate to infer some other sentence, the conclusion, on the basis of those premises. The consequence relation is the relation between the premises and the conclusion.

Typically, formal logic deals with formal languages rather than with natural languages. Not all aspects of a natural language play a special logical or inferential role. A formal language is an abstract shorthand that attempts to abstract away from features of the natural language that are deemed to be logically irrelevant. *Informal semantics* attempts to correlate particular words or phrases in a natural language with particular symbols or structures in a formal language. Informal semantics is essentially a crude mapping from the hoariness and sloppiness of the natural language to the crispness and precision of the formal language.

Consequence relations are usually formalized in two distinct ways: either by means of a proof theory or by means of a formal semantic theory. *Proof theory* identifies elementary inference patterns based on the syntactical structures of the conclusion and those in the premises. A derivation of a conclusion from a set of premises is some sequence of formulas of the formal language (satisfying particular conditions), such that each formula in the sequence conforms to one of the elementary inference rules.

Historically, *formal semantics* began as an attempt to formalize relationships between sentences by abstracting away from their actual content and concentrating on properties based on the concept of "truth". Thus formal semantics abandons the notion of linguistic meaning altogether. From the point of view of formal semantics, classical logic is based on two fundamental assumptions: (1) bivalence, and (2) truth-functionality. Bivalence is the view that every sentence is either true (T) or false (F), not both (no gluts), not neither (no gaps). Roughly, truth- functionality is the view that the truth value of any logically complex sentence is a function of the truth values of its simpler constituent sentences.

Bivalence and truth-functionality may be traced back at least as far as the Stoics. It seems that bivalence has always been somewhat controversial, having been opposed by the Epicureans and by several medieval logicians. One of the primary complaints about bivalence is clearly discussed in Aristotle. Briefly, the problem is that bivalence seems to commit us to determinism. If every sentence is either T or F, then even sentences about future events are T or F right now, and were so yesterday or 100 years ago. That suggests that in some real sense, the future is already determined, and that the notion of future contingency is meaningless. In [8] Łukasiewicz tells us that it was the problem of determinism which first led him to abandon bivalence and to formalize three valued logic.

Since Łukasiewicz's logical excursions were motivated by the problem of determinism, it was natural that he should try to formulate a theory of "necessity" and "possibility". As with bivalence, we find discussions of the modal operators as far back as the ancient Greeks and in the work of the medieval logicians as well. During the early 1900's, C. I. Lewis formulated a number of proof theories concerning necessity and possibility which seemed to be well motivated (see [7]). It was natural for logicians to investigate whether or not the Lewis proof theories could be married to the Łukasiewicz many valued semantics.

We can always find a truth-functional infinite valued semantics for any standard axiomatic theory for which a rule of substitution is admissible. We simply take the formulas themselves to be the semantic values. The provable formulas will be the "designated" values, or ways of being true. The unprovable formulas will be the undesignated values. However this "semantics" is somewhat unsatisfactory. It will not always be decidable which values are designated and which are undesignated. And without an independent way of determining theoremhood, this semantics is not autonomous from the proof theory; consequently, soundness and completeness results are completely circular and trivial. Such a semantics provides no philosophical illumination.

And it provides no tools (e.g., a disproof procedure) not already available through the poof theory.

So the interesting question is whether or not we can find a finite valued semantics for the Lewis (and similar) systems. The answer, as shown by Dugundji in [1], is sadly "no". The basic problem is that a finite valued truth-functional semantics can only distinguish between a finite number of formulas of a given fixed length. But for the Lewis systems (and many others), there are more proof theoretically distinguishable formulas of that given length than can be distinguished by the semantics. Thus truth-functional finite valued semantics proved to be incapable of accounting for important inferences using "possibility" and "necessity" and similar modal concepts. The problem seemed to be with the concept of truth-functionality. But logicians would only grudgingly give up truth-functionality. The failure of the bivalent and the many valued treatments of modal concepts led to only a partial rejection of truth-functionality and the development of "hidden variable" semantics for modal operators.

A parallel with standard universal and existential quantification is useful. The truth value of a universally quantified formula, such as $(x)Px$, is not a direct truth-function of the constituent parts of the formula. (Indeed, on many accounts, removing the quantifier leaves a syntactic structure, Px in our example, that is not regarded as a sentence, and hence does not even have a truth value.) Rather, the truth value of $(x)Px$ is a function of another semantic variable, the domain of the model. The domain is a "hidden variable" in the semantics. Formulas (and ordinary language expressions) do not carry an indication of the (size of the) domain with them. The domain of the model is something extra which the logician must supply. In some cases, simply changing the size of the domain, and no other semantic factors, will change the truth value of the formula. The truth value of $(x)Px$ is not a truth-function of Px; rather it is a function of the truth values of instances of Px over the domain of the model, the domain being the hidden variable in the semantics.

Medieval logicians had noticed the parallel between the modal operators and the quantifiers centuries before the development of modern model theory. The modal operator "necessarily true" was thought of as a sort of universal quantifier, perhaps "true in all possible worlds". The medieval suggestion in its modern incarnation as Kripke semantics makes the truth value of "Necessarily A" not a truth-function of the value of A alone, but rather a function of the truth values of instances of A over the set of "possible worlds" of the model. The set of possible worlds and the relationship of accessibility between the worlds become the "hidden variables" of the semantics. The truth value of "Necessarily A" is not a function solely of the truth value of A; rather it is a function of the truth values of A and some "hidden" variables, e.g., possible worlds and the accessibility relations between them.

But there are problems with the Kripke semantics. There are modal logics for which it is impossible to formulate a Kripke semantics with a binary accessibility relation (see [3] and [20]). If we are willing to arbitrarily multiply the number and adicity of accessibility relations, we can always formulate a possible worlds semantics for any propositional logic (see [18]). But such complications are without intuitive or philosophical justification and are all too reminiscent of astrophysical epicycles. There really ought to be a simpler way.

2 Reflections on reasoning

As a guide, let us reflect on the nature of our interaction with the world and the role of reasoning and inference. Any living organism at large in the world has a number of goals, such as reproduction, enemy avoidance, and obtaining food. We may identify three general categories of possible strategies: (1) random, (2) fixed response, and (3) predictive. A paramecium moves at random through its environment. Since food, enemies, and reproductive opportunities are randomly distributed in the environment of a paramecium, a more complex strategy would be of no advantage. On the other hand, a sunflower has limited sensory input and a relatively limited range of behaviours. So it tends to employ fixed response strategies, such as always turning its head toward the sun and growing its roots generally in the direction of gravitational pull. But fixed response strategies are not very successful if one wishes to be able to exploit a broad range of complex environments.

Most higher organisms with which we are familiar utilize strategies that involve the ability to predict the effects of their various possible behaviors on their environment. A cat hunting a mouse in a field needs to have some familiarity with the sorts of places a mouse is likely to inhabit. It needs to have some familiarity with the drift of scents, both its own and that of the mouse. It needs to be aware of its own visual presence, and that of its shadow, and how the mouse is likely to react to both. It needs to be familiar with the sort of noise it makes when it moves through various materials, and it needs to understand when being quiet is important (e.g. when stalking) and unimportant (e.g. during the final rush). An organism using a predictive strategy in effect has a model of its environment, a model that includes its self in that environment.

When predicting important aspects of their environment, the models which organisms employ are never completely accurate. Generally speaking, the greater the degree of precision of a prediction, the greater the amount of time and energy resources it takes to make the prediction. The more complex the phase space, the greater the amount of storage required and the more time consuming the calculations required. Greater storage and time requirements translate directly into greater energy requirements.

By employing complex enough models, we could be very accurate in computing tomorrow's weather; but if the computation takes longer than a day, then it is useless to us. The cat has no need to be able to predict the path of a running mouse to within an angstrom – being correct to within a centimeter or so will do. Being more accurate would simply cost the cat more time and energy to make the computation, with no compensatory return in terms of food (energy acquisition). Similarly, being able to predict a flight trajectory from the top of a tree is of no use to the cat; since it cannot fly, the cat cannot make use of the results of such a computation. Further, animal sensory organs are never arbitrarily accurate or completely reliable. In its natural state, a cat has no use for electromagnetic information in the radio wave range. Hence the energy expenditure to carry around and maintain organs capable of sensing radio waves would bring no positive return to the cat, and thus would be an evolutionary detriment.

Thus it is a truism that if the computation time exceeds the temporal utility of the prediction, then the energy resources employed are totally wasted. Similarly, if greater accuracy does not on average lead to a net gain in energy acquisition (or a reduction in energy expenditure) then the energy resources employed are totally wasted. And in a competitive environment, organisms which are more efficient in their utilization of energy are more likely to survive and have more reproductive opportunities. A model that is more complete, more detailed than is necessary for the prediction and control of matters important to the organism ... such a model is inefficient ... it takes more energy resources, slows the computational time ... So in terms of evolution, possession of such a model would be a detriment. Thus because of the tradeoff between (i) accuracy and (ii) time and energy costs, natural selection favors models that are less than completely accurate.

In summary, the briefest reflection suggests that organisms do not need a concept of "Truth" to survive in the world ... they need a model of the world that includes their place in it. By its nature, a model is an abstraction, a simplification of the real world. And to the extent that the simplification is computationally more efficient relative to the needs of the organism, the simplification is energetically and hence evolutionarily preferred.

Organisms which have the ability to store information in the environment and to systematically recover that information when needed will have a distinct advantage over organisms without such an ability. One of the advantages of environmental memory is that fewer internal states will be required for memory, so an organism with environmental memory can be physically simpler, and hence more energetically efficient. Or to put the matter another way, an organism with environmental memory will be able to remember a much greater volume of material for a fixed size of internal state space. Further, an organism with environmental memory will generally have greater computational capability than it would have without environmental memory. By analogy, we need only recall that a Turing machine is just a finite state machine brain controlling a read/write head on an unbounded tape. That is, a Turing machine is just a finite state machine with environmental memory. But Turing machines have much greater computational power than finite state machines.

At least one primary function of natural languages is to allow us to communicate, store, and retrieve information about our world. That information is an abstraction from, a simplification (not a reproduction) of, aspects of our environment. Natural languages evolved at least in part to allow us to predict and control aspects of the environment, under conditions of uncertainty. Truth seems to have little to do with it. Our use of natural languages then is based in part on their utility for modeling and predicting aspects of our environment under conditions of uncertainty. Our rational thought processes involve modeling and predicting aspects of our environment under conditions of uncertainty, and natural languages are involved in those processes.

One of the major problems with classical semantics is that uncertainty is not truth- functional. Rational elementary reasoning under conditions of uncertainty is often based on simple relative frequency schemes, as for example with decks of cards, throwing dice, or drawing items from an urn. But it is obvious that relative frequency does not combine in a truth-functional way. As a simple example, consider a deck of cards. Let us use "clart" for any card that is either a club or a heart. Then the frequency of clart cards is 0.5, the frequency of black cards is 0.5, and the frequency of cards that are both clart and black is 0.25. However the frequency of red cards is 0.5, the frequency of black cards is 0.5, but the frequency of cards that are both red and black is 0. So the relative frequency of "A and B" is not a function of the relative frequency of A and the relative frequency of B.

Hence we can expect a semantic theory based on truth-functionality, even truth- functionality relative to hidden variables, to provide a very poor account of many logical particles. In fact, if we just completely abandon truth-functionality for the modal operators, it is possible to find a bivalent characteristic semantics for any propositional modal logic whose proof theory can be axiomatised as an extension of classical sentence logic (see [9]). So with regard to modal logics at least, the problem is not the assumption of bivalence, but rather the problem is the assumption of truth- functionality. Many valued semantic accounts are not appropriate for reasoning about uncertainty, precisely because they maintain truth-functionality.

In summary, classical formal semantics abandons linguistic meaning and is instead based on the two principles of bivalence and truth-functionality. However, our rational thought processes involve modeling and predicting aspects of our environment under conditions of uncertainty. But uncertainty is not truth-functional. Further, the history of disputes concerning the concept of truth is much too long and well known to warrant even the briefest survey here. It suffices to say that the concept of truth is philosophically problematic. However, belief and evidential relationships seem to be more tractable.

We are often sure that we believe X without being able to be sure that X is true. So it would be reasonable to develop a formal semantics based on notions of rational belief and uncertainty rather than on notions of truth and falsity, abandoning truth-functionality. Early steps in this direction were taken in [16], [2], and [6]. In the next sections we will examine the prospects for such a theory. Some of the material from section 4 is adapted from [11] and [13], and some of the material from section 5 is adapted from [10] and [15].

Of course there is nothing new under the sun. When I was expounding these motivational considerations to a colleague, he remarked that they are very reminiscent of Nicholas of Aurtrecourt, who was writing during the 1300's. Nicholas rejected as inconclusive all reasoning that could not be reduced to the principle of non-contradiction. Thus he rejected virtually all deductive reasoning attempting to connect one physical event with another. He proposed that reasoning about every day affairs must be judged on the grounds of probability. However, he stated explicitly that he did not mean probability in the sense of relative frequency of past experience. Rather, he insisted on probability in the sense of the amount of weight an experienced and disinterested judge would accord the conclusion when presented with the premises, thus connecting probability with rational belief under conditions of uncertainty. See [21] for a brief description and more references.

3 Formal preliminaries

The primary focus of this paper is on formal semantics. But before we begin our discussion of formal semantics, it will be useful to say a few words about proof theory. We assume we are dealing with some formal language $\mathcal{L}$, which contains among its logical particles classical conjunction, disjunction, conditional, and negation, designated by $\wedge$, $\vee$, $\supset$, and $\sim$. We will assume that all of the logics we will be discussing can be formulated by means of a set of "inference rules" of the following sort.

IR If $\vdash C_1$, and ..., and $\vdash C_j$, and C and $C_1, \ldots, C_j$ satisfy conditions COND, then $\vdash C$.

We use COND for any sentence specifying special conditions which must be satisfied for the application of the rule. For example, the following rule is used in some formulations of first order predicate calculus.

EX.FO If $\vdash A \supset B$ and x has no free occurrence in A, then $\vdash A \supset (\forall x)B$.

Axioms are just inference rules with vacuous antecedents. The following is a standard example used in many propositional logics.

EX.PL $\vdash A \supset (B \supset A)$

Classical propositional logic will play a central role in our discussions, but the particular axiomatic development will be unimportant. Thus we adopt the following simple characterization, using "$\supset$" as the material conditional.

CPL If A is any instance of any classical tautology, then $\Gamma \vdash A$.

DET If $\Gamma \vdash A \supset B$ and $\Gamma \vdash A$, then $\Gamma \vdash B$.

The usual extensions of classical propositional logic may be formed by adding rules of the form IR to CPL and DET. Thus we include all the usual modal, deontic, epistemic, and temporal logics, and we include first and second order predicate calculus, with or without identity, and various versions of axiomatic set theory. In addition many logics which are usually thought of as "weaker" than classical logic can be formulated as extensions to classical logic. For example any finite valued logic can be formulated to include logical particles which satisfy standard conditions for negation, conjunction, disjunction, and conditional, in the way pioneered by Rosser and Turquette in [17]; and so any such logic can be formalized as an extension of classical propositional logic. We adopt the usual definition for theoremhood.

DF.PRF We say that a formula A is provable in the logic, written $\vdash A$, iff there is a finite sequence of formulas $A_1, \ldots, A_n$, such that A_n is A and each formula in the sequence follows from previous members of the sequence by an appropriate inference rule of the logic.

It is also possible to characterize derivations from background assumptions in this context. Where Γ is a finite set of sentences, we use the notation $\wedge\Gamma$ as shorthand for the sentence which results from taking a (classical) conjunction of the members of Γ in some arbitrary order; given our assumptions about conjunction, the order will not matter.

DF.DRV We say that A is derivable from the set of assumptions Γ, written $\Gamma \Rightarrow A$ iff $\vdash \wedge\Sigma \supset A$ for Σ some finite subset of Γ.

Given this definition, it is easy to verify the following about our consequence relation.

CR.1 If $A \in \Gamma$, then $\Gamma \Rightarrow A$.
CR.2 If $\Gamma \Rightarrow A$, then for some finite subset Σ of Γ, $\Sigma \Rightarrow A$.
CR.3 If $\Gamma \cup \{A\} \Rightarrow B$ and $\Gamma \Rightarrow A$, then $\Gamma \Rightarrow B$.
CR.4 If $\Gamma \Rightarrow A$, then $\Gamma \cup \Delta \Rightarrow A$

The notion of "universe design" will be useful in the material to follow. Intuitively, a universe design is a complete description of one possible way the world might be. We can give a formal characterization of universe designs using our consequence relation.

DF.CNST We say that a set of sentences Γ is *consistent* iff there is at least one sentence A such that it is not the case that $\Gamma \Rightarrow A$. A set is inconsistent iff it is not consistent.

DF.MXC We say that a set of sentences Γ is *maximally consistent* iff (i) Γ is consistent, and (ii) for all sentences B, if $B \notin \Gamma$, then $\Gamma \cup \{B\}$ is inconsistent.

A universe design is just a maximally consistent set of sentences. Of course as we change the logic (by changing the rules NPIR), we change what is to count as a maximally consistent set.

Using the standard proof and CR.1-4, it is easy to prove that any consistent set has at least one maximally consistent extension. In the standard proof, we assume that we begin with a consistent set, Γ_0. Then we step through an enumeration of all the sentences in the language, forming a sequence of nested consistent sets:

$$\Gamma i = \Gamma_{i-1} \cup \{E_i\}, \text{ if consistent}$$

$$= \Gamma_{i-1}, \text{ otherwise}$$

It is easy to show that the union of all the Γ_i is maximally consistent. Although the enumeration of sentences is arbitrary, it will be important in the material to follow that we use some single, fixed enumeration $E_1, E_2, \ldots$ for the construction of all maximally consistent sets. We call this enumeration the *default enumeration.*

DF.DME The *default maximal extension* of set Γ, written $DME(\Gamma)$, is: (1) the maximally consistent extension of Γ constructed with the default enumeration of sentences, provided Γ is consistent; or (2) the (inconsistent) universal set of all sentences, provided Γ is inconsistent.

Using our definitions, it is simple to verify the following.

CDME If $A \in DME(\Gamma)$, then $DME(\Gamma) = DME(\Gamma \cup A)$

Given our characterization, all of the logics under consideration are clearly monotonic. Non-monotonic reasoning is treated in detail in [14], where: (1) we prove that a nonmonotonic proof theory which corresponds to elementary principles of rational belief is logically impossible; and (2) we show that nonmonotonic reasoning can be seen to be the result of the way we use logic, rather than a result of the logic we use.

4 Numerical probability theory and formal semantics

Our most detailed theory about reasoning under conditions of uncertainty is formal probability theory. The most well known account of the classical theory of elementary probability is due to Komolgoroff in [5]. It consists of a set of constraints which any function P must satisfy in order to be a probability function. The constraints can stated as follows:

KP1 P is defined on a σ-field of sets.

KP2 $0 \leq P(\alpha)$.
KP3 $P(U) = 1$.
KP4 If $\alpha \cap \beta = \emptyset$ then $P(\alpha \cup \beta) = P(\alpha) + P(\beta)$.

A σ-field of sets is just a set of sets which includes the empty set $\emptyset$ and the universal set U, and which is closed under complements and finite unions and finite intersections. Intuitively, the field is a field of events, such as card draws of aces and dice throws of sevens. Komolgoroff takes *a priori* probability as basic and defines conditional probability as follows:

DF.KCP If $P(\beta) \neq 0$ then $P(\alpha, \beta) = P(\alpha \cap \beta)/P(\beta)$.

The expression "$P(\alpha, \beta)$" is read "the probability of α, given β". There is a well-known problem with the Komolgoroff account when the conditioning set has probability 0. We would like conditional probability functions to always be probability functions. But when the evidence set has probability 0, the conditional probability function is undefined; and because of KP4, we cannot admit the constant function $P(x, \beta) = 1$ for all x. Further, simple reflection on real examples shows that our reasoning about probabilities is never *a priori*. We are always making some sort of background assumptions, such as that there are six faces on a die, the numbers on the faces are positive integers from 1 through 6, the faces do not change when the die is thrown, etc.

There is a simple way of modifying the basic Komolgoroff theory to make it conditional and to get around the problem of conditioning on sets of probability 0. I have taken the central idea for the solution from Popper (1965). Although he did not formulate his probability theories in this way, I will refer to the following constraints as Popperian classical conditional probability theory.

PP1 P is defined on ordered pairs from a σ-field of sets.
PP2 $0 \leq P(\alpha, \beta) \leq 1$
PP3 $P(U, \alpha) = 1$
PP4 If $\alpha \cap \beta = \emptyset$ then $P(\alpha \cup \beta, \gamma) = P(\alpha, \gamma) + P(\beta, \gamma)$, unless for all δ, $P(\delta, \gamma) = 1$.
PP5 $P(\alpha \cap \beta, \gamma) = P(\alpha, \gamma) \times P(\beta, \alpha \cap \gamma)$

In this formulation, conditional probability is taken as primitive. So what was previously a definition, DF.KCP, now becomes a constraint, PP5, on a par with the others.

We should make a brief comment about PP4. The intuitive idea is that there are some sentences which are so bizarre that if asked to accept them as evidence, I would be unable to reject anything. Standard examples of such expressions are logically false sentences; but there may be similar sentences that are not logically false, e.g., the negation of the law of conservation of mass and energy or the denial of elementary principles of mathematics. At any rate, we adopt the following definition:

DF.PA A set γ is said to be P-abnormal, for Popperian classical conditional probability function P, just in case for all sets $\delta, P(\delta, \gamma) = 1$. A set that is not P- abnormal is said to be P-normal.

For simplicity, we will use the terms "normal" and "abnormal". Thus the Popperian constraints do admit the constant 1 function, unlike the Komolgoroff constraints.

Just as with the Komolgoroff formulation, we may wish to include a modified version of constraint PP4 to ensure that properties of an infinite sequence of sets determine properties of their infinite union. We will gloss over such details here.

On another point, in the Kolmogoroff theory, KP1-4, it is possible to prove that $P(\alpha, \alpha) = 1$ for all α such that $P(\alpha) \neq 0$. In the Popperian theory, PP1-5, one may prove a similar result. It is easy to show both of the following:

POP.1 If P is any function satisfying PP1-5, then for all $\alpha, P(\alpha, \alpha) = 1$ or $P(\alpha, \alpha) = 0$.

POP.2 If P is any function satisfying PP1-5, then for all α, if $P(\alpha, U) > 0$, then $P(\alpha, \alpha) = 1$.

But consider the simple sigma field of just U and $\emptyset$. The function which sets $P(U, x) = 1$ for all x, and $P(\emptyset, x) = 0$ for all x, satisfies PP1-5. However, we obviously do not have $P(\emptyset, \emptyset) = 1$. To ensure that $P(\alpha, \alpha) = 1$ is always true, we may simply add the following:

PP6 $P(\alpha, \alpha) = 1$.

Intuitively, our constraints capture the notion of "relative frequency". To be more precise, we begin with the definition of an elementary weighting function.

DF.WF An elementary weighting function w defined on a σ-field of sets is a function mapping the field into the real numbers, satisfying the following conditions:

(a) $w(\emptyset) = 0$

(b) $w(\alpha \cup \beta) = w(\alpha) + w(\beta) - w(\alpha \cap \beta)$

(c) if $\alpha \subseteq \beta$ then $w(\alpha) \leq w(\beta)$

For purposes of illustration, simple examples of elementary weighting functions can be had by assigning whole numbers to the cells of a Venn diagram. The weight of a set is then the sum of the weights in its cells.

We can use elementary weighting functions to construct examples of probability distributions satisfying our constraints. Let $\{w_1, w_2, \ldots\}$ be a well-ordered sequence of elementary weighting functions, all defined on the same σ-field. Then it is easy to verify that the function P defined below satisfies PP1-6.

PWF

$$P(\alpha,\beta) = 1, \text{ if } w_i(\beta) = 0 \text{ for all } w_i$$
$$= w_i(\alpha \cap \beta)/w_i(\beta) \text{ for the first } w_i \text{ such that } w_i(\beta) \neq 0$$

In order to motivate appropriate constraints for a probabilistic semantics, let us think in terms of "universe designs". Intuitively, a universe design is a complete specification for one way the world might be. In terms of proof theory, a universe design may be thought of as a maximally consistent set of sentences. Since our proof theory includes classical propositional logic, then there is a rather tight connection between the classical logical particles and set theoretic operations on universe designs. Using "$M(\Gamma)$" for the set of all maximally consistent extensions of Γ, it is easy to verify each of the following.

B1 $M(\{A\}) \cup M(\{B\}) = M(\{A \vee B\})$
B2 $M(\{A\}) \cap M(\{B\}) = M(\{A \wedge B\})$
B3 $M(\Gamma) \cap M(\Delta) = M(\Gamma \cup \Delta)$
B4 $M(\{A\})^c = M(\{\sim A\})$
B5 If $\Gamma \subseteq \Delta$ then $M(\Delta) \subseteq M(\Gamma)$

Now, consider a single elementary weighting function on the σ-field of maximal sets. If we think of $P(A,\Gamma)$ as being equal to $w(M(\{A\}) \cap M(\Gamma))$ $/w(M(\Gamma))$ for $w(M(\Gamma)) \neq 0$ and equal to 1 for $w(M(\Gamma)) = 0$, then using B1-5 and PP1-6, it is easy to extract the following constraints.

NP1 $0 \leq P(A,\Gamma) \leq 1$
NP2 If $A \in \Gamma$ then $P(A,\Gamma) = 1$.
NP3 $P(A \vee B,\Gamma) = P(A,\Gamma) + P(B,\Gamma) - P(A \wedge B,\Gamma)$
NP4 $P(A \wedge B,\Gamma) = P(A,\Gamma) \times P(B,\Gamma \cup \{A\})$
NP5 $P(\sim A,\Gamma) = 1 - P(A,\Gamma)$ unless $P(B,\Gamma) = 1$ for all B.
NP6 $P(A \wedge B,\Gamma) = P(B \wedge A,\Gamma)$
NP7 $P(C,\Gamma \cup \{A \wedge B\}) = P(C,\Gamma \cup \{A,B\})$
NP8 $P(A \vee \sim A,\Gamma) = 1$

In fact, these constraints would hold for any P determined by any sequence of weighting functions according to the scheme proposed in PWF. In spite of our motivational discussion, it should be noted that conditions NP1-8 do not depend on any notions from proof theory, and hence they are autonomous. Thus they are an acceptable basis for probabilistic semantics. Further, the constraints do not depend on any notions from classical formal semantics. And although we used the weighting function scheme to motivate the constraints, the constraints are completely independent of our discussion of elementary weighting functions. These constraints constitute what we call neo- classical conditional probability theory.

For future reference, we note that we can define "$A \supset B$" as "$\sim (A \wedge \sim B)$", from which it is easy to deduce the following constraint for the classical conditional from the constraints NP1-8.

NPCND $P(A \supset B, \Gamma) = 1 - P(A, \Gamma) + P(A \wedge B, \Gamma)$

Karl Popper in [16] seems to have been the first to try to formulate probability theory by a set of autonomous constraints, i.e., constraints independent of any proof theory and independent of any classical semantic or algebraic notions. Indeed one of Popper's complaints about the Komolgoroff formulation is that it obscures the algebraic characteristics intrinsic to the constraints themselves by presupposing a Boolean structure on the event space. In fact, the Boolean structure can be derived directly from the probability constraints.

For each probability distribution P, we may define a set equivalence classes for each background set Γ as follows.

DF.PEQC $[A, \Gamma] = \{B : P(A, \Gamma \cup \Delta) = P(B, \Gamma \cup \Delta) \text{ for all } \Delta\}$

We can define the meet and complement as follows.

DF.MT $[A, \Gamma] * [B, \Gamma] = [A \wedge B, \Gamma]$
DF.CMP $[A, \Gamma]^c = [\sim A, \Gamma]$

Given NP1-8, it is trivial to show that the following conditions are satisfied.

BA1 * is commutative
BA2 * is associative
BA3 $[A, \Gamma] * [B, \Gamma]^c = [C, \Gamma] * [C, \Gamma]^c$ iff $[A, \Gamma] * [B, \Gamma] = [A, \Gamma]$

But these are Stolls conditions for a Boolean algebra (see [19]). Stoll includes the condition that there be at least two elements in the algebra; but here we have no reason to exclude the single element algebra as a special case. Hence we have:

Theorem 1. *For each probability distribution P, for each background set Γ, the set of P determined equivalence classes under Γ is a Boolean algebra.*

In order to capture extensions of classical logic, we need to add extra constraints to the probability theory. For each inference rule of the sort IR, above, we need to add a constraint of the following sort.

NPIR If $P(C_1, \Delta) = 1$ for all Δ, and ..., and $P(C_j, \Delta) = 1$ for all Δ, and C and $C_1, \ldots, C_j$ satisfy conditions COND then $P(C, \Delta) = 1$ for all Δ.

Although the constraints NPIR are parasitic on the inference rules in an obvious way, they do not employ any explicit notions from the proof theory. Thus the extended probabilistic constraints remain autonomous.

The set of constraints picks out a class of functions, each function satisfying all the constraints. As we vary the constraints by adding or changing instances of the form NPIR, we will vary the class of functions. Thus each logic will be associated with a different class of probability functions. Given a set of constraints for a particular logic, we can define a corresponding notion of semantic implication.

DF.PSI We say that the set of sentences Γ *semantically implies sentence* A, in the sense of numerical probability, written $\Gamma| \vdash_{NP} A$, if and only if for all probability functions satisfying NP1-8 and the appropriate restrictions of the form NPIR, $P(A, \Gamma \cup \Delta) = 1$ for all Δ.

The intuitive notion behind our definition is the simple observation that if the argument from Γ to A is non-demonstrative, then there should be some additional evidence Δ that we could add to Γ that would lead us to reduce our confidence in A. Hence, if A is a demonstrative consequence of Γ, not only must our degree of belief in A, given Γ, be maximal, but no matter what evidence we add to Γ, our degree of confidence in A must remain maximal. Further, demonstrative inference is not relative to my particular set of beliefs; hence our definition must generalize over all probability functions. We are now in a position to prove soundness and completeness results.

Theorem 2. *(soundness) If* $\Gamma \Rightarrow A$, *then* $\Gamma| \vdash_{NP} A$.

Proof. Suppose $\Gamma \Rightarrow A$. Then by definition we know that $\vdash \wedge\Sigma \supset A$ for some finite subset Σ of Γ. Thus there is a finite sequence of formulas, $A_1, \ldots, A_n$, the last one of which is $\wedge\Sigma \supset A$, and such that each member of the sequence follows from previous members by one of the inference rules. By a simple inductive argument, we show that for each line, A_i, for all probability functions P satisfying NP1-8 and appropriate versions of NPIR, $P(A_i, \Delta) = 1$ for all Δ. Given Theorem 1 and NP8, the desired result holds for any line justified by CPL. The result for any line justified by DET follows from NPCND, NP1 and NP4. The result for any line justified by IR follows from NPIR. Thus we know $P(\wedge\Sigma \supset A, \Delta) = 1$ for all Δ. But by NPCND, NP1, and NP4 it then follows that $P(A, \{\wedge\Sigma\} \cup \Delta) = 1$ for all Δ. Hence by NP7, $P(A, \Sigma \cup \Delta) = 1$ for all Δ. For arbitrary Λ, substituting $\Gamma \cup \Lambda$ for Δ and remembering that $\Sigma \subseteq \Gamma$, we obtain $P(A, \Gamma \cup \Lambda) = 1$ for all Λ. Hence by DF.PSI, $\Gamma| \vdash_{NP} A$.

Theorem 3. *(completeness) If* $\Gamma| \vdash_{NP} A$ *then* $\Gamma \Rightarrow A$.

Proof. As usual, we will prove completeness by proving the contrapositive. So, suppose it is not the case that $\Gamma \Rightarrow A$. We must construct an appropriate probability distribution for which there is a set Δ such that $P(A, \Gamma \cup \Delta) \neq 1$. Recalling our previous discussion of default maximal extensions, we define the function $P*$ as follows:

$$P*(B, \Lambda) = 1 \text{ iff } B \in DME(\Lambda)$$

$$= 0 \text{ otherwise}$$

Our first task is the show that $P*$ satisfies all of the constraints NP1-8 and all appropriate constraints of the form NPIR. Most of the conditions follow trivially from classical properties of maximally consistent sets. NP4 requires CDME. We will explicitly deal with constraints of the form NPIR. From the antecedent of the constraint, suppose for each C_k that $P*(C_k, \Delta) = 1$ for all Δ and that conditions COND are satisfied. Using $\{\sim Ck\}$ for Δ, it follows that $C_k \in DME(\{\sim C_k\})$. Thus $DME(\{\sim C_k\})$ is inconsistent, so $\{\sim C_k\}$ is inconsistent. Then $\vdash C_k$ for each C_k. Thus by IR it follows that $\vdash C$. But then $C \in DME(\Delta)$ for all Δ, and so $P*(C, \Delta) = 1$ for all Δ. Hence NPIR is satisfied. Our next task is to show it is not the case that $\Gamma | \vdash_{NP} A$. Since it is not the case that $\Gamma \Rightarrow A$, from classical principles it follows that $\Gamma \cup \{\sim A\}$ is consistent. Hence $A \notin DME(\Gamma \cup \{\sim A\})$. Thus $P*(A, \Gamma \cup \{\sim A\}) = 0$, as required.

There is an interesting technical point about the proof of Theorem 3. Generally in Henkin style completeness proofs, we assume A is not a deductive consequence of Γ. We then try to construct a model in which Γ holds but A does not. For Kripke style semantics, a model consists of an ordered triple $\langle W, R, V\rangle$, where W is the set of worlds, R is the accessibility relation, and V is a valuation mapping each expression to a truth value at each world. If A is not a consequence of Γ, then we search for a model and a world w such that $V(A, w) \neq T$ but $V(B, w) = T$ for all $B \in \Gamma$. In fact, for normal modal logics, one can prove the existence of so called canonical models. In a canonical model, for any A and Γ such that A is not a consequence of Γ, there will be some world w in the canonical model such that $V(A, w) \neq T$ but $V(B, w) = T$ for all $B \in \Gamma$. Our proof of Theorem 3 demonstrates the existence of canonical probability distributions. In a canonical probability distribution, for any Γ and A such that A is not a consequence of Γ, there will always be some evidence one could find in addition to Γ that would cause one to doubt A; that is, there will always be a Δ such that $P(A, \Gamma \cup \Delta) \neq 1$. The existence of canonical models for modal logics does not imply that the possible worlds semantics is in any way trivial. Nor does it imply that models that are not canonical are not useful. On the contrary, the vast majority of models used for practical purposes when dealing with modal logics will NOT be canonical. Similar comments apply in the case of probabilistic semantics. The existence of canonical probability distributions does not imply that the semantics is trivial. Most of the probability distributions one actually uses, e.g. for decision problems, will not be canonical. Canonical models generally are infinite, much too complex to actually construct, but are useful in completeness proofs. And for philosophical reasons, Carnap was interested in canonical probability distributions, though he did not call them that. For more details, see [13].

Certainly NP1-8 are not bivalent. Neo- classical probability functions can take any values between and including 0 and 1. Nor are the conditions "truth-functional" in any sense. Note that for fixed evidence Γ, the probability of a negation, a conjunction, or a disjunction is not a function of the probability of the constituent parts. So we have totally abandoned truth-functionality. In doing so, we have completely overcome the incompleteness of the standard possible worlds approach. In this sense, probability theory is a universal semantics.

Probability theory allows infinitely many values. Thus, it cannot be autodescriptive; we cannot add to the language monadic operators corresponding to each semantic value, as we can do in finite valued semantics. But in a sense, even classical logic *permits* an infinite number of values, although only two values are actually required. Surprisingly, from the completeness proof above, we see that probabilistic semantics never *requires* more than two values. In short, we could have proved our soundness and completeness results if we had added another constraint requiring the probability functions to be two valued. This observation suggests that classical semantics with its emphasis on two values and truth-functionality may viewed as just a computationally more tractable abstraction from a more accurate model of our "belief environment"!

As observed earlier, standard formal semantics abandons concepts of linguistic meaning. Bivalent semantics takes the concepts of *truth* and *falsity* to be primitive and attempts to account for important semantic concepts in terms of truth and falsity. Probabilistic semantics also abandons concepts of linguistic meaning but takes as primitive the notion of *degree of evidential support.* In so far as we are willing to consider probability theory a formal theory of rational belief, we can characterize probabilistic semantics as the semantics of rational belief rather than the semantics of truth.

So we have seen the move from two values to three values and from three values to many values. And we have seen how abandoning truth- functionality allows the many valued semantics to be universal. We will now move to a non truth-functional semantics without any values.

5 Comparative probability theory and formal semantics

Certainly one view of probability theory is that it represents idealized degrees of rational belief. However, it is a matter of psychological fact that people find it very difficult, if not impossible, to assign precise numerical values to their degrees of belief. Most (I would say "all") humans would find it impossible to even linearly order their beliefs, much less assign numbers in any meaningful way. On the other hand, people generally find it easier to make comparative judgements concerning, for example, which theory is better supported by the evidence. These considerations led us in [10] to develop a very weak conditional comparative probability theory, minimally strong enough to serve as a formal semantic theory for classical logic. These results were extended in [15] and will be further developed here.

A conditional comparative probability distribution with sets of sentences in the evidence position is a type of quasi-ordering relation defined on ordered pairs from $\mathcal{L} \times \mathcal{P}(\mathcal{L})$. We represent such a relation by "LE"; we use the notation "$(A, \Gamma)LE(B, \Delta)$" which may be read in a number of ways, depending on the preferences and prejudices of the reader. The following are some standard readings: (1) the degree to which A is supported (confirmed) by Γ is less than or equal to the degree to which B is supported (confirmed) by Δ; or (2) the probability (likelihood) of A, given Γ, is less than or equal to the probability (likelihood) of B, given Δ. For convenience, we use the notation "$(A, \Gamma)EQ(B, \Delta)$" to mean that both $(A, \Gamma)LE(B, \Delta)$ and $(A, \Gamma)LE(B, \Delta)$ hold.

Our goal in this section is to formulate a conditional comparative probability theory that can serve as a formal semantics for all logics of the sort discussed above. In particular, we want to be able to use a reasonable notion of semantic entailment and obtain soundness and completeness results. In parallel with DF.PSI, we say that Γ semantically implies A just in case the argument from Γ to A is maximally strong, no matter what additional evidence is added to Γ.

DF.CSI We say that Γ *semantically implies* A, in the sense of comparative probability, written "$\Gamma| \vdash_{CP} A$", iff for all appropriate conditional comparative probability structures LE, all sentences B, and all sets Δ, $(B, \Gamma \cup \Delta)LE(A, \Gamma \cup \Delta)$.

Of course the semantic entailment relation will depend on the particular logic we are trying to capture. As we vary the logic, we will have to vary the defining constraints for the LE structures.

We now turn our attention to the formulation of the constraints. We will begin with those appropriate for classical logic. We will list our revised constraints and then discuss them briefly.

RCP1 The relation "LE" is reflexive and transitive.
RCP2 $(A \wedge B, \Gamma)LE(A, \Gamma)$
RCP3 $(A \wedge B, \Gamma)LE(B \wedge A, \Gamma)$
RCP4 If $(B, \Gamma \cup \Delta)LE(C, \Gamma \cup \Delta)$ for all Δ, then $(A \wedge B, \Gamma \cup \Delta)LE(A \wedge C, \Gamma \cup \Delta)$ for all Δ.
RCP5 $(A \wedge \sim B, \Gamma \cup \Delta)LE(C \wedge \sim C, \Gamma \cup \Delta)$, for all Δ, iff $(A, \Gamma \cup \Delta)LE(A \wedge B, \Gamma \cup \Delta)$, for all Δ.

If we think of LE as just numerical "$\leq$", then RCP1-5 are all just special cases of NP1-8. RCP1 is just what we would expect of any comparative notion. RCP2 is immediate from NP1 and NP4. RCP3 follows from NP6. RCP4 is just a weak version of the standard product rule for conjunctions, NP4. Certainly if the antecedent of RCP4 holds for all Δ, then it holds taking Δ to be $\Lambda \cup \{A\}$ for aribitrary Λ; applying the standard numerical product rule would then yield the consequent of RCP4. Similarly RCP5 can be derived, but perhaps some more intuitive justification is in order. If no matter what the evidence, the conjunction of A with $\sim B$ is no more likely than a contradiction, then in some sense, B must be implied by A. Thus in such cases, the assertion of A amounts to the assertion of both A and B.

Using just conditions RCP1-5, we can obtain an important result concerning the Boolean character of our structures, as with the numerical case.

For each comparative probability structure LE and each set of expressions Γ we define the LE determined equivalence classes under Γ as follows:

DF.CEQC $[A, \Gamma] = \{B : (B, \Gamma \cup \Delta)EQ(A, \Gamma \cup \Delta), \text{ for all } \Delta\}$

We then use DF.MT and DF.CMP for meet and complement. Note that for our purposes, we take the trivial algebra of only one element to be Boolean. An absolutely trivial modification of the proofs in [10] will then yield a proof of the following result.

Theorem 4. *For each comparative probability structure LE satisfying RCP1-5, for each background set Γ, the set of LE determined equivalence classes under Γ is a Boolean algebra.*

Using Theorem 4, it is not difficult to obtain a weak soundness result for classical logic. But we would like to be able to obtain a stronger result. Given our proof theory, we certainly know that no matter what the logic, if $A \in \Gamma$, then $\Gamma \Rightarrow A$ always holds. Note that our constraints RCP1- 5 do not place any restrictions on the way in which the evidence position relates to the conclusion position. The restrictions are all of the form: "the evidence must support X to at least as high a degree as it supports Y", or conditionals of such sentences. But given our definition of semantic entailment, we must have some constraint to ensure that any member of the premise set is a semantically entailed by that set. For this purpose we introduce the following constraint.

RCP6 If $A \in \Gamma$, then $(B, \Gamma)LE(A, \Gamma)$.

Again, because our original constraints were so minimal, virtually no requirements were imposed on the evidence statements. Thus, as noted in [10], it was perfectly possible for our constraints to be satisfied but for $(A, \{B \wedge C\})$ to be unrelated to $(A, \{C \wedge B\})$. By using sets for the evidence position, we can presume that order and repetition among our premises do not matter. But we still need a constraint that indicates that premise sets are thought of conjunctively. Thus we impose the next condition.

RCP7 $(A, \Gamma \cup \{B \wedge C\})EQ(A, \Gamma \cup \{B, C\})$

In order to obtain the theory appropriate for extensions to classical logic, we must also include for each additional inference rule of the form IR, above, a constraint of the following sort:

RCPIR If for all sets $\Delta, (B, \Delta)LE(C_1, \Delta)$ for all B, and ... and for all sets $\Delta, (B, \Delta)LE(C_j, \Delta)$ for all B, and C and the C_k satisfy conditions COND, then for all sets $\Delta, (B, \Delta)LE(C, \Delta)$ for all B.

In summary, for a specific logic defined by (i) a classical part consisting of CPL and DET, and (ii) a set of inference rules of the sort IR, we will take the appropriate constraints to be RCP.1-7 and a set of constraints of the form

RCPIR, one for each inference rule IR. Each set of constraints will pick out a set of admissible comparative conditional probability structures LE, namely all and only those satisfying the constraints. We think of these structures as analogous to probability distributions of the usual numerically based sort. Because our theory of comparative probability is so weak, we will have to do quite a bit of work to obtain soundness and completeness.

First we will need a few lemmas. (We use the designation "CPLn" for "comparative probability lemma n".) In the following we will use F for some arbitrary contradiction $A \wedge \sim A$, and we will use T for $\sim F$. From our result on Boolean algebras, Theorem 4 above, we immediately know each of the following holds.

CPL1 $(\sim\sim A, \Gamma)EQ(A, \Gamma)$
CPL2 $(F, \Gamma)LE(A, \Gamma)$
CPL3 $(A, \Gamma)LE(T, \Gamma)$

Further, a rather trivial modification of the proof of Theorem 12 of [10] will yield the following result.

CPL4 If $(A, \Gamma \cup \Delta)EQ(B, \Gamma \cup \Delta)$, for all Δ, then $(\sim B, \Gamma \cup \Delta)EQ(\sim A, \Gamma \cup \Delta)$, for all Δ.

With these lemmas, we can prove that semantic implication is preserved under the inference rule DET.

Theorem 5. *If $\Gamma | \vdash_{CP} A$ and $\Gamma | \vdash_{CP} A \supset B$, then $\Gamma | \vdash_{CP} B$.*

Proof. Suppose both $\Gamma | \vdash_{CP} A$ and $\Gamma | \vdash_{CP} A \supset B$. Then from DSE and CPL3, we know that for all Δ, both of the following hold:

$$(T, \Gamma \cup \Delta)EQ(A, \Gamma \cup \Delta) \tag{1}$$

$$(T, \Gamma \cup \Delta)EQ(A \supset B, \Gamma \cup \Delta) \tag{2}$$

From (2), the definition of "$\supset$", CPL1 and CPL4, the following holds for all Δ:

$$(A \wedge \sim B, \Gamma \cup \Delta)EQ(F, \Gamma \cup \Delta) \tag{3}$$

From (3) and RCP5, the following holds for all Δ:

$$(A, \Gamma \cup \Delta)LE(B \wedge A, \Gamma \cup \Delta) \tag{4}$$

But (4) with (1), along with CPL3 and RCP.2-3 guarantee the following for all Δ:

$$(T, \Gamma \cup \Delta)EQ(B, \Gamma \cup \Delta) \tag{5}$$

But by DF.CSI, (5) means $\Gamma | \vdash_{CP} B$.

With the help of a couple of additional lemmas, we will be able to use Theorem 4 to obtain a strong soundness result. A trivial modification of the proof of Theorem 7 from [10] yields the following result, which simply says that conjunction is the greatest lower bound.

CPL5 If $(C, \Gamma \cup \Delta)LE(A, \Gamma \cup \Delta)$ for all Δ, and $(C, \Gamma \cup \Delta)LE(B, \Gamma \cup \Delta)$ for all Δ, then $(C, \Gamma \cup \Delta)LE(A \wedge B, \Gamma \cup \Delta)$ for all Δ.

We can now use CPL5 to prove our next lemma.

CPL6 If $A \in \Gamma$, then for all $\Delta, (B, \Gamma \cup \Delta)EQ(A \wedge B, \Gamma \cup \Delta)$.

Proof. Suppose $A \in \Gamma$. Then by RCP6, we have for all sets Δ:

$$(B, \Gamma \cup \Delta)LE(A, \Gamma \cup \Delta)$$

But then by reflexivity (i.e., RCP.1) and CPL5, we have for all Δ:

$$(B, \Gamma \cup \Delta)LE(A \wedge B, \Gamma \cup \Delta)$$

Then the desired conclusion follows by RCP.2-3.

We have used the conditional in our definition of "$\Gamma \Rightarrow A$". The next lemma will allow us to relate the conditional to our definition of semantic entailment as well.

CPL7 If $A \in \Gamma$, then for all sets $\Delta, (A \supset B, \Gamma \cup \Delta)EQ(B, \Gamma \cup \Delta)$.

Proof. From Theorem 4 we know that for all Δ:

$$(A \wedge (A \supset B), \Gamma \cup \Delta)EQ(A \wedge B, \Gamma \cup \Delta)$$

Then the desired conclusion follows from the assumption that $A \in \Gamma$, CPL6, and RCP.1.

Now that we have both syntactic and semantic entailment characterized in terms of the classical conditional, we are finally in a position to prove both strong soundness and strong completeness.

Theorem 6. *(soundness) If $\Gamma \Rightarrow A$, then $\Gamma | \vdash_{CP} A$.*

Proof. Suppose $\Gamma \Rightarrow A$. Then by definition we know that $\vdash \wedge\Sigma \supset A$ for some finite subset Σ of Γ. Thus there is a finite sequence of formulas, $A_1, \ldots, A_n$, the last one of which is $\wedge\Sigma \supset A$, and such that each member of the sequence follows from previous members by one of the inference rules. By a simple inductive argument, we show that for each line, A_i, for all structures LE satisfying RCP1-7 and appropriate versions of RCPIR, the following holds for all Δ and all B:

$$(B, \Delta)LE(A_i, \Delta)$$

For any line justified by CPL, the result follows from Theorem 4. For any line justified by DET, the result follows by DF.CSI from Theorem 5, taking the evidence set to be empty. For any line justified by an inference rule IR, the desired result follows from RCPIR. Hence we know that for all structures LE, all sentences B, and all sets Δ:

$$(B,\Delta)LE(\wedge\Sigma \supset A,\Delta)$$

So considering sets Δ of the form $\Gamma \cup \{\wedge\Sigma\} \cup \Lambda$, we have for all structures LE, all sentences B, and all sets Λ:

$$(B,\Gamma \cup \{\wedge\Sigma\} \cup \Lambda)LE(\wedge\Sigma \supset A,\Gamma \cup \{\wedge\Sigma\} \cup \Lambda)$$

Applying CPL7 and RCP1, we then have for all structures LE, all sentences B and all sets Λ:

$$(B,\Gamma \cup \{\wedge\Sigma\} \cup \Lambda)LE(A,\Gamma \cup \{\wedge\Sigma\} \cup \Lambda)$$

Finally, by elementary set theory, RCP7, and RCP1, we obtain for all structures LE, all sentences B and all sets Λ:

$$(B,\Gamma \cup \Lambda)LE(A,\Gamma \cup \Lambda)$$

Hence by definition, $\Gamma | \vdash_{CP} A$.

Theorem 7. *(completeness) If $\Gamma | \vdash_{CP} A$, then $\Gamma \Rightarrow A$.*

Proof. For our proof, we will adopt the usual approach of establishing the contrapositive. Suppose it is not the case that $\Gamma \Rightarrow A$. We must show that it is not the case that $\Gamma | \vdash_{CP} A$. Thus we need to find an appropriate structure LE, some sentence B, and some set Δ such that the following does **not** hold:

$$(B,\Gamma \cup \Delta)LE(A,\Gamma \cup \Delta) \qquad (6)$$

To this end, we will define a single structure $LE*$ which has the desired property no matter what A and Γ may be. For all expressions C and D, and all sets Λ, we initially let $LE*$ be defined by:

$$(C,\Lambda)LE*(D,\Lambda) \text{ iff there is some set } \Sigma, \text{ a finite subset of } \Lambda, \text{ such that } \vdash \wedge\Sigma \supset (C \supset D). \qquad (7)$$

Then to ensure that RCP7 is satisfied, for all C, D, E, and Λ, we extend $LE*$ as follows:

$$(C,\Lambda \cup \{D \wedge E\})EQ*(C,\Lambda \cup \{D,E\}) \qquad (8)$$

Since it is not the case that $\Gamma \Rightarrow A$ and we are dealing with extensions to classical logic, there can be no finite subset Σ of Γ such that the following holds:

$$\vdash \wedge\Sigma \supset (T \supset A) \qquad (9)$$

But given the failure of (9) for all finite subsets of Γ, the definition of $LE*$ guarantees that the following does not hold:

$$(T, \Gamma) LE * (A, \Gamma) \tag{10}$$

So as required we have a case in which (6) fails. To complete the proof, we need to show that $LE*$ satisfies the constraints for logic L. But it is easy to show that given the definition of $LE*$, all of RCP1-7 correspond to elementary theorems of classical logic. Further, all constraints of the form RCPIR correspond to trivial variants of inference rules of the form IR. Hence $LE*$ satisfies the constraints appropriate for the logic. Thus it is not the case that $\Gamma | \vdash_{CP} A$.

Our proof of Theorem 7 actually establishes the existence of canonical comparative structures, similar to canonical probability distributions and canonical models in modal logic. As discussed before, the existence of canonical models in no way trivializes the semantics. Canonical models are generally too complex to be of much practical use; they are useful in completeness proofs and are of interest for technical and philosophical reasons. Models constructed for real world applications or to solve decision problems will not generally be canonical, and there is no reason they should be. See [13] and [15] for more discussion and references.

The astute reader will note that we did not need the usual theory of maximally consistent sets to establish completeness. The real reason we did not need such sets is that our comparative structures LE are permitted to be quite sparse. For example, consider some atomic sentence Q. In real life, if Q talks about some exotic situation in some exotic place, I may have no idea whether or not Q is true, nor will I have any estimate of its probability. But in classical model theory, we **must** assign Q some value; and in classical probability theory, we **must** assign some number to the probability of Q. But in our comparative structures, we need say very little about Q. Basically, if my logic does not connect Q with sentence A, on the basis of my background information Γ, then it is perfectly permissible to have Q and A be non-comparable on the basis of Γ. Such non-comparability is not permitted by either classical model theory or by numerically based probability theories. Thus our comparative structures may be a better model of belief than numerically based probability theories.

However, a little reflection suggests that our current constraints may permit structures that are too sparse. In their present form, our constraints place almost no restrictions on the degree of support a given expression may have when the evidence set is varied. (The only exception is for sets that are set theoretically equal, modulo conjunctions of their members.) We do not even require that (T, Γ) bear any relation to (T, Δ). However, there seems to be no good reason for not requiring $(T, \Gamma) EQ (T, \Delta)$ for all Γ and Δ. If we do so, then we can consider the set of all LE determined equivalence classes to be one algebra, with meet and complement defined as in DF.MT and DF.CMP, generally treating $[A, \Gamma]$ as incomensurable with $[B, \Delta]$ for $\Gamma \neq \Delta$. Under these conditions, the resulting structure will be a partial Boolean algebra, as defined in [4] for the treatment of quantum logic.

Another obvious technical lacuna in our current account is the complete lack of any conditionalization scheme. When new information is acquired, or at least entertained, how are beliefs conditional on the combination of the old and added information to be related to beliefs conditional on the original information? Our current theory places no restrictions on such conditionalization. In classical probability theory, it is trivial to derive the following condition, which is independent of any logical particles.

$$P(A, \Gamma) \times P(B, \Gamma \cup \{A\}) = P(B, \Gamma) \times P(A, \Gamma \cup \{B\})$$

This equation will give rise to many constraints on the *LE* structures by considering what happens as each of the multiplicands takes maximum and minimum values. The peculiar observation is that such constraints do not seem to be required by the logics, as is indicated by our soundness and completeness results.

I would like to mention one last problem, which is directly related to the practical utility of comparative probability distributions. Although our constraints permit LE structures that are extremely sparse compared to normal numerical probability distributions, they still require a degree of specificity which makes them unreasonable as an account of actual belief structures. Consider the strong soundness theorem for example. If we were attempting to actually build up an *LE* structure for a robot in a real world situation, we would have to at least be able to correctly determine, for every set Γ and every sentence A, if (A, Γ) is maximal. But this level of logical omniscience is beyond us in most cases, and certainly does not correspond to actual belief structures. It seems that for practical applications we need some account of partial *LE* structures. We would not take lack of logical omniscience as an indication of irrationality. But we might accept the principle that a belief structure that cannot be extended to a full *LE* structure is indicative of irrationality. What conditions can we impose on a partial structure to ensure that it can always be extended to an acceptable *LE* structure? At present I have no answer.

6 Concluding remarks

We have briefly traced the movement of formal semantics from two values, to three values, to many values, and finally to no values. In the process, we have seen good empirical, philosophical and technical reasons for abandoning truth-functionality. The resulting formal semantics has a number of theoretical benefits. Certainly it presents a uniform semantic account suitable for all extensions of classical sentence logic, including many logics not normally presented as extensions of classical sentence logic. Thus probabilistic semantics does not suffer from the incompleteness problems of standard many valued and Kripke style semantics. Thus probabilistic semantics is considerably more general.

Standard formal semantics abandons notions of linguistic meaning and attempts to define logically important semantic concepts in terms of a primitive notion of *truth*, either bivalent or multivalued. By contrast, numerically based probabilistic semantics abandons the notion of truth and attempts to define logically important semantic concepts in terms of a primitive notion of *degree of rational belief* - given evidence Γ, it is rational to believe A to degree r. But many feel that numerical probability theory is too determined in its details to serve as a good model of real belief systems. Thus our semantics based on comparative probability structures abandons the notion of degrees of belief and defines logically important semantic concepts in terms of the primitive notion of comparative belief: *the rational support given A by Γ is no greater than the rational support given B by Δ.* Certainly from a formal point of view, comparative rational belief semantics is at least as robust as classical truth valued semantics. Further, elementary observations and reflections of the sort advanced in section 2 provide very strong empirical evidence that an account of semantic concepts should be based on notions of rational belief under conditions of uncertainty, rather than on notions of truth and falsity. Thus we suggest that comparative rational belief semantics is a step forward from truth value semantics, both philosophically and scientifically.

In section 1, we criticized the trivial many valued semantics in which the values are the sentences themselves. One difficulty with that semantics is that it is crucially dependent on the proof theory. In the trivial semantics, the designated values will often not even be decidable, but there is no such problem with the probabilistic semantics. In sharp contrast, both the numerical and the comparative probability semantics are autonomous; they do not require knowledge of the theorems of the logic, and consequently they may be used to solve decision problems, among other things. We also criticized the trivial semantics for providing no philosophical illumination. Such a complaint cannot be made against the numerical nor comparative probabilistic semantics. For example, in the trivial semantics, we would be told something like "Sentence A is logically certain because it is logically certain"; by contrast, in probabilistic semantics, we would be told something like "Sentence A is logically certain because your beliefs have such and such a structure". Relating formal semantic concepts to rational belief structures can hardly be said to be devoid of technical, empirical, and philosophical interest.

There remains the question of how probabilistic semantics can be applied to problems in computing science. One area in which it has already been applied with very interesting results is the theoretical analysis of nonmonotonic reasoning (see [14]). No doubt there are other areas as well. For example, I have yet to see any development of probabilistic semantics for programming languages, but from our presentation here, it should be obvious that it is possible to do so. Perhaps we will gain insights into particular programming languages by relating them theoretically to principles of probability and rational belief. And of course there are standard applications of probability theory to connectionism. Perhaps looking at the proposed probability structures and updating procedures used in connectionist models and extracting matching proof theories will tell us something interesting about such models.

Coming at the matter from the opposite direction, we have already mentioned that with the abandonment of truth-functionality, we have given up a great deal in computational tractability. Truth tables are generally finitary, but probability distributions generally are not. How can we make probabilistic semantics and probabilistic models more computationally tractable so that the formal semantics will be more useful in computer applications? For numerically based theories, the weighting function construction P.WF is very useful; examples may be found in [11]. And if the number of logically distinct expressions is finite, tabular techniques are sometimes possible; see [12] and [16] for examples. Further, our soundness and completeness results show that for many applications, we can restrict our attention to simple two valued functions, which would greatly simplify many applications. But the comparative structures remain computationally difficult. Of course part of the problem is that at present it is still easier to compute with numbers than with abstract algebraic structures. The situation might be greatly improved if we could develop a useful theory of partial comparative structures LE.

References

1. Dugundji, J.: Note on a property of matrices for Lewis and Langford's calculi of propositions, Journal for Symbolic Logic, vol. 5 (1940) 150–151
2. Field, H.H.: Logic, meaning, and conceptual role, Journal of Philosophy, vol. 74 (1977) 379–409
3. Fine, K.: An incomplete logic containing S4, Theoria, vol. 40 (1974) 23–29
4. Kochen, S., and Specker, E.P.: Logical structures arising in quantum theory, Proceedings of the 1963 International Symposium at Berkeley on the Theory of Models, North Holland Publishing Company, Amsterdam (1965) 177–189
5. Komolgoroff, A.: Foundations of the Theory of Probability, Chelsea Publishing Company, New York (1950)
6. Leblanc, H.: Probabilistic semantics for first-order logic, Zeitschr. f. math. Logik und Grundlagen d. Math., vol. 25 (1979) 497–509
7. Lewis, C.I.: A Survey of Symbolic Logic, University of California, Berkeley (1918)
8. Łukasiewicz, J.: Philosophische Bemerkungen zu mehrwertigen Systemen des Aussagenkalküls, Comptes rendus des séances de la Société des Sciences et des Lettres de Varsovie, Cl. iii, 23 (1930) 51–77
9. Morgan, C.G.: The truebeliever, the unbeliever, and a new modal semantics, Essays in Epistemology and Semantics, Leblanc, H., Stern, R. and Gumb, R. eds. Haven Publications, New York (1983) 33–53

10. Morgan, C.G.: Weak conditional comparative probability as a formal semantic theory, Zeitschr. f. math. Logik und Grundlagen d. Math., vol. 30 (1984) 199–212
11. Morgan, C.G.: Logic, probability theory, and artificial intelligence - Part I: the probabilistic foundations of logic, Computational Intelligence, vol. 7 (1991) 94–109
12. Morgan, C.G.: Conditionals, comparative probability, and triviality: the conditional of conditional probability cannot be represented in the object language, Topi, vol. 18 (1999) 97–116
13. Morgan, C.G.: Canonical models and probabilistic semantics, in Logic, Probability and Science, Poznań Studies, vol 71, Niall Shanks et al., eds. (2000) 17–35
14. Morgan, C.G.: The Nature of Nonmonotonic Reasoning, Minds and Machines, vol. 10 (2000) 321–360
15. Morgan, C.G.: Les probabilits comparatives comme fondements de la logique, in Sémantique et théories de la vérité: l'héritage carnapien, Lepage, F. and Fivenc, F. (eds.) Vrin/Bellarmin, Paris/Montréal (in press) (2001)
16. Popper, K.R.: The Logic of Scientific Discovery, Basic Books, Inc., New York (1959)
17. Rosser, J., and Turquette, A.: Many-Valued Logics, North-Holland, Amsterdam (1952)
18. Routley, R.: Universal semantics?, Journal of Philosophical Logic, vol. 4 (1975) 327–356
19. Stoll, R.R.: Sets, Logic, and Axiomatic Theories (2nd ed.) W.H. Freeman and Company, San Francisco (1974)
20. Thomason, S.K. An incompleteness theorem in modal logic, Theoria, vol. 40 (1974) 30–34
21. Weinberg, J.: Nicolas of Autrecourt, Encyclopedia of Philosophy, Edwards, P. (ed.), The Macmillan Company, New York, vol 5, (1967) 499–502

www.ingramcontent.com/pod-product-compliance
Ingram Content Group UK Ltd.
Pitfield, Milton Keynes, MK11 3LW, UK
UKHW021857190726
13853UKWH00003B/1313

* 9 7 8 3 6 6 2 0 0 2 9 9 5 *